W0080162

EVALUATING NEW TELECOMMUNICATIONS SERVICES

NATO CONFERENCE SERIES

I Ecology
II Systems Science
III Human Factors
IV Marine Sciences
V Air–Sea Interactions
VI Materials Science

II SYSTEMS SCIENCE

EVALUATING NEW TELECOMMUNICATIONS SERVICES

Edited by
Martin C. J. Elton
New York University and
Communications Studies and Planning
London, England

William A. Lucas
Rand Corporation
Washington, D.C.

and

David W. Conrath
University of Waterloo
Ontario, Canada and
Institut d'Administration des Enterprises
Aix-en-Provence, France

Published in coordination with NATO Scientific Affairs Division

PLENUM PRESS · NEW YORK AND LONDON

Library of Congress Cataloging in Publication Data

NATO Telecommunications Symposium, University of Bergamo, 1977.
Evaluating new telecommunications services.

(NATO conference series: II, Systems science; v. 6)
"Proceedings of the NATO Telecommunications Symposium held at the
University of Bergamo, Italy, September 1977, sponsored by the NATO
Special Program Panel on Systems Science."
Includes index.
1. Telecommunication—Social aspects—Congresses. I. Elton, M. C. J. II. Lucas,
William A. III. Conrath, D. W. IV. Nato Special Program Panel on Systems
Science. V. Title. VI. Series.
HE7631.N2 1977   384   78-4684
ISBN 978-1-4757-0177-7   ISBN 978-1-4757-0175-3 (eBook)
DOI 10.1007/978-1-4757-0175-3

Proceedings of the NATO Telecommunications Symposium held at the University of
Bergamo, Italy, September 5—8, 1977, sponsored by the NATO Special Program
Panel on Systems Science

© 1978 Plenum Press, New York
Softcover reprint of the hardcover 1st edition 1978
A Division of Plenum Publishing Corporation
227 West 17th Street, New York, N.Y. 10011

All rights reserved

No part of this book may be reproduced, stored in a retrieval system, or transmitted,
in any form or by any means, electronic, mechanical, photocopying, microfilming,
recording, or otherwise, without written permission from the Publisher

Preface

This book contains the proceedings of the first international symposium devoted to research on the evaluation and planning of new person-to-person telecommunication systems. It was sponsored by NATO's Special Programme Panel on Systems Science and took place, in September 1977, at the University of Bergamo in the north of Italy.

Telecommunication systems which provide for communication between people, rather than computers or other instruments, are of two kinds. There are mass communication systems (broadcast radio and television) and interpersonal systems (for example, the telephone and Telex) which join together individuals or small groups. Here we have included in the interpersonal category certain systems for retrieving information from computers, essentially those systems in which the role of the computer is primarily to act as a store and to identify that information which best fits a user's request. (This excludes management information systems in which the computer performs important transformation functions.)

Distinctions between interpersonal and mass communication systems, and between these two and data communication systems, are increasingly breaking down for those who provide the services. (In the U.K. broadcasters are piloting information retrieval services and the British Post Office is competing with a more sophisticated system which could also be used for the exchange of messages. Elsewhere computer data networks are increasingly employed for the exchange of personal messages. And in the United States there are various experiments in the use of cable television systems for interpersonal communication.) Nevertheless, the distinctions remain meaningful in terms of the different uses to which the systems are put. And it is a common characteristic of all current research in our field that it is explicitly concerned with use.

In the laboratory and in the field there are a variety of new telecommunication services. They range from simple extensions to the basic capability of the telephone - allowing it to serve more

than two locations and more than one person per location – to picture
telephones and two-way color television systems using satellites or
lasers to connect health-care establishments.

They are seen as making possible new solutions to problems of
major social concern. Applications of the technology, which are
addressed in this volume, include: reducing the burden of business
travel; dispersal of office work from city centers to the suburbs,
smaller towns, rural areas, and "neighborhood work centers"; provision of health care, personal social services, and educational opportunities to those who are relatively underserved by reason of
physical handicap or geographical location; public participation in
local government; and improved coordination between the parts of
large organizations.

While promising help in alleviating some problems, the new
technology threatens to exacerbate others. There is, for example,
concern about the dangers of dehumanization, invasion of privacy,
and information overload. There is the risk of unintended side
effects: maybe the reinforcement of undesirable trends in the
balance between centralization and decentralization, or the possibility of increasing energy consumption by encouraging more dispersed
working and living patterns. Then there are the perplexing problems
of regulation and the development of policy at national and international levels. These grow ever more complicated as the computer
industry increasingly penetrates the telecommunications industry,
and as these two penetrate the economically fragile postal services.

Nor is it easy to predict whether, in a particular context,
people will actually use some new telecommunication service.
Confravision, a European public studio videoconferencing service,
has fallen far short of its market targets; picture telephones have
not lived up to their early expectations; and many early uses of
telecommunications for the delivery of health care have been disappointing.

Considering all this it is scarcely surprising that, once started, research on the use and usefulness of new interpersonal telecommunications systems has grown rapidly. It is, however, somewhat
surprising that (outside the military arena) it came into being only
about seven years ago. One might have expected that at an earlier
stage it would have provided a modest complement to the enormous
efforts of technological development which have made the new systems
possible.

Today's worldwide telephone system is a remarkable triumph of
systems engineering. The systems science research which new telecommunication services require is not, however, a simple extension
of that which guided the development of the telephone system. Uncertainties regarding individual and organizational users are far

harder to treat. No longer can one consider only service time characteristics (how long calls last), or the ergonomic design of the
telephone instrument, or the necessary acoustical standards, according to the problem at hand. And the concept of marketing has become
relevant, as it used not to be, for the provision of telephone
service by statutorily protected monopolies.

There are the communication problems one would expect of a young
and fast growing area of interdisciplinary endeavor. How best can
new entrants, especially from countries not previously involved,
make connections to past research and to practitioners with like
concerns? What are the methods for weeding out the false starts
encouraged by demands for "quick fixes"? Where are the forums for
the exchange of ideas and the challenge of one's peers? Are research
conclusions reaching and being understood by those they are intended
to influence? Do the latter consider researchers to be in touch
with reality?

The symposium was designed to address such problems: to enable
researchers, together with some business managers and administrators,
to learn from one another. To judge from a questionnaire survey
of participants it certainly succeeded.

In soliciting and selecting papers we encouraged authors to
present research on which comment would still be useful to them; many
of the papers describe work in progress. We also encouraged contributions which would be helpful to newcomers to the field; several
of the papers contain useful reviews. We did not restrict ourselves
to papers which portrayed systems scientists' wares, but included
thoughtful discussions of aspects of the environments in which they
must operate. We emphasized the need for papers to communicate successfully across cultural and disciplinary frontiers; most of the
authors met this challenge without, we believe, trivializing their
work.

Certainly we were demanding in our relations with the authors:
with few exceptions, the editors insisted that papers be produced
for circulation before the meeting, rigidly enforced limitations
on their length, and in some cases required them twice to be totally
rewritten. We are most grateful for the long suffering goodwill
shown by the authors. Two, it may be noted, swallowing their own
medicine, presented and discussed papers from their offices in North
America, using a decidedly ad hoc audioconferencing system.

For presentation here papers are grouped into eight sections.
The first of these provides some introductory overviews. Next come
two sections which deal with the delivery of health care, education
and community services. The fourth section comprises contributions
from the field of scientific and technical information (STI). This
is followed by a group of papers concerned with teleconferencing and

computer conferencing services; some report upon trials of new
services and others look more deeply into communication processes
at the level of individuals and of organizations.

While almost all the papers deal with new services, four approach particular services in such a way that they do not fall
naturally into any of the preceding sections. They provide the
sixth section, entitled New Services. Three of the papers in the
seventh section view developments in the field of telecommunications
from different perspectives regarding society's use of information
technology. The fourth paper in this section considers developments
in the field of electronic funds transfer (EFT). Finally comes a
group of papers concerned with aspects of planning and design. The
last of these is concerned with planning one aspect of a field trial;
the others address much more wide-ranging concerns.

To guide the reader through the book each section is introduced by a short summary of the papers it contains. Some changes
have been made in the way papers were ordered and grouped at the
symposium so as to make for easier reading here. The changes have
been made in such a way that it remains meaningful to conclude the
sections below with a summary of points raised in the corresponding
discussion sessions at the symposium. For the final session of the
symposium four participants were invited to relate points raised
during the meeting to the twin themes of policy development and
methodology which ran through all the sessions. The book concludes
with edited transcripts of their observations.

An international conference in an emerging field is at best a
complex undertaking, and contributions are made in a variety of
forms. Without the generous support of NATO's Special Programme
Panel on Systems Science these proceedings would not exist. And
without the further support of the U.S. National Science Foundation
there would have been a substantially less weighty contribution
from the U.S.A. In addition to the editors, the organizing committee for the symposium included Michael Tyler and Dieter Kimbel.
Others assisted the committee in the review of papers.

The book represents the work of around sixty authors, as well
as the efforts of the discussion leaders and of the two rapporteurs,
Barbara Lucas and Hilary Thomas. We have also made use of cartoon
drawings of some of the participants by P.G. Holmlöv.

It is a pleasure to acknowledge all these contributors.

Contents

SECTION 3

SECTION 4

SECTION 5

SECTION 8

CONCLUDING DISCUSSION

APPENDIX

Section One

AN OVERVIEW OF RESEARCH ISSUES

Michael Tyler was asked to provide a paper drawing upon the critical review of methods of forecasting demand for new telecommunications services, which he and his colleagues recently completed for the British Post Office. His paper is concerned with methodology, primarily with questions of how to achieve results which are reliable and will provide useful input to those responsible for the introduction of new services. It aims to provide criteria which will be useful in assessing some of the papers in following chapters.

The paper by Robert Chapuis takes the form of the speech which he offered in the opening session. His attention is turned inward upon the enormous organizations responsible for providing telecommunications services in European countries ("whales swimming merrily....practically unnoticed......sending up a little water spout to indicate their presence at a few infrequent public relations events" or "dinosaurs" with skeletons "too ossified to enable them to adapt to the changing environment"?). One of his main concerns is to introduce their endoeconomics as a rewarding subject for research. The other is to draw attention to their need to adapt to rapidly changing conditions.

In the light of recent developments in Britain, France, Germany and Sweden this is a particularly timely paper.

Two other papers are included in this chapter although they were presented later in the Symposium. Their authors come from the two telecommunications agencies which have, we estimate, undertaken and published, from very different perspectives, appreciably more policy-relevant research than their counterparts in other NATO countries: Bell Canada and the British Post Office. (The papers are, of course, written in an individual, rather than a corporate capacity.)

Larry Day examines the role of "telecommunications policy analysis" in the planning and development of new services. As context he describes the service planning process. By way of illustration his paper presents a case study of the technology assessment of the substitution of telecommunication for travel. It concludes with hypotheses as to the current "rules of the game" in the development of national telecommunications policy, and looks ahead, perhaps somewhat optimistically, to the day when large groups of users go ahead in using new capabilities as they see fit, regardless of the restrictions of policy.

From a European perspective, Jim Cowie's paper describes current policy issues in the development of new services, together with the actual and potential roles of the different actors in the process. Like Day, he draws attention to the rapidly increasing points of intersection between computers and telecommunications.

No technical knowledge of telecommunications is required to follow any of the four papers. While an interest in and some understanding of methodologies of applied research are assumed in Tyler's paper, none of the other contributions make even these demands.

USER RESEARCH AND DEMAND RESEARCH: WHAT'S THE USE?

An enquiry into the how and why of telecommunications studies

Michael Tyler

Communications Studies and Planning Ltd.

56/60 Hallam Street, London W1N 5LH England

'Entia non sunt multiplicanda praeter neccessitatem'. William of Occam.

PREFACE

This paper is an introductory contribution designed to stimulate discussion at the NATO Symposium on the evaluation and planning of telecommunications systems. It argues the case for a searching reappraisal of the quality and relevance of the growing volume of social, economic, behavioural and policy research into the implications of new telecommunications services and information technologies. The need for such research to underpin policy-making and planning through economic analysis, rational design, demand forecasting and many other approaches is generally acknowledged. A reappraisal of the field along the lines sketched in the paper would, it is suggested, show that the practical impact of the work has so far been slight and that serious deficiencies of strategy and method remain, despite considerable research achievements. An 'agenda' for the effort to develop this field of research and enhance its impact is suggested.

INTRODUCTION

Over the last ten or fifteen years, it has become almost a truism that high priority should be given to the study of user "needs" - a notably vague concept[1] - for telecommunications and information technology. Few would deny the importance of studying the behaviour and preferences of users and the "human/machine interface". Indeed, my observation is that the number of

engineers, planners and managers who would openly dismiss such
work as 'bunk' - after the manner of Henry Ford's legendary
dismissal of the historians - has diminished steadily. No
engineering conference is now complete without its solemn if
superficial special sessions on user needs or social impacts,
going far beyond the narrow - though methodologically deep -
concerns of the long established tradition of 'Human Factors'
studies dealing with such questions as the optimum presentation
of visual displays.[2]

Numerous research groups were established during the 1960s
and 1970s to develop the new fields of work. They ranged from
groups which initially focused mainly on the 'microscopic' study
of human communications and the effectiveness and acceptability
of media - for example the Communications Studies Group at
University College, London - and Professor Chapanis' group at
Johns Hopkins University - to those whose orientation emphasized
public policy questions (e.g. the two Annenberg Schools of
Communication in the U.S.A., or the Harvard Program on
Information Technology and Public Policy). Another orientation
involved the combination of research with the implementation of
communication innovations as part of a process of social change -
an approach exemplified by the Alternate Media Center at New York
University.[3] Recently, groups within - or closely associated
with - telecommunications carriers have become increasingly
active in the same fields: leading examples have been the
Business Planning Group of Bell Canada and the Long Range Studies
Divisions in the British Post Office, as well as groups within
some of the European PTTs - such as the user research group at
the FTZ in Darmstadt, West Germany[4] - and in Bell Laboratories
and Bell Northern Research. Some government research bodies such
as the Communications Research Centre (CRC) of the Canadian
Department of Communications have also made an active
contribution.

Moreover, the growing awareness of telecommunications as a
matter of wide public interest and concern has led in a few
countires - notably the U.S.A. - to financial support from public
research funding agencies such as the U.S. National Science
Foundation, though on the whole European countries have lagged
behind in this respect. The total result of the expansion of
activity and support is that the field of work covered by the
NATO symposium on the planning and evaluation of telecommunications
systems can no longer be considered an 'infant industry'. It must
justify itself as a major contender for policy influence and for
resources.

The premise of this introductory paper for the NATO Symposium
is that the time has come for a searching appraisal of the value
of the burgeoning field of work which I have labelled 'user research
and demand research'. I exclude not only planning methods which
rely on the more or less sophisticated extrapolation of past
trends[5], but also those macroscopic demand models that are based
on economic theory but whose specification deals only crudely with
user behaviour, most of the underlying structure of the problem
being subsumed in the measured - and, hopefully, stable - values of
parameters such as price- or income-elasticities of demand. As
B. Cartwright and I have argued elsewhere[6], such methods are
effective where the system of interest has been highly stable over
time (as in POTS - Plain Old Telephone Service) and no major
structural changes are expected. Where radical changes or the
introduction of entirely new technologies, services or applications
are involved, analysis must dig deeper into the broader and more
fundamental regularities of human behaviour if it is to find some
bedrock on which to build generalisable findings and predictions
about effectiveness, acceptability, benefits or demand.

Work conceived within this perspective must be judged from
two distinct angles: what its authors were trying to do and
whether that was the right aim to try for. Knowing that the
magic of words has robbed much of our descriptive terminology
of its meaning (the terms 'evaluation study' or a 'field experiment'
can label many different sorts of activities) it seems best to
work instead in terms of the sorts of questions that are posed.
Most of the literature relevant here attempts in one context or
another to answer one or more of the following questions:

- does a particular innovation[7] perform the function expected
  of it (for example, does a teleconference system really
  allow business to be transacted effectively and acceptably?)

- will people or organisations want and use the innovation?

- How should the innovation be designed to maximise its
  utility to users and its acceptance by them? Is there
  a better way - new or old - to achieve the same result
  as that offered by the innovation?

- How many people will use the innovation and how much?
  How much will they (or some 'social fund' acting on their
  behalf) be willing to pay?

- How should the introduction of the service be arranged,
  in economic and managerial terms? For example, how should
  it be priced and how should it be introduced to potential
  users?

● Should 'society' (by which we usually mean public policy
  implemented through legislation or the executive actions of
  governments) favour or oppose the innovation? What should
  be done to obtain the socially preferred outcome?

Each of these questions corresponds more or less closely to
the areas of research designated by such labels as 'studies of
effectiveness and acceptability', 'demand forecasting', 'evaluation'
or 'technology assessment'. In each case, I have sought to
stimulate critical discussion with a first attempt to suggest

● Why we might need research (if indeed we do)

● What are the consequences of <u>not</u> doing the research?

● How useful has the research, in its various forms,
  been so far?

● What have we learned that will help us design better-
  directed and more useful research in future?

ASSESSING SYSTEM EFFECTIVENESS AND ACCEPTABILITY

Why we try

It has long been recognised that there are different ways of
arranging technological systems to make them more or less comprehensible and acceptable to their human users, and to make the
human/machine interface more or less 'efficient' in various
respects. The original focus of 'Human Factors' or 'Ergonomics'
was fairly narrowly restricted to design questions related to
services with an established demand: it was only in the 1960s
and 1970s that researchers - stimulated by innovative work such
as the studies of teleconferencing conducted at the Institute of
Defense Analyses and at the Communications Studies Group and
work on scientific and technical information at MIT[8]- recognised
that the net could be cast wider to try to reduce the risks
associated with technological innovation. Behaviourally-oriented
research prior to the large-scale introduction of a new product
or service could, it seemed, cast light on its utility - and the
utility of its various attributes - in such a way as to enable
improved decisions in design and planning. For example, the
British Post Office - subsequently joined by six other European
PTT Administrations through CEPT - decided to undertake such a
work programme, largely building on fundamental research at the
Communications Studies Group, before becoming more committed to the
provision of video-telephone or teleconferencing (including
conference call facilities).

Can we Assess Effectiveness and Acceptability?

The 'effectiveness and acceptability' approach has probably generated more user-oriented research in telecommunications than any other. It has made use of a variety of social-science instruments such as controlled psychological laboratory experiments, monitored field trials and quasi-experiments, and survey research.

This paper is not the occasion for a review of these lines of research, especially since several excellent review articles now exist.[9] It is sufficient for my purpose to note that these lines of research have been extremely productive of research findings - though perhaps somewhat less productive of deep general insights into behaviour. The US work on computer conferencing (at the Institute for the Future, SRI, U.S.C., the New Jersey Institute of Technology and elsewhere), and the British work on the relative strengths and limitations of face-to-face meetings and of audio-only or audio-visual teleconferencing, provide good examples.

There is however, no room for complacency. Distinguishing between those approaches that rely heavily on field-trial data, and those that rely on theoretical models and controlled laboratory tests, we find that each faces severe problems:

● The field-trial approaches often:

- lack the suitability for the derivation of generalisations from the field test to a wider universe of users, or even fail to collect the socio-economic data that would be a precondition for this

- lack the 'quasi-experimental' design that would allow the key hypotheses to be tested. For example, do information-retrieval systems or teleconference systems replace existing use of other communication media (the 'diversion effect'), generate entirely new information-gathering activity, or cause a widespread reorganisation of users' communication habits? "Before and after" surveys, comparisons with a control group of non-users, or comparisons between the effects of different experimental 'treatments', would be needed in order to answer these questions[10]

- are too small to give the innovation being tested a real chance of winning user loyalty

- compromise unsuccessfully between the need to stimulate usage before the trial can yield any scientific data and the paradigm of the "objective" and detached experimenter.

- The laboratory and model based approaches tend to suffer from:

  - problems in achieving realism

  - restrictive conceptual 'paradigms'. For example, much of the psychological work on teleconferencing has taken face-to-face communications as a 'norm' of effectiveness, and emphasised the substitution of teleconferencing for face-to-face communication; but both these perspectives - while they can be defended as useful interim devices for structuring the research - are open to challenge

  - lack of opportunity for verification and test.

This last remark calls for some explanation. Within its own terms, an experiment on, say, the outcomes of information searches or business meetings embodies its own tests of hypotheses. The question, however, is how far the hypotheses are the right ones and how far the findings have a relevant predictive value in the real world. Will a system which is, according to the experiments, "effective", actually be favoured by users? To answer this calls for an entirely new kind of verification effort that is still in its early stages.

What If We Don't Do the Research?

The starkest argument in favour of user research and demand research in the context of business planning is the exceedingly high 'death rate' among innovations in telecommunications. The range of services available to most residential and business users is in most areas not substantially different today from the offerings of twenty years ago - apart from the incremental improvements such as the replacement of operator connection by automatic dialling for various classes of long distance calls, virtually all of the formidable technological development of those years has gone into providing the same services cheaper or slightly better. Along the way we have seen numerous cases where high hopes were dashed by experience in the market place: for example the major efforts to market facsimile, in the 1940s and 1950s; Picturephone; most teleconferencing services and probably many of the 'advanced services and facilities' (e.g. short code dialling or 'camp on busy') now being offered by some telephone companies as the result of the introduction of advanced Stored Program Control switching equipment.

Should we expect to be able to greatly improve the success rate of innovations by means of user-oriented research? I believe so, but as the next section shows, this is still something of an 'act of faith'.

The Track Record

A rather comprehensive review of our field and reading of its literature - stimulated by a critical review study recently carried out for the Long Range Studies Divisions of the Post Office Telecommunications in the UK[11] - has convinced me that while 'human factors' research has sometimes - but not always - influenced the detailed design of telecommunications systems in a favourable way, the wider research on effectiveness and acceptability has had remarkably little practical impact so far.

The experience with teleconferencing - comprehensively catalogued in a report by Hough[12] - is instructive in this respect. Not only do the fundamental choices - between audio and video; studio or desk-top terminals; public or private terminals; voice switched, manually switched or unswitched audio and video - seem to have been little influenced by the research, but we even find that the avalanche of field-trial activity has inadvertently occurred in such a way that most of the research findings on these points can scarcely be tested. In general, audio-only, in-house systems set up for specific organisational purposes and with highly accessible terminals have been the most successful. But since all these variables are so strongly associated together (most video systems are in general-purpose public studios) their effects cannot be disentangled.

Nor is the practical impact of the more detailed 'evaluation' studies in such fields as telemedicine, scientific and technical information, teleconferencing or computer mediated communication yet established clearly.

Some Suggestions for the Future

Most of the prescriptions I suggest for discussion flow naturally from my diagnosis of the problems of our field of study. After the traumatic experiences many telecommunications administrations have had with service innovation in recent years, the inherent attractiveness of evaluative research as a means of improving planning and reducing risks is very great. The task facing the researcher or research planner is to demonstrate that his magic really works in this respect... and to persuade decision makers to give him the opportunity.

In particular, he or she needs to:

• improve the verifiability and verification of the research results in policy-relevant frameworks

• test and improve the relevance of the research 'paradigms' and priorities to real-world problems

- improve the entrepreneurial effectiveness <u>and</u>
  scientific quality of field studies, and especially
  their generalizability.

Where field evaluations are concerned, the future credibility
of the work makes two aspects especially important:

- Assessment of the economic value placed on the
  benefits of an innovation by individuals or
  institutions willing to pay for those benefits.

- An honest investigation of how far the problem being
  tackled in the field demonstration is really a
  communication problem, and of how far the desired
  result could be obtained by other - possibly non-
  communication - means. For example, is deficient
  secondary education in a particular case really due in
  some degree to inadequate access to information
  materials? If so, is the answer interactive
  television; or perhaps the production and distribution
  of video cassettes, or some other means altogether?

QUANTITATIVE DEMAND ESTIMATION

Demand Estimation in Recent Practice

Quantitative projections of demand for new telecommunications
services are a commonplace of planning and management in telecommunications. They are often produced by methods requiring
little more than the capacity to draw lines on logarithmic graph
paper. It is essentially only within the last five years that we
have seen serious attempts to build quantitative assessments of
the potential demand for various innovations in telecommunications
upon a substantial foundation of research findings.

At the most general level, the structure of these studies is
as indicated in Figure 1. A set of hypotheses are set up, dealing
with the behaviour of the user (an individual or business
establishment, say) in specific circumstances. This hypothesis
might deal with, for example, the circumstances under which a bank
would be willing to entrust a particular flow of transactions
documents to an electronic medium; or the cases in which business
executives would be willing to conduct a particular meeting via
teleconferencing. These hypotheses are checked by reference to
the kinds of fundamental research discussed briefly in the previous
sections. Data are acquired from large-scale surveys of
communication activity which measure the size of communication
flows and the frequency of occurrence of the different
circumstances considered in the hypotheses dealing with user
behaviour. The application of these hypotheses to the survey data
then makes it possible to derive general quantitative results

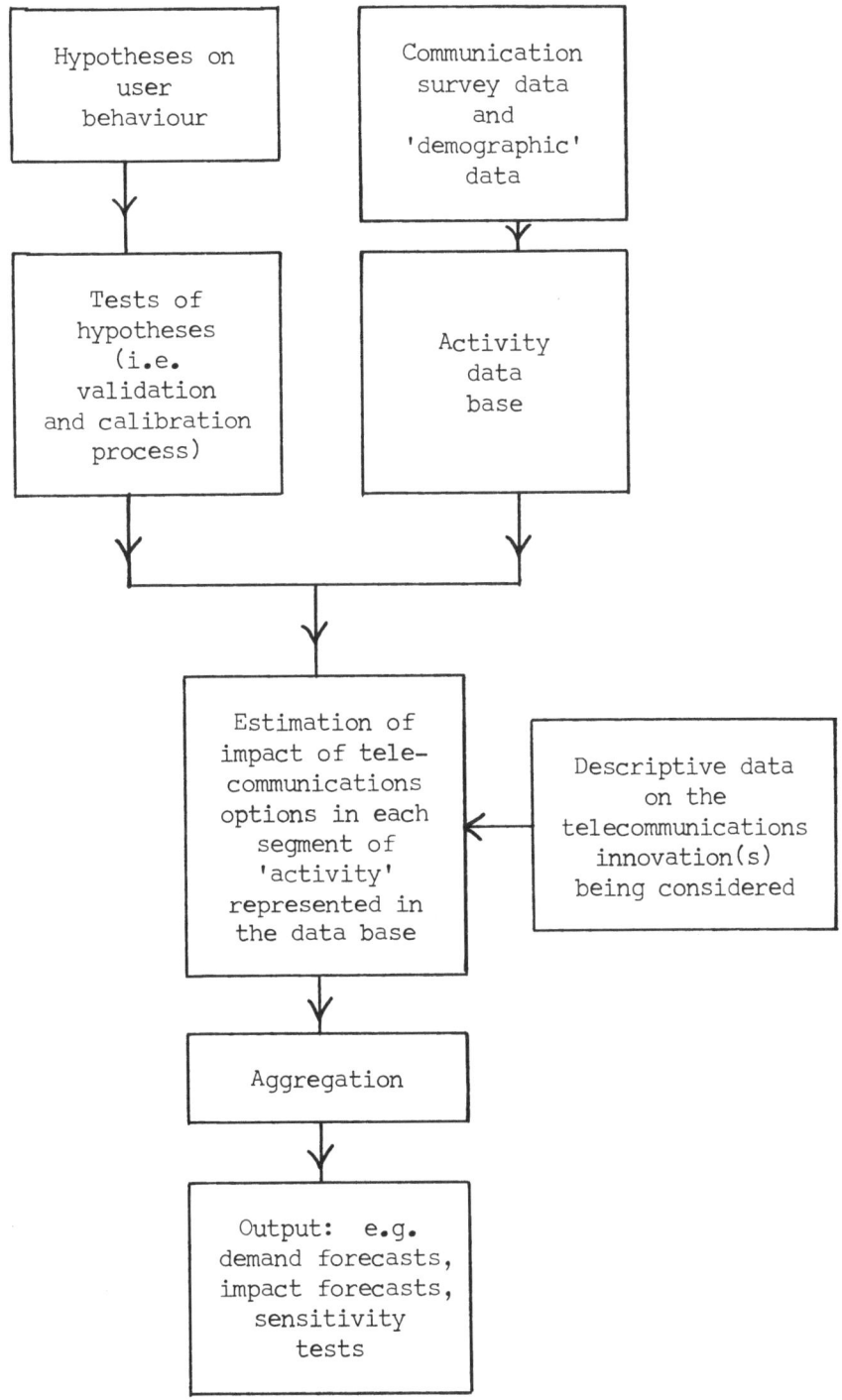

Figure 1   General conceptual structure of quantitative demand
            models

about the likely volume or pattern of use of a particular telecommunications innovation. This can be done by means of
established techniques of mathematical modelling or computer
simulation, provided the researcher is willing to supply the
necessary 'heroic assumptions' in the many instances where
research results are <u>not</u> available to indicate or test key
parameter values or structural features in the model.

Notable examples of recent attempts at quantitative
estimation have included:

- The Eurodata demand-estimation study for data
  communications, commissioned jointly by the European
  PTT Administrations through CEPT; and similar and
  earlier - but unpublished - studies undertaken for
  DATRAN, a U.S. specialized common carrier that has now
  ceased operations. [13]

- The European Business Communications Study, carried
  out by 7 European PTTs and co-ordinated by Post Office
  Telecommunications in the U.K. This study, building a
  basic research work carried out by the Communications
  Studies Group, involved extensive surveys of business
  communications activity and the use of mathematical
  models to investigate potential demand for conference
  telephony, teleconferencing, video telephones and
  other business communications media. [14]

- The analyses of demand for text-message services being
  carried on by Kalba Bowen Associates in the U.S.A.,
  Long Range Studies Divisions of Post Office Tele-
  communications in the U.K.;[15]Communications Studies and
  Planning Ltd. in the U.K., the KtK commission in
  W. Germany,[16]and others.

The modelling strategies they represent have certain common
features, and fall into two main categories. The common features
include the emphasis on a <u>disaggregate</u> approach, focusing on the
actions of individual decision-makers, and the emphasis on the
<u>measurement of communication flows</u>. Alternative approaches
involve:

- The 'diversion/generation' approach, in which the
  scale and composition of total communication flows are
  represented in the model, and the flows are allocated
  between the competing media according to formulae or
  algorithms reflecting the cost, effectiveness and
  acceptability of the different media, with
  supplementary representation (if any) of the
  generation of totally new traffics as a result of
  introducing the new media. The paper by Dormois,
  Fioux and Gensollen in the present symposium[17] is an

example of the diversion/generation approach.

- The 'leading edge' approach, applicable where substantial and established use of the new service exists among some sub set of users. Here the model predicts future demand according to a 'diffusion' algorithm which reflects the rate at which additional decision-makers (households or firms) join the 'club' of users, and the rate at which usage <u>within</u> the club intensifies over time. Most demand models for data-communications have been of this kind.

Does It Work? The Validity of Quantitative Models

At this early stage, there has been little work in the calibration, validation and retrospective testing of the models: this in itself is a serious deficiency in the present state of this field of work although, as we noted earlier, the situation is not much better where the 'qualitative' findings of behavioural research are concerned. Our knowledge of how well the various concepts and models predict behaviour is exceedingly sparse.

While the traffic forecasts for data communications in Europe seem to have stood the test of time relatively well,[18] it is still unclear how much the complex structure of the models contribute to their predictive usefulness, since relatively simple syntheses of trend projections and 'leading edge' extrapolations - as applied by Hough in many studies in the U.S. and Canada[19]- also appear to have been similarly successful.

Still less is known about the predictive performance of the diversion/generational models. There are two ways to test models - comparing the output with reality, or testing the structural assumptions. Where the model is intended for long-range studies of demand for new services, or for the evaluation of a range of design or policy options - most of which will of course never be adopted - the former approach is simply not available. Some recent work on teleconferencing sponsored by the Long Range Studies Divisions of the U.K. Post Office has used the second approach.[20] It suggested that many - but not all - of the behavioural assumptions of the CEPT model are justified. While the sample was small (45 respondents) and the results must there-fore be considered provisional, this study illustrates the feasibility and value of such 'validation and calibration' studies.

Why Do It? The Need for Quantitative Models

The necessity for quantitative forecasting of demand for service innovations (new services or radical changes in services) seems, on a little reflection, less obvious than at first sight. Is it really necessary to predict the quantitative pattern of

demand in order to make good decisions? A recent study of
research needs carried out for Post Office Telecommunications in
the U.K.[11] suggested that quantitative information _is_ often needed,
but only selectively, for example:

- Where a new service places new demands upon the public
  network, knowledge of certain attributes of demand
  (e.g. the call holding time associated with
  information retrieval services or message services)
  may be much more important, because of the
  implications for the design and dimensioning of
  switching centres, than good estimates of total
  demand. Uncertainties about the latter may be
  accommodated by changing the timing of installation of
  extra capacity.

- Where indivisibilities in supply or major 'steps' in
  the cost curve make an incremental approach to
  planning insufficient, it may be necessary to estimate
  total demand. It may, however, be sufficient to know
  how likely it is that demand will exceed a certain
  critical figure in order to decide whether, for
  example, a Viewdata computer would be justified, or
  whether the spatial density of mobile radiotelephone
  calls would make it necessary to go over to a cellular
  radio system.

The Impact of Quantitative Demand Forecasting

It is surprising - and in some ways worrying, when one
considers the embryonic 'state of the art' - that the various
quantitative demand-modelling exercises undertaken so far,
especially in the data communication field, appear to have had a
considerable impact on policy and planning. To take two examples:

- The rapidly increasing interest in automatic dial-up
  telephone conference-call services in Europe, is in
  part a consequence of the findings of the CEPT study,
  with its emphasis on the market advantages of audio-
  only, high accessibility desk-top systems.

- Extensive use has been made of the Eurodata studies in
  planning data networks and services.

While it is gratifying to see the research being used in this
way, there is clearly a risk of a 'fallacy of misplaced
concreteness' based on the sometimes spurious authority of
computer print-out. Where the model really has a substantial
behavioural content and is well-validated, the qualitative
findings of the quantitative model - for example about market
structure or the sensitivity of demand to certain key policy
levers - may be more important than the exact quantitative

estimates. The usefulness of the CEPT study lay, to a substantial extent, in providing an order of magnitude distinction between likely levels of demand for:

- audio-visual and audio-only conferencing
- desk-top versus studio-based systems
- point-to-point versus multi-point systems.

Directions of Development

The experience so far suggests that if the quantitative demand models are to earn themselves a permanent place in the planners' tool kit, the model-builders' practice must change in certain respects:

- Models need to be constructed more in anticipation of specific information needs for policy and planning, rather than as general-purpose machines for simulating markets. Given that only limited skills and resources are available for model-building they should be directed in a planned way.

- Much more effort needs to be devoted to testing, validation and calibration (i.e. the statistical estimation of the parameters of the model).

- Model design should, in many cases, focus on the 'policy sensitivity' of the model as well as other aspects of its ability to model the impact of changes in parameters, policies, or other exogenous variables.

- The high degree of uncertainty inherent in forecasting demand for new services should be recognized, and the role of the model should be seen primarily as reducing this uncertainty.

- In view of the high degree of uncertainty, there is much to be said for modelling approaches that embody cross-checks on their results using more than one method.

A particularly interesting possibility of the latter kind is to combine 'bottom up' analysis which predicts total demand by aggregating data on communication activity and mode choice for individuals or groups, with a 'top down' approach in which estimates of total demand are controlled by reference to an independent analysis of total activity as indicated by overall activity levels, time - or money - budgets or other aggregate measures. The macro-economic work pioneered by Porat[21] and represented in this symposium by the paper by de Chalvron and Curien,[22] is highly relevant here.

THE INTRODUCTION OF INNOVATIONS: CAN YOU GET THERE FROM HERE?

Why Analyze the Process of 'Take off'?... and Can We Do So?

The discussion in the previous sections emphasized the role that knowledge about users plays in the life or death of innovations. Clearly, an extensive knowledge of how a particular innovation might be used once it achieves very widespread acceptance will not be very useful if the innovation itself never 'gets off the ground'. Nevertheless, the more one examines the literature the more one is aware that most of the explanations and models developed so far are –

- 'Static', in that they ignore the dynamic processes by which users and society generally adapt – or maladapt – to the innovation, and the providers of service adapt – or maladapt – to their users.

- 'Equilibrium oriented' in that they deal almost entirely with a situation where all users are fully aware that the innovation is 'available', and knowledgeable about it; and where the 'snowball' problem of conditional decisions-to-purchase and decisions-to-use does not significantly impede its use.

The reductio ad absurdum can be neatly indicated by a 'thought experiment'. Suppose telephones were not yet available, but laboratory studies indicated potentially high effectiveness and acceptability to users, and related demand models pointed to the existence of a large potential market. Even so, it would seem irrational for any one particular person to decide to acquire the first telephone, since it would obviously be an entirely useless instrument when no one else is connected. Of course the first subscriber might join in the <u>expectation</u> that others would soon subscribe to the telephone service: expectational phenomena are clearly a vital issue here.

These are not academic quibbles: an elegant proof of the existence of Eldorado is surely of little value without a map of how to get there. But few means are yet available for making such a map for new services. Take the question of pricing, for example: though some economists have begun to investigate the 'take off' problem[23] none of the economists' prescriptions for pricing policy currently influential in telecommunications administrations – long run marginal cost pricing, for example – depart from the usual assumption of equilibrium. 'Promotional' pricing of innovations usually rates – at best – a dismissive footnote. Michael de Smith and I are using economic models that indicate where varying strategies for cross-subsidy and 'promotional' pricing in the early stages of an innovation can maximize both the Net Present Value of net income, and social

welfare; and make it possible to identify other cases where the
application of conventional pricing rules at the outset kills off
an innovation that is potentially able to contribute positively to
business revenues and economic welfare.[24] Parallel research areas
exist in behavioural research - where learning processes and the
diffusion of innovation are fundamental.

Why Bother to Do Research on 'Take Off' Phenomena?

In earlier sections, I emphasized the high death rate among
telecommunications innovations, and suggested some reasons for it.
Now another reason can be added... the case where everything about
the innovation is sound except the strategy used in introducing
it. In the absence of research into the 'natural history' of
innovation in telecommunication[25] it is dangerous to be dogmatic
about the causes of success and failure. Nevertheless, it may be
instructive to contrast the success rate and pace of <u>service</u> or
<u>performance</u> innovation in the computer industry - where
promotional pricing and strategies designed to guarantee an
initial 'critical mass' for new systems are commonplace - with
that in the telecommunications services.

History and Future Prospects

This field, unlike most of those discussed in the paper, has
virtually no 'track record'. In future, however, I believe it
must become a major part of our research field if users of the
research are to be justly convinced that it can help them innovate
more successfully.

NORMATIVE RESEARCH, TECHNOLOGY ASSESSMENT AND SOCIAL PLANNING

The Value and Feasibility of Technology Assessment

It seems appropriate to complete this paper by shifting focus
from the 'positive' research which seeks to ascertain the facts of
user response to telecommunications innovations, to 'normative'
research dealing with the social gains and losses which they
cause. Efforts to assess technological innovation from a broader
social perspective than the 'verdict of the marketplace' are
hardly new: writers like John Stuart Mill and Karl Marx had a
good deal to say about the matter. Still, technology assessment
studies for telecommunications date back only a few years.
Perhaps the subject would have attracted more attention if the
operation of telephone switching equipment was typically
accompanied by black smoke belching from tall chimneys, or some
other nuisance of that sort. As it is, no telecommunications
equivalent of the innumerable social, environmental, economic and
energy-economy studies of the transportation sector - for example
- existed until very recently. And because the 'social impacts'
associated with telecommunications tend to be less obvious, more

pervasive, and more subtle than those of - say - transportation, the methodological problems have proved to be severe.

The recent work largely dates from the establishment of the Office of Technology Assessment (OTA) by the U.S. Congress, and the start of the Technology Assessment funding by the U.S. National Science Foundation's RANN programme. Studies that have been carried out include work on rural telecommunications (OTA), electronic funds transfer (EFTS) (Arthur D. Little) and inter-actions between transportation and telecommunications (Stanford Research Institute and Communications Studies and Planning Ltd.).

From these early studies, it appears that

- There is a dearth of basic research on the inter-actions between telecommunications and the remainder of the economy, so that it is extremely difficult to assess the implications of changes in tele-communications for such topics as overall economic efficiency and productivity, locational patterns, or transportation demand. A partial exception can be found in the work on the economic value of EFTS, and the work of Pye and others on the location and costs of office activity.[26] While considerable work has been devoted to the analysis of interactions between transportation and telecommunications,[27] much of the theory and model-building involved has yet to be validated by observation of telecommunications-induced changes in economic behaviour.

- Given a judgemental analysis of the likely pattern of these responses of the economy to telecommunications innovation, sufficient data exist to allow useful estimation of the secondary consequence for employment, energy consumption, pollution, urban settlement patterns, transportation demand and other 'physical' variables. An extensive discussion of such impacts will be found in the SRI Technology Assessment study.[27]

- It seems likely that the most important impacts of innovations in telecommunications and information technology will be relatively intangible, involving very fundamental matters such as changes in the organization of work and the pattern of use of time and social contracts away from work, with repercussions not only for economic behaviour but also for fundamental values and habits of thought and behaviour. Moreover, the kinds and magnitudes of 'quantitative' impacts - such as travel substitution or locational change - that may come about are likely

to depend very much on these 'qualitative' changes.
John Clippinger[28] and others are seeking to develop
rigorous research methods that cast light on the ways
technology and people interact at this fundamental
level. The approach draws on basic theories of
linguistics and human information processing. The
work is highly significant, although the
methodological difficulties involved are severe.

It is clear that to do 'Technology Assessment' for telecommunications and information technology usefully is very
difficult, and we are only at the very early stage of learning
how.

Does It Matter?

Society appears to have coped pretty well without such
Technology Assessments in our field for many years, and does not
seem to have suffered by their absence - by stark contrast with
other fields, such as food and drug regulation and transport
planning, where it is clear that a better mechanism for 'social
foresight' could have avoided many costly and damaging errors.
But such a complacent point of view about telecommunications and
information technology leaves one with an uncomfortable feeling.
The lack of appropriate historical data and the ubiquity of the
telephone in the business world make it very difficult to
assess, in the absence of a 'control group' for comparison, just
what changes its adoption brought about in habits of life and
work, in the location of business and so on. One persuasive view
is that we live in a century of massive, often inadvertent and
sometimes unnoticed change - typified, for example, by the social
consequences of the expansion of universal compulsory schooling to
later and later ages; broadcasting; or the development of the
contraceptive pill. The case for attempting to be less
'inadvertent' in the future seems compelling.

The Future

It seems to me that Technology Assessment, harmed by
excessive early expectations and dogged by methodological
problems, is falling from favour - as Cost Benefit Analysis did a
few years ago - in one of those cycles of intellectual fashion
that are becoming all too familiar. This line of work could, and
I believe should, be saved. One way to do this may be to move
away from the 'grand overview' type of Technology Assessment study,
regarding the existing work as essentially a general 'clearing of
the ground' for detailed research, and set about investigating in
depth certain questions that the general studies have highlighted
as important. For example -

- Changes in working patterns, and the psychological consequences for workers, could be investigated in depth in organizations that have adopted advanced office information-handling and communication systems. A wide range of communications-survey, linguistic and psychological research techniques could be adapted for this purpose.

- The actual changes in location decision and travelling patterns... 'working from home' for example... could and should be observed in organizations which are 'leading edge' users of the new technologies.

CONCLUSIONS

A few strong general themes emerge from the particular arguments of this paper, and will I hope provide a useful and provocative starting point for the discussions at Bergamo:

- User research and demand-research for tele-communications is a field of work that, despite its great promise, has so far had a disappointingly small practical impact. It is essential to understand the reasons for this and to seek to correct them.

- Much of the fundamental work on communication behaviour has yet to be accepted widely in practice as a means of evaluating the usefulness of communication systems - or indicating preferred design features. The 'missing link' is research to test the <u>predictive</u> value of such concepts as 'system effectiveness and acceptability'.

- To be credible, research studies of telecommunication innovation - such as field evaluations of new applications - must take full account of alternative means of accomplishing the same 'mission'.

- Similarly, much more emphasis on the testing, validation and calibration of models, as opposed to the elaboration of the models themselves, is needed.

- A shift of focus is needed away from analyzing the long-run equilibrium situation and towards the analysis of how usage of an innovation can 'take off' initially.

- Normative or evaluative research is worth planning, but must move from its present highly general level to deeper analysis of important impacts.

NOTES AND REFERENCES

(1) Some authors write 'need', but mean 'demand' more or less
precisely in the sense defined by economic theory: the amount of
a service or a product that can be sold in the market under some
given pricing arrangement. Others - including a significant but
diminishing band within public telecommunications administrations
- have in mind an approach that can be termed 'paternalistic' or
'socially planned', according to ideological preference. Yet
others leave the meaning of 'need' totally unclear.

(2) Ederyn Williams (1974). Is 'Human Factors' answering the
important questions about telecommunications? In: Human Factors
in Telecommunications, Proc. of the Seventh International
Symposium, Montreal.

(3) The work of the CSG in the formative period up to 1975 is
well summarized in a report, 'The effectiveness and acceptability
of person-to-person telecommunications systems' republished by
Post Office Telecommunications in the U.K. as Long Range Research
Report 3 in May 1975. The Annenberg Schools, the Harvard Program
and the Alternate Media Center all publish informative Annual
Reports.

(4) The Fernmeldetechnisches Zentralamt (Central Technical Office
for Telecommunications) of the Bunderministerium fur Post und
Fernmeldewesen (PTT Ministry), Federal Republic of Germany.

(5) Notably the Box-Jenkins time series methods which are used
extensively in telephone companies and administrations for fore-
casting connections and traffic. S.R. Brubacher, 'Forecasting in
Bell Canada: Applications and Extensions of Box-Jenkins Model
Building Techniques'. Proc. Int. Forecasting Conference,
Windermere, 1977.

(6) Tyler, M., and Cartwright, B. (1974). Forecasting long-term
for Telecommunications Services: methods and problems. Paper
presented at the Aston Conference on Telecommunication Economics.
Long Range Studies Divisions, Post Office Telecommunications
(mimeo).

(7) I use the word innovation as shorthand for any addition to -
or fundamental change in - the range of telecommunication or
information-processing services offered to the user, their
detailed attributes, or the policies, prices and conditions
associated with their use.

(8) A comprehensive review of the teleconferencing work is given
in Tyler and Cartwright, op. cit., and in the CSG report referred
to in note 3. The MIT work is typified by: T.J. Allen and

P.G. Gerstberger (1967), Criteria for selection of an information
source. Report No. 284-67. Alfred P. Sloan School of Management,
Massachusetts Institute of Technology, September 1967.

(9) See for example R.R. Panko, 'The outlook for computer mail',
and R. Pye and E. Williams, 'Teleconferencing: is video valuable
or is audio adequate?', both in Telecommunications Policy, Vol. 1,
No. 3, June 1977.

(10) A useful discussion of some of these issues can be found in:
B. Stapley (1974), A comparison of field trials of tele-
conferencing equipment. Communications Studies Group, University
College, London. Working Paper P/74244/ST.

(11) Tyler, M. et al. (1977). Review and recommendations on the
assessment of demand for new telecommunications services:
Interim Report. Communications Studies and Planning Ltd. in
association with PA International Management Consultants Ltd.,
PO/77001/TY, January 1977.

(12) Hough, R.W. (1976). A state of the art survey and
preliminary analysis. Prepared for the National Science
Foundation by the Telecommunications Science Center, Stanford
Research Institute, Menlo Park, California.

(13) PA International Management Consultants Ltd. (1973).
Eurodata, 1972-1985. Copyright/publisher: Televerket (Swedish
Telecommunications Administration) Farsta, Sweden.

and

Data Transmission Company, Inc. (DATRAN) (1970). Comments of Data
Transmission Company, FCC Docket No. 18920 (special carrier
enquiry), 1st October 1970. Federal Communications Commission,
Washington D.C., U.S.A.

(14) There is an extensive set of reports on the CEPT Business
Communications Studies, but this has not yet been published. The
CEPT Rapporteur for studies of demand for new services is Mr. B.
Cartwright, Post Office Telecommunications/TSS6.1, 88 Hills Road,
Cambridge, England.

(15) Collins, H.A. (1977). Long range forecasting of telecommunications demand: message services in the U.K. Proc. Public
Utilities Forecasting Conference, Bowness-on-Windermere, March
1977.

(16) Federal Ministry of Posts and Telecommunications (1976).
Commission for the development of the telecommunication system.
(Kommission für den Ausbau des technischen Kommunikationssystems:
KtK.) Telecommunications Report. Federal Ministry of Posts and

Telecommunications, Bonn, Federal Republic of Germany.

(17) M. Dormois, F. Fioux, and M. Gensollen. Evaluation of the potential market for various future communication modes via analysis of communication flow characteristics.

(18) See the KtK report: Federal Ministry of Posts and Tele-communications, Federal Republic of Germany, op. cit.

(19) Many of these reports are not publicly available, but the approach is typified by an early and publicly available report for NASA: A study of trends in demand for information transfer. SRI, Feb.'70.

(20) Williams, E., and Young, I. (1977). The choice to travel or teleconference amongst loudspeaking telephone users. Working Paper E/77077/WL, Communications Studies Group, University College London.

(21) See for example: Porat, M. (1976). The information economy. Institute for Communication Research, Stanford University, report no. 27, August 1976.

(22) J.G. de Chalvron, and N. Curien. Information, energy and labour force.

(23) See for example: Rohlfs, J. (1974). A theory of inter-dependent demand for a communications service. Bell Journal of Economics and Management Science, Vol. 5, No. 1, Spring 1974.

(24) Appendix E, 'Pricing and the Dynamics of Demand Growth', in 'Review and recommendations on the assessment of demand for new telecommunications services', op. cit.

(25) An historical analysis of the causes of success and failure in innovation would be extremely valuable. Some work of this kind has been undertaken by the Science Policy Research Unit (SPRU) at Sussex University in England, with very interesting results summarized in Appendix B of the CS & P report to the U.K. Post Office cited in note 11. Two key points emerge: that a qualitative understanding of the nature and requirements of the market may be more crucial than the possession of a sophisticated quantitative model of demand; and that demand research prior to the extensive introduction of a new product or service is necessary since an 'ad hoc' process of modification of a product or service after introduction is more often associated with failure than with success. This should certainly not be interpreted as implying that services should not be modified as necessary subsequent to their introduction. It does suggest that such a pragmatic process on its own is unlikely to converge quick-ly and reliably enough as the 'right' service configuration or

marketing strategy to ensure success: prior research is clearly
necessary.

(26) See for example: Pye, R. (1976). The effect of telecommunications on the location of office employment. Omega: the
International Journal of Management Science. Vol. 4, No. 3,
pp. 289-300; Goddard, J.B., and Pye, R. (1977). Telecommunications
and office location. Regional Studies, Vol. 11, pp. 19-30.

(27) See: Stanford Research Institute (1977). Technology
Assessment of telecommunications/transportation interactions.
(3 volumes), prepared under contract NSF-C1025 for the National
Science Foundation (Office of Exploratory Research and Problem
Assessment) by Stanford Research Institute, Menlo Park,
California, in association with Communications Studies and
Planning Ltd.

(28) Research on the impact of technology on workers' value-
systems is being funded by the U.S. National Science Foundation
and directed by Dr. John Clippinger at Kalba Bowen Associates, 12
Arrow Street, Cambridge, Massachusetts 02138.

TECHNOLOGY AND STRUCTURES - MAN AND MACHINE

R. J. Chapuis

Senior Counsellor, CCITT

Geneva, Switzerland

In our developed countries, society is now tending to describe itself as "post-industrial". The main distinguishing features of this society, which have been the subject of many studies, may be defined as follows (Ref. 1):

1) economic sector: the change from a goods-producing to a service economy;

2) occupational distribution: the pre-eminence of the professional and technical class;

3) axial principle: the centrality of theoretical knowledge as the source of innovation and policy-making for society.

Now, telecommunications

- are service-providing activities;

- essentially depend on the decisions of a professional and technical stratum of society; and

- are based on centralized decisions arrived at in accordance with the principles of technological and econometric research.

Telecommunications thus posses all the main attributes characterizing an activity of the post-industrial society.

Another point that should be stressed in passing is that,
amid the stagnation of the present-day economic scene, the telecommunications sector stands out as one which continues to flourish
vigorously: the growth rate of telecommunications rises
inexorably, and now stands, for instance, at the level of 20% a
year for international traffic.

In contemplating telecommunications we are by definition
contemplating the post-industrial society, thereby engaging in
what amounts to a sociological analysis.

Since the post-industrial society as defined by the analysts
is still in the stage of gestation or, at most, infancy,
reflections on telecommunications are automatically reflections on
the future of our contemporary society, on the society of the
future. This is the whole point of this Symposium.

Telecommunications may be considered from many angles. For
present purposes, we shall first eliminate everything pertaining to
technological and scientific literature, but shall bear in mind
three points. First, the order in which these two adjectives have
been placed, "technological" first and "scientific" second, reflects
historical developments. For instance, if we consider the century
during which telephony has existed, we see that during its first
half developments were essentially the work of artisans, draughts-
men and technicians. It was only during and immediately after the
First World War that pure scientific research began to supersede a
technology based on quasi-craftsmanship and to serve as the
driving force for the development of telephony.

Second, technical and scientific literature on telecommunications is proliferating apace. The Central Library at the
International Telecommunications Union subscribes to over 700
periodicals devoted exclusively to telecommunications. Third,
against the background of this explosion of technical publications,
it is only very recently and on a very modest scale that the
economic and sociological aspects of telecommunications have begun
to arouse any interest.

These economic and sociological aspects provide the reason
for our meeting in Bergamo. If one looks around to see what has
been published in this area, one finds: among the innumerable
technical journals, only one specializing in the economics of
telecommunications, the Bell System Economic Journal, and most
recently the new review "Telecommunication Policy"; and a variety
of articles scattered in the varied reviews of different speciali-
zations - often telecommunications, sometimes sociology.

Once this bibliographical check has been completed - and the

task is relatively difficult because of the wide dissemination of
the source material - I think it can confidently be stated that
nearly all the studies conducted fall into two main categories:

   i)   the first is microsociology or microeconomics, or, to use
        a neologism that I find most apposite, picoeconomics or
        picosociology. This area is largely covered by the
        concept "human factors", that is to say, the study of
        relations between man and machine, between the subscriber
        and his network.

   ii) the second, at the other end of the size scale, is macro-
        economics, conducted at a national level. The yardstick
        applied here is the Gross National Product.

   The content of this paper will, however, fall outside the
area defined by these two categories. It will be concerned with
what I propose to call the "endoeconomics" of telecommunications,
to coin another neologism. It will consist of some comments on
the internal structures of the organization which directs and
manages the machinery **or, rather, the great machine of telecommuni-**
cations.

THE ENDOECONOMICS OF TELECOMMUNICATIONS

   I refer to a great machine because an essential feature of
telecommunications is that, technically speaking, all the equipment forms a unified whole. I know of no other area of human
activity in which everything depends as much as it does in telecommunications on the complete interlocking of machinery and
equipment. Consider the case of a manufacturing company, even a
multinational enterprise: it will have workshops and factories
spread over an area which may be vast, but each of its production
units will be relatively autonomous, limited at most to the
dispatch and exchange of goods and products between the factories.
In the case of railway or air transport, the train or aircraft
will carry its cargo as far as it can go, until the gauge is
changed for the train, or until the load breaking point ... But
in telecommunications no connection can be established without
perfect standardization of the essential parameters all along
the circuit.

   The organization responsible for telecommunications,
whether an administration or a private company, thus manages -
plans, constructs, organizes and operates - a single machine:
under the twofold influence of an unprecedented technological
expansion which has caused production costs to plummet, **and**
correspondingly, of ever-expanding demand, the machine has now
become enormous.

In its early days, the telephone was a little fish which, in
the European countries, had great difficulty in insinuating itself
into the wake of its older brother, the telegraph. But this little
fish of the years 1880 - 1920 has now grown into a whale or a
mastodon.

To pursue these zoological similes, a question arises with
regard to the telecommunications authority. Is it really at the
"whale" stage of evolution? The simile seems to fit a number of
these telecommunications organizations perfectly. Do we not see
them swimming merrily in the ocean of our modern society, most
often practically unnoticed and ignored by the public, at most
occasionally sending up a little water spout to indicate their
presence at a few infrequent public relations events? Do we not
see them thriving on the ample food provided by the nourishing
plankton and small fry represented by the ever-growing density
and numbers of their subscribers? "Whales, full of well-being
and glad to be alive" is the optimistic view of the situation.

But of course another, pessimistic, view can also be taken,
that of the telecommunication organization as a mastodon, which
may be assuming the form of a diplodocus or a dinosaur.
Palaeontologists claim that these beasts of the Tertiary period
became extinct because their skeletons were too ossified to
enable them to adapt to the changing environment...

The study of the internal behaviour of organizations responsible for telecommunications and of the way in which their structures adjust and evolve in accordance with the development of
society and technology is thus a topic of major importance.

It could already become a special branch of study, and with
your permission I would suggest designating it as

the endoeconomics of telecommunications

i.e., "the internal economics".

This endoeconomy would be a most fascinating subject for
research at this time because it is:

   i)   a virtually unexplored subject;

  ii)   a problem area which should be of crucial importance to
        telecommunication organizations;

 iii)   a highly topical subject.

It is unnecessary to stress this last point to American participants in this Symposium, who are well aware of the upheavals and even clashes caused in the United States this year by the Congress's current review of the Communications Act of 1934 which governs the terms of reference of the various US carriers and Government agencies.

HUNTLEY'S LAW

We now enter into the substance of our theme and provide a first and fully characteristic example of these studies of internal economics and structural analysis.

There exists in telecommunications economics a relationship which is well known and which was designated "Huntley's Law" in 1975 (at the ITU Symposium held in connection with the Telecommunication Exhibit in Geneva) (Refs. 2 to 5).

This is the key relationship between the annual turnover and the fixed assets of a common carrier telecommunications organization.

This relationship was first brought out in 1967 by Mr. H. R. Huntley, formerly Chief Engineer of the American Telephone and Telegraph Company (AT&T), who subsequently acted as Consultant to ITT for several years (the period covering his publication of 1967). Many articles, especially those published between 1972 and 1974, have demonstrated the full validity of Huntley's Law on the basis of actual figures.

This Law fixes at roughly 3 the ratio of the invested capital value of the plant of the telecommunication agency* (numerator) to the gross annual income (turnover) of the telecommunication agency (denominator). There are critical upper and lower thresholds: deficit financial operation is associated with values greater than 4 and inadequate service quality with values less than 2.

Huntley's Law is fundamental in telecommunication economics. It brings out the structural difference between telecommunication undertakings and ordinary industrial and commercial companies. In the latter the situation is just the opposite, capital invested in the plant being more often than not, less than one-third of the annual turnover. In other words, in relation to turnover the fixed capital of a telecommunications operating agency is about ten times that of an ordinary industrial undertaking.

*More accurately: "existing investment, i.e., accumulated annual investment less accumulated annual depreciation".

It is the size of the capital invested in the network and
installations of a telecommunications operating agency which
justifies describing it as a heavy industry, while industries
manufacturing consumer goods and equipment (including telecommunications equipment) are correctly designated as light industry.

On recognizing the validity of Huntley's Law, not without
some initial surprise, one question naturally arises. How is it
that in examining the Annual Financial Reports received in the
ITU Library from nearly all countries of the world, one always
finds, over successive years, this same mean value of 3 for the
financial ratio?

This can of course be explained as an Act of God: it is so
because it is not otherwise. Such an explanation will not
however appeal to the rational mind. Personally, I believe that
there is a historical explanation which serves as a basis for
Huntley's Law.

At the beginning of the century, when the development of
telecommunications really started, two completely different types
of structure began to emerge in North America and in Europe.

In the United States, the home of free enterprise, telecommunications were the concern of private companies. These private
companies, which held a virtual monopoly in their sector, were
very soon subjected to regulation by Government departments at
both the state and federal levels. This regulation of telecommunications was similar to that exercised over, for example, means
of transport; it was modelled on the regulation of railways, which
was the cause of so much political turmoil in the USA in the
period 1860-1900.

In Europe, after a brief period of telephone development
under the auspices of private agencies, the all-powerful State -
as a result of persisting Napoleonic concepts or the influence of
Hegelian philosophy - drew all telecommunications activity into
the purview of its Posts and Telegraphs Administration, incidentally to the great detriment of telecommunications development.

In European countries, as in other countries which have more
or less modelled their administrations on the European ones, no
separate financial accounting for PTT administrations existed for
a very long time. Expenditure was entered in the debit account,
merged with all the other expenses of the national budget.
Income was lumped together with all other state income, in the
same way, for instance, as revenue from state lands and particularly with income from taxation. There was no correlation
between income and expenditure for telecommunications; the same
applied to the postal services.

It was not until the 1920's, largely owing to the growing financial importance of telecommunications, that national parliaments, one after another, began to grant their PTT administrations a degree of financial autonomy, such as the "Budget annexe" for the French PTT. A framework for this autonomous structure had then to be established, and for telecommunications the accounting model followed was more or less that provided by the financial rules of the American "common carriers", especially those of the American Telephone and Telegraph Company.

Thus, it is osmosis with the American "regulated industry" model that has served as the pattern for the financial structures of telecommunication organizations throughout the world. Huntley's Law is therefore firmly based. It would very probably be possible:

1) to take as a basis the financial constraints imposed by Government regulations of the American type,

2) to set up "in abstracto" a financial model of a regulated company subject to such constraints,

3) and to prove by simple accounting calculations that the end result is Huntley's Law,

The entire theoretical model could be constructed in legal and financial terms without any reference whatsoever to the activity - in this case, telecommunications - to which the model related.

This, incidentally, would be an ideal subject for theoretical studies at the most advanced level by young chartered accountants seeking higher qualifications.

SOME CONSEQUENCES OF REGULATION

Now, what are these famous constraints resulting from governmental regulation? It is difficult for a non-American to speak with any authority on such a delicate subject, which has caused so much controversy and has consequently given rise to so many reports and so much other documentation of a legal rather than a technical nature. With apologies to the Americans for my limitations in this respect, I shall try to give a very rough sketch of what these constraints may be, mainly for the benefit of the non-American participants. To this end and to be as specific as possible, I shall compare (if such a comparison is not regarded as sacrilegious, since there are certain taboos in this regard for those who are not involved in these matters) the financial balance-sheets of two companies: the first, an American common

carrier and the second, a free undertaking not subject to the
former's governmental constraints.

To give honour where it is due, they are two household
names, ATT and IBM. Their financial status is comparable and,
being among the world's leaders, they are both in the vanguard of
electronic progress; it is hardly necessary to enlarge any
further on a comparison which has already been made on innumerable
occasions. For our purposes, I shall merely refer to their annual
financial reports and particularly to the rules governing the
establishment of their balance sheets at the end of the financial
year.

In the case of the ATT, as in that of any other American
common carrier, an accountant would say that the balance-sheets
are drawn up in a very conservative manner:

- all expenditure relating to new plant (including labour
  costs, even for staff engaged in installing the equipment)
  is charged to a capital investment account, from a very
  low threshold;

- the investments charged to this account are amortized,
  not in accordance with legal provisions allowing accelerated amortization, but linearly over relatively long
  periods matching as closely as possible the estimated
  life of the equipment installed.

For IBM, as for any other "non-regulated" undertaking, the
service life of equipment assumed for purposes of amortization
is much shorter and a considerable proportion of medium-term and
even long-term production costs are charged to current costs;
that is to say they are charged to annual operational costs,
whereas the administrative stringency imposed by "regulation"
would require them to be charged to investment costs in the case
of common carriers.

Regulation thus has a damping effect on consumer charges,
since any investment cost, however small, is ultimately spread
out in time and is carried over to subsequent financial years.
This damping effect is greater in periods of inflation which are
so characteristic of our times.

For a telecommunication agency - whether American or not,
since the accounting rules described above have become standard
practice in telecommunications - the investment amortization
period, more generally the amortization system, and the system
whereby any expenditure with medium or long-term effects is
carefully assigned to investment, results in:

i)   intensive capitalization of the telecommunication agency,
     as expressed by Huntley's Law,

ii) an increasing appetite for fresh capital, which is bound
     to cause numerous difficulties for telecommunications
     development,

iii) but, on the other hand, lower operating costs and hence
     lower rates for users.

Cynics are bound to sneer when regulation - whether imposed
by the government or built into the procedures followed in
compiling the budgets of telecommunication administrations - is
extolled as the most practical means of protecting the consumer's
interests. They will start out by pointing to the disastrous
situation of the American railroads and the regulated companies
operating them, their thesis being that a regulated company
meeting the full blast of competition from free enterprises is
automatically faced with bankruptcy. That was indeed the case
of the American railroads: competition from road and air transport has practically put an end to passenger transport by rail
in the United States.

According these criticisms, only the existence of a de facto
or de jure monopoly could ensure the existence of governmental
regulation of activities and the survival of regulated undertakings. In fact, the terms "monopoly" and "regulation" are by
definition inextricably linked. The two concepts have been
connected from the outset of American legislation on regulated
companies, as in the case of concessionary companies in other
legislations.

We thus see the gradual emergence of the basic components
which define the structures of a telecommunication agency, that
is to say, the vectors of our telecommunication endoeconomies:

1)   extreme centralization of technical, operational,
     financial and therefore administrative decisions

2)   considerable development of the telecommunication
     agency from a technical and hence a financial point
     of view

3)   monopoly

4)   financial constraints which result from explicit or
     implicit regulation and which run counter to the
     financial rules of free enterprise

5) as a consequence of 4), "overcapitalization" (without
    attaching any pejorative meaning to this word)

6) as a consequence of 5), all the difficulties arising
    from the need to find capital for long-term investment.

Each of these vector characteristics could alone constitute
a chapter heading or even a treatise in book form. I mention
them to open up some horizons. Two points, however, are particularly worth emphasizing at this stage:

- the last-named vector, the need to find the necessary
  capital

- and the inertia effect resulting from vectors 1) and 2),
  extreme centralization and the considerable size of
  telecommunication agencies.

RAISING CAPITAL FOR INVESTMENT

Finding the large amount of capital needed for a telecommunication agency is a major concern of management. There are
several strategies, each depending on the conditions specific to
the country concerned.

In countries where telecommunication services are provided
by private companies, with the United States as the outstanding
example, capital is raised on the money market. But there is
one essential condition for stimulating the flow of this capital,
namely the confidence of investors or shareholders. This
confidence requires "an appropriate rate of return", i.e., an
appropriate rate of benefit for the company over the years. The
profit rate depends on tariffs; any tariff increase must be
authorized by the regulating government department, and it will
be seen that, as in any regulating system, there is a feedback
loop between the telecommunications operating agencies and the
regulating government department. It is even possible to develop
a "system analysis" of this macroeconomic machinery.

In countries where telecommunications are the concern of
states and administrations, a distinction should be made between
two situations:

- that of countries with developed economies (that is to
  say, the European model);

- that of countries with underdeveloped economies (third
  world countries).

The European model itself covers a whole range of situations, generally arising out of the legislation and administrative regulations peculiar to each country which, incidentally, by their rigidity, have served as the decisive factors for the development rate of the national network of each of these countries.

i)   In some European PTT administrations the considerable amount of capital collected in Post Office Savings Funds and postal cheque accounts – which are very widely used in Europe and are favoured by a large number of depositors – is available to the administration for the development of its telecommunication services (at an appropriate rate of interest, of course). In other words, the telecommunication agency uses its own administration as a bank. Switzerland provides an excellent example of this practice.

ii) Other less fortunate administrations do not have this possibility for financing, as the public treasury appropriates the funds collected by the financial services of its PTT administration. Hence, loans have to be raised, first of all by the PTT administration and, since the various State services compete among themselves for loans, other sources of financing have to be sought, more or less based on the American model, if not in structure, at least in their results. An example of this is provided by France.

In countries with an underdeveloped economy, the essential capital can only come from outside financing, which may be:

-    bilateral between two countries – with government loans, action by export/import banks or long-term credits from equipment suppliers;

-    multilateral, with financing by international bodies such as the World Bank, the European Bank, the OPEC Bank and so forth.

Apart from the American model and the European model with internal sources of capital coming from the PTT administration, models which can generally be claimed to be successful, the financing of telecommunication investments will call for a great deal of imagination and also a vast fund of energy and determination when it comes to negotiating the maze of administrative regulations and the variegated combinations of aid which may be provided by outside organizations.

The fundamental requirement for telecommunications development is thus <u>unequivocal determination</u> of the country's policy-makers. There must be:

- first of all, a thorough grasp of the problem, since the
  essential role of telecommunications in the economic
  development of a country must be duly recognized from
  the outset. An excellent formula to be borne in mind in
  this connection is: "Telecommunications represent the
  key opening the door of modernity to an economy".

- stubborn persistence, since cutting the Gordian knot of
  the administrative and legislative bonds which hamper the
  access of telecommunications to the capital required for
  their development presupposes action which must be
  decisive and which must be sustained by consistent
  activity over many years.

To cite typical examples here would be to engage in a political analysis outside the scope of the Symposium, but there can
be no doubt that changes in the rate of development of a country's
telecommunications facilities can safely be attributed in each
case to the unequivocal determination of a handful of men, heads
of government or state, who are clearsighted about the requirements of a modern economy and the part that telecommunications
can play in it.

SOME STRUCTURAL PROBLEMS

Great strength of character is needed to manipulate the
unwieldy legislative, administrative and financial machinery
which weighs down the structures of a telecommunications agency.
As we have already pointed out, this unwieldiness is the result
of the inertia generated both by the extreme centralization and
by the considerable size, especially where financing is concerned,
of a telecommunication agency. Examples of this abound and I
shall confine myself to citing only two:

- the first, which is a problem of vital importance, is
  provided by the administrative structure of the telecommunication agency;

- the second is quite a minor problem, but one should not
  always be thinking at the macroeconomic level without
  sometimes descending to more mundane realities. It is
  the problem of establishing international accounts.

The forms and structures of telecommunication agencies and
their degree of autonomy vary from one country to another.
Agencies may be schematically classified according to three basic
criteria:

i) whether the agency takes the form of a company, private or nationalized, or of a state administration;

ii) whether the agency is concerned only with telecommunications or is a branch of a PTT administration (hence associated with postal services);

iii) generally, but not necessarily, depending on the first two criteria, the degree of freedom of action and autonomy for the telecommunications agency or for the telecommunications department of a PTT administration.

The CCITT, through its economic Working Party known as GAS 5, conducted a whole series of studies on this subject, published in 1972, in an attempt to reply to the question, "Is there a relationship between

- the structure of a telecommunication agency, on the one hand, and

- the profitability and quality of the service it provides, on the other hand?"

Reference may be made to the relevant chapter (Chapter IX) of the CCITT Handbook "Telecommunications - Economic studies, 1972". The conclusions of this Capter are expressed very diplomatically, in somewhat evasive terms, while of course suggesting a certain degree of autonomy and freedom of action for telecommunications agencies.

It is important to note that this subject is very topical: the parliaments of many countries have had to consider more or less radical changes which would provide telecommunications with some autonomy and would set them up as state enterprises combining the interests of a public service with the efficiency of an industrial organization. The case of Australia presents a typical example and could be singled out as a model.

In many other countries, however, where the administrative and economic situation is comparable to that of Australia, we find concrete examples of the phenomenon of unwieldiness and inertia which we have already criticized. These countries seem to lack the strength to break out of an administrative mould dating back over a century, to the time when telecommunications consisted of a few telegraph lines. There are a number of explanations for the lack of interest which these countries display in restructuring their telecommunication agencies:

i) the public is poorly informed; users know what services

are available to them and only a very few people have any
idea of what telecommunications represent in other
countries or of the potential offered by the state of the
art;

ii) the influence of the press in these matters may be
described as blunting the public's awareness and is often
decisive. PTT administrations which give their newspapers
clearly preferential postal rates are sacred cows. To
take the example of activities of equivalent economic
importance, it will be noted that the space allocated to
telecommunication subjects in the press is only one-tenth
of that given up to topics more calculated to stir up
passions on the ecological or political front;

iii) the labour unions which are so strong in PTT administrations are generally horizontally structured in these
countries. Between the postal service which is a big
employer and the telecommunication service, which is much
less labour-intensive, union positions very often represent the majority opinion of postal workers rather than
that of telecommunication operatives;

iv) many other no less cogent reasons could also be found.

It is therefore in the nature of things for the structures of
a bygone era to be perpetuated, even when the sociological and
technological environment is completely transformed - one more
proof of the resistance to change of human institutions, it
applies completely to the telecommunications structures.

My second example of administrative inertia is, unlike the
first which is fairly well know, a completely obscure little
story.

It concerns accounting between administrations for
international services, in this case the telephone service. When
international telephone operations began in Europe in the 1920's,
the service was manual with preparation, with a long waiting
period for the subscriber and intervention by operators at both
the outgoing and incoming ends of the international line, not to
mention all the other operators who might have to take part in
setting up the call. It will be recalled that the international
circuits between two European countries could then be counted on
the fingers of one hand. Telephone communication was therefore a
rare commodity, dealt out, so to speak, with an eye-dropper and
with maximum precision. Its unit price per three-minute call was
arrived at by adding up the shares of the terminal and transit
countries. The duration of the call was measured by the international outgoing operator and was entered on the ticket used for

charging the calling subscriber, but the administration of the country of arrival also noted the duration and in the evening the call tickets were collated between the international outgoing and arrival exchanges.

With the increase in traffic, this checking of the duration of calls became more and more cumbersome and expensive. The international accounts were therefore based on the recording of the international operator of the country of departure, because it was she who made out the ticket for billing the caller. The administration of arrival however retained the right to check the duration of calls by means of certain traffic spot checks and to ask the administration of departure to match up the duration of the monitored calls.

As the traffic continued to increase, these sampling checks became an obsolete administrative rule and it was recognized as a general principle that "international accounting is performed at the discretion of the country of origin of the call".

After the Second World War, the traffic became semi-automatic, using numerous and multifarious channels and routes. This situation prevailed for over 10 years before a radical reform was introduced in 1968: the amounts paid by administrations for the reciprocal use of telecommunications facilities outside their territories were thenceforth dissociated from the exact total of the amounts charged to subscribers with respect to each call. The 1968 reform also systematized the use of metering devices – call time meters – installed on the international circuits for measuring the traffic exchanged between administrations and providing the data base for setting up the international accounts.

These metering devices are operated by the signals – answer signal by the called party and clear forward signal from the calling party – exchanged on international circuits and present at both the incoming and the outgoing ends of the international circuits.

There were several excellent technical reasons – which I shall not go into here – why the metering devices were installed at the incoming end rather than at the outgoing end of the international circuit. The latter would have made it possible to conform with the standard commercial practice, whereby it is the provider of the service (that is to say, the administration of arrival or transit) who sends the bill. It is indeed unusual for the client and user (in this case, the administration of origin), after receiving a service, to draw up the accounts and the bill for what he has to pay.

Yet to depart from a tradition, however much it may conflict

with the elementary rules of commercial practice, and to introduce a change which would have been technically advisable and
relatively practicable at that time, seemed to be as insurmount-
able a problem as the switch from left to right-hand driving.

Rigidity of the structures is due to:

- traditions often going back to outdated historical
  contexts,

- administrative unwieldliness due to centralization,

- the effect of inertia engendered by the size of the
  undertaking,

and this rigidity must be denounced as a characteristic symptom
of the "dinosaur" effect I have already mentioned.

It must be realized that science and technology are progres-
sing by leaps and bounds. The machine is now hurtling at such
a speed that its driver can no longer control it. As an example,
let us take switching equipment, which is in a state of
extremely rapid evolution.

The technological development of this equipment is advancing
at an ever accelerating rate. In many countries, such electro-
mechanical systems as Strowger have been the only standard systems
for 40 years, if not more. The crossbar generation will have
lasted for 15 to 25 years. And now that a new generation of
Stored Programme Control - SPC - exchanges is coming into being,
the period of development and introduction of a new system is
about 10 years, corresponding to the advent of new technologies.

SPC exchanges have emerged in the wake of computers. The
computer industry, has thus been an essential driving force in
the development of modern switching systems and the spectacular
advances of modern telecommunications are ultimately due in
part to the dynamism of those free enterprises, the computer
manufacturers. Digitalization and the large-scale advent - or
breakthrough - of digital techniques, with integrated switching
and transmission, are the result of the increasing pace at which
the performance of electronic components has been stepped up
over the past ten years, at the same time leading to a drastic
reduction in, for instance, the price per function of LSI
components.

CONCLUSION

It seems to me that evaluation of the sociological and
technological environment of our times and scrutiny of the
structures of telecommunication agencies and of the ways in which
they should be modified - either by radical reforms or by gradual
measures - fully warrant objective, scientific studies of this
subject, which we have christened the "endoeconomy of
telecommunications", and large-scale national action.

REFERENCES

1. D. Bell   -   The Coming of the Post Industrial Society -
                         Basic Books, New York, 1976.

2. R. H. Huntley   -   "Some Ideas Regarding Economics of
                         Telecommunication" - Electrical
                         Communication, Vol. 42, No. 1, 1976,
                         pp. 6-21.

3. L. L. Bower   -   "Telecommunication Market Demand and
                         Investment Requirements" - Telecommunication
                         Journal, Vol. 39, No. III, pp. 177-178.

4. R. J. Chapuis   -   "Common Carrier Telecommunications In the
                         World Economy" - Telecommunication Journal,
                         Vol. 39, No. X, October 1972, pp. 601-620.

5. R. J. Chapuis   -   "Telephony Is a Heavy Industry" -
                         Telecommunication Journal, Vol. 42,
                         No. XI, November 1975.

THE ROLE OF TELECOMMUNICATIONS POLICY ANALYSIS IN SERVICE
PLANNING

LAWRENCE H. DAY

BELL CANADA

ROOM 800, 2001 UNIVERSITY STREET, MONTREAL, P.Q.

INTRODUCTION

This paper is an examination of the role of Telecommunications Policy Analysis in the planning and development of new
telecommunications services. The focus here is from the point
of view of a common carrier which must plan and develop new
telecommunications based services in a policy driven environment
rather than that of a government body which must develop various
policy positions. The author's observations are based both
upon practical experience as a long term service planner in a
large telecommunications carrier and as an observer of the role
of policy development in his position as General Editor of the
journal: Telecommunications Policy.

The paper begins with an overview of the total service
development process. The various steps are reviewed as a
means of illustrating both the process itself and the role that
telecommunications policy analysis plays in this process.
Several illustrations are offered throughout the paper. The
concluding section reviews the impacts that this form of analysis
can have on service planning and development.

The telecommunications services development process follows
these steps: New service concepts are derived from two basic
sources: 1) user needs analyses and/or 2) technological
innovation. These concepts are then tested against a number of
market analysis criteria. The concepts that survive this
preliminary market analysis should then undergo analysis to
determine their potential social impacts, both positive and
negative. Services that meet all of the tests presented to

this point should be then examined against the current and
forecasted telecommunications policy environment.   Then, and
only then, should these new concepts be tested in the real market place.

This review will only examine the role of telecommunications
policy analysis in the development of new service concepts.
The impact of telecommunications policy upon existing services
will not be examined even though it is an important field in its
own right.   This focus is on the innovation process in the
telecommunications arena and the position of policy analysis in
that process.   The paper will not only review the theory but will
also note several key areas where future interactions of service
and policy development are likely to have a significant impact.

THE SERVICE PLANNING PROCESS

The Development of Service Concepts

User Needs Analysis.   The analysis of user needs is an ideal
means of identifying new telecommunications services designed to
serve those needs.   The concept is widely accepted but the
reality often falls far short of the theory.   There is not any
accepted "user science" which has an acceptable track record of
identifying unmet user needs.   Conventional market research techniques are quite useful in identifying evolutionary steps in the
existing product lines of telecommunications carriers.   However,
in this paper we are concerned with more ambitious steps in service
development which lie beyond the development of normal product/
service evolution.

User needs analysis falls into two rough categories.   One
can be called "behavioural analysis"and the second is normally
termed "technological forecasting".   The former approach deals
with analysis of individual and group interactions with existing
or simulated telecommunications services under controlled
scientific conditions.   This normally involves the use of techniques adapted from behavioural psychology and other social
sciences.   The findings may result in the development of new
telecommunications service technology designed to meet unmet
needs identified through the behavioural analysis.   Existing
services may also be modified or withdrawn as a result of this
form of research.   Best known examples of this type of research
are the studies of the Communications Studies Group of University
College (London)(1)and at John Hopkins University (2).

Another form of behavioural analysis is the use of survey
research.   Survey research can be used to capture user attitudes

in their "real world" environments.   Survey research has been
used to gain basic information on user's communications activities
(as with contact-record diary studies), on users' attitudes
towards existing communications media, and on users' attitudes
towards new or proposed media.

The field of technological forecasting contains a number of
techniques that can be used to help identify new telecommunications service opportunities.   These range from trend extrapolation
through to "expert opinion analyses" via the DELPHI technique and
its derivatives.   In each of these cases, the focus should be
to identify new service possibilities.

Technological Innovation.   The discussion in the section on
user needs analysis implies what many would describe as an ideal
model.   An understanding of user needs leads us to develop various
technological means of offering services that satisfy latent
demands for telecommunications services.   In reality, technology
itself has often been the driving force in telecommunications
services development.   This is particularly true as the computer
and telecommunications fields have merged into an uncomfortable
alliance.

There are few who would argue that technology alone should
be the mainspring in the development of new telecommunications
services.   The often-quoted experience of AT&T with PICTUREPHONE®
is public testimony to the dangers of a technology-push approach.
The best approach is a symbiosis of both techniques with frequent
interactions and feedback between the two.   In some cases, basic
user needs should direct technology research efforts.   In other
cases, behavioural research techniques can be used to evaluate
the service concepts that emerge from the technological research.
In any case, the identification of a new service concept is only
the first step in a long process.

Market Analysis and Trial.   Service concepts that appear
viable should undergo extensive cost/benefit analyses.   These
analyses are both from the viewpoint of the common carrier and
the potential user communities.   Much of this work is, by
necessity, "desk research".   Desk research which deals with
secondary data may be supplemented with various forms of field
research.   The extent of the original research that is conducted
by the marketing or business development organization is determined
by a number of factors: budget, the amount of time available
(original research is often an intensive time consumer!), the
perceived value of the additional data, and the competitive climate
(field research often warns competitors of the service/product
directions that the carrier is planning to take).   This point
on competition would have appeared absurd 10 years ago. However,
Canada and the U.S. now have several levels of direct competition

between retail and wholesale communication common carriers.   In
addition, the competition between the telecommunications and the
computer industries raises new areas of concern for those planning
new telecommunications services.

Field research may include customer surveys or interviews.
However, market researchers argue that the most exact form of
market analysis is a product/service trial.   Prototype services
are developed and exposed to user communities in a variety of
environments.   The feedback from user trials can lead to service
modification, cancellation of the concept, or the development of
specific plans to introduce it into the marketplace.

The market analysis and trial process weeds out the weaker
service concepts.   In a more simplistic time, the successful
concepts would be introduced with a close eye to the new service
concept's acceptance in the marketplace.   However, in recent
years, many organizations have become concerned with the social
implications or impacts that arise from the use of their services
and products.   This form of analysis is often known as "Technology
Assessment".   The next section of the paper reviews this area
in some detail.   The concept of technology assessment is very
important in placing the role of policy analysis in perspective.
The line between the two activities is somewhat blurred and
arbitrary distinctions may have to be made.

SOCIAL ASSESSMENT (TECHNOLOGY ASSESSMENT):   A CASE STUDY

Introduction.   The material in this section of the paper
will introduce the concept of technology assessment and note
several studies that the Business Planning Group (BPG) of Bell
Canada has conducted.   One case example will be traced through in
some detail in order to illustrate the scope of technology
assessment work and its overlapping relationship with telecommunications policy analysis.   This example cuts across a number of
studies which have dealt with the inter-relationships between
travel and telecommunications.   It will also serve as a partial
case study to illustrate the role that telecommunications policy
analysis can play in a private carrier.

The definition of technology assessment that has been used
is one of the accepted ones in the technology assessment
"profession".   Our shorthand definition is that technology
assessment is the advance identification of the secondary impacts
that often arise from the uses of technology.   Our more formal
definition has been the one presented by Vary Coates:

"Technology assessment is the systematic identification, analysis, and evaluation of the real and potential impacts of technology on social, economic, environmental, and political systems and processes. It is concerned particularly with the second and third order impacts of technological developments; and with the unplanned or unintended consequences, whether beneficial or detrimental, which may result from the introduction of new technologies or from changes in the utilization of existing technologies. Technology assessment seeks to identify society options and clarify the trade-offs which must be made; this approach is designed to provide an objective and neutral input to public decision-making and policy formulation with regard to science and technology." (3)

The Business Planning group has undertaken T.A. activities in five areas. The actual work ranges from projects that follow a rigorous T.A. methodology to those that are involved with exploring fundamental interactions between telecommunications and other sectors of society. These activies are:

A) The study: A Technology Assessment of Computer-Assisted-Instruction Use in Colleges.

B) Exploring the societal impacts of proposed "wired city" services.

C) Exploring future trade-offs between travel and communications services.

D) Sponsoring research into the impact of new telecommunications services upon native populations of the Canadian North.

E) Participation in a U.S. National Science Foundation funded study of travel/communications interactions. This study was conducted at the Stanford Research Institute.

The first three of the above studies were conducted using internal professionals with outside support where required. The fourth study was contracted with Queen's University of Kingston, Ontario and was only monitored by a BPG professional.

T.A. of Travel/Communications Substitution

Introduction. This area is one that has captured con-siderable attention in the BPG during the past five years. A

variety of projects have addressed issues associated with the
potential of future communications services to substitute, supplement, or interact in some way with the transportation sector.
In total, these activities add up to a multi-faceted technology
assessment of the various social, political, environmental, energy,
behavioural, and corporate impacts of the interaction of travel
and communications.

The work in this area has been concentrated in a number of
projects.   The key areas are outlined below:

A continuing assessment of the technologies that are the
driving forces in the telecommunications fields.

A detailed behavioural evaluation of the individual
traveller's perception of travel and communications and
his or her attitude towards the substitution of certain
types of travel with telecommunications alternatives.

An examintation of the energy implications of travel/
communications substitution.

A review of the policy implications of substitution
across a wide spectrum of activities.

Participation with the Stanford Research Institute's
recent NSF sponsored technology assement of Travel/
Communications Substitution.

The phrase "susbtitution" is used here as a shorthand
expression that refers to very complex, mostly unknown (to date),
relationships between the transportation and communications
sectors of our society.   This relationship is not new, of course,
as these two sectors have been intertwined in a maze of relationships since the development of postal, telegraph, and telephone
services.   Research on the impacts of these old communications
services upon personal travel has been extremely limited to date.
The simultaneous rapid growth in the use of modern communications
and transportations systems during the last few decades in North
America has masked the development of interrelationships between
these two sectors.   Studies have indicated that those who travel
a great deal also use communications systems frequently.(4) Thus,
existing communications and transportation systems appear to be
mutually reinforcing.   However, many argue that the rapid
proliferation of new communications technologies when combined
with the current crises, congestion, and negative side-effects
of many transportation systems will lead to a new era of
substitution.

Susbtitution: Will It Happen?

An adequate analysis of the intercity substitution question must examine the fundamental reasons why people travel.   There are obvious stated occupational reasons for many travel activities; however, there are a host of unstated social and personal factors at work when travel decisions are being made.   An understanding of the travel/communications substitution issue requires research into these behavioural factors that would also underlie any future decisions on substitution.   The other key variable involved in determining whether or not substitution will occur in the future is the financial cost-benefit trade-offs that must be determined between the costs of travel and the proposed communications substitute.   This section of the paper considers both these variables on an inter-urban basis.

Behavioural Analysis.   The Business Planning Group has undertaken an analysis of these issues.   This survey research was concerned with business travel in Canada between the cities of Montreal, Toronto, Ottawa and Quebec City.   Business travellers between these cities utilizing air, rail, auto or bus modes of transportation were given a questionnaire to obtain the types of information shown below:

a)   basic trip statistics

b)   purpose(s) of meeting(s)

c)   information carried to or acquired at meetings

d)   reasons for not substituting existing communications media for this trip

e)   indirect personal activities associated with the business trip

f)   an assessment of the most satisfactory and unsatisfactory aspects of the current trip

g)   the potential of various future communications capabilities to replace the type of trip the traveller is currently on

h)   organizational, personal, statistical data on the individual respondents

The questionnaire was given to 30,000 business travellers during October 1973.   Approximately 9,600 usable replies were received which permitted detailed sub-analysis of the substitution question by mode of transportation, particular intercity corridor,

organizational characteristics of the traveller's employer, executive level, ethnic group, and potential substitute capabilities.
The response to the survey was a much higher level of returns than
expected.   This in itself may indicate that interest in the
substitution question is becoming widespread among travellers.

An overview analysis of the survey results indicates several
interesting findings.   Presentation of these comments should be
prefaced with the reminder that these questionnaires were
distributed before the Arab-Israeli war and its subsequent impact
on energy supplies, travel convenience, and public consciousness
regarding the negative societal implications of transportation
systems.   The overall results of the sample indicated that 20%
of the business travellers would have substituted the existing trip
they were on, if appropriate communications substitutes had been
available.   This finding is not offered here as an indicator of
the overall average potential for substitution but it appears to be
a representative look at the short trip, commuter travel often
experienced on the travel corridors studied.

Cost Benefit Analysis.   Behavioural research helps determine
if people are willing to substitute in "free decision" environments.
Often this is not a relevant factor in governmental, business and
educational institutions.   Telecommunications systems that meet
cost-benefit criteria definitely are more easily acceptable to
managers.   Various studies have shown that audio and augmented
audio teleconferencing systems usually turn out to be cheaper than
travel for defined trip patterns.   The reverse is almost always
true for video based systems.   All of these calculations involve
an assumption of the cost of the time of the traveller (i.e. it
costs "x" dollars per hour for an employee whether he is travelling or in the office;   hence, travel time saved equals dollars
saved).   The problem with these forms of calculation is that the
institution has to spend more money on telecommunications systems
in order to optimize the existing expenditures in salary charges.
Illogical or not, many managers would rather have employees underutilized than spend more money to optimize a "sunk" cost, namely
salary.   If this attitude can be overcome then many non-video
teleconferencing systems can result in net savings for the using
organization.

One operation instance of detailed cost-benefit analysis is
the experience of NASA during the Apollo program.   A series of
teleconference networks were created, ranging from audio teleconferencing through to high speed facsimile (50 kilobits per
second).   These networks were used to replace certain types of
travel during the course of the program.   A series of analyses
using various assumption patterns indicated that the use of the
networks saved from $1.4 to $4.1 million per year.   An examination
of travel costs before and after the introduction of the telecon-

ferencing systems indicated that the average travel cost per professional assigned to the Apollo program dropped from $860 per year to $650 per year.(5) These latter figures ignore the costs of time spent in travel whereas the earlier calculations assume an allocated cost of salary against travel. The study also noted that the use of teleconferencing resulted in many informal contacts and faster decision making than could occur if travel had been the main form of long haul interaction between the various groups involved. No attempts have been made yet to try and quantify these types of factors in a cost-benefit analysis. However, it is interesting to note that rigorous application of cost-benefit analysis in the major institutional environment can result in considerable substitution of travel through the use of telecommunications.

Energy and Environmental Issues

Introduction. There are a host of environmental and energy issues related to travel and transportation systems. These have become increasingly important in public and private policy determination in the past decade. One forecast is that these factors may lead governments to promote or encourage communicating rather than travelling in the future. This could be through a wide variety of administrative mechanisms, including ones that may alter the economic cost/benefit ratio in favour of communications alternatives.

The environmental costs associated with transportation systems have become identified in considerable detail in the past decade. Current research is expanding information on these issues at a rapid pace. The environmental considerations associated with communications systems have been virtually unknown although recent interest had been expressed on the subject. Analysis that has been undertaken to date leads us to believe that these costs are far less on a per capita user basis than those for transportation systems.

Energy Impacts of Travel/Communications Substitution. Several studies have tried to examine the impact of substitution for specific trips using defined technologies. Similar research has been conducted by the Business Planning Group and by the Communications Studies Group (CSG) of University College, London and the London School of Economics in conjunction with their work with the U.K. Post Office.

The approach is to calculate the energy consumed in particular journeys using various forms of transportation. This requires estimates of the number of people travelling to a meeting which is a key variable for the transportation/energy consumption

calculations.   One of the difficulties in this form of estimate
is that the number of people at one location for a teleconference
is not a cost or energy consumption variable.   The incremental
cost of adding people on to an audio or video teleconference is
vitrually zero and hence more people may attend a teleconference
more cheaply than in person.   On the other hand, the length of
a meeting within a reasonable period (1 day) is a key variable for
the energy consumption of telecommunications alternatives and a
fixed cost item for transportation systems.   Another key variable
in the telecommunications side of the equation is an assumption
on the source of electricity for operation of the system.   Electricity derived from hydro dams has a much more efficient conversion ratio (85%) between primary and secondary energy (the ratio
of the raw energy available at the source compared to the energy
actually obtained for end usage) than say coal/oil (35%) or nuclear
(30%) sources.   This is particularly important in the Canadian
situation since some provinces are highly dependent on coal/oil
and nuclear sources (Ontario) while others have large supplies of
hydro power (Quebec).

The result of these types of calculations result in trade-
off curves for various city pairs using defined transportation
and telecommunications systems.   This form of micro analysis can
be built up to national estimates of macro savings.   The two
groups have found in independent studies for Canada and the U.K.
that approximately 2% of national energy consumption could be
saved through the reduction of a moderate amount of business intercity travel.(6) This does not include infrastructure savings or
the reduction of intra-urban transportation.

The energy elements of the substitution equation are certainly
incomplete at this time.   We can obtain an idea of the relative
impact of various trade-offs through the preparation of scenarios
assuming levels of substitution.   This is only one step towards
a greater understanding of this issue.   However, it is certainly
a move away from merely assuming that these trade-offs will favour
the communications half of the substitution issue.

                              Social Issues

The material reviewed to date has revealed that the substitution issue has several dimensions.   This last section overviews some of the other questions and issues that will have to
be examined in any comprehensive examination of travel/communica-
tions substitution.   They also indicate the type of questions
that enter into an extensive policy analysis.

Potential <u>negative impacts.</u>

    Loss of Privacy

    Loss of Interpersonal Interactions

    Disruption of Life at Home

    Sector Unemployment (in Transportation Related
                              Industries)

    Significant Disruption Potential by Strikers, Disasters,
                              Etc.

There are certainly other potential negative implications of travel/communications substitution. This short list is merely presented to indicate the range of possibilities and the agenda for technology assessment and policy analysis.

THE ROLE OF TELECOMMUNICATIONS POLICY ANALYSIS

IN SERVICE PLANNING

The paper has followed the course of service innovation in the telecommunications field from concept creation through to an analysis of the potential social impact of the proposed service. Each of these steps can be visualized as a filter that eliminates many concepts and/or options. Only the hardy concepts should reach this stage in the innovation process. The service must now be measured against the current and forecasted telecommunications policy environment.

Our hypothesis is that a service may be technologically feasible, meet both the supplier's and the potential user's cost-benefit criteria, not be perceived as socially harmful, and may still never have the opportunity to exist in the "real world". The final filter is telecommunications policy.

Telecommunications policy has become very complicated in the last decade and a half, both nationally and internationally. Most of the conventional wisdoms have fallen or have been modified radically. There are new "winners" and some old "losers" here. A few examples of these changes include:

- terminal competition and interconnection in both the traditional voice markets and in the computer and visual communications markets.

- the development of "Special Service Common Carriers" or new carriers based on a particular technology (e.g. Satellite Carriers).

- interconnection of these competitive networks to the
  Telephone network (U.S.) or pressures for this (Canada)

- the emergence of "Value Added Carriers (or valued added
  networks - VANS) who resale communications services after
  buying the "raw" capability from an existing carrier and
  adding some new capability to it (e.g. Packet Switching
  Carriers).

- the movement of telecommunications issues into the political
  arena (e.g. The Consumer Communications Reform Act in the
  U.S.)

The service planner must take this new environment into
account.   The general trends dictate that the planner must anticipate a fully competitive situation.   This competition comes from
the new entrants in the market place and competing technologies.
More importantly, the current or forecasted environments may favour
some activities and make others very difficult.

Supportive Policy Issues

Telecommunications policy development may be triggered by
events in the broader environment.   For example, as noted above,
the energy shortage in the industrialized nations is becoming more
than a superficial concern in many nations.   Energy conservation
is becoming a new ethic with considerable and mounting government
support and pressure.   In this type of environment, the energy
issues associated with teleconferencing take on a new meaning.
Telecommunications policy developers may find that it is quite
attractive (politically and socially) to promote telecommunications
alternatives to travel in order to conserve energy.   The savings
may not be large but they have a certain attractiveness to them,
especially from a media viewpoint.   The "Conserver Society" movement is socially attractive to many and cutting back on some
business travel will not be viewed by the population-at-large as
a difficult sacrifice by business.   Thus, the communications
carrier has a vested interest in monitoring something as seemingly
unrelated to new service development as energy conservation.   This
is a case where the carrier can market a service profitably and be
regarded as socially responsible in the broadest sense.

New telecommunications policy developments also can be the
impetus to major new service development efforts.   IBM's entry
into the satellite communications business via its participation
in the Satellite Business Systems (SBS) venture has been the
result of new supportive telecommunications policy developments
in the U.S.   A venture of this type would have been unthinkable
a decade ago and, today, SBS is forecasted by many to be a power-

ful force in the telecommunications market place of the 1980's.
The lesson in this case is that changes in the telecommunications
policy environment can product new service and business opportunities that would not exist in another policy environment.

Restrictive Policy Issues

A particular telecommunications policy (or in some cases, a
lack of a policy) can also lead to restricted opportunities for
communications service development.   A case in point has been the
field of computer conferencing and the closely related computer
mail area.   Both of these services use computers to act as
intelligent intermediaries between individuals who communicate
on a delayed basis via a computer system and/or network.(7)

The key policy question is whether or not computer conferen-
cing/mail systems constitute data processing or telecommunications.
This is not a minor point since in North America, the former is
an unregulated business while the latter is regulated.   Businesses who are unregulated do not want, as a rule, to become regulated
and it is illegal for U.S. regulated common carriers to enter into
unregulated businesses except through separate subsidiaries.
Computer conferencing/mail systems are on the fuzzy border between
the two areas and exist in a form of limbo.   Communications
carriers have not offered these services to date although several
of the new VANS are planning to do so(8) in the U.S.   On the other
hand, the providers of time-sharing systems and VANS let the
services exist on their systems but do not acknowledge their
existence or promote their use.   Hence, we have a case of a
potentially new collection of "Electronic Mail" services that
exist technically, appear to serve a collection of user needs
(if their extensive "underground" usage is any guide), do not
appear to cause any significant social dis-benefits, and yet,
have not been aggressively introduced into the communications
market place as a result of a telecommunications policy vacuum.

This issue was touched upon tangentally in the first FCC
Computer Communications Inquiry and appears to be subject to
consideration in the current Second Inquiry.   However, many
knowledgeable observers of the Second Inquiry do not hold out any
hope for significant resolution of the key issues as they relate
to computer conferencing/mail systems.(9)

The computer conferencing/mail question also illustrates the
international impacts of telecommunications policy. The
international record carriers and PTT's perceive this new service
concept as a threat to the international TELEX business (the
fastest growing component of the international telecommunications
field) and have adopted policies that restrict the use of computer

conferencing/mail services or apply price surcharges that
eliminate most of the user cost benefit advantages.   The actions
of the British Post Office in this regard have become somewhat
of a cause celebre.   Hence, the lack of a policy regarding this
innovation in the U.S. case and the presence of a discriminatory
policy as in the British case has resulted in very little development of computer conferencing/mail services for users even
though all other indicators point to a positive future for the
service.

CONCLUSIONS

Observation and analysis of social and telecommunications
policy impact indicates that a few hypotheses or "rules of thumb"
would apply in a variety of cases.   These are summarized briefly
below:

1. Developments in the policy environment should be
    viewed mainly as filters that block out new service
    concepts or slow down the pace of innovation.   This
    should not be interpreted to imply that policy
    development is always negative but the overall direction
    is towards reducing service development options rather
    than increasing the options.

2. The main "supportive" element in the telecommunications
    policy environment is the encouragement of more
    competition in the telecommunications service area.
    Competition speeds up the innovative process in both
    the existing common carrier industry and between the
    new competitors.   This drive towards competition
    in the telecommunications field is the main counter-
    vailing force to the negative filtering process
    noted above.

3. The policy process will virtually always lag the
    innovation process in the telecommunications technology.
    The policy development process will often slow down the
    provision of new services to advanced users.

4. Many policy activities are concerned with defining or
    defending the status quo.   Existing actors use the
    process to maintain their position of dominance in
    the field.

5. Those who create the telecommunications policy precedents
    may not bear the fruits of their labours.   (Datran,
    Thomas Carter, and Packet Communications Inc. may have

broken new ground but it is IBM, RCA and ITT who will
benefit most from the precedents.)

6. Non-traditional subjects such as energy budgeting and
    analysis may become important driving forces in the
    telecommunications policy environment.

7. As trends are developing in the policy area, individual
    events may appear to be quite unrelated and contra-
    dictory.   Short term developments may also appear to be
    quite "illogical".

8. U.S. policy determination will likely end up in the
    legal system for final resolution.   This, of course,
    extends the time scale required for policy development
    and hence, the uncertainty that may be surrounding any
    particular service development.

9. Telecommunications policy determination will be inter-
    facing more and more with the political environment
    in the industrialized nations.   Recent developments
    in Canada and the U.S. show an accelerating intersection
    between the political environment and the telecommuni-
    cations policy one.

10. The service planner must monitor the policy environment
    on a continuous basis.   The planner must attempt to
    project this environment through the planned lifetime
    of the proposed service concept.   The planner will
    likely have to use the technique of having several
    scenarios and contingency plans to deal with each of
    them.

These hypotheses are not iron-clad or exclusive.   However,
they do point out a few of the "rules of the game".   The final
few paragraphs in this paper deal with one of the most vexing
telecommunications policy questions.

The issue of "enforceability" is going to vex policy
developers and analysts more and more in the future.   It would
appear futile for regulatory bodies to develop policies that
cannot be enforced in any practical way. A further example
may clarify the point.

The policy issues associated with computer conferencing/mail
systems were discussed above.   In reality, any computer network
can offer this service or any computer on a network can offer it
as well.   Any policy that attempts to restrict the development
of computer conferencing/mail systems will only lead to widespread abuse and "underground" use of various user-built systems.

To a computer and a network, messages are just another bit stream
from a remote terminal that must be transmitted and (perhaps)
processed.   Technical limitations placed upon time-shared computers to eliminate computer message traffic would almost destroy
the utility of time-sharing for most "normal" applications.   Users
and providers could not stand for such a limitation and we would
expect to see widespread legal/political intervention to prevent
technical limitations.

The key issue is that computer mail systems are mainly
"user-driven".   The users are defining the real utility of
the systems and are developing services based upon the raw
computer and telecommunications capabilities being placed at their
disposal by system providers.   This is similar to the development
of CB radio service or the widespread use of office copiers to
duplicate all forms of copyright material (against the well known,
but totally unenforceable, copyright laws).   Regulators are also
finding that CB radio is a similar user based phenomenon with
virtually no opportunity to guaranteeing that the rules and
regulations will be followed.

This question is going to become the centrepiece of many
telecommunications policy issues.   This is going to challenge
both the policy maker and those who must try and anticipate
the policies and develop services for the perceived environment.
The author believes that this will cause more delays in service
innovation than have been seen to date.   The process will likely
be resolved only when large groups of users decide to ignore the
policy issues and go ahead and use the new capabilities as they
see fit.   The only option will then be to accept the new
"policyless" environment as the status quo and find a means of
accommodating to it.   The process may also be stimulated by a
few innovative or well financed service providers who decide to
go ahead and offer and promote borderline services in a policy
vacuum.   In either case, the policy arena of the future is
going to be populated by a new cast of issues, a new cast of
actors, and a new intensity of interest by all concerned parties.

FOOTNOTES

1. This work has been summarized in the recent book by: John Short, Ederyn Williams, and Bruce Christie, The Social Psychology of Telecommunications, John Wiley & Sons, New York, N.Y., 1976.

2. Alphonse Chapanis, "Interactive Human Communications", Sciencific American, March, 1975.

3. Vary R. Coates, Technology and Public Policy: The Process of Technology Assessment in the Federal Government, Summary Report, Program of Policy Studies in Science and Technology, The George Washington University, Washington, D.C., July, 1972. p 1.

4. James H. Kollen, Transportation & Communications substitu-tability: A Research Proposal, Business Planning Group, Bell Canada, Montreal, Quebec, February, 1973. (Revised Edition)

5. Samuel W. Fordyce, "NASA Experience in Telecommunications as a Substitute for Transportation", NASA Memo, April, 1974.

6. This is reviewed at greater length in Michael Tyler, Michael Katsoulis and Angela Cooke, "Telecommunications and Energy Policy", Telecommunications Policy, (Vol. I, No. 1) IPC Science & Technology Press, Guildford, U.K. pp 21-32

7. This has been described in detail by Raymond R. Panko, "The Outlook for Computer Mail", Telecommunications Policy, (Vol. I, No. 3) IPC Science & Technology Press, Guildford, U.K. pp 242-253.

8. Panko, p 252.

9. However, it is interesting to note that the FCC has recently acknowledged the issue and is in the process of funding some contract research to explore the policy questions associated with electronic mail.

COMMUNICATIONS POLICY - THE NEED FOR RESEARCH[1]

J. B. Cowie

Long Range Intelligence Division

British Post Office

INTRODUCTION

Until recently the major task of most PTTs was to work towards a a target of installing a telephone in each home and to supply the basic telephone needs of business. In this situation there was rarely major discussion of telecommunications policy unless it involved questions of capital investment and relative priorities between competing needs. Recently there has been a change in the situation. Developments in technology have meant that new issues have arisen in relation to the impact on Computing and Communications. There is now - and will be increasingly in future - a widening of the area of public debate.

This paper reviews the challenge facing those responsible for Communication Policy with brief comments on those likely to be involved and suggests aspects requiring more research.

COMMUNICATION POLICY - THE CHALLENGE

It is now widely recognised that the fields of telecommunications and computing are converging. Improvements in basic technology are having a significant impact on both fields. Cheaper, more powerful processing and storage devices can be used directly for controlling traffic flow in networks and for their more

---

[1] The views expressed are personal observations after a brief time exposed to such policy issues, following many years in computing. They are not necessarily the views of the British Post Office.

efficient management. The trend towards distributed processing
will change the characteristics of network traffic and a growing
range of products and services for data and information processing
is creating additional communications demand. Provision must be
made not only for continuing expansion of traditional services on
a world-wide basis, but for new demands in such areas as data
processing, message services, mobile communications, cable television and information services.

As pressures for new services build up, and as the distinction
between computing and communication services and broadcasting
becomes increasingly blurred, there will be debate in some countries
on which services should be provided on a monopoly basis, which
should be available from open competition, what constraints if any
should be imposed on each type of organisation and what choices
should be made between mutually exclusive alternatives, eg use of
parts of the frequency spectrum for wireless broadcasting or mobile
communications. Governments will have to decide on what is in the
national interest while faced with conflicting evidence from the
interested parties.

In trying to decide between claimants to be the provider of
services, careful consideration must be given to the strengths
stemming from the capabilities and experience of different organisations and the constraints which have shaped their development
and orientation.

The PTTs have been mainly concerned with provision and protection of network services. All points in the network should
connect with all other points and attachments must not endanger
the service through misoperation. Some enhancements do not become
attractive until the facility is widely available. The need for
standardisation at international level also retards the flow of
new telecommunication services. The scale of investment inhibits
the early withdrawal of substantial amounts of old plant and new
technology must co-exist with the old for decades.

Despite these constraints, the experience from developing
services on a national and international basis through mechanisms
for international cooperation should not be underestimated.

In contrast, computer firms have in the past been much less
constrained by commitment to existing plant and standard practices.
They have thrived by persuading customers to invest in products
and services and to replace these with improved facilities at
typically seven year intervals. Applications were initially often
self-contained, without much apparent need to be developed to
recognised standards. An important factor in success was to be in
close touch with user needs and to be persuasive that the needs

would be met by their offerings.

As a result, computer firms have created a base of customers who are more sophisticated in their awareness of the potential of new technology and more aggressive in their demands for supporting communication facilities. The computer firms will be more dependent for their future growth on the provision of timely communication facilities and may wish to be able to offer some of these themselves.

The further market penetration of computing could be constrained to some extent by the investment in existing systems and greater user insistence on ease of transition to new arrangements - a constraint only too familiar to PTTs.

Hence, compared to computer firms, the PTTs are strong in providing network capabilities extending to an international basis but less experienced in providing support arrangements which keep them in close touch with emerging user requirements.

For both the PTTs and computer firms, research emphasis has been on the technology side, with much less investigation of the potential implications of the developments either in advance of, or after, facilities are provided.

It is envisaged that, as the growing impact of computers and communications on all walks of life (eg in business, industry, the professions and on leisure) becomes more widely recognised, pressure will mount for more of this second type of research with conclusions influencing policy makers and eventually the law. It should also be noted that sophisticated research in this field is likely to use systems analysis, computer modelling, information retrieval and other computer packages, ie many of the techniques and products whose potential for good or evil is the subject of the research.

THE INTERESTED PARTIES

The interested parties and the stronger links between them are indicated in a generalised way in figure 1.

The extent to which each box is currently well established will vary from country to country, eg Government Agencies, Research Units and Pressure Groups are more strongly represented in the United States than in the United Kingdom. However, neither an abundance of uncoordinated agencies nor the lack of them are likely to lead to the emergence of coherent national communication policies. Policy issues are likely to be stimulated first by the activities of suppliers - both national and multi-national. Three inter-

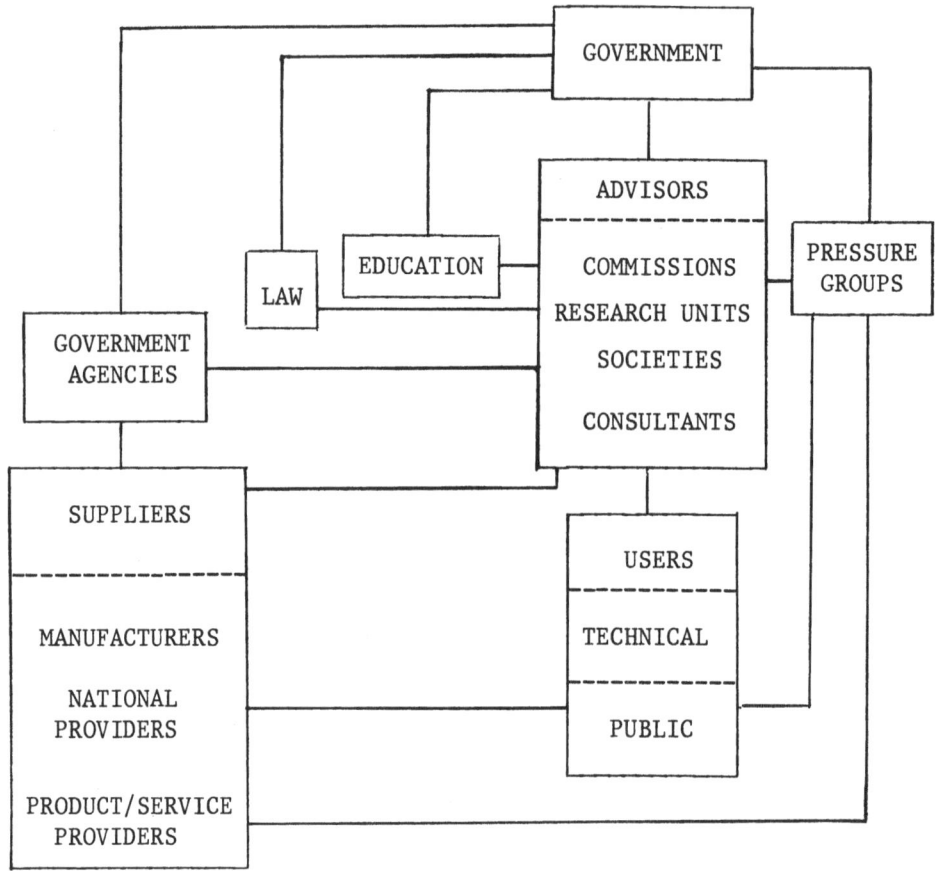

Figure 1

Notes

'Government' is the entities which pass legislation

'Government Agencies' included Government Departments and other
have a direct involvement in administering Communications Policy

'Advisors' are those organisations or individuals who could be
invited to provide advice on policy

'Pressure Groups' are those who attempt to influence the lawmakers.
A member of the public can be a user and a member of a pressure
group (eg as a constituent of his local representative)

'National Providers' include PTTs, Broadcasting Authorities and
Press owners.

related policy themes are:

a)   industrial policy for the communications sector,

b)   who should be allowed to provide which services
      and products

and   c)   how can the new opportunities be best exploited
            and abuses avoided.

Users, and particularly the minority defined as 'technical', ie
those who have a professional interest in the provision of communication services, will be pressing for an expanding range of
capabilities provided in a convenient and cost effective manner to
meet their requirements. If they feel their needs are not being
met adequately they will be looking for policy changes, but their
concept of what they want and what is feasible is likely to be
heavily influenced by what suppliers tell them or what is becoming
available elsewhere.

     Pressure groups, and in particular those who wish to defend
the interests of society, tend to be set up when there is already
significant evidence of a threat which it may be too late to deal
with effectively. However, having come into existence and become
more aware of some of the implications of past developments, such
groups should be able to respond more effectively to future
challenges.

     Advisors may be gathered together temporarily in a commission
to consider some aspects of national policy and then disband.
Others in research units, learned societies and consultancies may
dedicate themselves more permanently to communications policy issues
but as yet the area is not well established as a professional
specialism with a distinct methodology based on research findings.

     Policy recommendations accepted by the Government may lead to
legislation and the law may be involved in adjudicating on the basis
of existing statutes.

     The educational sector could do more to help increase the
general awareness of the implications for society of developments
in computing and communications.

     Each of the parties listed will have some interest in research
relevant to communications policy issues, although naturally each
will be seeking to use the results to protect its interests. For
example, the 'National Providers', which include the Broadcast
Authorities and Press owners as well as the PTTs will be seeking
to extend their services with innovations which they are favourably
placed to provide. Strategies will be required to cope with any

threats which have serious implications for their organisation and
convincing arguments need to be developed that their proposals are
in the national interest. Well conducted research may add credence
to such arguments but perhaps the greatest challenge to the research
community will be to develop methods which aid policy makers by
more clearly separating the special pleading of vested interests
whoever they may be from the intrinsic merit of their case.

RESEARCH NEEDS

Research tasks in the field of communications policy include:-

1   Estimation of demand for new services

   (a)   Past practice

   (b)   Environmental factors

   (c)   Identification of basic characteristics of user needs

   (d)   Study of factors which inhibit the adoption of
         new technology

   (e)   Ways of exploiting new technology

   (f)   The relationship of new and existing services

2   Study of potential abuses in use of new technology

3   Study of suitable national control arrangements

4   Study of the need for institutional change

1   Estimation of Demand for New Services

Past Practice. In the past, PTTs, for example, have not
looked to the research community to contribute significantly to
their future planning of services. Research results have influenced technical choices but have had less impact on telecommunications system strategy. However since the extending range of
customer requirements can be met in various ways, more research is
now needed to select the best mix of services and their order of
introduction.

The methodology for forecasting future demand needs further
development. Projections of past data, customer expressions of need

or preferences via questionnaires and delphi techniques are all useful but do not produce reliable forecasts of future demand. One approach is to provide a capability and try to make the service successful by marketing effort. If the ability to generate demand has been misjudged, this can prove expensive both in direct losses and in opportunity costs for excluded services that might have been more successful. If the start-up costs of a new service are relatively inexpensive, it may be appropriate to provide it and reduce the commitment or withdraw it, if it is not a commercial success. Test marketing of a new service is desirable but difficulties arise if it only becomes attractive to users after there is an extensive and expensive commitment to the service.

Environmental factors. Developments in the environment can influence the take-up of new services and can be researched; eg economic developments at national, organisational and individual levels affecting investment policies, individual disposable income, etc; changes in industrial structures, size and locations of firms; changes in social patterns and values and trends in attitudes to consumer information products. These are only a few examples of factors whose trends have market implications.

Identification of basic characteristics of user needs. The prospects for a new service depend on a complex mix of technical, economic and behavioural factors. Potential new services must be considered in terms of basic characteristics and compared with existing services for suitability and value for money. A taxonomy of user requirements if developed might identify gaps in the provision of service. Such a taxonomy could then be mapped onto technical functions to be supplied by various technical facilities.

Research into the demand for, and effective use of, new facilities should contribute to their further development, their marketing and training support; eg a recent investigation of the use being made of loud speaking telephones in one area, gave pointers to the way the device could be improved, its current selling points and the advice needed to raise use towards the level of those who exploited it to the full.

Study of factors which inhibit the adoption of new technology. Inhibitors to the take up of new technology can include institutional barriers, poor instructional support, and a failure to trigger a snowball effect by not reaching a critical mass in a timely manner. In particular there is a need for a better understanding of human reactions to new technology. For example, if a new service is perceived, by those whose attitude is important to the success of the innovation, as involving them in a loss of

status or in less interesting work or in reduced job opportunities,
they are unlikely to give it their full support. This perception
may be accurate, in which case the effect should have been identified
at the planning stage and where possible steps taken to ameliorate
the problem, or it may not be, which should lead to more careful
thought about how to remove such misunderstandings and avoid them
in future.

A counter example of an innovation which has been universally
welcomed is the copier. The negative side effects in this case
tend to be economic, which does not induce strong reactions if the
users are not paying for the service themselves. Laboratory
experiments, field trials and modelling can provide useful insight
into these factors but behaviour of subjects in test environments
may not give a complete or even accurate picture of how devices
will be viewed and used in the normal workplace.

Ways of exploiting new technology. In the example of the
use being made of loud speaking telephones, further research
could provide guidance on the most effective way of exploiting
the device, eg if being used for audio conferencing, the methods
for effective chairmanship and membership may vary from conventional face-to-face meetings. Advice on this as well as the
circumstances in which audio conferencing was likely to be most
successful could stem from research rather than leaving individuals to form their own views by trial and error. As terminals
become more widespread at work and at home, many innovative uses
become possible, eg for voting or information retrieval. Research
should clarify the more promising of these applications but
pitfalls, would need to be avoided, eg the danger of regarding
present terminal owners as being a representative sample of the
total population.

The relationship between new and existing services. Even
if one achieves the difficult task of assessing the basic strengths
of a new service and relates these to user requirements, consideration must also be given to their relationship to existing
services. Are they significantly in competition or are they
independent? Alternatively, does the existence of one service
have a positive impact on the market for the other, perhaps
because they can share support arrangements or because one service
may be a natural extension of an existing service?

These displacement and generation factors must be taken into
account in assessing the potential of new services.

2   Study of Potential Abuses in Use of New Technology

Some abuses have been well publicised such as the misuse of
data gathered for other purposes. Other problems have been less
fully considered such as the confusion of responsibilities which
could follow the introduction of an on-line management information
system if senior management intervened at stages that would normally
have been dealt with by junior or remote management. Here, the
apparent advantage of more powerful control systems using sophisticated technology may be more than cancelled out if subordinates
react against encroachment on their responsibilities and authority
in their domain. The dissatisfaction may take some time to manifest
itself if their concern is not made clear to their superiors. Similarly, there is a thin dividing line betwen the efficient use of
mobile communication devices and an invasion of the privacy of those
carrying them. The feeling that the bearer's working time is being
substantially controlled by a device is similar to the impression
created by assembly lines.

These examples relate to side effects from the use of new
technology which could affect the quality of working lives. Private
lives could also be disturbed, eg by the misuse of telecommunications
access. Research into these real or potential abuses would be of
benefit to Government in setting policy and creating legislation,
to suppliers in influencing their product and marketing policies,
to users in alerting them to misuses which could reduce the return
on their investment in the new facilities, and to society who could
then respond to threats to the quality of life.

3   Study of Suitable National Control Arrangements

Communications media make a significant contribution to the
development of attitudes and values in society. New communication
services, including a number first developed abroad, could affect
the existing boundaries between telecommunications, broadcasting,
press, publishing and library functions.   Issues will arise which
require national consideration. For example, in various countries,
the proper responsibilities of the Broadcasting authorities, the
PTTs, private companies and other bodies involved in the provision
of TV, cable TV, and Pay TV could be a recurring theme which provokes lively debate in future.

4   Study of the Need for Institutional Change

Developments in computing and communications could lead to
three types of institutional change. Firstly, it will be argued

that new bodies should be established in some countries, eg
Government agencies to administer communications policy. Secondly
the technical possibilities could encourage new structures, eg in
industry, firms could break up into smaller units connected by more
sophisticated communications facilities. Thirdly, the emphasis on
this field could grow as its importance becomes more widely recognised, leading to adjustments in established practice, eg educational
establishments, particularly universities could become more active
in research and education on the implications of these developments.
Research interest to date has been limited and mainly concerned with
the technology which can be studied more easily within traditional
departments. However, effective study of the implications is best
undertaken by interdisciplinary teams. If the universities are to
respond, they will need to break down departmental barriers and
offer more hybrid education programs and research effort. Again
research could help indicate desirable changes and evaluate their
implementation.

NATIONAL AND INTERNATIONAL CONSIDERATIONS

In deciding who should be allowed to offer communication
services, a government will hear valid arguments in favour of monopoly provision and conversely, for a wider range of suppliers.

If interconnection, particularly on an international basis, is
a significant factor, the well established arrangements for cooperation among the PTTs will be an advantage. Their cooperation also
extends to joint research into methods and prospects for new services.
However, a wider range of suppliers leads to a broader range of services available more quickly which is advantageous if interconnect
arrangements are less important.

Where choices have to be made between alternatives, eg in the
use of the spectrum, or for or against Pay TV, national preferences
may differ from country to country.

The pressure for change will often come from capabilities
available elsewhere and research has a significant role to play in
enabling each country to assess the potential implications of these
developments within their culture and environment.

SUMMARY

The theme of this paper is that computing and communication
services are likely to have an increasing impact on society and
that more research is needed to help suppliers, including PTTs,
choose wisely the services to be provided, to enable users to
exploit them effectively and to enable Governments to control these
developments in their national interest.

The research community is not yet well equipped to deal with these problems and closer collaboration with their sponsors could help develop their capability and make their results more useful, eg by being available when key decisions are due to be taken. However, despite progress in research, advisers on communications policy, like economists, are still likely to be influenced by basic political beliefs in their assessment of what is best for their nation.

M.T.

J.C.

DISCUSSION OF PAPERS BY CHAPUIS, COWIE AND DAY

Discussant and moderator: Dieter Kimbel

Kimbel opened the discussion saying he hoped it would serve to bring together the divergent views on regulation presented in each of the papers.

Chapuis then began by discussing the difficulties of comparing research questions and results on policy questions when the policies considered are so diverse. He suggested the need for a multi-disciplinary study which addresses and compares the motivational factors behind internal regulatory structures of different countries. Such a study is needed, he said, to draw together the common threads of interest in regulatory policy for telecommunications. Chapuis then spoke of some of the problems resulting from the absence of a telecommunications policy in Europe which transcends national boundaries. He suggested that the structure of the international network is currently depending more on accountants' national views than on an internationally defined policy.

Mandelbaum suggested that even if there were no national boun-daries, the distribution and centralization of telecommunications systems would not look much different than they do today because decisions tend to be based primarily on economic considerations.

A general discussion followed on regulation and the entreprenuer. Goldstein differed with these portrayals of the development of telecommunications services and again pointed out the discussion was neglecting the role of the entrepreneur in executing policies and delivering actual systems. Decision on telecommunications, he said, result from choices by either the private or public sector--decisions which are not necessarily based on abstract need. He suggested that differences in systems lie more in the different nature of the decisionmakers than in an abstract conception of policy. Cowie suggested that policy researchers need a better understanding of which aspects of entrepreneurial activities and objectives could produce communications difficulties which are not in the national interest, for example, interconnect problems, and which aspects could produce benefits. Wells mentioned a dual role for the entrepreneur--to provide facilities and to use these facilities to provide services. What's lacking, according to Wells, is research on the effects of new services on people. People are more vulner-able than organizations and policy research must look to the effects of policy on minority interests. While not necessarily objecting

to this view, Goldstein pointed out that we need to recognize when
regulation contributes more problems than it solves. Regulatory
effects, he said, need to be weighed so that the problems caused
by regulations don't outweigh the benefits.

Baudazzi suggested that the discussion tended to overemphasize
the importance of people's interest research. Research cannot
deal with all problems. While it may be impossible for a citizen
to recognize what's better between two choices, once the user
learns something is better through experience, he doesn't need
research. In his opinion, the different views of regulation among
entrepreneurs or between entrepreneurs and government is merely
a difference of opinion which is perhaps based on economic matters,
but is not something which can be resolved by research.

Moss asked how can research be designed to address the
user-generated telecommunications developments which emerge
outside of the regulatory environment? Day responded that in
discussing research, one is talking about a variety of problems and
a variety of research methods. Just because research may not have
an impact, he said, doesn't mean we shouldn't try. He went on to
suggest that perhaps researchers should try to understand things
besides the newest technologies. There should be more concern with
the technologies--new or old--which are widely used, e.g., pocket
computers and their impact.

Cowie referred back to Goldstein's remarks on the potential
problems which may be caused by regulations. He said that regulations are accepted when they are seen to have a purpose which is
beneficial to society, for example, speed limits. There will always
be those who ignore the regulations and they must expect to be
penalized if found out. In special circumstances, regulations may
be strengthened, such as in an oil crisis. Society expects that the
temporary or longer standing regulations will be adjusted when there
is a consensus that they are no longer operating in the best
interests of society overall.

With respect to entrepreneurial freedom, one speaker asked the
group to consider as a specific example the problems which arise when
a leased circuit is set up between, say, New York and Paris, and when
the organization leasing the circuit uses it to pass information
to the remainder of other European countries.

Day returned the discussion to the problem posed by citizen
disregard of regulations and said that when telecommunications
regulations are broken, nothing can be done about it. In his
opinion, regulations exist to protect certain vested interests.
Users will adapt technology to their own purposes and there is
nothing regulators or entrepreneurs can do about this. He also

suggested that unanticipated developments can have a major impact.
EFT evolved for one reason, but it may evolve further for another
set of reasons.

Goldstein pointed out that his earlier remarks on entrepreneurs has been misunderstood. Policy research has ignored not the
right of the entrepreneur but his role. It has ignored the decision
of the entrepreneur to do or not to do something and how he does it.
He felt that the papers presented by Chapuis, Cowie, and Day ignored
this area. He also commented that much of policy research is not
useful because it has concentrated on the past--not what is likely
to occur in the future. Day suggested that one reason the entrepreneurs role may be ignored is that it is the funders who decide
which questions will be researched, and the entrepreneurs don't
fund research.

This led to a question to Cowie on the types of policy research
activities in Britain. Cowie responded that there was recognition
in Britain as elsewhere that the convergence of various fields involved in information processing and transfer generated policy
issues which needed to be studied. Some issues had global implications, others were more localized to individual countries. It
would benefit researchers if views and information could be exchanged
as far as possible but it had to be recognized that this is a
sensitive area.

Elton remarked that while policy research should help formulate
better policy, there's a lag time between research and decisions.
He stressed that unless the research results are exposed to a broad
number of people, one must be cautious about relying on them.

The discussion again turned to the function of policy research
and the factors it should consider, particularly the entrepreneur's
role. Cowie remarked that independent research groups might provide government with more objective analysis of important issues
and suggest policies which are in the best interests of the nation.
Jull suggested that policy research could be viewed as a "chess
game," to complement the "poker game" between the government and
entrepreneur.

Next the discussion truned to the differences stressed by Yerrell
between policy research (or research into policy) and policy-
oriented research. Someone suggested that what's lacking is research on policy, and expressed the opinion that this should (a) be
done outside the government and (b) take into consideration the
views of all sectors.

Lucas pointed out that in his organization, successes often come
from conceputalizing the problem in different terms than can be done

by the government organization charged with making policy. His
organization's contribution to policy is often reconceptualizing
the problem. He went on to say that stability in funding and a
close affiliation with a government agency is very helpful in
making policy research both relevant and useful. Cowie agreed
with Lucas' remarks and suggested that since the major payoff of
policy research is often embedded in the conceptualization at the
beginning of the studies, research groups should spend more time
than at present on conceptualization and less time on lengthy
implementations. Lucas pointed out that in order to provide this
conceptual work researchers need the remainder of the project time
to replenish their store of intellectual investment.

Section Two

PUBLIC SERVICES: THE DELIVERY OF HEALTH CARE

"Telemedicine" or "telehealth care" (the jargon is still at a formative stage) has become a significant field of research for North America and Japan; it is becoming one for certain developing countries. Though it is beginning to receive some attention in Italy and Sweden, by and large it has been of relatively little concern in the more industrialized European countries. Appropriately then we have papers from the USA, Canada and Turkey.

Maxine Rockoff has been responsible for managing the US Department of Health, Education and Welfare's very substantial program of research in telemedicine. Her paper, written in collaboration with Art Bennett of the Mitre Corporation, presents a general model for assessing the performance of different "manpower-technology combinations" for providing health care to isolated rural communities. They describe briefly the criteria that such a model must meet and some of the difficulties arising in meeting them. The model, based on the concepts of decision points and probabilities of transition from one node in the health care system to another, is illustrated using hypothetical data.

The model has been used as the foundation for a computer simulation of the flow of patients through a primary health care facility. The authors describe this process and the results obtained. The latter suggest that priority should be given to the use of narrow-band technology (i.e., systems which, unlike interactive television, are modest in the capacity they require for transmission). Finally they discuss the limitations of the methodology described.

Anna Casey-Stahmer's paper is quite different. After providing some background information regarding the Canadian context, she describes three recent experiments which made use of the Hermes satellite. Their approach to evaluation and, where possible, their preliminary results are summarized. The last quarter of the paper is devoted to a discussion of coordination between telecommunications and social service agencies, and of other issues arising in the evolution from research and development to operational systems.

The objective of the final paper is to propose a framework for the design and assessment of telemedicine systems for developing countries. Unver Cinar draws attention to distinctive features of the latter: their demographics, transport and telecommunications infrastructure, and distribution of existing health care resources. These are illustrated with a description of the present situation in Turkey. An outline design is then presented for the organization of a Turkish telehealth system and some preliminary consideration is given to its component telecommunications systems.

The three papers demand no prior knowledge of the technology, nor of particular analytical methods. Familiarity with basic OR techniques will, however, make it considerably easier to follow and reach a position on Rockoff's and Bennett's paper.

THE "PATIENT TRAJECTORY": A MODELING TOOL FOR PLANNING AND EVALUATING RURAL TELEMEDICINE SYSTEMS

Maxine L. Rockoff, Ph.D.

National Center for Health Services Research

3700 East West Highway, Hyattsville, MD 20782

Arthur M. Bennett, M.B.A.

The METREK Division of the MITRE Corporation

1820 Dolley Madison Boulevard, McLean, VA 22101

This paper presents an analytical tool developed and applied to plan for the use of telecommunications technology to link a non-physician health care provider in a rural satellite clinic to a central source of medical expertise. An implicit underlying hypothesis was that high technology plus lower skills (non-physicians) would be "equivalent" in some to-be-defined sense to low technology plus higher skills (physicians). The measurement tool developed to compare the performance of such different manpower/technology combinations is based on the concept of a "patient trajectory," the sequenced set of interactions that take place between a patient and the health care system from the time the patient perceives a need for medical care for some problem until that problem is resolved. For a fixed non-physician manpower level (defined by medical protocols), three independent physicians estimated the ability of different telecommunications technologies to avert patient travel at decision points within the protocols that called for physician consultation or referral. These judgments were used as inputs to a computer simulation to assess prospectively the overall effects on reducing patient travel that might be expected from each of several telecommunications technologies. The major result obtained was that *telephone-compatible narrowband technologies (such as slow-scan television) could avert nearly two-thirds of the travel that could be averted with broadband technologies.*

INTRODUCTION

The National Center for Health Services Research has been investigating since 1971 the use of telecommunications technology in the delivery of health care services.

As a first step, seven experiments were funded in June of 1972 to explore the utility of two-way visual telecommunications in the delivery of health care.[1, 2] The purposes of these exploratory projects were to: (1) gain "clinical impressions" of the utility of this technology in a wide variety of health care settings and applications; (2) develop methods for assessing the utility of the technology; and (3) develop a framework for further research on the logistics of health care delivery.

One of the underlying problems in the health care delivery system that led to the telecommunications research initially is the provision of health care services in communities that are unable to attract and retain a fulltime physician, either because they are too small, too isolated or both. Among the more promising solutions posed for such communities is that a specially trained non-physician be stationed in the community with backup made available through telecommunications technology to provide primary care services (i.e., a range of diagnostic and therapeutic services for uncomplicated cases on an ambulatory or walk-in basis).

Recognizing the importance of this problem, two of the seven exploratory experiments involved the use of two-way interactive television to provide backup and supervision to nurse-practitioners providing primary care in neighborhood health centers. One was in Cambridge, Massachusetts[3] and the purpose of the experiment was to compare telephone and television as a medium for providing physician backup and consultation to three nurse-practitioners in satellite health clinics in moderate- to low-income neighborhoods. The second was in New York City and it explored the feasibility of providing physician coverage from the Mt. Sinai School of Medicine to a neighborhood pediatric clinic in an East Harlem housing project.[4]

The data from the Cambridge and Mt. Sinai experiments were in conflict on some important issues. For example, the data from the Mt. Sinai project appeared to indicate that nurse-practitioners handled an increased percentage of patients without in-person referral to a physician when the two-way television was available, suggesting that this modality _increased_ the health system's productivity by substituting specially trained nurses for physicians. In contrast, the Cambridge study appeared to indicate that television consults took longer than telephone consults and in addition led to _more_ in-person physician referrals, suggesting that television _decreased_ the productivity of the total system. However, both of these experiments were done in urban areas where good hospital facilities were only a short taxi ride away. The technology was not stressed because the alternative of an in-person visit was relatively easy; this was especially so in Cambridge. Hence, it was decided to focus the next phase of study on remote sites where the transportation alternative was not as attractive as it is in urban settings.

The question to be addressed was what telecommunications technology would be required to support a non-physician primary care provider in an isolated rural area? This paper reports on part of a research program undertaken to answer that question.

We report first on our general approach and introduce the concept of the "patient trajectory" in which we follow patients as they move from site to site within the health system. We selected this concept because we believed that the introduction of telecommunications technology would improve patients' trajectories. We then discuss a method by which we developed prospective theoretical estimates of what the impact on patient travel would be if different levels of telecommunications technology (ranging from telephone alone, through telephone augmented with various narrowband add-ons, to full two-way broadband television) were to be added to the health system. We report that our analysis showed that almost two-thirds of the in-person physician referrals that could be averted with _any_ telecommunications technology could be averted with _narrowband_ technology. Next, we mention briefly the laboratory research and field trials that we are now undertaking as "next steps." Finally, we discuss some of the limitations of the analytic approach that we have undertaken.

DEVELOPMENT OF THE PATIENT TRAJECTORY

The first requirement in the research program was to develop a methodology for assessing the performance of an arbitrary "manpower/technology combination" providing health care services in a rural, isolated community.[5] This methodology would be used in three different, but interrelated roles, as follows:

1. The methodology must be capable of assessing prospectively the effects of various manpower/technology combinations on the health care system's performance. This is essential for purposes of _selecting manpower/technology combinations_ that

appear, analytically at least, to offer greatest promise for
improving the health care system. The selected manpower/technology
combinations would then be tested in field experiments.

2.   The methodology must be suitable for assisting in the
selection of sites for the field experiments and for matching
specific manpower/technology combinations to the sites' characteristics in order to define the field experiments, including the
design of site-specific telecommunication systems. That is, the
methodology must be usable by designers of health care communication networks. It must provide information as to the nature
and suitability of various manpower/technology combinations in
the context of a specific community's problems.

3.   The methodology must provide an analytic framework for
the field experiments, both in terms of specifying the types of
data to be obtained during the experiments and in terms of
specifying an analysis plan for these data.

In meeting these three requirements, the methodology must be
oriented to the measurement of the health care system's performance, rather than to the measurement of the telecommunication
system's performance.

When we say "health care system" we refer to the set of
health care resources required to deliver comprehensive health
care to a defined population. In assessing and comparing alternative systems we recognize that health care system performance
must be measured along three major dimensions in which improvement
is desired. These are: access, quality, and cost. From a
system's perspective we recognize that these may in many ways be
incompatible and even competitive. Thus, a centralized health
care system may provide high quality health care efficiently (at
low cost) and yet be inaccessible to its target population and
hence it would perform poorly on the access dimension. Also, the
patients' perspective of the system may be different from the
providers'. Indeed, the exploratory experiments suggested that
some of the important benefits of improved communications are
seen by focusing on what happens to patients rather than what
happens to providers. For example, one of the benefits of television as opposed to telephone that came out of the Cambridge
experiment[3] was that although more in-person referrals to
physicians were required following television consultations than
telephone consultations, many of these could be delayed without
harm to the patient. Thus, the ability to avoid an immediate in-
person consult by waiting until a physician would be at the
remote site appeared to be a benefit from the patients' perspective (although not from the providers' perspective) that was
worth further quantification and investigation. Ideally, the
methodology to be developed would allow simultaneous consideration

of all of the system's performance characteristics from all
perspectives.

Measurement Issues

Our first question asked: What are the measurable effects
of utilizing different manpower/technology combinations in the
provision of health care services?

Of the three measurement dimensions, access, quality, and
costs, the most difficult to deal with is the effect of manpower/
technology combinations on the quality of health care. We would
expect, *a priori*, that the effect would be large, in that specialized skills that would otherwise not be available could be
brought to a remote site via telecommunications technology. The
problem is that it is difficult to measure the quality of care.
If we view the quality of care as being measured primarily in
terms of the outcome of services, we must acknowledge that the
majority of primary care received in ambulatory settings will
have little effect that can be measured in terms of outcome
differences. Most of the services provided are of a relatively
straightforward nature and a large part of the health conditions
dealt with are self-limiting. A wide variety of approaches may be
utilized in the delivery of care for a specific condition and
result in the same, generally satisfactory, outcome. Kessner's[6]
tracer studies have generated a limited number of indicators
that may be utilized as partial measures of changes in the
quality of health care in ambulatory situations. The staging
approach of McCord[7] may be used for a limited portion of cases
presenting in an ambulatory setting when satisfactory recognizable
outcomes do not occur as defined by progressive stages of a
disease or condition. Therefore, because of the difficulties in
measuring quality of care, we will assume, at least for our initial efforts, that it is approximately constant.

The measurable effects of different manpower/technology
combinations on accessibility and cost are somewhat more amenable
to systems analysis. Further, these were the major effects where
benefits were anticipated and where we felt, both intuitively and
based on the exploratory experiments, that important tradeoffs
were available. Hence, we undertook to develop measures for
these.

Development of the Model

The analytic construct that we decided to pursue is that of
a "patient trajectory" by which we mean the path that a patient
follows as he interacts with the health care system. We believed

that it would be possible to "characterize" a particular health
care system (i.e., manpower/technology combination) by looking at
the overall set of trajectories of patients using that system.

We anticipated that trajectories would be sensitive to
telecommunications technology inputs primarily in being able to
increase the level of care that could be provided by a remote
non-physician provider through consultation, i.e., that in-person
patient referrals to a physician-staffed central clinic would be
reduced.

If all possible steps that might be taken by a patient in
obtaining health care are assembled, and these steps are inter-
connected in every possible order of occurrence, any given interaction of a patient and the health care system (i.e., episode of
illness) could be represented by the patient's path through these
interconnected steps. When this concept of patient's path is
combined with the time required to traverse each step of the
path, the result is the generation of a "patient trajectory" as
the representation of the measurable effects of changing manpower/
technology combinations.

It follows that if we know the health care system's and the
patient's fixed and variable costs for each possible step in the
process of obtaining health care services, the cost of acquiring
and/or providing health care can be calculated directly from the
knowledge of the patient trajectory.

Since the set of possible paths and the associated times and
costs will be dependent on the nature of the medical problem
being addressed, the patient trajectories must be differentiated
by problem type.

The actual procedure for determining a patient trajectory
can be streamlined by first identifying all possible steps, and
then utilizing the following empirical data: (1) the probability
of transition from each node to each other node and (2) the
probability distribution of time spent in each node.

It is apparent that the transition probabilities and times
for various steps (arcs linking system nodes) will change due to
direct and indirect effects of manpower/technology combinations.
Using the trajectory approach, the accessibility and cost dimensions of the health system's performance can be evaluated by
monitoring the transition probabilities and the transition times
associated with various system configurations.

An Illustration of Patient Trajectories

    Figure 1 illustrates a simplified interaction diagram for a
system consisting of six nodes. These nodes include the patient's
"home" (starting point for seeking care), a non-physician provider
(NPP) staffed local clinic, a local capability for teleconsultation (TC) with a remote physician (MD central), the periodic
availability of the physician at the local clinic (MD local), and
entry into the secondary care level of the regional health care
system (secondary care). These nodes represent different levels
of capability within the system. The basic patient trajectory
classes associated with this system are displayed in Figure 2.
It can be observed that there are are five basic classes. All
trajectories involving entry to the secondary care system are
grouped in class 5.

    As mentioned earlier, the identification of all possible
patient paths is the first step in the development of a data base
for patient trajectory analysis. The next step involves the
estimation of the transition probabilities for transition from
any node to each other node in the system. Figure 3 displays a
matrix of transition probabilities for a given problem type.
Cells with no entries represent infeasible transitions. By
identifying the transitions associated with any trajectory, and
utilizing the appropriate transition probabilities, the probability of a patient with a given problem type following that trajectory can be estimated. Since these transition probabilities will
depend on the health care system being monitored, the probability
distribution of various classes of trajectories for a given
problem type will also depend on the system being observed.

    Figure 4 illustrates a hypothetical comparison of the trajec-
tory class probabilities for two different health care systems, A
and B, for cases involving one problem type. For example, such
results might be obtained in a situation where the capability of
the more elaborate technology (telemedicine) employed at site B
compensates for the lower degree of independent capability of the
NPP at site B. This hypothesis could be generated by comparing
the sum of Class 1 and Class 2 trajectories at the two sites.
Such a sum represents the ability for the local NPP to handle
problems of the given type without external in-person referral.

    Matrices similar to that shown in Figure 3 can be prepared
for recording the transition times and costs. Using appropriate
combinations of transition probabilities and cost and time
estimates, measures of accessibility and cost of care can be
developed.

    Since significant differences can exist between the charac-
teristics of the initial treatment phase (diagnosis and treatment

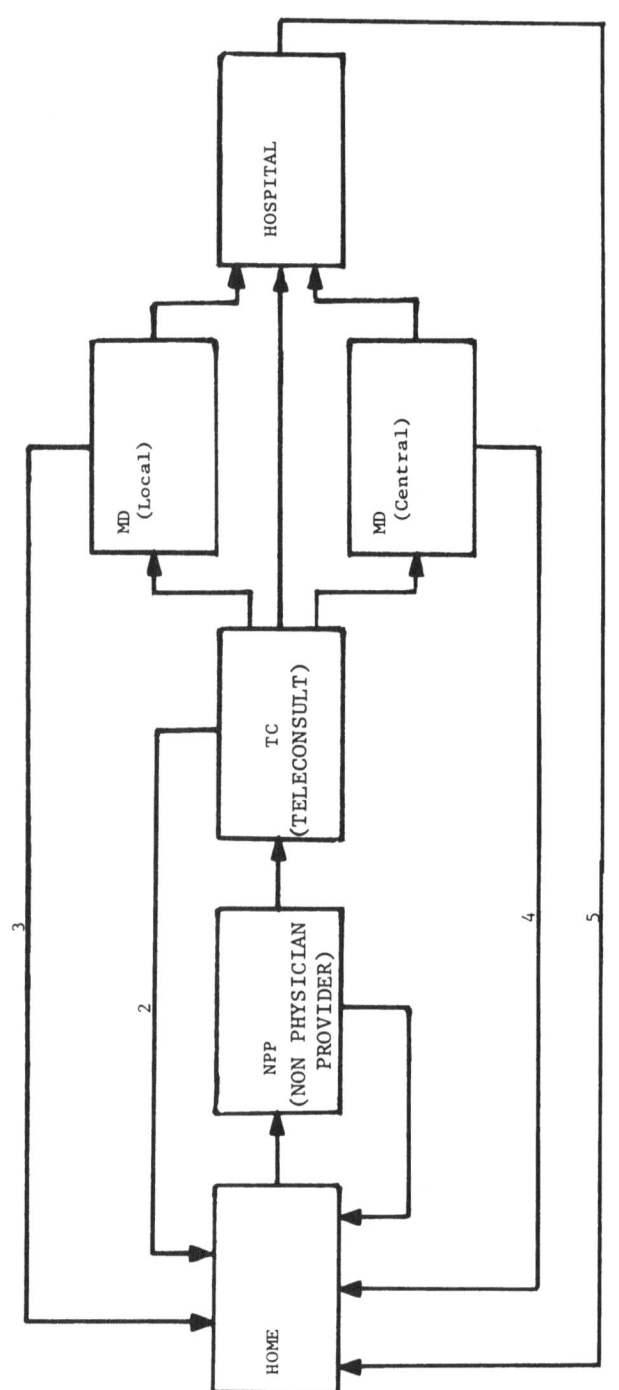

Simplified Interaction Diagram

FIGURE 1

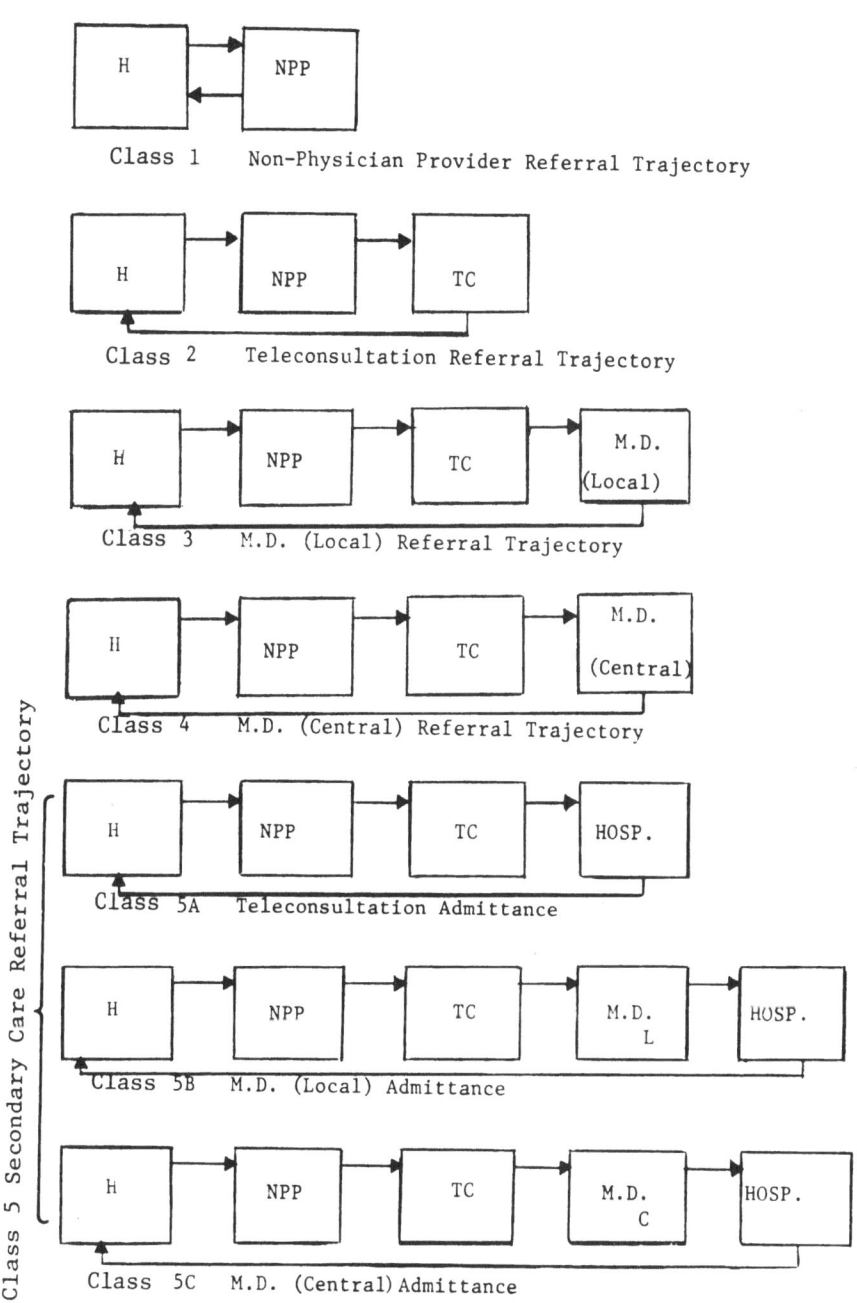

Basic Patient Trajectory Classes

FIGURE 2

	HOME	NPP (INIT)	NPP (UFU)	NPP (SFU)	TC	LAB	MD (LOC)	MD (CENT)	HOSP
HOME		.90					.02	.08	
NPP (INIT)	.60		.02	.15	.10	.10	.02	.01	
NPP (UFU)	.50		.02	.15	.15	.15	.02	.01	
NPP (SFU)	.40		.01	.25	.05	.25	.03	.01	
TC	.60		.02	.20		.08	.05	.04	.01
LAB		.60	.01	.25	.05		.05	.04	
MD (LOC)	.70		.01	.15		.05	.05	.02	.02
MD (CENT)	.70		.01	.15		.05	.05	.02	.02
HOSP	.75		.01	.20			.03	.01	

NOTATION

INIT: INITIAL VISIT
UFU: UNSCHEDULED FOLLOW-UP
SFU: SCHEDULED FOLLOW-UP
TC: TELECONSULT
LAB: LABORATORY
HOSP: HOSPITAL

Example Matrix of Transition Probabilities For A

Given Health Problem Type

FIGURE 3

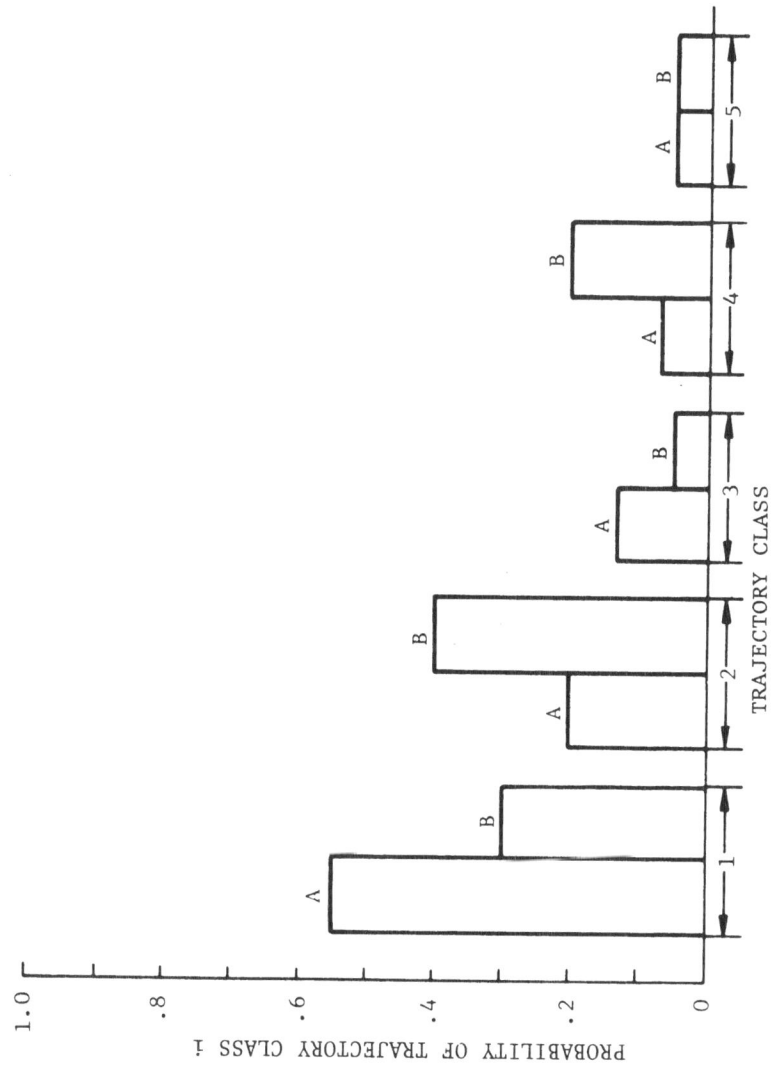

Comparison of the Probabilities of Each Trajectory

Class for Two Hypothetical Health Systems, A and B

FIGURE 4

plan development) and the follow-up phase (monitoring and management of treatment to conclusion of episode), it is desirable to consider complete episodes in order to distinguish between different systems in terms of the cost and time characteristics of initial and follow-up phases. That is, during the follow-up phase, we are concerned with how the design of the system has affected the number and nature of the follow-up visits in an episode of illness involving a given problem type.

A hypothetical comparison of two alternative health care systems in terms of their follow-up visits characteristics is shown in Figure 5. Review of such results could generate a number of hypotheses. For example, the large number of three follow-up visit cases for site B may reflect a rigid return visit policy which may be indicated by the nature of the actual problem severity. The fact that trajectory classes three and four are the major ones involved in these visits would further support such a hypothesis.

These figures represent an example of some of the significant observations that could be expected to emerge from a well-executed patient trajectory analysis. However, the execution of a patient trajectory analysis may become unmanageable when the set of problem types is large. If the system being analyzed consists of a large number of nodes (and consequently a large number of transitions and patient trajectories), even a single problem type may involve an enormous computational effort. Willemain[8] has observed that utilization of commonly used techniques for network analysis (including the use of the Markovian formulation) can simplify the computations, and yield some extremely revealing statistics concerning system performance. Specifically, Willemain has shown that even though individual patient trajectories may be non-Markovian, the overall system performance measures can be obtained by analyzing a Markovian model provided we only seek the average trajectory costs (or times) and do not attempt to compute the likelihood of any particular trajectory.

THE USE OF PATIENT TRAJECTORIES IN ESTIMATING TECHNOLOGY IMPACTS

In order to utilize the patient trajectory concept for purposes of predictive analysis, we must determine a method for estimating the effect on specific transition probabilities of different manpower/technology combinations. The patient trajectory concept itself gives us some assistance in this regard by focusing our attention on transition probabilities that are meaningful to the patient's progress through the health care

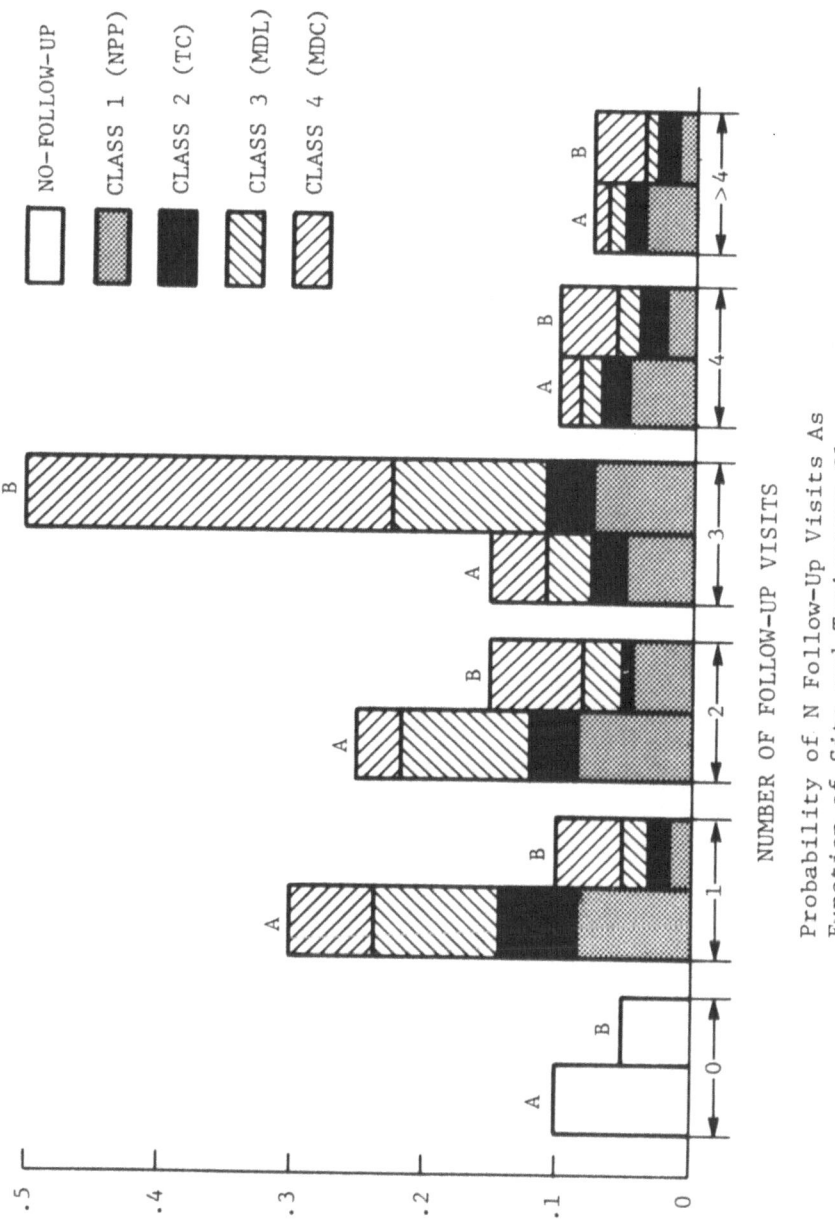

FIGURE 5

Probability of N Follow-Up Visits As
Function of Site and Trajectory Class

system, rather than attempting to estimate quantities such as
differences in resolution, dynamic range, etc.

   In terms of our major objective, to analyze the potential
for various technological alternatives to extend the performance
of non-physician providers in isolated rural settings, a measure
of cost/access effectiveness applicable to arbitrary manpower/
technology combinations and health care settings is required.
The measure selected for the technology comparison is the proportion of patients for whom an in-person physician referral from
the non-physician provider is recommended in order to have their
medical problems resolved. Simply stated, we are interested in
the capability of technology to alter the probability of transition from the satellite non-physician provider-staffed clinic
to the central physician.

Estimation of the Referral-Averting Potential
of Different Technologies

   In order to determine the effects of particular technologies
on the probability of transition from satellite to central clinics,
we concentrated on the ability of the specific technology to permit
a remote consultation to take place between the physician and the
non-physician provider instead of requiring an in-person referral.
This was accomplished[9] by analyzing a set of clinical protocols
that had been developed at the Dartmouth Medical School[10, 11] and
the Beth Israel Hospital[12, 13] to guide non-physician providers
through the identification and treatment of specific presenting
health problems. These protocols were developed both to train non-
physician providers and to be used by them in actual health care
situations.

   Clinical protocols were used as the basis for the technology
comparison analysis because they provide a detailed description
of tasks that a non-physician provider is typically trained to
carry out in order to arrive at a diagnosis and treatment plan
for addressing acute and chronic primary care problems. They
also specify those conditions called "decision points", for which
a consultation between the non-physician provider and the physician, or an in-person referral of the patient to the physician,
is recommended.

   The decision points are logical candidate situations for the
use of telecommunications technology that can aid in consultation,
thereby avoiding patient referral. Figure 6 shows a portion of a
protocol for treatment of Upper Respiratory Infection (URI). As
indicated in this Figure, there are two decision points in this
portion of the protocol. One, stemming from an undescribed abnor-
mality, requires an MD referral and the second, stemming from the

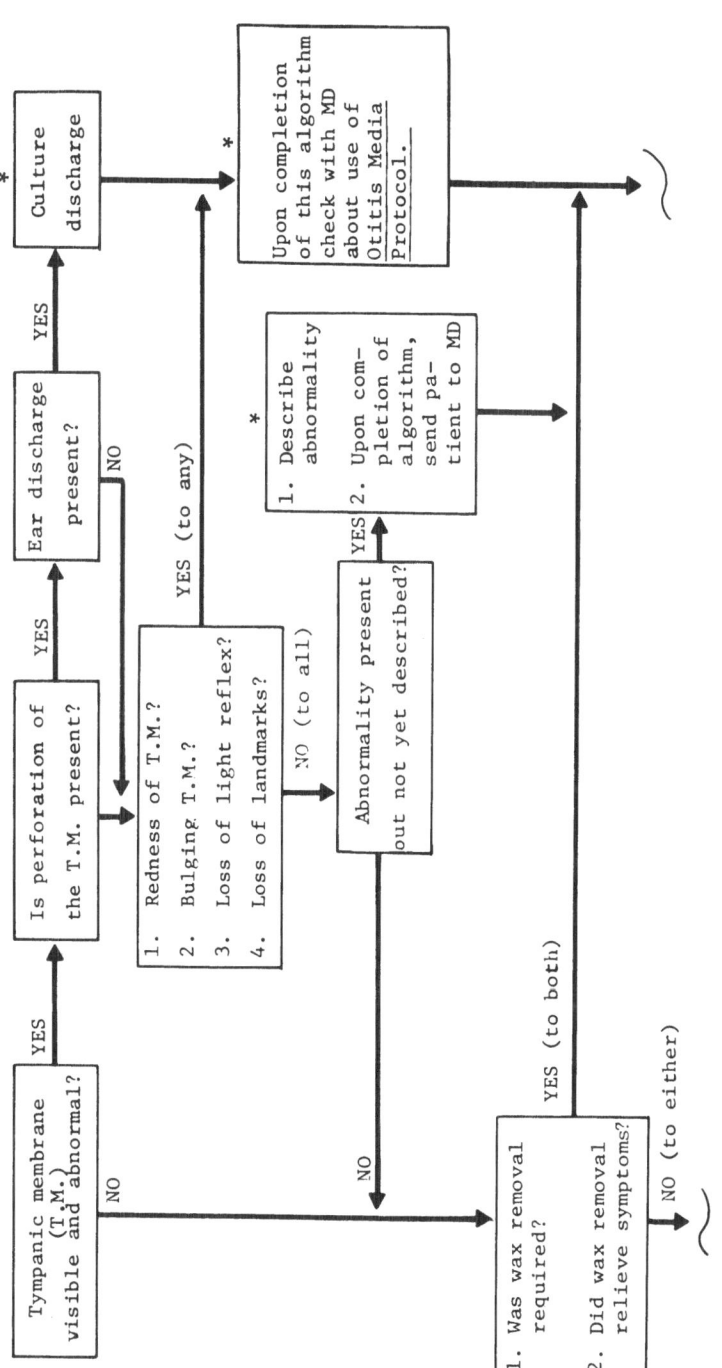

*Potential areas of expansion through technology incorporation.

Portion of Dartmouth URI Protocol

FIGURE 6

performance of a culture procedure, requires an MD consultation.
The question to be answered is, can telecommunications-based technology play a role in "extending" the protocol and thereby avoiding
the in-person referral at these two decision points?

The ten protocols used in the analysis cover the following
presenting problems: back pain; cough; diabetes/hypertension/
heart disease; ear problems; male genitourinary and female geni-
tourinary/gynecology; laceration; nausea, vomiting, diarrhea,
abdominal pain; prenatal; and upper respiratory infection (URI)
for which protocols from two different sources were analyzed.
These protocols were selected because utilization data were available reflecting actual field experience relative to the frequency
of patients accessing various decision points. These protocol
utilization data are of key importance since they provide information on how often patients access particular decision points in
the protocols, thereby indicating how often particular technologies could be used in avoiding referrals.

Three physicians independently estimated the probability that
a particular technology at a particular decision point could avert
a referral at that point. Two of the physicians had been involved
in developing the two sets of protocols utilized in the analysis
(Beth Israel and Dartmouth) and the other acted as an independent
reviewer. The independent reviewer was familiar with the development and implementation of protocols but had no clinical experience with the Dartmouth or Beth Israel protocols. Each protocol
set was reviewed by both the developing physician and the independent physician.

The major steps taken by the reviewing physicians in developing the protocol extension decisions were:

o Screening of decision points
o Selection of technology that could be used to avert a
   referral at each decision point, and
o Estimating the post-consult referral probability.

The decision point review process gave the physicians the
opportunity to screen the decision points, eliminating those that
they felt were inappropriate or that, based on utilization data,
were found to have been encountered so infrequently that they
could be dropped from further consideration.

The next step in the physician review process involved each
physician stating whether, in his opinion, each decision point
could be handled via a teleconsult and if so, what technology (or
combination of technologies) could be used. Based on preliminary
review and analysis of alternative technologies, seven different

telecommunications-based technologies were considered for potential
use in the analysis. They were:

o Telephone
o Electrocardiogram (ECG)
o Electronic Stethoscope
o Slow-Scan Television (including X-ray Transmission)
o Interactive or "Live" Television
o Patient Viewing Microscope
o Video Microscope

Referring to this technology list, the telephone represents
an audio link provided by the standard telephone network or a
point-to-point radio system. The electrocardiogram and electronic
stethoscope are devices for monitoring/transmitting/receiving ECG
and cardiac sounds respectively. Slow-scan television enables
the monitoring/transmitting/receiving of video information over
a narrowband (voice) channel. Interactive television represents
the standard two-way (black and white) broadband video system.
The patient viewing microscope is a fiber-optics device that can
be connected to a television system, enabling the remote viewing
of body parts including orifices, and the video microscope encompasses a standard optical microscope coupled to a television
system to permit remote viewing of slides.

Following the decision point screening and technology
assignment activities, each physician was asked to estimate the
probability of the need for an in-person patient referral at each
decision point given that a teleconsult using the assigned technology had taken place; this is called the "post-consult referral
probability". For those decision points that were deemed un-
addressable by teleconsult, the post-consult referral probability
was placed at 1.0.

Table 1 illustrates this decision point data collection
activity for the Dartmouth Upper Respiratory Infection protocol.
This Table also incorporates the protocol utilization data that
gave actual field experience information relative to the frequency
of patients accessing various decision points in the protocols
analyzed. Thus, for example, out of 2,882 cases in which the
protocol was used, 23 patients accessed the "Enlarged or swollen
epiglottis" decision point.

However, the frequency of the patient referral as indicated
by the decision points does not, unfortunately, correspond
directly to the number of patients referred because a patient may
have more than one cause for referral (accesses multiple decision
points) but make only one visit to the doctor. Referral data
obtained from one source, Beth Israel, showed that the number of
patients having no referrals, one referral, two referrals, etc.,

Decision Point	Frequency	Protocol Tests/Procedures	MD(A) Technology	MD(A) Add'l Tests/Procedures	MD(A) Referral* Prob.	MD(A) Referral* Number	MD(C) Technology	MD(C) Add'l Tests/Procedures	MD(C) Referral Prob.	MD(C) Referral No.
Enlarged or swollen epiglottis	23		(Referral to MD)				PVM		0.8	18.4
Swollen tonsil on one side only	44	Throat culture	Telephone				PVM		0.4	17.6
Pharyngeal or tonsillar exudate, swollen tonsils or red throat	2124	Throat culture	(Drop)				(Drop)		-	-
Sore throat	2198	Throat culture	(Drop)				(Drop)		-	-
Tender sinus	234	Nasal discharge culture,	(Drop)				(Drop)		-	-
Periapical abscess	4	Nasal discharge culture, Sinus X-ray	(Referral to MD)				X-ray		0.9	3.6
Runny nose	1416		(Drop)				(Drop)		-	-
Foreign body in the ear	9		PVM				PVM		0.3	2.7
Otitis externa	97	Irrigation of ear canal	(Drop)				(Drop)		-	-
Wax removal not done or done but not succseeful	129		(Referral to MD)				Telephone		0.1	12.9
Undescribed abnormality	206		PVM				PVM		0.3	61.8
Red or bulging tympanic membrane, or loss of light reflex or landmarks	380		Telephone				(Drop)		-	-
Ear discharge and tympanic membrane perforated	15	Ear discharge culture	Telephone				PVM		0.1	1.5
Tender mastoid	32		Telephone				X-ray		0.2	6.4
Tinnitus or loss of hearing present for more than two weeks	28		Telephone	Audiogram			PVM		0.1	2.8

*These data not available for MD(A)

DECISION POINT DATA COLLECTION
DARTMOUTH PROTOCOL FOR UPPER RESPIRATORY INFECTION
(2882 CASES)

TABLE 1

Decision Point	Frequency	Protocol Tests/Procedures	TELECONSULT								
			MD(A)				MD(C)				
			Technology	Add'l Tests/ Procedures	Referral*		Technology	Add'l Tests/ Procedures	Referral	
					Prob.	Number			Prob.	No.
History of infectious mononucleosis and severe neck swelling	95	WBC, Differential, Heterophile, Throat culture	(Referral to MD)				(Drop)		–	–
Enlarged spleen or axillary nodes enlarged	128	WBC, Differential, Heterophile	Telephone	CBC			Telephone	CBC, Monospot	0.1	12.8
Localized chest abnormality	195	PA and lateral chest x-ray	TV				X-ray, Steth		0.2	39.0
Temperature > 101°	105	PA and lateral chest X-ray	TV				Telephone		0.1	10.5
Chest pain	290	PA and lateral chest X-ray	TV				TV, X-ray, Steth.		0.1	29.0

*These data not available for MD(A)

TABLE 1 (CONCLUDED)

for the Diabetes protocol followed a Poisson distribution. The
Poisson distribution was used to represent patient decision point
access patterns for the other protocols used in the analysis
since no other data were available that described this phenomenon.

The third analysis activity involved the development of a
computer simulation to model patient flow through a primary
health care facility. Patient flow was examined in terms of the
percentage of patients entering the primary care clinic that:
(1) are treated and return home without need for physician consultation, (2) require a teleconsult in the treatment process and
then return home, (3) require a referral to a physician without a
prior teleconsult (other than telephone contact to set up an
appointment), and (4) require a teleconsult and then are sent to a
physician. This simulation, based on the physician technology
assignment data, provided estimates of the number of patients
referred when using different technology alternatives for various
medical problem categories.

RESULTS OF ESTIMATING TECHNOLOGY IMPACTS

The patient simulation results served as inputs to the final
analysis activity, the technology impact assessment. This assessment enabled conclusions to be drawn regarding the impact on
averting patient referrals for specific problem types (the ten
listed earlier) as a function of various combinations of technology
used to support non-physician personnel. Following the development
of the alternative technology referral avoidance capacities for
the different protocols, these results were scaled to a representative patient population for which data indicating patients'
presenting problems were available. This step enabled the results
of the application of technology for individual protocols to be
extrapolated to a typical patient population in which the incidence
of the various problem types was known. Thus, the number of
averted referrals for this population could be calculated for
each of the alternative technology combinations.

The patient simulation results reflected the patient referrals in the ten protocols that could be averted with the utilization of alternative technologies. These results were then
scaled to the appropriate number found for the corresponding
problem type in the representative patient population.

The overall impact of telecommunications-based technology on
the representative population is illustrated in Figure 7. As
indicated, more than half of the visits would have required a
physician referral but with the use of telecommunications-based
technology, approximately half of these referrals could be avoided.

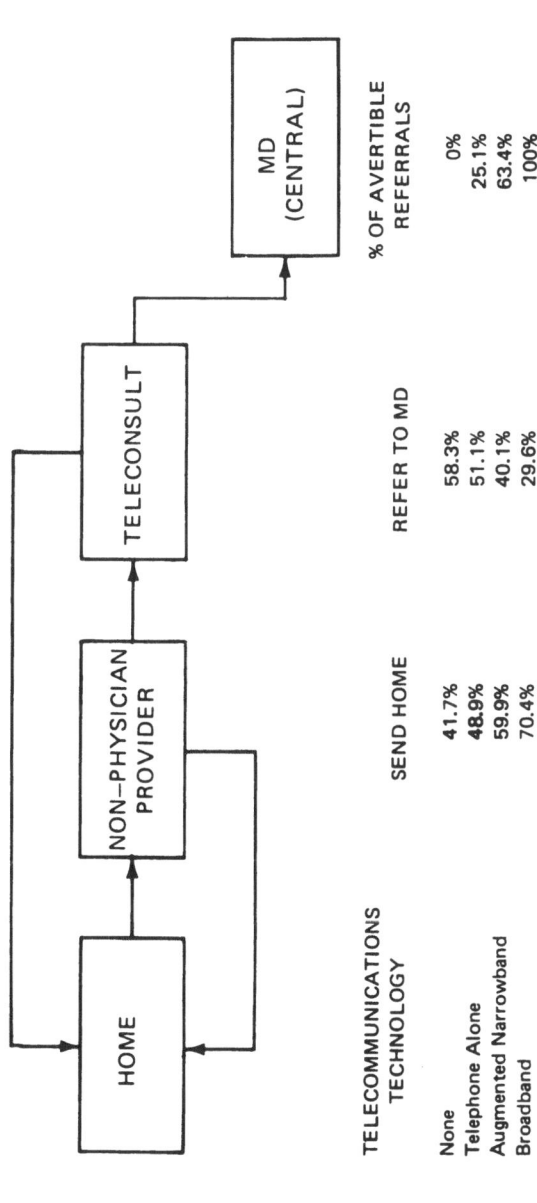

TELECOMMUNICATIONS TECHNOLOGY	SEND HOME	REFER TO MD	% OF AVERTIBLE REFERRALS
None	41.7%	58.3%	0%
Telephone Alone	48.9%	51.1%	25.1%
Augmented Narrowband	59.9%	40.1%	63.4%
Broadband	70.4%	29.6%	100%

Results of a prospective, theoretical analysis of what the impact of alternative telecommunications technologies would be in averting patient travel from a remote, non-physician-staffed clinic. Significantly, narrowband technology can avert nearly two-thirds of the avertible referrals.

FIGURE 7

Summarizing the results of this analysis:

1. Based on a total of 236,624 patient visits for medical
    problems in the representative population for which data on
    presenting problems were available, 137,870 (58.3% of all
    patients) referrals to or consultations with a physician
    were indicated.

2. Of these, 70,035 were to be referred no matter what telecom-
    munications technology was available; this is 51% of the
    patients with decision points or 29.6% of all patients.

3. Of the 67,836 referrals (28.7% of all patients or 49% of
    patients with decision points) that could be averted through
    use of telecommunications technology, telephone alone
    accounted for 17,000 (25%); telephone plus ECG, electronic
    stethoscope and X-rays transmitted via slow-scan television
    averted 25,000 (37%); all of the above plus a patient-
    viewing microscope transmitted via slow-scan television
    accounted for 43,000 (63%); all of the above plus two-way
    black and white television plus slide video transmission
    accounted for the full 67,836 (100%).

Thus it can be seen that narrowband technologies can avert
almost two-thirds of all of the referrals that could be averted
with broadband technologies.

Cost Performance Analysis

        Further analyses were performed(14) to assess the impacts of
these "averted referrals," in terms of the total time involved in
patients utilizing a rural health care system and in terms of
the cost per episode of illness (and average cost per visit).
This analysis was performed by simulating a patient population
having the characteristics (in terms of incidence of various
problem types) of the representative population utilized in the
protocol extension analysis in a hypothetical setting involving a
local, non-physician staffed clinic separated from a distant
physician-based clinic. Patients "presented themselves" at the
local clinic and were either served directly by the non-physician
provider, served after the use of teleconsultation, served by a
physician who visited at regular intervals at the local clinic,
referred to the central physician clinic, or referred to a source
of secondary care. Different manpower/technology combinations
were "utilized" in the satellite (local) clinic as represented by
the transition probabilities determined by the previous analysis.

        In addition, the sensitivity of the results obtained in
these simulation analyses was investigated by means of varying

the values of critical parameters. The performance of the different manpower/technology combinations was analyzed with respect to variations in the value of patient time, the number of patients visiting each of the clinics annually, the proportion of physician time spent at the remote clinic, the required return on investment in telecommunications equipment, and the effectiveness of technology options in assisting with remote physician consultation.

The first major result of the cost performance analysis was that a significant difference in cost per episode of illness (and cost per visit) existed between a centralized system requiring direct patient visits only to the distant physician and the various "distributed system" options (alternative systems utilizing the satellite clinics with different manpower/technology combinations). It was found that distributed systems offered significantly reduced costs (patient travel costs were included in the analysis) in comparison to centralized systems which required rural residents to receive their health care only through direct trips to distant physician-based clinics. The second major result of the cost/performance analysis dealt with the relative cost savings achieved through the use of different technology options. These analyses showed that an option designated as the "augmented narrowband technology" option showed superior cost performance relative to other technology options which could be employed in the rural health care delivery system investigated. However, even though this option which utilized the maximum medical instrumentation compatible with a single voice channel communication system showed significant cost improvements over other technology options, the degree of improvement was not as great as that shown by choosing a distributed system over a centralized system.

Similarly, the augmented narrowband technology option provided minimum patient time involved in receiving health care over any other technology option. Once again, the distributed system approach provided significantly improved patient time characteristics over the centralized system, when one-way trips to the physician required approximately 45 minutes or greater.

The results of the sensitivity analyses indicated no change in these two basic results for variation of any of the parameters indicated. However, extreme variations did change the degree to which the augmented narrowband system offered improved capability for a few parameters. In particular, increasing the proportion of time spent by a physician at a local non-physician staffed clinic significantly reduced the advantage of any technology options over a standard telephone.

NEXT STEPS

As a result of the analytic activity and results obtained
and reported above, the following "next steps" are being undertaken at the present time. Under funding from the Health Underserved Rural Areas Program of the Bureau of Community Health
Services, DHEW, an Augmented Narrowband Telehealth Laboratory is
underway. The technologies that appear to be cost-effective
based on the above analyses are being aggregated at the METREK
facilities in McLean, Virginia. These include stethophone,
electrocardiogram, facsimile, slow-scan television and patient-
viewing microscope. A team including physicians and engineers is
developing a set of operating procedures to guide users of the
technology when it is placed in the field. Questions being addressed include the best ways for lighting an object whose image is
to be transmitted and the best views to be transmitted in order to
obtain medically-useful information. These procedures are being
developed to be problem-specific, i.e., different procedures are
needed for dermatology, radiology, etc. Field trials of the technology are being undertaken in three sites at present, and additional sites are being studied for their potential as telehealth
demonstration projects.

DISCUSSION

The patient trajectory methodology as described above has
several limitations and shortcomings as a tool for planning and
evaluating field trials in telecommunications technology for
health care delivery.

First, the trajectory concept trades off patient travel
costs and telecommunications costs. These are not dollars that
come from the same source so the decision to transfer them from
one to the other cannot be made by a single decision-maker. In
the private sector, patient travel costs are not reimbursed as
part of the costs of obtaining health care services. In Federally-
sponsored projects, grant dollars may be used for the installation
of telecommunications equipment and this is being done in the
field trial projects. However, it is clear that for wide-spread
dissemination of this technology into the health care system, a
revision in the financing mechanisms will have to be found
(unless telecommunications technology can be justified on other
grounds than saving patient travel). One possibility is to seek
reimbursement for services provided *via* telecommunications
technology and use some of the funds to defray the telecommunications costs; another is simply to absorb these costs as overhead.

Second, the trajectory methodology ignores psychological
issues. The utility of telecommunications technology may be most

marked in reducing the sense of isolation experienced by health care providers in rural sites. Also, patients may be more comfortable than they would otherwise be if they believe that they are part of a health care system that has a full array of resources available to them. The benefits in telecommunications technology coming from these sources are not calculated in the trajectory analysis described above, although patients' increased confidence might be measured by observing that more complicated cases present at a remote clinic with telecommunications technology than at a comparable clinic without such technology. That is, patients may triage themselves and go around a non-physician provider-staffed clinic when they consider themselves really "sick". If patients are unwilling to use the non-physician provider in such cases, the data upon which the trajectory methodology was developed will be inapplicable in a real setting.

Third, the trajectory methodology only looks at the use of telecommunications technology in the medical care process and ignores its potential with respect to administration and education. Both of these application areas have been cited as important reasons for installing telecommunications equipment.[1]

Fourth, the trajectory methodology is not "outcome-oriented" in that it does not look at the results of using telecommunications technology to see whether the use of the technology made any difference in terms of what actually happened to patients.

Fifth, there is the potential for both overuse and underuse of telecommunications technology that is not addressed by the trajectory methodology. The judgment of the non-physician provider as to whether a teleconsultation is needed will certainly be important in a field trial in determining whether the telecommunication technology is used. The protocols as developed do not allow any such judgment. Also, there is the factor having to do with the ease of telecommunicating which may affect the use of the technology. The laboratory effort will not tell us whether in actual use the technology will be cumbersome and irritating; this information on time and convenience will be sought in the field trials.

Finally, the protocols on which the analysis presented in this paper was based were developed for use in clinics in which the non-physician providers were co-located with physicians. As discussed above, our analysis of these protocols indicates that 58.3% of patients presenting require physician consultation or referral. Data from a recent University of Washington survey[15] of 13 non-physician providers in remote rural clinics showed telephone consultations ranging from 0% to 17% with an average of 4% and physician referrals ranging from 5% to 27% with an average of 8%. Thus, the impact of telecommunications technology on

reducing patient travel to physicians from these sites could be
anticipated to be considerably less than the impact projected
from the theoretical analysis presented in this paper.

Nevertheless, the decision to pursue the investigation of
telephone-based technology still appears sound. This direction
is consistent with the findings of Conrath, et al.[16, 17] who
found few differences between television and hands-free telephone
for remote diagnosis. In fact, Conrath and his co-workers have independently reached the same conclusions that we have reached and
they are undertaking a major demonstration of slow-scan television
for health care delivery in rural Canada.[18]

In conclusion, this paper has presented an analytic approach
to planning for the use of telecommunications technology in health
care delivery. The results of the analysis have strongly suggested
that telephone-based technologies should be vigorously and systematically explored.

REFERENCES

1. M.L. Rockoff, "An Overview of some technological/health-care system implications of seven exploratory broad-band communication experiments," *IEEE Trans. Commun. (Special Issue on Interactive Broad-Band Cable Systems)*, vol. COM-23, pp. 20-30, 1975.

2. J.J. O'Neill, J.T. Nocerino, and P. Walcoff, "Benefits and problems of seven exploratory telemedicine projects," MITRE Corp., McLean, Va., Tech. Rep. MTR-6787, 1975.

3. G.T. Moore, T.R. Willemain, R. Bonanno, W.D. Clark, A.R. Martin, and R.P. Mogielnicki, "Comparison of television and telephone for remote medical consultation," *New England J. Med.*, vol. 292, pp. 729-732, 1975.

4. C. Muller, C.L. Marshall, M. Krasner, N. Cunningham, E. Wallerstein, and B. Thomstad, "Cost factors in urban telemedicine," *Medical Care*, vol. 15, pp. 251-259, 1977.

5. A.M. Bennett, "Assessing the performance of rural primary health care systems," MITRE Corp., McLean, Va., Tech. Rep. MTR-6788, 1974.

6. D.M. Kessner, C.E. Kalk, and J. Singer "Assessing health quality - the case for tracers," *New England J. Med.*, vol. 288, pp. 189-194, 1973.

7. J.J. McCord, J.S. Gonnella, and D.Z. Louis, "The staging concept - an approach to the assessment of outcome of ambulatory care," *ORSA/TIMS Joint Meeting*, San Juan, P.R., October, 1974.

8. T.R. Willemain, private communication

9. A. Doermann, D. MacArthur, and P. Walcoff, "Extending the capabilities of non-physician providers in isolated rural areas: an investigation of the potential impact of telecommunications-based technology," MITRE Corp., McLean, Va., Tech. Rep. MTR-7063, 1975.

10. H.C. Sox, C.H. Sox, and R.K. Tompkins, "The training of physician's assistants. The use of a clinical algorithm system for patient care, audit of performance and education," *New England J. Med.*, vol. 288, pp. 818-824, 1973.

11. R.K. Tompkins, "Computer-based paramedic support and audit,"
     Final Report Contract No. HSM 110-73-403, National Center
     for Health Services Research, 1976.

12. A.L. Komaroff, B. Reiffen, and H. Sherman, "Problem-oriented
     protocols for physician-extenders," in: *Applying the
     Problem-Oriented System*, J.W. Hurst, H.K. Walker, and
     M. Woody, eds., MEDCOM Press, N.Y., 1973.

13. A.L. Komaroff, W.L. Black, M. Flatley, R.H. Knopp,
     B. Reiffen, and H. Sherman, "Protocols for physician assistants:
     management of diabetes and hypertension," *New England J. Med.*,
     vol. 133, pp. 294-299, 1974.

14. H. Dhillon and A.M. Bennett, "A cost-performance analysis of
     alternative manpower technology combinations for delivering
     primary health care," MITRE Corp., McLean, Va., Tech. Rep.
     MTR-7068, 1975.

15. University of Washington, "Utilization of new health
     practitioners in remote practices," *Proc. of Conference on
     New Health Practitioners in Remote Settings, October 20-22,
     1975*, pp. 22-50, 1975.

16. D.W. Conrath, E.V. Dunn, J.N. Swanson, and P.D. Buckingham,
     "A preliminary evaluation of telecommunication systems for
     the delivery of primary health care to remote areas," *IEEE
     Trans. Commun. (Special Issue on Social Implications of
     Telecommunications)*, vol. COM-23, pp. 1119-1126, 1975.

17. E.V. Dunn, D.W. Conrath, W.G. Bloor, and B. Tranguada, "An
     evaluation of four telemedicine systems for primary care,"
     Health Services Research, vol. 12, pp. 19-29, 1977.

18. D.W. Conrath, private communication.

TELEHEALTH CARE IN CANADA

A discussion of projects, research and policy considerations

Anna Casey-Stahmer

Social Policy and Programs Branch
Department of Communications
Ottawa, Canada

This paper reviews Canadian experiments and projects in the
health services field utilizing telecommunications based delivery
systems. It reports on evaluative findings of several projects.
It discusses research and development activities which might be
necessary to ensure that the benefits of these ongoing activities
will be applicable to the existing telecommunications and health
care delivery systems, but which at the same time take into consideration future developments in both areas.

INTRODUCTION

Ever since its introduction the telephone system has been
used by health care providers to support various aspects of their
work. A recent survey of a sample of nursing stations[1] in
Northern Canada shows that the eleven nursing stations called the
base hospital on an average of 52 times a week. The nurses called
mainly for medical advice (33%), and for administrative support
(46%), including ordering of supplies and to check up on
hospitalized

What is then novel when we talk about telemedicine or telehealth care? Did it develop to promote high technology, e.g.,
one and two-way television transmission facilities or satellites?
This aspect may be present in some of the current surge of
research and development in the field. However, telemedicine and
telehealth care derive their primary impetus from the recognition
that remote health care delivery could utilize telecommunications
more effectively. Basically, the concept of telehealth care
should include three conditions:

(1)   the organization of the health care system makes tele-
      communications an integral part of its delivery, e.g.,
      nursing education by telephone, administrative tele-
      conferences, routine "doctor calls"'

(2)   the network configuration, including the telephone
      system, integrates special needs of the health care
      system, e.g., emergency access, teleconferencing net-
      works, slow-scan TV or interactive TV;

(3)   medical and technical standards are developed jointly,
      so that the operational telecommunications network can
      provide many of the special transmission services that
      might be required, e.g., medical and technical standards
      for the transmission of x-rays via slow-scan are speci-
      fied to make optimum use of the existing telecommunica-
      tions system.

Is there a role for the Canadian federal and provincial
governments in the exploration and development of these issues and
the subsequent implementation of recommendations? Different
levels of governments have vast responsibility for the health care
delivery systems in Canada either by means of policy making
powers, (e.g., insurance schemes), or through the mandate to provide
medical care to remote areas and native peoples. In the communications area, provincial and federal governments have various
policy responsibilities as well as regulatory authority over the
common carriers. These structures do not imply, that there is
no room for private industry or for private health care organizations. However, governments are required to maintain an
overall view of the field and facilitate its adequate exploration
in technical, economic, legal, policy and regulatory matters.

                              BACKGROUND

In large areas of Canada the delivery of health care, of
education, and of other social development services is faced with
the vast distances between communities and relatively low population density. Canada is the second largest country in the world,
but has only about 22 million people. A high percentage of these
live in a 100 mile belt north of the United States border. The
northern 40% of the country is inhabited by 58,000 people.

Generally, a three-tiered health care system is in existence.
At the primary level a registered nurse, nurse practitioner or a
physician provides the care. At the next level one finds small
base hospitals with clinical and laboratory facilities and a limited
number of staff members in   different medical specialities. At the

third level large research and teaching hospitals provide various specialized services.

The public telecommunications systems are owned and operated by common carriers with provincial or federal charters. They provide a mix of terrestrial and satellite services. The common carriers lease domestic satellite services from Telesat Canada, which owns and operates the system. The shares of Telesat Canada are almost entirely owned by the federal government and the common carriers. Anik I, the first domestic satellite, was launched in 1972. Anik II and III were launched in subsequent years. In early 1979 a satellite (Anik-B) will be launched to replace Anik-I. Plans for an expanded satellite system include the launch of three new satellites (Anik-C series) in the early 1980's.

The mission of the Department of Communications (DOC) in the area of communications satellites is limited to a R&D role. The Hermes Communications Technology Satellite launched in 1976 is an example of such experimental work. Overall, it is the mandate of DOC to ensure the orderly development of the Canadian telecommunications system. Apart from enforcing technical standards, DOC policies aim to guide the implementation of telecommunications systems which allow socially equitable access to such services and which support social and cultural aspirations of the Canadian people. The Hermes project is one tool with which the desirability and utility of various social applications can be initially explored.

TELEMEDICINE PROJECTS IN CANADA

In the late 60's and early 70's research and development related to telemedicine mushroomed in North America. Telemedicine applications in urban, rural and remote settings were explored. Telecommunications technologies tested for these purpose included simple telephone systems, interactive TV-systems, and computer applications for diagnostic support.

In Canada, telecommunicatons systems to support health applications can be found in almost all provinces. Sixteen projects have come to our attention which were conducted or are underway. Some are experimental systems, others provide ongoing services and others are research studies. Out of these projects six are in the area of medical, nursing and continuing medical education. Four deal with the transmission of medical data, including x-ray and EKG, four support remote medical diagnosis and management decisions. Two are broader in scope and try to identify future medical communications needs for a given health care system.

Sponsorship and funding for these projects comes from a variety of sources. Some of the projects developed out of personal interest and initiative of individual researchers and practitioners. Others were developed by provincial government departments. The major activities of the federal government in this area started with the Hermes experiments project. The DOC made satellite time and terminals available to the experimenters. The DOC and the Department of Health and Welfare participated in the Hermes project by providing funding, by monitoring and by evaluating the experiments. The following description and discussion deals with only those experiments in telehealth care that were sponsored by the DOC. The effect and impetus of other projects on the development of telehealth care in Canada is not to be underestimated, however.

THE HERMES EXPERIMENTS

In 1972, the DOC invited interested groups and agencies to participate in the experimental Hermes satellite project. A major objective for the department was to obtain feedback concerning future demands for satellite services. Sixteen experiments were accepted in the areas of education, community development, administration and health care.

Three of the Hermes experiments relate to health care delivery. The experiments are designed to support the present type of health care system. They do not attempt to assess completely different systems in which, for example, community aides could replace regsitered nurses.

One experiment is primarily concerned with the provision of continuing education for physicians and nurses in remote hospitals. It also has public health education components. The other two experiments are primarily providing second medical opinions and support in a variety of medical desciplines.

The Technology

The experimental Hermes Communications Technology Satellite is the first communications satellite to be operated in the 12/14 GHz frequency band and at a very high transmitted power level (200W). It has been used to explore various technological systems features which may be part of the next series of operational satellites in Canada. The Anik-B and Anik-C series of satellites will utilize the 12/14 GHz frequency bands.

The Hermes experiments have used three main categories of small satellite terminals.

Two-way voice links   – ten terminals, manufactured in
                                                     Canada. Antenna diameter 1 m.

TV-receive; two-way voice   – eight terminals, manufactured in
links   Canada. Antenna diameter 2 m.

Two-way TV; two-way voice   – two terminals, manufactured in
links   Canada. Antenna diameter 3 m.

The limited number of terminals available were shared by all
experimenters. The large number of experimenters required
frequent transport of the terminals. Additionally, the satellite
was designed to have a two-year life. This necessitated the
limitation of the duration of individual experiments. Also, the
Hermes satellite time is shared between the two countries on an
every other day basis.

EXPERIMENT 1 – GOVERNMENT OF ONTARIO

The provinical Ministry of Health is a participant in an inter-
ministry experiment, which is coordinated by the Telecommunications
Services Branch of the Ontario Ministry of Government Services.

During the period June to August, 1976 the Ministry of Health
tested the utility of audio-conferencing for discussions between
physicians located at isolated hospitals and specialists at a
large specialized hospital. The conference subjects included
radiology, cardiology, orthopedics, and hospital administration.

The ministry also undertook a short but impressive demonstration which may have great future potential in the development of
ambulatory care in remote areas. In this demonstration a medical
emergency evacuation was simulated from a remote mine site.
Medical data were transmitted via a small ground station from
hand held radio units in an isolated area, from a ground ambulance,
from a small nursing clinic in the mining town, and from an air
ambulance to an urban medical centre. During this experiment
the EKG of a patient was continuously monitored by specialists
while in transit from the remote site to a hospital. Doctors at
the medical centre and at the base hospital monitored the EKG
and were able to discuss the medical case via satellite. For this
demonstration special equipment was developed by the Ministry of
Health to relay signals from ground locations and from the plane
to the satellite terminal.

EXPERIMENT 2 - UNIVERSITY OF WESTERN ONTARIO

From October 1976 through February 1977, Norhtern Ontario was the site of another telemedicine experiment. This experiment linked a remote nursing station, a northern base hospital and the specialized resources of the Hospital of the University of Western Ontario. The base hospital and nursing station are both under jurisdiction of the Federal Department of Health and Welfare. The University is on contract to provide various specialist consultant services to the base hospital. This includes regular travel to the remote hospital by specialists, the placement of residents at the base hospital as well as the receipt of referral cases from the base hospital.

Objectives

In support of the above arrangements, the experiment aims:

- to test alternatives and additions to the existing system of providing medical specialist support to remote base hospitals,

- to assess if the scope of support to a base hospital can be increased without major strain on the resources of the university hospital,

- to identify the uses of the telemedicine system which are of benefit to the base hospital,

- to describe the use and effects of the communications technology employed in support of medical specialist consultation,

- to assess the acceptance of the system by the care providers and allied health personnel,

- to analyse the effect of reliable communications links between the nursing station and the base hospital upon patient management and professional satisfaction of the nurse providers,

- to describe the acceptance of the telemedicine system by the communitites at large, i.e. patients, relatives, village leaders.

Description

The experiment used three satellite terminals. Via a 3m terminal the base hospital communicated with the 2m terminal at the university hospital as well as with the 1m terminal at the nursing station. All three sites were linked with two-way voice communications, with facsimile and with EKG transmission. The base hospital was also linked to the university hospital by one-way video transmission. Black and white cameras at the base hospital could be controlled remotely from the university hospital. One colour TV unit was used for the transmission of microscope images.

The link between the nursing station and the hospitals was in use approximately one half hour per transmission day. Uses of the link included: obtaining second medical opinions from the hospital; handling administrative matters; transmitting EKG for interpretation; transmitting patient documents; exchanging informaiton on hospitalized patients; and keeping families in touch with a hospitalized family member.

The video link between the base hospital and the university hospital was available for about 250 hours. It was used for a wide variety of specialist consultations including radiology, psychiatry, pediatrics, pathology, obstetrics, and cardiology. These consultations accounted for about 11% of the available video time. The main usage categories break down as follows: x-ray interpretation - 11.8%; ultrasound reading - 6.1%; morphology - 5.2%; anaesthesia - 4.9%; respiratory technology - 2.4%; physiotherapy - 1.7%.

Generally, the different programs, (e.g. pathology, patient consultation, x-ray reading) were scheduled a day or two in advance of the transmission day. Some programs were planned on a regular basis. On the morning of each transmission day, or on the preceding afternoon the transmission schedule was distributed among participating sites. Any relevant patient history and other information was transmitted prior to the actual consultation.

Evaluation

The experiment is being evaluated from various viewpoints, including technical and management assessments. The overall evaluation was undertaken by an independent team on contract to the DOC and in consultation with the federal Department of Health and Welfare.

The main data collection efforts were designed to provide a description and assessment of the effect of the system on criteria such as perceived changes in patient diagnosis, treatment and management, acceptance of the system, the types of

uses of the system and perceived adequacy of the system for various
tasks.

During the experiment each participating health care provider
responded to a series of questions gathering his/her reaction to
individual sessions. Short patient questionnaires were designed
and implemented towards the end of the experiment. Post experiemnt
interviews have been conducted for more global reactions to the
experiment, its acceptance and perceived future effects on
health care delivery in the North. Interviews with the communities
at large are part of the post experiment data collection.

Some Preliminary Findings

Final reports by the various participants are being prepared
as of the time of writing this paper. The following information
was obtained in discussion with the participating groups and by
scanning the evaluation data.

The final report by the technical contractor[4] states that the
system worked without difficulties. Recommendations for future
research as well as project planning include: the need to further
define engineering and medical standards to develop appropriate
systems; the desirability to have input from all user groups in the
technical design so that the engineer has a good idea of how and
why certain facilities will be used; the need to develop effective
communications scrambling devices to ensure medical privacy; the need
to test the equipment in simulations; and the need to train the users
in the utilization of the technology prior to the actual commence-
ment of the project.

As of this time the independent project evaluation team has
just completed the post-experiment interviews and the data have not
yet been analysed.[5] It seems justifiable, however, to say that the
participants thought that the experiment was a useful undertaking:

  - About 90% of the participants in the experiment rated their
    overall impression of individual transactions as excellent
    or good. Similar positive ratings were given to the
    quality of the technical system for medical purposes and
    to the ease with which the equipment could be handled.

  - Over the experimental period, the acceptance of various
    features of the system increased as users became more
    familiar with it. For example, the technical system only
    provided a one-way video link from the base hospital to the
    University Hospital. The other links were audio-only. The

acceptance and utility of the asymmetric links was in doubt at the outset of the experiment. From the available data it appears, however, that the participants at the university hospital as well as those at the base hospital had overcome any possible difficulties due to the use of asymmetric channels toward the end of the experiments. In physiotherapy, psychiatry and some continuing education programs, however, the need for 2-way video was voiced quite strongly.

- Also, many participants felt at the beginning of the experiment that they could have achieved the same by using the telephone. Toward the end of the experiment more participants indicated that the experimental system offered a considerably improved service over that which the telephone system could provide.

- The perception of the medical usefulness to the patients of the transactions also changed over time. At the end of the project over three-quarters of the transactions were seen as medically useful to the patient by the care providers at both hospitals.

- A series of questions was asked in regard to changes in patient diagnosis and management plans. In general, some changes in both occurred as a result of the remote consultations. The highest percentage of respondents felt that the remote consultations were helpful, even when no changes in patient management occurred. This was felt to be so particularly by the people at the base hospital. One explanation could be that the respondents generally appreciated the transactions as a learning experience.

- A popular, albeit questionable, measure in telemedicine research is the change in the number of patient transfers. In this experiment the participants at the base hospital stated that in over 80% of the cases presented, patient transactions were felt to have prevented patient transfer; and some transactions resulted in patient transfer which otherwise would not have occurred.

All of the above data will be analysed in more depth in the upcoming months and different or new trends will become apparent. Specific recommendations will evolve. Overall, the experiment demonstrated that telehealth care can gain acceptance of care providers and consumers alike, and that it can serve a useful function in supporting the delivery of health care to remote areas.

EXPERIMENT 3 - MEMORIAL UNIVERSITY

From March to June, 1977 four hospitals in Newfoundland and Labrador are linked via the Hermes satellite by live television to the Educational Television Centre of Memorial University in St. John's, Newfoundland. All sites were linked by interactive voice channels. The links were mainly used for continuing education for health care professionals. For several years, CME Programs from Memorial University have been given at the four sites by a variety of traditional methods relying on face to face methods.

The objectives of the CME experiment included the following issues:

- acceptance of the experimental program by all participants,

- cost estimates and comparisons of Hermes programs and traditional programs,

- software development for an interactive education system, and

- acceptance of the Hermes program in public health education for special interest groups, e.g. diabetics, prenatal classes, social workers.

Description

The project addressed different topics in continuing education for all health care professionals, public health education and medical consultation. Public health education played a limited part in the experiment. The medical consultation and data transmission component of the experiment were also of limited scope.

The continuing education program took up about 80% of the projects' resources and time. The program is divided into three components: 1) Continuing education for physicians including topics such as communication disorders in children, cardiology, therapeutics and anaesthesia. Course material covered between eight to ten hours per topic. 2) Another eight to ten hours were devoted to continuing nursing education. 3) The Provincial Hospital Association developed another 10 hours for allied hospital workers.

Program development for these programs was based on a great deal of interaction between the learners and teachers in order to ensure that the program content was relevant and reflected the information needs of the learners. For example, for the communications disorders in children course, local tapes were filmed at various sites throughout Newfoundland and Labrador. Whenever relevant, footage was used in the course to illustrate problems and treatment approaches.

The present experiment by Memorial University grew out of the work of a provincial committee which was established as a result of DOC's announcement of the opportunity to participate in the Hermes project. During its existence the telemedicine committee provided valuable liaison between the experiment, different medical institutions as well as the provincial government. Such a structure may be very beneficial in the long run by ensuring a continual involvement of agencies with operational mandate to implement/fund similar services.

Evaluation

An evaluator was attached to the project team. The main emphasis of the evaluation was on the CME component. Descriptive information will be provided on the other two components.

The evaluation of the CME component focuses on satisfaction measures along the following dimensions: attitude of participants, cost and organizational factors, and technical performance.

Evaluation data are being collected from all groups of participants including learners, teachers and instructional technologists. Pre-broadcast data included the participants' perceptions of the properties of the course, the local organizational climate and project background. During broadcast time all five sites were monitored periodically. At given intervals questionnaires were administered to capture views on the technical qualities as well as responses to instructional format and content. Major interviews were conducted with all participants in the post-broadcast period. The focus was on reactions to the delivery system, didactics, subject matter and comparison with alternate continuing education programs.

FUTURE ACTIVITIES

What next? The Hermes social experiments are successful in that they have raised awareness and interest in satellite communications in various sectors of social development services. The translation of this interest into the identification of communications needs, demands and requirements which can enter communications systems planning is a demanding task.

Sharing of Responsibilities

Many social, health care or educational agencies which might benefit from utilizing telecommunications systems, and particularly satellite systems, have only limited technical or engineering

know-how. This limits their capability to input into the development or expansion of telecommunications systems by the common
carriers. Also, their knowledge of the regulatory processes in the
telecommunications field is limited. This limits their capability
to influence regulatory decisions in their favour.

On the other hand, the various bodies in the telecommunications
sector, i.e. policy and regulatory authorities and common carriers
cannot carry out these tasks for the social development sector.
First, they might appear as "pushing technology as a solution in
search of a problem." Second, they do not have the tools (or the
mandate) to identify the areas in which telecommunications are
genuine solutions to problems of service delivery and service
access.

Different approaches have been taken by governments to overcome this dilemma. One approach is to establish telecommunications
offices within social development agencies. The other is to
establish social policy and programs divisions within telecommunications agencies. The third is to disregard any governmental
involvement and leave the interests of social development services
to the market system in the same way as the business or private
sectors. All aproaches have obvious advantages and disadvantages.

The second approach has guided developments in Canada and will
continue to do so – with mounting input and co-ordination from an
increasingly informed group of potential user agencies. For
example, several joint activities including workshops, studies and
planning committees are underway or anticipated bringing together
the interests of the federal Departments of Health and Welfare and
of Communications. Similar types of activities are necessary in
the further exploration of the other areas which are being tested
in the Hermes project, i.e., administration, education and community development.

One advantage of the Canadian approach is that it facilitates
the aggregation of user needs over such different service and
interest areas, because the DOC is concerned to foster the development of responsive and effective communications systems for all
areas. Although specific developmental tasks differ for each area,
a few common guidelines exist. First, program responsibilities
must be shared increasingly with user agencies. (In the Hermes
project the DOC carried much of the project funding and other
support). Second, the DOC has an important role to play in facilitating the development of Hermes-type projects from experiments-
demonstrations to operational applications. The following discussion will limit itself to considerations in the health services
field.

From Demonstrations to Operational Systems

The planning for satellite applications to the health services field can be seen as consisting of three steps. The first step constitutes the experiment - demonstration phase in which technical and medical service possibilities are being explored and from which wider interest is generated. This phase is presently underway in the Hermes telehealth care experiments. In the next phase, pilot projects of major size and closely co-ordinated with all groups that have potential interest in the outcome will be required. A rigourous evaluative research component would assess utility, feasibility and effects of satellite communications in remote health care delivery. Such a program is under consideration for the Anik-B satellite. The DOC may support the program by accepting the costs for satellite capacity and ground stations. The national Department of Health and Welfare would be responsible for the program definition, development and implementation. This step needs to be preceded by a needs and demands assessment to help in the design of pilot projects which address priorities in health care delivery. Some preliminary work is underway in this area. The third phase would be the operational implementation of desirable systems and organizational features. It is anticipated that technical res- ponsibilities for operational telehealth care systems will largely remain with the common carriers. Many activities in preparation for the third phase will therefore have to be undertaken jointly with the common carriers and the provincial and federal departments of health.

Research and Development: Some Lessons

The Hermes experiment evaluations have identified research and development (R&D) tasks relating to the development of joint standards in the medical and technical areas. For example, the medical utility, resolution requirements and gray scale variations of equipment for the transmission of x-rays needs to be explored further. Also the Hermes experiments did not explore interactive television systems, or television links to nursing stations.

A particular concern in regard to R&D is to develop medical standards and technical specifications in such ways that the findings can be applied as much as possible to existing or planned telecommunications systems. In other words, the development of devices for medical information transmission should take into account the capacities of the existing telecommunications systems. If this is not done the usefulness of these equipments could be limited to applications where private networks are desirable. Also, system implementation and operation costs will be considerably higher as the shared use of such systems will be limited.

In the non-technical areas of telehealth care, the Hermes
experiments can give also directions to future R&D. Basically,
they can be seen as a formative stage, providing a basis upon which
research issues can be based regarding long-range social, medical
and clinical effects of telehealth care. For example, the Hermes
experiments explored a variety of health related applications;
should an operational telehealth care system integrate all of these,
are some services of marginal importance or are additional services
desirable? What institutional and administrative structures would
make best use of such systems? The Anik-B program should exemplify
this type of research.

It appears that the health services field would best be served
if this R&D were initiated and sponsored by the government. This
should ensure that medical requirements are the main rationale and
guiding principle for the work. Also, in order to ensure that the
right questions were asked in the Hermes projects, the DOC was
careful to have the social research work carried out by groups with
expertise in the social applications and not with strong telecommunications experience.

The Hermes experiments themselves came into existence mainly
because of technological R&D factors related to communications
satellites. The decision to make the satellite available for
experimentation by social user groups was well considered. Although
it gave rise to some frustrations and required consolidation of
different requirements in the technical and social sectors, the
positive features are overriding. It gave the social sector the
opportunity to explore and develop applications which would not
have existed otherwise. The area of social uses of satellite
systems would have been little explored in Canada without the
project. The long-range implications of the project make this step
a useful one.

Research and Development or Operational Systems?

Research and Development in the exploration of telehealth care
systems is guided by immediate requirements and future potential.
The two must proceed in parallel. The exploration of future
potential, often more exciting and creative, must not overshadow
the need for 'action-now'. For example, the support of the northern
health care system by interactive television may well be feasible
or desirable at some point in the future and should not be overlooked. This future potential should not detract from the urgency
to explore and implement systems which can more immediately support
the present health care system.

Costs and the overall development of the telecommunications system in a given area must be major determinants for present activities. Unfortunately, broadband systems are costly and their superiority in regard to other strategies for health care delivery is not established. Even if health care planners are so convinced of their effectiveness as to invest in such services, the overall telecommunications situation needs to be assessed. For example, many communities have not yet access to live television broadcasting. In this context broadband services for the nursing station may not appear socially desirable at present.

Not detracting from the explorations of the Hermes project or from considerations of future F&D, a project is underway to specify a telephone system[6] for delivery of health care services. The analysis of telephone systems and the interface requirements for health care systems can be undertaken now. Findings and recommendations can be negotiated and implemented. The project identified present technical or health care organization problems which prevent effective use of the telephone system. It will identify services which the system could provide and desirable system configurations. A field trial to test some recommendations is planned in the near future. If the trial proves successful, the common carriers would have standards by which to develop health care specific requirements for their operational services.

SUMMARY

Various experiments, projects and studies have created interest and awareness of the potential applications and utility of telecommunications and space technology within various user sectors, particularly in the health services field. In order to assess operational feasibility as well as desirablility of satellite based delivery systems, however, large scale pilot testing and further research and development in the technical as well as applications sectors will be required. These further efforts must be designed to satisfy present operational requirements, respecting the existing state of telecommunications coverage, and telecommunications service costs, but at the same time the efforts should anticipate future developments of communications technologies and equipments, as well as related changing cost and rate structures. Any effort in this area, however, can only succeed if the health services sector and the telecommunications sectors undertake them jointly. At present the state of development is such that governments have an important role to guide and promote these developments in order to assess their viability in operational systems.

ACKNOWLEDGMENTS

The author wishes to thank the evaluators and experiment leaders of the Hermes telehealth care experiments without whose enthusiasm and hard work these projects would not have been possible. They include Robin Roberts, Sally Skene, Dr. Earl Russell, Dr. Russ Kempton, Borys Koba and Emerson Johnson of the University of Western Ontario experiment, under the leadership of Dr. Lewis Carey; Dr. J.A. Baldwin, Glen Chung-Yen and J. Butler of the Government of Ontario experiment; Alan Pomfret, Judy Roberts and Ken Hauschild of the Memorial University experiment, under the leadership of Dr. Max House and Craig McNamara. Special thanks also go to my colleagues at the Department of Health and Welfare, Dr. Norm Fraser, Dr. Ken Butler and Dr. James Bagnall who have actively participated in analysing and guiding overall project development. My colleagues at the Department of Communications have each made their valuable contribution throughout the development of the project and are always open to discuss issues and problems, they include Terry Kerr, George Davies, Doris Jelly of the Space Branch and Jacques Langlois and Juris Silkans of the Social Policy and Programs Branch.

REFERENCES

1. "Baffin Zone Telemedicine Study" by Jacques Langlois, Anna Casey-Stahmer, Dr. H.J. Bagnall, D.O.C. Ottawa, June 1977.

2. Information compiled by Jacques Langlois and Anna Casey-Stahmer, Department of Communications, Ottawa, April 1977.

3. Information obtained from Robin Roberts and Sally Skene, McMaster University, Hamilton, Ontario.

4. 'CTS Telemedicine System Engineering Evaluation Report', by Bell Northern Research Ltd., DOC contract No. OST76-00043, May 1977.

5. The following information is based on preliminary data analysis of the experiment evaluation and was obtained in discussion with Robin Roberts and Sally Skene.

6. "Baffin Zone Telemedicine Study" by Jacques Langlois, A. Casey-Stahmer, Dr. H.J. Bagnall, DOC Ottawa, June 1977.

M.R.

A. C.-S.

A METHODOLOGY FOR DESIGN OF ADVANCED TECHNOLOGY-BASED HEALTH CARE SYSTEMS IN DEVELOPING COUNTRIES

Unver Cinar

SHAPE Technical Centre

The Hague, Netherlands

I. INTRODUCTION

The objective of this paper is to outline the methodology for designing a telehealth system for developing countries. By definition, developing countries have limited health care system resources and lack adequate transportation and telecommunication facilities for direct support of the health care system. Increasing demand for health care, high rates of population growth, and slow progress in supplying health care resources in most developing countries suggest that alternative forms of technology should be used in order to remedy the situation, both in the short and long-term provision of health care.

There are vast differences in the organisational structure, allocation of resources, resource composition and financing of the health care systems in various countries. One of the theses of this paper is that although the technological principles are general in nature, their utilisation in different environments varies widely. The existing socio-economic structure, health care resources, telecommunication and transportation systems and their spatial distribution are important variables to be considered in the design of a telehealth system.

Most efforts in the field of telehealth are either in the experimental or developmental stage but a few are operational (1, 2). These efforts are confined to providing various forms of medical care support in an urban environment or in rural (remote) areas (3, 4). The level of technological sophistication also varies from one study to another ranging from simple audio-link (telephone) to interactive television system including special diagnosis equipment (e.g., ECG, X-ray transmission, video microscope etc.). Most of the studies are conducted in developed countries where technology is either available or there is provision for research and development in the area of tele-health (5).

In this paper, the provision of a health care system for Turkey is taken as a case study. A methodology is proposed for the design of a telehealth system primarily for the rural areas. An evolutionary approach is suggested in order to provide some level of integration with the overall health system. Preliminary conclusions are provided to enable the selection of a transmission technology.

II. A BRIEF DESCRIPTION OF THE METHODOLOGY

The methodology described here is of a conceptual nature and does not include the analytical approaches or techniques to be used in its various stages. The objective is to provide a logical framework for an overall design, derive a set of alter-native configurations, assess their economic and technical feasibility, provide cost estimates and recommend configurations and/or solutions.

In this context, a configuration is defined as a telehealth system which consists of an organisational structure, set of functions to be performed, level of technology to be utilised, manpower and facility requirements, interfaces with other medical organisations which support the system and the anticipated population coverage.

The design of a telehealth system for rural areas requires two major investigations. The first is data collection and analysis of rural area characteristics. This effort should concentrate on those areas with no health care facilities of their own and limited access to others. This would lead to the collection of statistics and an initial assessment of demand for health care.

Although demographic data are available in developing coun-tries, the general health statistics of rural areas are either not available or very limited. A subset of this field of study is the accessibility of the rural population to a health care facility. One important element is availability of road network. Another set of data required relates to the impact of socio-economic factors. This analysis would include an assess-ment of the attitude of the rural population to support such a system. The phenomenon that supply creates demand in the health field may have consequences in the design and operation of a telehealth system.

Accessibility to a health care facility in a rural area is a function of the road network as well as the availability of means of transportation. Therefore the study should investigate the present and planned road network nodes and availability of means of transportation. Provision of a telehealth system requires a telecommunication system. In developing countries rural areas lack telecommunication facilities; therefore an assessment of present and planned capabilities of the telecommunication system should be an integral part of this analysis.

The second major investigation required is a study of the distribution of health care resources which could support rural areas. This would also cover the spatial distribution in terms of medical staff and health care facilities. The type and capacity of facilities and composition of medical staff at present, and their projection into the near future, would determine the viability of a telehealth system in the respective regions.

In the above section, two major requirements for data collection and analysis were briefly described. The next phase of the study is concerned with the derivation of alternative configurations for a telehealth system.

A configuration basically describes a telehealth region composed of a set of remote health centres staffed by non-physician medical staff, a set of hospitals that provide medical, administrative and technical support including means of transmission and computer technology. The set of alternative configurations could be evaluated using various criteria, including their costs. Maximisation of population coverage, maximisation of accessibility, minimisation of travel time or distance, and minimisation of total system costs are a few examples of selection criteria that could be used in an analysis of this nature. The main categories of cost include investment in facilities and equipment, organisation and maintenance, and research and development costs. The major components of the methodology are represented schematically in Figure 1.

The availability of health care and telecommunications resources would have a direct impact on the design of a telehealth system. The more such resources are available and the better balanced their spatial distribution, the fewer will be the facilities which a telehealth system will require for its support. On the other hand, it could be anticipated that the introduction of a telehealth system would improve the utilization of available resources. Thus, the aggregation of the existing capability and telehealth facilities would broaden the population coverage and drastically improve accessibility in remote areas. Another anticipated impact of the introduction of a telehealth

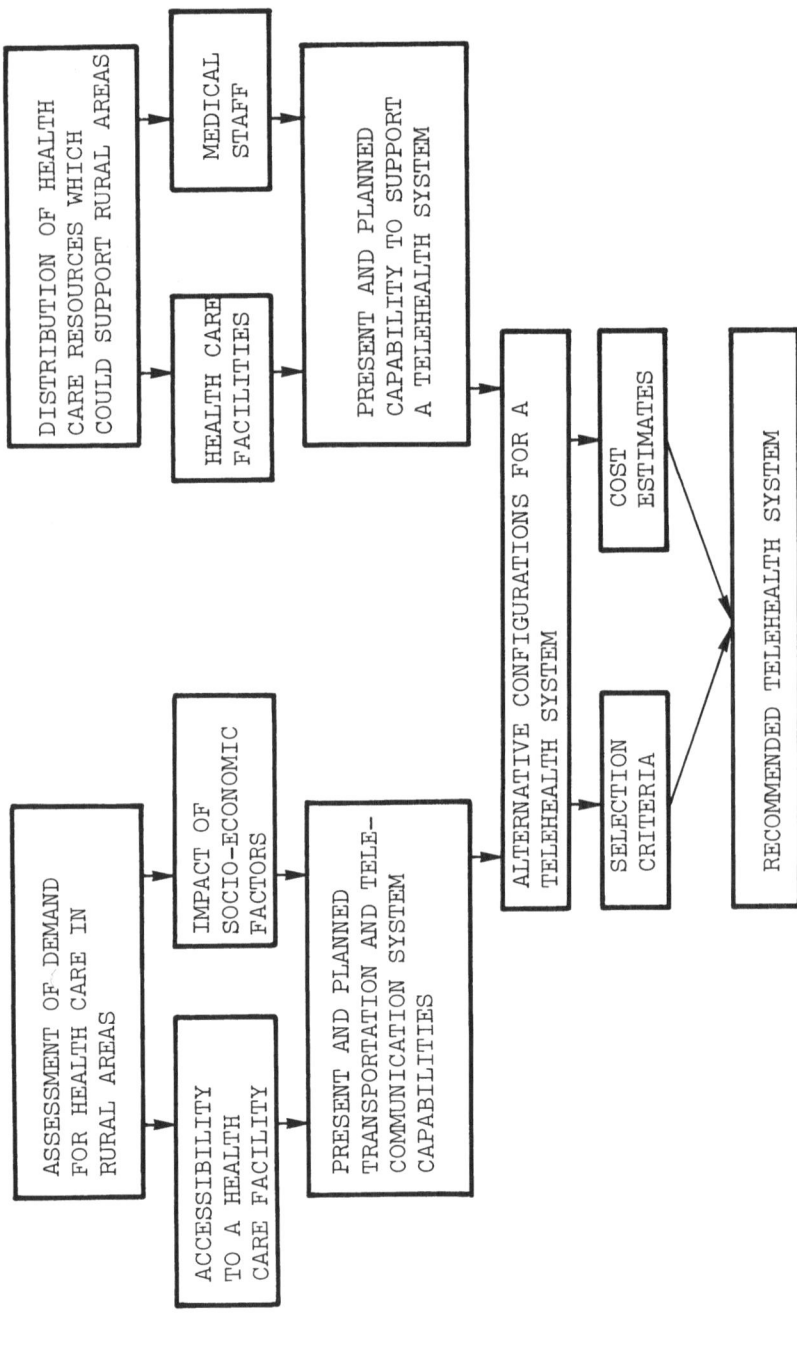

Fig. 1: A Conceptual Model for the Design of a Telehealth System

system is the provision of data collection for health planning
and the capability for detection of epidemics.

III. HEALTH CARE SYSTEM IN TURKEY

3.1 Background

Turkey is a developing country with a population of forty
million of whom 42% are less than 14 years of age.   If the current
rate of population growth continues, it is expected that the
population will exceed seventy-five million by the year 2000.
The infant mortality rate in the period 1965 - 1970 was about
one hundred and fifty per thousand.

The country has an area of 776,000 km^2 with mountainous
regions and there is a prolonged winter in the eastern part of
the country.   The annual gross national product per capita is
about one thousand US dollars and illiteracy rate is 40% (6).

3.2 Health Care System Resources

The Ministry of Health and Social Welfare is responsible
for the provision of health care delivery in Turkey and controls
all health and health-related organisations.   In each of the
sixty-seven provinces in the country, it is represented by an
office of health director.   In addition to the Ministry, medical
schools, various state enterprises and social welfare organisations
operate hospitals and clinics.   In the period 1960 - 1974, the
funds allocated for health services have not exceeded 5% of the
national budget.

Although there has been a substantial improvement in health
manpower and facilities in the last fifteen years, the available
resources are far below the standards of developed countries.
Between 1960 and 1972 the following manpower increases were
attained:

Physicians	+ 79%
Nurses and auxiliaries	+ 344%
Medical technicians	+ 160%
Midwives	+ 285%

During the same period the population increase was over 25%.

Out of 821 hospitals in the country with a bed complement of
95,300, the number of general hospitals is 729 with a bed comple-

ment of 72,900.   Nearly half the beds are in hospitals operated
by the Ministry.   The total number of physicians is about 25,200,
nearly half of whom are general practitioners.

The distribution of health care delivery resources, both
manpower and facilities, is extremely unbalanced.   About 68%
of the physicians practice in the provinces of Istanbul, Ankara
and Izmir, which have a total population of about five million.
The percentage distribution of hospital beds by provinces is as
follows:

No. of Provinces	Beds per 10,000	Percentage of total no. of beds	Percentage of population
51	less than 15	37	65
10	less than 25	13	15
6	more than 25	50	20
67		100	100

A health care delivery programme initiated in 1961 aimed
at the establishment of health centres in remote areas.   It had
planned to provide a health centre staffed by a physician, a nurse
and five medical auxiliaries to provide health care to a population
of between 7,500 and 8,000; and a health unit staffed by a midwife
(primarily in charge of mother and infant care) for a population
of between 2,000 and 2,500.   The so-called "socialization of
medicine" programme was initiated in twenty-five provinces in the
eastern and south-eastern parts of the country.   However, substantial difficulties were encountered in staffing these
facilities (7).

IV.   TELECOMMUNICATIONS IN TURKEY

The installation and operation of all public telecommunications
in Turkey is under the authority of the General Directorate of PTT.
This organisation provides telephone and telegraph services as well
as channels for Turkish television.   According to 1974 statistics,
there are about 600,000 main telephone subscribers and the
telegraph network is country-wide.   There is a substantial
demand for telex terminals; the network had nearly 4,000 terminals
in 1974.   The facsimile transmission networks which interconnect
major cities serve mainly the press which owns the terminal
equipment; the PTT provides only the transmission channels.

The existing telephone system is not capable of direct distance
dialling except between a few metropolitan areas.   Furthermore,
the grade of service is very low since each telephone is used more

frequently than in developed countries. The problems encountered
include absence of dial tone, loss of dialled number, wrong
connections and difficulty in reaching certain subscribers even
within the same metropolitan area.

The telephone density in Turkey is 1.5%; in the eastern and
south-eastern parts of the country this density is far lower and
in rural areas there is no telephone service at all. However,
there is potential demand for this service. The expected demand
for telephones is 2.3 million subscribers in 1982.

A number of institutions carry out research in communications.
"At the PTT Research Laboratories, 1) development work of high-
channel-capacity FDM equipment and high-speed telegraph systems
for a national network is being carried on and 2) a 24-channel
VHF and a 60-channel UHF radio systems are under study" (8).

V. PRELIMINARY CONSIDERATIONS FOR DESIGNING A TELEHEALTH SYSTEM FOR RURAL AREAS IN TURKEY

The high rate of population growth, low levels of health
resources, unbalanced distribution of available health manpower
and facilities, low income levels and limited national funds are
the major factors that lead to the consideration of using advanced
technology to support the provision of health care in rural areas.
Because of geographical and environmental conditions and the
paucity of transportation and telecommunication facilities, a
near-future health care improvement is only feasible through the
application of telehealth.

Although there is an accelerated rate of urbanization in
Turkey, the number of metropolitan areas are few. About 90% of
all residential areas (cities, towns, and villages) have less than
10,000 inhabitants. In certain rural areas accessibility to a
health facility is made extremely difficult by the lack of proper
road networks and/or means of transportation (9). This situation
makes the use of telehealth technology a viable and attractive
alternative.

An overall analysis of the organisational structure, manpower, facilities, health services rendered and financial aspects
of the Turkish health care delivery system is contained in
Reference (7).

In the two following sections some preliminary considerations
arising in the design of a telehealth system for rural areas in
Turkey will be briefly described.

5.1 Organisational Aspects

The projected telehealth system is composed of a hierarchically interrelated system (Figure 2) consisting of four levels of organisation. Conceptually, all levels of the system should be able to communicate with each other and higher levels should have information storage and retrieval and associated interactive software.

The first level comprises remote health care facilities each staffed by one or more non-physician staff (paramedicals). Such units will have direct access to the other levels and will form the primary interface with the public. A large number of these facilities will be set up, and located in small population centres in rural areas. One of the possibilities is to have mobile units in addition to static facilities to serve areas of scattered population. These units will be administratively connected to small local hospitals.

The small local hospitals, which form the second level of the structure, will have responsibility for providing consultation, care for cases referred, and administrative and logistic support. Where such a hospital is not easily accessible, the remote unit may be administratively linked to a general hospital which would have similar responsibilities.

General hospitals constitute the third level, and are assumed to contain a wider spectrum of medical knowledge and facilities. Such hospitals could provide medical and administrative support for the second and first level requirements.

Medical schools, the fourth level, are envisaged as organisations primarily responsible for education, training and research and development (including the field of telehealth).

The primary and secondary functions considered for these four levels of organisations are outlined in Table 1.

Two of the items included in the outline of functions are particularly noteworthy. It is suggested that the system be utilized for the establishment of individual medical records, and also for health screening purposes so as to provide information for early detection of epidemics and for health planning.

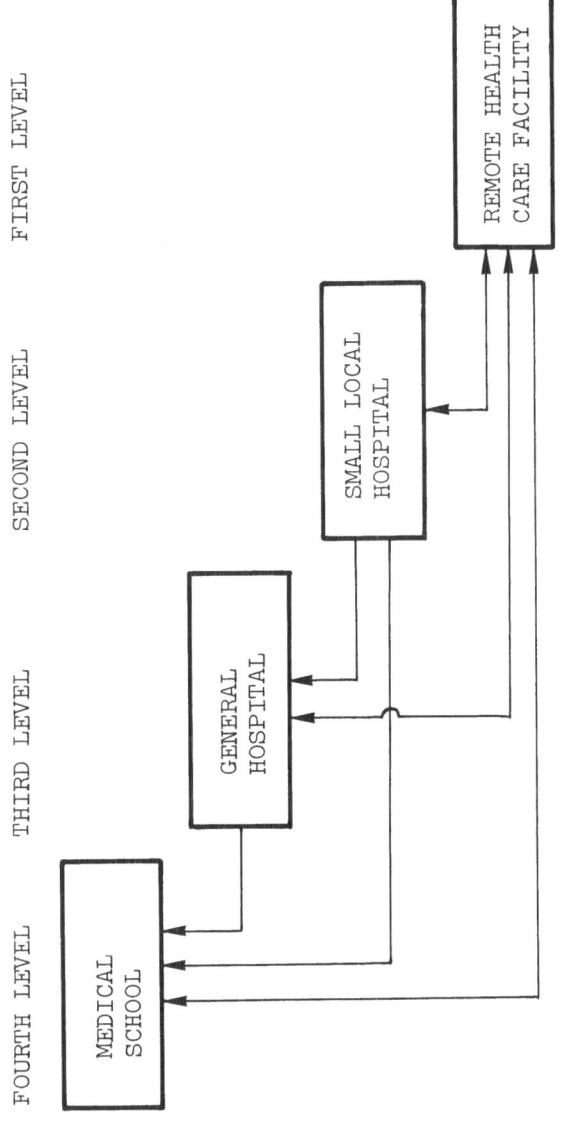

Fig. 2: Proposed Hierarchical Structure for a Telehealth System

LEVELS	MEDICAL SCHOOL	GENERAL HOSPITAL
PRIMARY FUNCTIONS	- Consultation and support for medical decision process - Provision of specialist know-ledge and facilities for complex cases and those critically ill - Education and training of physicians and paramedics in the area of telehealth	- Consultation and support for medical decision process - Provision of specialist know-ledge and facilities for acute cases - Centralised medical records management
SECONDARY FUNCTIONS	- Development of software to support telehealth system - Re-training of personnel	- High-level administrative support and managerial control

Table 1:   Primary and Secondary Functions of the Four
           Levels of Organisation in a Telehealth System

LEVELS	SMALL LOCAL HOSPITAL	REMOTE HEALTH CARE FACILITY
PRIMARY FUNCTIONS	– Consultation and support for medical decision process – Provision for in- and out-patient facilities	– Diagnosis – Treatment after consultation – Referral to MD – Conduct of simple clinical tests – Dispensing
SECONDARY FUNCTIONS	– Administrative support for first-level routine operations	– Medical records – Health screening

Table 1: Primary and Secondary Functions of the Four (cont'd) Levels of Organisation in a Telehealth System

5.2 Telecommunication Aspects

It is assumed that, ideally, all organisational units, whether at different levels or at the same level, should be capable of communicating with each other. These requirements could hardly be accomplished without the use of a dedicated system.

In order to be able to carry out all of its tasks, a remote health care facility will have to communicate with a medical centre of higher level using audio, visual and data transmission modes.

Audio information will consist of human voice and stethoscope sounds. The quality of transmission for voice will be measured in terms of intelligibility, while in case of stethoscope sounds fidelity will be more important. These requirements could be met by transmission quality equal to that of a radio broadcast channel. Commercial quality colour television is satisfactory for the transfer of visual information. Since the picture to be transmitted will be stationary or slow moving, the speed at which the picture is scanned can be very low. Therefore, facsimile transmission from the remote unit to higher levels for purposes of diagnosis may be an alternative.

Data will be transmitted both from and to the remote unit and other levels. They will consist of diagnostic and clinical data, prescriptions and administrative data in accordance with the specified functions, decision and information flow requirements within the system.

In view of the preceding sections, it is considered that the use of **satellite** communications will best meet a number of needs of the telehealth services. A satellite system will offer the following advantages:

 i   it can be designed according to user requirements

 ii the system's configuration can be easily modified
     to meet new requirements

 iii it can be used for broadcasting and, as a con-
     sequence, teleconferencing will be simple

 iv the space segment, which is the expensive com-
     ponent of the system, can be leased from inter-
     national organisations or from developed nations.
     Therefore, a large capital investment is not
     required (10)

v a national telehealth system utilizing satellite
   communications can easily be integrated into a
   regional or national system and be used by other
   industries or the government (e.g., energy,
   banking, manufacturing).

As stated in the description of the methodology, however,
an economic and technical analysis is required for the selection
of the proper type of transmission technology.

VI. CONCLUDING REMARKS

The methodology described here sets out a broad framework
for designing a telehealth system for a developing country.   In
its present form it is rather crude and further refinement is
required.

In Turkey most of the data on present and planned health
care and telecommunication system capabilities is already available;
data collection in some rural areas is required.

The need for improvement of the health care delivery system
in Turkey is quite obvious.   The discrepancy in resource allocation between metropolitan and rural areas is great.   In three
metropolitan areas there is a physician for every 300 inhabitants;
in rural areas it is 3,500.   The introduction of telehealth
technology is needed to overcome the shortcomings of such dis-
crepancies as well as to improve the overall level of health care
delivery.

One of the objectives of this paper is to identify the
scope of such an effort and prepare a project description.   A
second step would be to implement the proposed studies through
various health agencies including the medical schools.

Acknowledgement

The author greatly appreciates the cooperation of Dr M Celebiler
of NICSMA.   His contributions in the area of telecommunications
enabled the preparation of the later part of this paper.

REFERENCES

1. R. Bashshur, "Telemedicine's History", IEEE Spectrum,
            December 1976, p. 33

2. R.B. Munson, "New Mexico's Proposed Physician-Monitored
            Remote Areas Health Program", Proceedings
            of the IEEE, 57, 11, 1969, pp. 1887-1893

3. R. Allen,   "Coming: The Era of Telemedicine", IEEE
            Spectrum, December 1976, pp. 31-35

4. W. Anthony and P. Walcoff, "Telehealth: The Application
            of Technology to Health Care Delivery",
            Proceedings of the International Conference
            on Communications, Philadelphia, 1976,
            p. 39.14-39.19

5. J.H.U. Brown, "Communications for a Health Care System",
            IEEE Transactions on Systems, Man and
            Cybernetics, January 1977, pp. 61-64

6. Third Five-Year Development Plan (1973-1977), State
            Planning Organisation, No. 1272, Ankara,
            1973 (In Turkish)

7. U. Cinar,   Health Care System in Turkey, Chapter 5,
            Operations Research in Health Care Systems,
            Rehabilitation Thesis, Middle East Technical
            University, Ankara, 1975 (In Turkish)

8. G. Bayraktar and H. Abut, "Present State and Future of
            Telecommunications in Turkey", IEE Trans-
            actions on Communications, COM-24, 7, 1976,
            pp. 684-686

9. U. Cinar,   "An Operations Research Study of the Location
            of Health Service Facilities in Rural Regions",
            Proceedings of the NATO Conference on Cybernetic
            Modeling of Adaptive Organisations, Porto,
            Portugal, 1973

10. I. Goldstein, "INTELSAT and the Developing World", IEE
            Transactions on Communications, COM-24, 7,
            1976, pp. 742-748

DISCUSSION OF PAPERS BY CASEY-STAHMER, CINAR AND ROCKOFF

Discussant and moderator: David Conrath

The discussion focused primarily on some critical questions which underly telemedicine designs.

1. Interdependence of the Telemedicine System and Organizational Structure

One of the first questions raised was: What is the interface between telemedicine and organizational structure for health care? Several speakers including Cinar and Rockoff focused on the idea that telemedicine cannot be considered in isolation from the organizational framework in which health care is delivered. In this regard, Conrath pointed out that, while the examples presented in the papers involved sophisticated technology, designers of health care systems cannot assume that technology is the answer. Cinar expanded this point, indicating that the technology should not only be considered in terms of other components of health care delivery (e.g., training, marketing, etc.) but also in terms of the entire spectrum of social and economic resources outside health care. Social variables and transportation networks are especially important factors which must be weighed. Another point raised with respect to organizational structure is that tele-medicine can have a centralizing or decentralizing bias and planners must be aware of the potential for changing the locus of decisions in an organization when a telecommunications system is introduced.

Rockoff indicated that her presentation focused on whether one could substitute people (for example, nurse practicioners) with lower skills utilizing high technology for people with higher skills (for example, doctors) and low technology. In his commen-tary, Conrath suggested one might also consider creating new forms of organizations in order to use both *low* skills and *low* technology. Cinar noted that systems designers have often tried to impose technology on existing health care organizations rather than seek to adapt the organization and the technology to each other. In response to a statement that developed nations which have develop-mental and operational experience in telemedicine should provide assistance to developing countries, Cinar said that technology transfer across national boundaries has its own special problems. These include problems associated with utilizing foreign personnel unfamiliar with local conditions, short duration of assignment,

lack of participation in phases of implementation and lack of
organizational continuity.

2.   Impacts of the Project Experience

A second line of discussion focused on the impacts of the
project experience. Casey-Stahmer had suggested in her presentation that a positive impact of the satellite communications
experiments in Canada was increased interagency dialogue on health
care delivery. Conrath, on the other hand, expressed concern that
increased high-level interest could also be detrimental given the
propensity of policy makers to base decisions on one or two
experiences. Perhaps, Conrath suggested, experimental results
should maintain a low profile so that conclusions on policy
won't be made on one or two visible demonstration projects which
may or may not be representative. In response, Casey-Stahmer noted
that alternatives to highly visible demonstration projects may
not be available and that in fact such experiments are useful
because they catch the attention of policy makers. Project details
are still usually left for analysis and recommendation by staff.
It is important, however, that the staff keep level-headed in their
analysis and in their recommendation to the policy makers.

A second point regarding the impact of demonstration projects
was voiced by several participants who expressed concern that these
projects may raise false expectations about health care in the
sample population. Ohlman pointed to the Satellite Instructional
Television Experiment in India. It was clear from the beginning
that NASA would only provide its advanced communications satellite,
ATS-6, for a period of one year. The experiment certainly stimulated, informed, and educated thousands of villagers during this
period, but it was a let-down when they confronted the blank television screens the day after the satellite was pulled away. Casey-
Stahmer noted that false expectations did not appear to be a
problem in the projects she had studied. The areas of Canada in
which telehealth care is being explored are the northern parts,
where health care is generally a governmental program (i.e., with
salaried staff, etc.), and it is the government which pays the
operational costs. It is really the government (or its
employees providing the health services) who are the *users*.

Conrath mentioned that the funding group can help meet the
problem of raised expectations by paying operational costs of the
technology so the users can continue the services if they desire.
Some participants indicated, however, that even with continued
availability of the technology, demonstration projects seldom
continue after the initial demonstration period has ended.

Section Three

PUBLIC SERVICES: EDUCATION AND COMMUNITY

A diverse set of educational experiments has recently been
undertaken in Canada using the Communications Technology Satellite.
Individual experimenters are responsible for evaluating their own
projects; in addition an overall assessment is being made by John
Daniel and his colleagues. The latter is concerning itself with
the wider problem of institutional assimilation of new technology,
as well as issues of educational effectiveness. The overall assessment was designed prior to the implementation of the experiments;
hence a series of instruments were developed for the individual
projects to draw upon.

Their paper describes the experiments. It also presents some
observations and practical interim conclusions arising in the
evaluation process, (this was still at an early stage at the time
of writing).

The second paper is also concerned with Canadian research in
the educational field, in this case the Public Service Commission's
professional training for civil servants. Nicole Mendenhall and
Rene Lortie present the results of field and laboratory studies
used to investigate the applicability of videoconferencing to
educational and administrative functions. They go on to propose a
teaching/learning model for adult tele-education and to describe
its testing in a laboratory simulation.

Peter Zorkoczy describes the needs for communication which
arise within the context of the British Open University. He than
describes current activities directed towards meeting them with
the use of interactive telecommunication systems: e.g., audio-
conferencing for tutorials, computer-based systems and an electronic
blackboard which shares the telephone line used for speech. The
Open University is developing some of its own low-cost technology.

The last three papers derive from the US National Science
Foundation's research program concerning the two-way potential
of urban cable television systems for the delivery of public
services. On of three experiments conducted by the Rand Corporation
in Spartanburg, South Carolina investigated the use of a simple
terminal to allow students, being taught by television in their
homes, to signal back to the teacher. After outlining the experiment,
Bill Lucas and Suzanne Quick describe an instrument they developed
to provide process information on the instructional dynamics in the
cable and conventional (control group) classes. The instrument was
designed to focus upon the distribution of classroom activities
and upon the frequency of classroom interactions.

The authors also report how results provided by use of the
instrument were fed back to instructors; statistics are provided
which compare activity and interaction in earlier and later classes
(i.e., before and after the information was fed back).

Mitchell Moss describes the experiment conducted by New York
University in Reading, Pennsylvania. The cable television system
there was used to provide interactive television connection between
three Neighbourhood Communication Centers for older citizens and,
on some occasions, a fourth mobile unit. Initially the proceedings
("programs") were carried live to a further 125 or so senior citizens
in their homes; subsequently they were made available to all cable
subscribers. He points out that the value of the system turned out
to lie more in community communications than, as had been expected,
in the delivery of public services. He notes that the system is now
allowing users to play an important part in overcoming problems of
inadequate coordination of services. Some interim conclusions are
presented concerning the use and impact of the system.

The paper by Red Burns was presented in a session on methods
of evaluation and implementation. It is now placed here since it
too deals with the Reading experiment. Her concern is to show how
a process was set up by which an emerging community could involve
itself in the design and implementation of its own communication
system. The paper deals with variables such as trust, comfort
simplicity, and flexibility. Some conclusions are drawn about the
nature of the interactive television medium and the process of implementing community-based telecommunications projects.

All the papers in this section are easily accessible, both in
describing how interactive telecommunications may be used for educational purposes and in the community, and in raising a variety of
issues of more general theoretical and procedural importance.

EDUCATIONAL EXPERIMENTS WITH THE COMMUNICATIONS TECHNOLOGY

SATELLITE: A MEMO FROM EVALUATORS TO PLANNERS

J.S. Daniel, M.L. Côté, M. Richmond

Tele-universite

3108, Chemin Ste-Foy, Québec, Canada   G1X 1P8

SUMMARY

A series of educational experiments are being conducted in Canada on the Communications Technology Satellite. The description of
these experiments reveals a diversity not previously encountered in
educational applications of satellite systems. In evaluating these
experiments, the authors adopted Stufflebeam's Context-Input-
Process-Product model in order to include in their investigation,
as well as issues of instructional effectiveness, the wider problem
of institutional assimilation of new communications technology. To
date the project has shown that organizational constraints were
more severe than experimenters had expected and that multipoint
conferencing is very sensitive to technical quality. Although some
of the experiments were very short the project has demonstrated
that satellite communications have applications in education which
go beyond the classical telelecture.

1. INTRODUCTION: THE CONTEXT

1.1 An experimental satellite programme

Made in Canada and launched by the U.S. on January 17th 1976,
the Communications Technology Satellite is a joint venture between
the two countries. Intended as an experimental rather than an operational satellite its novel features are firstly high power (200W)
and secondly the frequency at which it receives and emits (12-14
GHz). High power in the satellite means that smaller and cheaper
ground terminals can be used, whereas communicating at a high
frequency part of the radio spectrum means less congestion and

less interference with terrestrial facilities. In concrete terms
this means that terminals can be placed in cities (e.g. on the
roofs of buildings).

   The objectives of the Canadian Department of Communications in
the CTS project were stated as:
1) To enhance in Canada a capability in the design and manufacture
   of spacecraft subsystems for domestic use and export.
2) To maintain a Canadian capability to specify, assess and construct space application systems for domestic use.
3) To develop and flight test spacecraft subsystems and components
   for use in future communications satellites.
4) To conduct communications experiments to explore the use of
   future high-power communications satellites in Canada to low
   cost ground stations in the 12 and 14 GHz bands.
5) To explore by means of communications experiments the social,
   cultural and economic impact of the eventual introduction of
   services that might be provided.

   Clearly the CTS project was justified to the Canadian Government principally on the basis of the technical developments and
experiments which it would permit. Important components of the
experiments were the study of the attenuation of 12-14 GHz signals
under different weather conditions and of the use of wave polari-
zation to increase channel capacity. However, these and other
technical experiments required relatively little satellite time so
because of a growing realization that communications satellites
were not an end in themselves, the government invited institutions
and organizations across the country to propose experiments in
social communications to be carried out on CTS. About forty
proposals were studied in the areas of education, telemedecine and
social interaction. The present article is limited to the educational experiments which have been carried out in Canada since the
satellite was launched.

   1.2 The institutionalization of new educational technology

   From his experience with educational innovations in the US
and the UK Hooper 1) has listed factors which facilitate the assimilation of new technology in educational institutions. Some of
these factors are:
1) The technology answers real needs linked to the primary objectives of the institution.
2) The technology blends in with technology already being used.
3) The team responsible for implementing the technology is
   integrated into the institution and not a peripheral tack-on.
4) Serious attention is given to training personnel to use the
   technology.

5) The institution puts up some of its own resources in order to use the technology.
6) The technology has high visibility both within and without the institution.
7) The use of the technology is evaluated.
8) Different departments and institutions collaborate on the project.
9) The technology favours reforms and developments already under way in the institution.
10) The technology has high transferability.
11) Existing equipment and courseware can be used with the technology.

Many of these conditions were fully or partially met by the institutions participating in the CTS project. Carleton University, the first experimenter, had already acquired, in its Wired City Laboratory, experience of the use of telecommunications in education whereas the second experimenter, the University of Quebec, uses teleconferencing on a routine basis and had already taken part in satellite exchanges.

However, despite these apparently favourable circumstances the highly important first condition in Hooper's list could only be partially met because of the experimental nature of the CTS project. The length of the periods allotted to experimenters (a few months) and the timetabling of satellite time within these periods (after-noon and evening slots on alternate days) were not designed to fit university sessions and course schedules. Furthermore the uncer-tainty which properly surrounds an experimental project meant that any institution using the satellite for a vital function had to have a back up system ready.

For these reasons the CTS educational experiments were viewed within participating institutions as an opportunity to acquire familiarity with the technical and organizational constraints of satellite communications and to explore various applications. Before contemplating any permanent operational satellite system all experimenters would wish to run pilot projects for longer periods than was possible with CTS.

2. THE EDUCATIONAL EXPERIMENTS

Before discussing the evaluation of the experiments and their implications for planners we shall describe the aims and configu-rations of the experiments which will have been completed by the time this symposium is held. Four types of terminal were involved:

Diameter	Location/mobility	Function
1m	Portable	Emit and receive audio
2m	Transportable	Receive video; emit and receive audio
3m	Mobile (truck mounted)	Emit and receive video and audio
9m	Fixed (Ottawa)	All of above plus control and telemetry

In the text these terminals will be designated by their diameters (e.g. 3m).

Two institutions, Carleton University (Ottawa) and The University of Quebec, carried out experiments between the summer of 1976 and the spring of 1977. Two other projects, those of Memorial University (Newfoundland) and the Federal Public Service Commission, began in spring 1977.

2.1 The Carleton-Stanford curriculum exchange

This project combined both technical and educational objectives. The basis aim was to use the satellite to allow students at Stanford University in California to take courses originating in Ottawa and vice versa. In one mode one class was simultaneously originated at each end and communications to the remote students were provided by means of one way video and two way audio. In the other mode two way video was used to enable a fuller interaction between the ends of the link. In addition to satellite technology the exchange used experimental methods of television signal processing. These digital processing methods, which permitted transmission of video signals from 2m. terminals, were developed by the NASA Ames Research Centre, a partner in the project. In simple terms they doubled the capacity of the video channel.

The first course exchanges by satellite began on October 19th 1976 and for the rest of the term five courses were available at both universities. These were:

Originating University	Course Title	Hours per week	Number of students Carleton Stanford	
Carleton	Digital Systems Architecture	3	25	5
Carleton	Computer Communications Systems	3	30	15
Stanford	Management of Research Institutions	2	20	30
Stanford	Statistical Methods in Signal Analysis	3	6	20
Stanford	Seminar course	1	Variable	

In the winter term, only one course was offered in each direction.

2.2 The University of Quebec Omnibus Network

The only multi-campus university in Canada, the University of Quebec, already had experience of satellites through its involvement in France/Quebec exchanges in the Comminsat Frabec and Symphonie programmes. The University's CTS proposal differed from those submitted by other institutions in that the content of the experiment was purposely left undefined. The satellite would be used to create a communications network between the campuses and local centres of the university and the whole university community would be invited to submit projects. Of the thirty experiments submitted fifteen were carried into the planning stage and the nine experiments of which descriptions follow have been conducted.

Another of the original features of this project was the requirement to move terminals to set up different networks. Thus the 3m terminal was located successively at Montreal, Trois-Rivières, Quebec and Rouyn and the smaller terminals were moved correspondingly.

2.2.1 Electron Microscopy (3m: Montreal; 2m: Trois-Rivières, Quebec, Rimouski). The satellite allowed interactive research discussions between bacteriologists in Montreal, Quebec and Trois-Rivières. The other sites received live images from an electron microscope in Montreal and discussion centred on identifying viruses and bacteria. Videotapes, reprints etc. could be called up at will and the electron microscope adjusted in real time to provide the field and magnification requested by the distant participants.

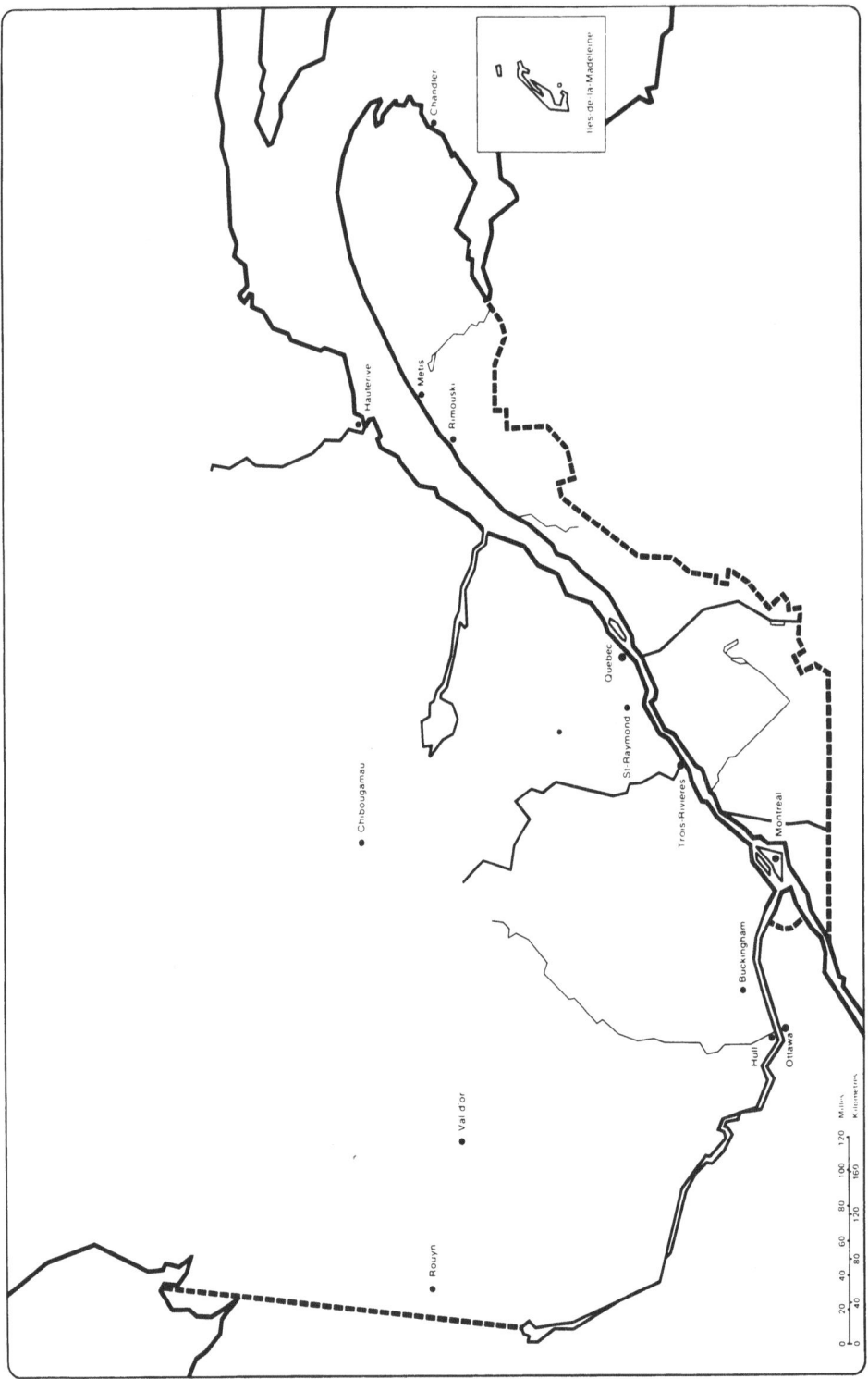

Figure 1. University of Quebec CTS Project – Terminal Sites

2.2.2 Teledocumentation (2m: Rimouski; 1m: Chandler; tele-phone: Hauterive). Since library services pose a problem for off campus courses the satellite was combined with rapid facsimile transmission equipment to make documents available for students in a Business Management course organised by the Rimouski campus in the towns of Hauterive and Chandler. Students used the audio link to discuss their needs with the Rimouski librarian who sent out the appropriate texts by rapid facsimile transmitter through the satellite.

2.2.3 Telereference (3m: Trois-Rivières; 2m: Rimouski)· Although somewhat similar to 'Teledocumentation' this project, which involved the University's libraries in Rimouski and Trois-Rivières, included a one way video link. Students in Rimouski could see the reference librarian in Trois-Rivières to whom they explained their needs. The video link was also used to show tables of contents, abstracts etc. to help the student zero in on the most appropriate reference text. The chosen text was then transmitted by satellite using the rapid facsimile equipment.

2.2.4 Centre for the Development of Learning Environments (CDES) (3m: Trois-Rivières; 2m: Quebec)· At any given time the CDES is involved in innovative projects with teachers in a number of schools around the province. The satellite was used (one way video - two way audio) for consultation between CDES staff and teachers in the Quebec City area. The objective was to study the effect of a consultation where each participant is in his every-day milieu. The traditional form of consultation implies that each participant is either visited or a visitor and it was hypothesized that this influences the consultative interaction.

2.2.5 Graduate course in public administration (3m: Quebec; 2m: Hull). One of the only classical teleteaching projects carried out in Quebec, this experiment brought three sessions of a masters course from the University's Graduate School of Public Administra-tion in Quebec City to students in Hull. The one way video link from Quebec to Hull was used to transmit images of the instructor and his transparencies etc. and discussion took place over the two way audio link. The objective was to study the changes in beha-viour which such an arrangement both requires and provokes on the part of both the instructor and the students.

2.2.6 Radio Orbital (2m: Montreal; 1m: Iles-de-la-Madeleine). An ambitious project to link the Iles-de-la-Madeleine with emigré islander communities in Montreal and Chicoutimi for a Christmas communications jamboree, this experiment unfortunately fell foul of numerous technical problems. One multipoint radio session took place over landlines but the more intimate telephone sessions had to be abandoned.

2.2.7 <u>Oceanology (2m: Rimouski; 1m: Métis)</u>. This was a technical experiment to use the satellite for transmitting oceanographic data on wave heights, conductivity, salinity, temperature (and gradients), depth and turbidity from the probes to the computer.

2.2.8 <u>Western Quebec (3m: Rouyn; 2m: Hull (link with Ottawa 9m); Val d'Or, Chibougamau)</u>. The University's centre in Western Quebec, CEUOQ, serves a sparsely populated area of 200,000 square miles. Already involved in various teleteaching and telemanagement projects, the CEUOQ used some 50 hours of satellite time for telework sessions, teleconferences, and a telelecture course. Some seven sites were involved.

2.2.9 <u>Intercom (3m: St-Raymond, 2m: Buckingham (link with Ottawa 9m)</u>. Scheduled for the end of the University's experimental period, this project linked community TV stations in two widely separated small towns. Some 500 citizens took part in exchanges on a variety of topics and the stations' ratings hit an all time high.

2.3 Memorial University
(3m: St. Johns; 2m: Stephenville, St. Anthony, Labrador City, Goose Bay-Happy Valley)

Starting in March 1977, this project linked St. John's with Stephenville, St. Anthony, Goose Bay-Happy Valley and Labrador City. The satellite was used for four types of activity.
- Continuing medical education for physicians.
- Continuing nursing education.
- Community health education.
- Consultation services and transmission of medical data.

2.4 The Public Service Commission
(3m: St. Johns; 9m: Ottawa)

This experiment has both technical and educational objectives. Four groups in St. John's were linked with Ottawa for a course in long range planning. Using a multiplexing technique developed by Miller Communications Systems Ltd., the TV images from the four rooms in St. John's were sent to Ottawa over a single video channel. From the educational point of view interest centred on the non-directive mediated learning model which was used in the course. This experiment ran from April through June 1977.

3. EVALUATION OF THE EXPERIMENTS

Each experimenter is responsible for evaluating his own project and the present authors have been charged with conducting an
overall assessment of the CTS educational experiments with a view
to developing models for the future use of satellites in Canadian
education. In the University of Quebec experiment, for example,
this meant that three groups of people took part in the evaluation,
the individual experimenters, a University of Quebec team reviewing the project overall, and the CTS educational evaluation group.
Although such a structure obviously works against the rapid publication of results since each group interprets the raw data before
passing it on to the next, it has the advantage of involving many
people in an evaluative appraisal of the project.

The present authors proposed to the experimenters an evalua-
tion procedure 2) based on Stufflebeam's Context-Input-Process-
Product model 3) and drew up a series of instruments covering the
following aspects of the experiments:
- Technical and interactive aspects (questionnaires for participants
  and site coordinators)
- Participant satisfaction (questionnaire)
- Project staff data (questionnaire)
- Checklists for interviews with experimenters, instructors,
  facilitators and participants)

Experimenters could either use some or all of these standard
questionnaires as provided or modify them to increase their
relevance to a particular experiment. Naturally institituions
also had their own special evaluative interests. Thus Carleton
University used a questionnaire on attitudes to educational
technology developed over the years in its Wired City Laboratory
and the University of Quebec placed strong emphasis on interviews
and observation with the aim of analyzing the type and quality of
the interactions taking place.

Although the present authors had proposed a standard data
processing format this has not been used by experimenters to date.
Each experimenter is the first to compile and interpret his
evaluation data and this process is under way at the time of
writing. Since little of this data has yet been passed to our
team the comments and conclusions which follow are based largely
on interviews and observations which we have made ourselves over
the preparatory period and during the first six months of
experiments.

3.1 Context Evaluation

We began this article by adumbrating the context of the CTS

project. Having described the experiments themselves some further
comments are in order.

3.1.1 Experiments or demonstrations? If an experiment necessarily implies an experimental or quasi-experimental research
design then the CTS project is better described as a series of
demonstration applications of satellite technology. There were no
control groups and the experimenters themselves seemed rather
uninterested in the issue of educational effectiveness although
this question seemed to loom large in the government planners'
minds at the start of the project. In this respect the experimenters were probably right since the results of the now famous
Dubin and Tavaggia 4) survey on comparative studies of instructional
media revealed 'no significant difference' with great regularity.
Furthermore a recent conference devoted exclusively to the
evaluation of educational TV and radio 5) came to the conclusion
that ten years activity in this area had produced little of
lasting value. In this context the experimenters were sensible to
concentrate on 'having a go' with satellite communications in order
to test their acceptability to both users and the institution and
determine whether this technology was worth exploring further.

3.1.2 Objectives or placebos? Since objectives became an
obsession in the world of education there has grown a tendency for
the writing of them to become an art form divorced from real life.
The CTS project did not always escape this tendency and some of the
objectives identified by the experimenters beforehand now have a
faint air of pomposity. The University of Quebec was lucid enough
to realize that the objectives it stated for the overall Omnibus
Network experiment were not necessarily shared by individual sub-
experimenters. In general, the somewhat rarified metalanguage
employed by experimenters and sponsoring agencies at the beginning
of this type of project creates confusion and misunderstandings
once the project gets under way.

3.1.3 Evaluators or participants? Evaluators usually complain
that they were involved in the project too late. We are glad to
say that our team began work in plenty of time - long before the
experimenters had even begun to think about evaluation. No
attempt was made to maintain the stance of detached observers for
one of our roles was to help experimenters design their project
and its evaluation and to advise on instructional issues at the
planning stage. However, we do not believe that this strategy
significantly contaminated any of the experiments!

3.2 Input evaluation

3.2.1 <u>Dollars or dimes</u>? The educational component of the CTS
programme was not generously funded. This was not necessarily a

bad thing although it would have been better had the experimenters been able to employ elsewhere the considerable energies they had to devote to raising the meagre (by US standards) funds they got. Hooper 1) has noted that educational projects which depend on generous outside funding do not have a high survival rate once the external source of money dries up. It already seems clear that the necessity of blending the CTS experiments into normal institut-ional activities and structures so as to affect resources to them will ensure that CTS leaves a more permanent trace than a more lavishly conceived programme might have done.

3.2.2 <u>Organization or improvisation?</u> We are grateful to the experimenter who explained patiently to us that universities always did things at the eleventh hour and made a success of them all the same. It is easy to say that the experiments could have been better planned but more difficult to prove that this would have made a difference. Some indication will be obtained by comparing the first two experiments with the later and better prepared projects of Memorial University and the Public Service Commission.

As with objectives, organizational charts can become an independent art form. In one experiment in particular the impressive organization of the project on paper bore no relation to the way the operation was really run. A possibly laudable attempt to both respect the hierarchy and encourage widespread participation in planning and execution actually caused considerable confusion as people tried to discover where the real lines of communication and authority were.

3.2.3 <u>Men or supermen?</u> It was impossible to follow this project without gaining considerable admiration for the people who made it work. From the instructor who spent twenty hours a week preparing his satellite courses to the site coordinator who could get up the three flights of stairs from his control booth to the terminal on the roof in a matter of seconds, CTS involved a large number of people working long hours on an overload basis. Whether the effort involved in educational satellite exchanges could be reduced to workaday proportions in an operational educational system is a major question.

3.3 Process evaluation

3.3.1 <u>On or off?</u> Problems with the power supply of the satel-lite itself caused uncertainty just before the experiments began and again in the winter of 1977 when the decision to continue experiments during the eclipse was taken at the last minute. Both events increased experimenters' difficulties.

3.3.2 <u>Mobile or transportable?</u> The ground terminals used in the project were not sufficiently rugged and reliable. This particularly affected the University of Quebec experiment which involved moving terminals several times and into some pretty remote places. An operational system would require better equipment.

3.3.3 <u>Analog or digital?</u> The technology of digital TV being in its infancy some problems were expected with the Carleton Stanford link. Although there were momentary image break-ups with increasing frequency as time went on the overall success of this spectrum saving system is a tribute to the engineers at the Nasa Ames Research Centre.

3.3.4 <u>Business or housekeeping?</u> During the Carleton-Stanford course exchanges administrative details (e.g. set books, exams) took more time to organize than had been expected and during the University of Quebec project sub-experimenters often complained they didn't know what was going on. The experimenters advise those involved in similar enterprises to set aside at least 10% of satellite time for housekeeping purposes. Similarly in the University of Quebec experiment telephone contact by landline was often maintained throughout sessions to ensure efficient and unobtrusive communication between site coordinators.

3.3.5 <u>Two or more?</u> The University of Quebec project provided a sharp reminder of the difference between multipoint interactive conferencing and the more traditional one to many (broadcast) and one to one (telephone) communication modes. In multipoint confe-rences technical problems, nearly always associated with the sound and frequently with echos, were often obtrusive. It seems that neither the experimenters nor the CTS programme control foresaw these problems sufficiently to install the efficient audio teleconferencing equipment that already exists 6). However, the existence of even a one-way video channel makes poor sound more tolerable (by reducing uncertainty) than in a solely audio system.

3.4 Product evaluation

Although it is too early to make any conclusive statements about the value of the projects of the CTS experiments a few comments can be made.

3.4.1 <u>Telelectures and tele-education.</u> Although the tele-lecture courses conducted over the CTS have been successful the project has shown that the possibilities of tele-education by satellite do not end there. The electron microscopists' research seminars, the telereference library consultation experiment

and several of the teleconferences generated considerable
enthusiasm amongst participants.

   3.4.2 <u>Audio vs video</u>. Since nearly all the Canadian educational CTS experiments involve asymmetrical communication links the
data which has been gathered should permit comparison of participant attitudes to different communication environments carrying
a common content. For the moment, we hypothesize that for tele-
lectures the one way video/multipoint audio mode produces less
interaction than the multipoint audio mode, simply because with
video instructors make less effort to implement the interactive
telecommunications teaching techniques described by Parker 7).
The addition of one way video to a multipoint audio system does
prove useful for research seminars, teleconferences and telework
sessions.

   3.4.3 <u>Classical vs romantic</u>. Wedemeyer 8) has suggested that
satellites should not and will not provide a "highway in the sky
for conventional teaching and learning". In the CTS project
although the institutions given time on the satellite were all
part of the conventional educational establishment the activities
conducted were not all classical in nature. It is unfortunate
that the most unconventional (romantic) educational experiment,
the University of Quebec's Radio Orbital, fell foul of technical
problems. However the Intercom project provided a foretaste of a
new type of educational communication.

   3.4.4 <u>Satellites in education: the cost</u>. We have mentioned
(para 3.2.1) that the CTS experiments were largely absorbed into
the financial bloodstream of participating institutions. This
makes exact costing of the experiments impossible, a situation
they share, incidentally, with most educational activities. A
separate cost study is being conducted 9) on the space, ground,
infrastructural and instructional costs of two hypothetical
educational satellite networks.

4. CONCLUSIONS: A MEMO TO PLANNERS OF FUTURE SATELLITE PROJECTS

   Our experience with the CTS educational experiments in Canada
indicates that the Hooper checklist given in para 1.2 provides
useful guidelines for planning an experimental or operational
satellite system for educational purposes. Particular attention
should be given to the following items.

   4.1 Go for Real Needs and Primary Objectives

   Ensure that the pompous metalanguage of experimental objectives
does not detract attention from using the satellite for useful

purposes. Do not mix technical and educational experiments unless
you are sure that the technical innovation will work.

4.2 Involve Your Own Resources and Organize Them for the Job in Hand

A few people full time on the project will give better results
than a larger number of part timers. When planning satellite
classrooms or studios consider air-conditioning an essential, not
a luxury.

4.3 Train Your Personnel

Technical staff, instructors and evaluators all need training
before the project starts. This was a weak aspect of the CTS
project which probably explains why participants in the experiments
(e.g. students, researchers) appear to have been more satisfied
with the project and interested in continuing it than were the
staff running the experiments.

4.4 Work at Circulating Information

Somewhat ironically for a communications experiments, a
common complaint within participating institutions was that nobody
seemed to know what was going on. Organize bulletins and information sessions so that the project has high visibility and
participants can see the whole picture.

4.5 Involve Experimenters in Evaluation

Make sure that roles and duties within the overall evaluation
plan are clear. Be prepared for the tendency of experimenters to
hoard their data.

5. ACKNOWLEDGEMENTS

We acknowledge financial support from the Department of
Communications (Canada) under contract OSU5 - 0169. The friendly
cooperation of CTS project personnel and participants in the four
experimenting institutions is greatly appreciated.

6. REFERENCES

1. Hooper, R., "Two Years On" National Development Programme in Computer Assisted Learning, Council for Educational Technology, London (1975).

2. Daniel, J.S., Côté, M., and Richmond, M., "CTS Evaluation Education – Evaluation Model and Instruments 1976-77" Télé-université, Québec (1976). See also "CTS Evaluation Education – Bulletins 1, 2, 3, 4 and 5.

3. Stufflebeam, D.L., "The relevance of the CIPP evaluation model for educational accountability", Journal of Research and Development in Education, $\underline{5}$(1) pp. 19-25 (1971).

4. Dubin, R., and Taveggia, T.C., "The Teaching-Learning Paradox: A Comparative Analysis of College Teaching Methods", Center for Advanced Study of Educational Administration, University of Oregon, Eugene (1968).

5. CEREB International Conference on Evaluation and Research in Educational Television and Radio, Open University, (1976) Proceedings to be published.

6. Braun, D., Gilbertson, D., and Hansen, H.C., "ETN: A Technical System" in L. Parker and B. Riccomini, Eds. "The Status of the Telephone in Education" University of Wisconsin Extension (1976) pp 51-74.

7. Parker, L.A., "Humanizing Telephone Based Instructional Systems" in L. Grayson and J. Beidenbach, Eds. "Proceedings Fifth Annual Frontiers in Education Conference" IEEE/ASEE pp 354-360 (1975).

8. Wedemeyer, C.A., "Satellite and Cable – – No Highway in the Sky for Conventional Teaching and Learning", in L.A. Parker and B. Riccomini, Eds. "A report on University Applications of Satellite/Cable Technology", University of Wisconsin Extension, pp 2-7 (1975).

9. Daniel, J.S., George, D.A., and Miller Communications Systems Ltd. "A Cost Study of Satellite Systems in Canadian Education", Télé-université, Quebec, 1977.

J.D.

P.T.

EVALUATIONS OF INTERACTIVE TELE-EDUCATION

IN THE PUBLIC SERVICE COMMISSION

Nicole Mendenhall and René Lortie

Public Service Commission

Ottawa, Canada

This paper summarizes the communication needs requirements in the Public Service Commission. It presents results of laboratory and field studies undertaken to investigate the application of teleconferencing to educational and administrative functions. It describes the learning model developed for the adult learner in an interactive educational situation, i.e. the principles underlying the model, the role of the resource person, and learner, and some information exchange strategies. Results obtained from the simulation of the model are also presented as well as its future application in a satellite-mediated learning situation.

INTRODUCTION

The Public Service Commission (PSC) acts as the central training agency for the Canadian federal government providing professional, managerial, and linguistic training and development courses to more than 20,000 civil servants a year. These courses are specifically designed to answer the overall training needs of all government departments as distinct from the particular need of a given department. To effectively carry out its mandate throughout Canada, the PSC decentralized its training services into six regional centers with Ottawa as its headquarters.

This paper will first describe the telecommunications experiments undertaken in view of applying teleconferencing to both administrative and educational purposes. It will then provide a description of the learning model developed for mediated interac-

tive learning, the simulation results of this model and its future
application. Research issues relevant to the possible implementation of interactive telecommunications are also addressed.

Rationale for Telecommunications Research

In Canada, the PSC provides more than 90% of the language
training and 15% of the professional and managerial training for
the Public Service. Nearly seventy-five percent of federal
public servants are located outside the national capital area.
In 1975, the PSC offered a total of 428 courses (excluding language courses), of which 139 were offered by the six regional
offices to some 2,547 federal employees. These figures indicate
the vast amount of interpersonal communications and training activities required to:
  • provide equal access to training and development courses;
  • access resource persons, trainers, and learners;
  • provide courses to meet the training needs in regional
    urban centers.
These figures also are suggestive of the costs implied in current
face-to-face training. Presently, federal employees and resource
persons have to gather in major regional centers, or in Ottawa
itself, to attend courses. As a consequence, residential living-
in expenses and costs of time of participants in courses add considerably to tuition fees. The fees incurred by the employee are
paid by each attending government department.

In view of the high direct and indirect costs to meet these
training requirements, it was decided to investigate alternate
methods of delivering courses to regional centers from headquarters and among regional centers.

THE EXPERIMENTS

Research on teleconferencing was undertaken with the following
objectives:
  • to provide information as to the feasibility of telecon-
    ferencing for both administrative and educational purposes;
  • to determine the educational content and methodology ap-
    propriate to the medium;
  • to identify the psychological and technical variables
    related to man-machine interaction using various techno-
    logies.
Both field studies and laboratory experiments using interactive
video systems were undertaken to achieve these objectives. Interactive video systems rather than broadcast video systems were
chosen on the basis that these favoured learning/teaching behaviors
comparable to face-to-face training. On the one hand, interactive

systems provide learners with the capability of becoming involved
in the educational session through questions, discussions etc.
(Cyrs, 1976). On the other hand, these systems provide a means
for participants to act as resource person and offer each member
of the group equal opportunity to interact with any other member.
The video component was chosen on the basis it would permit an
immediate feedback as to the comprehension of the content material
and would promote a greater sense of group unity. As will be
seen, interactive video systems were used in both two and four-
nodes training situations.

Field Studies

     To investigate educational teleconference applications, two
experimental sessions were undertaken: a financial seminar and
a language teacher training seminar. Both field trials utilized
the Bell Canada Video Conferencing System offering complete audio-
video facilities between two locations. The financial seminar
designed for PSC personnel consisted of a one-day session on
"Services Contracts within PSC". These three one-day sessions
held on consecutive days had respectively six, eight, and nine
participants. The teaching methodology used was content presentation followed by dialogue with the participants. Results
obtained from questionnaires indicated that:
   ● the content and the methodology were both suited to the
     technology;
   ● the degree of satisfaction was inversely proportional to
     the amount of knowledge prior to the educational session,
     and
   ● one-day training activities were perceived to be better
     suited for teleconferencing than face-to-face.
According to the participants, the following differences were observed when they compared teleconference to face-to-face. They
felt teleconferencing was more businesslike and more serious, and
that it inhibited aggressiveness and imposed a greater feeling of
constraint. (Lortie, 1977).

     The second field experimental session, concerned with the
training of language teachers in a new French language teaching
methodology, also took place between two locations. Similar technical arrangements were made available by Bell Canada. Six specialists of Dialogue Canada II, the new Public Service Commission
language program, participated from Ottawa with 14 language teachers
from Montreal. These were divided into sub-groups with seven language teachers and three to four specialists. Each session consisted of two half-days with actual teleconferencing time of about
two hours and fifteen minutes per half-day session. The first
session content was a review of the older Dialogue Canada I (DC 1)

while the second session was an introduction to Dialogue Canada II
(DC II). Thirty minutes of theory was spent on DC I while the
remaining time was for discussion and questions. However, for DC
II only twenty minutes of formal presentation took place and subsequent time was left for discussion. On the whole, results obtained were similar to those of the financial seminar:

- Participants preferred the format of DC II involving a
  higher ratio of interactive communication versus formal
  one-way presentation (DC I).
  In the case of DC I, there emerged a need for planning
  the session content more adequately, that is, by using
  proper amount of subject matter per unit of time allocated.
- Communication was largely content-oriented and little interaction was possible within the same group. A high percentage of the participants felt that a face-to-face session
  would have been more lively, more intimate and less tense.
- Participants of the first and second sessions felt that
  they had achieved their goals which were to be informed
  about the teaching methodology. (Frenette, 1976).

For both of these seminars, the learning/teaching methodology
was not specifically adapted for the audio-video medium. The
course content and presentation format were those of similar face-
to-face courses. Although results obtained indicate participants
as being satisfied with the content and methodology, they stressed
the fact that the use of the system would have been more conducive
to learning if more group interaction had evolved. This is not
surprising since the participants have a solid educational background, managerial experience and expertise, and felt that they
had an active role to play in these educational sessions by bringing
pertinent information and insight into the subject matter. Thus
students felt that the quality and quantity of interactions among
themselves would have been greater in a face-to-face situation.
It has been shown (Weston, Kristen, O'Connor, 1975) that the number
of messages transmitted in face-to-face is greater than in any
other communication mode for the same given time.

Laboratory Experiments: Administrative and Educational Applications

The Public Service Commission and Department of Communications
jointly set up a two-node interactive audio and video laboratory.
The laboratory experiments performed replicated the government
working environment as much as possible. Subjects were government
employees who performed tasks related to their managerial functions.
Groups of four to five managers were located at each node to replicate the regional nature of Canadian decentralized government
departments. Data obtained under these conditions provided knowledge of the attitudes, feelings, satisfaction and performance of
government users to teleconference. (Ryan and Craig, 1975; Ryan,
1976; Jull and Mendenhall, 1976; Mendenhall and Ryan, 1977). An

example of such a study was the laboratory experiment undertaken in 1975 which compared two-node audio, two-node video and face-to-face. Fifty-one managers, divided into three and five-persons groups, performed managerial tasks, i.e. problem solving, brainstorming, information exchange. Half hour limits were imposed in the accomplishment of these tasks. These results were obtained:

- Audio conferencing was perceived more negatively than either video and face-to-face.
- Information given over audio-only systems elicited more skepticism.
- Audio and face-to-face discussions were rated more private than video.
- Face-to-face and video discussions were found more active, more positive and more influential than audio discussions.
- Face-to-face was considered the medium most satisfactory for discussion. Although audio teleconferencing elicited some negative reactions, it is important to note that the managerial tasks accomplished over teleconference did not differ significantly from face-to-face.

Results obtained co-incide with other comparisons of media studies performed in a group-to-group situation (Stapley, 1974; Christie and Elton, 1975; Institute for the Future, 1976).

Another recent laboratory experiment was aimed at exploring the use of teleconference in a four-node, audio-video, interactive, non-directive, learning environment. The environment was non-directive, in the sense that participants and the resource person would decide, over three mediated educational sessions, what procedures would be used to teleconference effectively in a learning situation. Twenty participants, staffing officers from various government departments, were divided into two groups of four and two groups of six with the resource person located in one group. The course content was "Staffing" and the teaching methodology introduced did not differ from that of face-to-face session. Sessions lasted approximately 67 minutes, 47 minutes and 59 minutes. The three sessions were evaluated in terms of the educational activity, the relationship among the nodes and feelings about the session. No significant main effects were observed. Briefly summarized the results indicated a significant longitudinal effect ($F = 15.58$; df = 2,51; $P > .05$) in that sessions became progressively more task-oriented over time. A significant difference among nodes was also observed ($F = 3.63$; df = 3,50; $P > .05$), the third node in which the resource person was located, showed a more static rating on a static-dynamic scale ($\bar{x}$ of 2,42) than did the other nodes ($\bar{x}$ of 2.82). This may tentatively be explained by the lack of participation by the learners due to the dominance of the resource person in her own node. Objective speech data analysis shows the resource person's node took up the greater part of the

total speech time. (Treurniet, 1977). Speech patterns indicate
that most interactions occurred between the resource person and
students. This is consistent with the fact that participants did
not come to a consensus on procedures until the third session at
which time they implemented traditional classroom procedures:
raising hands to speak, having the teacher acknowledge speaker
and self-identification.

These field and laboratory tele-education experiments showed
some of the major deficiencies of applying the traditional learning and teaching approach, and confirmed the need to develop an
educational learning model more closely matched to the needs of
the adult learner in a mediated situation. It became evident that
one of the key elements for successful mediated learning is interaction, in particular, the capability of exchanging information
among learners themselves. The model would have to take into
account some of the following preconditions for interaction: the
traditional role of the resource person and of the learner would
have to be redefined; the responsibility of the learning would
have to be in the hands of the learner, the learning behavior would
take the medium into consideration. From these elements evolved
the following mediated adult learning model.

DESCRIPTION OF MODEL

A learning model can be defined as a proposed set of coherent
steps or measures aimed at applying a philosophy or a given set of
educational principles. According to those principles, certain
roles and behaviors can be recommended to the educator and learner.
These roles lead to behaviors appropriate for the role and this is
shown in the following diagram:

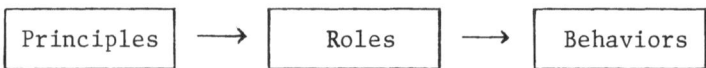

This adult learning model was elaborated for application in a mediated learning context over a multi-way satellite channel and
included modifications appropriate for that context.

The principles upon which this model is based follow:
● Man is basically self-directing (however this characteristic undergoes varying degrees of actualization in actual
  practice).
● As an adult learner, this self-directed man is characteristically oriented toward solving some problems that he is
  being confronted with in real life situations (Knowles,

1973). He tends to organize his learning under the form of a self-directed project (Tough, 1971).

- A clear distinction is made between learning and teaching. Learning is a basic internal process involving the whole person; it is highly individualized and directed from within. Teaching is an external intervention which is incidental to a person's learning. The model is a <u>learning model</u> as opposed to a teaching model.

These principles elicit certain patterns of behaviors (i.e. roles) to be performed by people participating in an educational activity.

Some of the major components of the role of the adult educator and learner can be briefly stated as follows. The educator has, among others, the following responsibilities:

- He is an animator and a facilitator.
- He is a resource on the content and the process of the educational session.
- He is a change agent.
- He promotes a greater awareness and personal growth.
- He assists learners to take charge of their own training.
- The educator becomes a learner.

Similarly, some of the main roles of the learner include the following:

- He takes the responsibility for his learning.
- He is responsible for determining his needs, goals and educational objectives.
- He develops his skills for self-directed learning including greater awareness of needs, potential, and interests.
- He is an active inquirer.
- He contributes his own resources to the learning activity.
- He takes charge of his own evaluation.
- The learner takes up the role of a resource person as appropriate.

These roles complement each other and contribute to self-directed learning.

The behaviors described below operationalize the principles and roles of the model in the context of the five location audio-video learning environment specific to the Public Service Commission's Staff Training by Satellite project.

<u>Preparation for the course</u>. The ten suggested steps or behaviors mentioned below will take place prior to the mediated session; they are intended to provide:

- <u>involvement</u> of learners in their self-directed learning activity;
- an early <u>interaction</u> among participants;
- some degree of <u>acquaintance</u> among learners prior to the tele-education sessions;

- some <u>common</u> and familiar <u>grounds</u> on which to base further discussion and interaction.

Each of the following is necessary to meet the principles underlying the model:

1. A needs survey is conducted which offers a number of potential participants an opportunity to indicate their professional development requirements.
2. The educator selects a course topic which satisfies the needs of most potential participants.
3. Needs survey respondents who requested the topic are offered the desired course and register.
4. With the assistance of the educator, learners are asked to define their learning objectives relative to the chosen topic in successive steps. These become more specific over time.
5. Learners inform the educator of their own potential resources concerning the topic. They may also mention other resource-persons that they know.
6. The educator assembles information on the learners' addresses, phone numbers, objectives, and each person's expertise, in order to distribute it to every participant.
7. Learners are informed that a number of preparatory documents are made available to them that seem to be relevant to their purposes. This pre-documentation serves to:
   - answer potential expectations;
   - increase learners' self-confidence;
   - stimulate a self-directed learning process;
   - facilitate the clarification and specification of one's own learning objectives prior to the session.
8. Learners are invited to further this process of pre-documentation by finding documents and references of their own, and exchanging information among themselves and the educator.
9. The educator tries to follow the evolution of the individual learners' objectives as they become more clearly defined and specific. A few weeks before the session begins, he compiles those objectives and draws up a proposal of tentative session plan which is sent to the learners to be discussed and negotiated among all involved.
10. On the basis of the learner's objectives as well as the subject matter's requirements, the educator progressively prepares for a given content and learning activities. These include namely identifying, contacting, and possibly contracting, internal or external resource people. This is a continuous process which really starts with the needs survey. It intensifies towards the end of the preparatory period (last few weeks). It will culminate on the first mediated meeting when a common agreement on session plan, schedule and content is "finalized". There is always a possibility of renegotiating at any point in time.

These preparatory steps increase the likelihood that the learner
will be engaged in his own learning process prior to the first
mediated learning session.

    First meeting. This meeting will need particularly careful
preparation in view of the prime effect. A logical sequence of
events would be the following:
    1. People introduce themselves, socialize briefly, get ac-
       quainted with this new environment.
    2. The educator proposes that participants first use the
       audio-video learning system for a while and then decide
       on the necessity and nature of a procedure for interaction.
    3. Participants hold a group planning session for the whole
       learning activity. This event seems to be a repetition
       of a type of activity (negotiation and planning) which
       has taken place previously. However it seems warranted
       because, the very first 30 minutes or so, many partici-
       pants will be distracted by the medium itself, however
       unobtrusive it may be. Thus, the period may be practically
       wasted if totally new information is exchanged.
Right from the start, and especially then, the educator adopts
certain behaviors likely to promote "self-directed" learning, such
as
    • listening attentively to learners' interventions;
    • reformulating to check for understanding;
    • refraining from elaborating too much in order to stimulate
      personal inquiry;

    After the first meeting. Basically the same type of attitudes
and behaviors as described above will repeat themselves over the
rest of the session. Whatever the task being performed, the educator remains an animator and a facilitator, promoting among the
learners an increasingly stronger sense of, and capabilities for,
self-directed learning.

    Between mediated meetings. Work in small groups as well as
a period of time in-between mediated meetings has often been mentioned as being desirable (Wigren, 1976). Optional 2 hour-long
periods for face-to-face meeting among members of a sub-group are
offered before each mediated session. Learners can also work by
themselves, or in smaller groups, or not take advantage of this
period at all.

    Types of tasks over the multi-way satellite channel. For
obvious reasons, tasks for transmission over the satellite channel
are those which require substantial interaction among the largest
number of participants at a same time. This means, for example,
that we would hold over the satellite channel those discussion involving potential participation of everyone. Discussion is likely

to be on of the major activities over this channel. In this respect, the desirable role of the discussion leader (Brilhart, 1974),
can be summarized as:
- guiding;
- stimulating thinking;
- facilitating communication;
- promoting cooperative interpersonal relations.

The above provides an idea of the principles, design and purpose
of the self-directed learning model. In practice, however, the
model has to be adapted to particular situations. The general
validity of this model for the present application has been confirmed by related research (Wigren, 1976).

Model Simulation. This educational model was tested in a laboratory experiment during a four-day session on long-range planning.
Sessions alternated between the face-to-face and mediated context
and lasted approximately two and a half hours each. The educational model was presented under three different formats:
- session I: a formal lecture presentation;
- session II: 10 minute presentations followed by questions
  and answers;
- session III: discussion on evaluation of the educational
  session among participants.

Twelve participants and two resource persons were divided into
four groups. The audio-video teleconferencing system used was
equipped with an open microphone system and four-way video.

Although the sample size was quite small, preliminary results
supported the usefulness (applicability) of the learning model.
The analyses showed significant differences among the three sessions
on the following dependent variables: the educational session, the
satisfaction and the team participant evaluations. Scheffé Post
Hoc Analyses showed that the interactive sessions (II and III)
elicited more positive evaluations to the educational session as
well as perception of greater team participation. The analyses
also showed that the satisfaction was greater with the presenta-
tion-discussion session than with the formal lecture session.
These data seem to imply for this group that the more students
can play an active role in their learning, the more positive they
are towards the educational session.

It is interesting to note that results obtained on the Attitude
Towards the Medium Questionnaire did not differ significantly over
the three sessions. This may imply that the medium used was transparent and unobtrusive.

A rigorous test of the applicability and usefulness of this
educational model will be undertaken in late April, 1977. At this
time, the Communications Technology Satellite (CTS) will provide

five-way interactive audio-video tele-education facilities to 25 Canadian federal civil servants located in four locations in Newfoundland and one location in Ottawa.

The long-range planning course will be offered twice in the mediated satellite context and once in a face-to-face context. This course will be offered on a split school/work day basis, that is, Tuesdays and Thursdays for two hours and for a period of four weeks.

The results of this CTS study will help educational decision makers to assess the benefits and/or disadvantages to be derived from interactive educational delivery systems. It will also provide information as to the extent to which these delivery systems can replace or substitute face-to-face teaching. Furthermore, the results will also indicate whether these public servants are willing to take on a more active role in their learning, and to what extent this type of learning satisfies the organizational needs. However, although this study is expected to produce a wealth of data on interactive tele-education, there are other major issues to be considered in more detail prior to any decision on the wide scale implementation of these telecommunication facilities for training:

- What are the learning styles that should be addressed in the application of interactive delivery systems?
- What are the characteristics of the learner which will most profit from mediated learning situations?
- Will interactive audio-video systems be useful for dual training functions: to better equip the individual to accomplish his present job responsibilities, and to better provide knowledge and skills to an individual in view of career advancement?
- What would be the major forseeable consequences and impact of the implementation of educational delivery systems on government departments in general and on PSC in particular?

SUMMARY AND CONCLUSIONS

This paper has shown that the traditional learning/teaching models needed to be reconsidered in view of the adult learner, the interactive medium, and the multiple group learning requirements. The self-directed learning model developed gives promise of being particularly well-suited to adult education over interactive audio-video channels. The disadvantages of broadcast methods of instruction with limited interaction between students and predominance of the teacher are overcome. The use of interactive audio-video systems seems to respond to an adult learner's need to become more actively engaged in his learning process.

This paper has also shown that the use of interactive audio-video channels has definite implications in satisfying the training and administrative needs of the PSC such as providing wider access to regional and central human resources. This possibility may in itself provide a greater exchange of information among regional federal employees and as such a greater awareness of diverse regional problems. The use of these systems makes different configurations possible, that is, two-node, four-node or multiway links such as required to satisfy regional training needs. It is now possible for example, with audio-only systems to offer a course to three people in Vancouver, ten in Halifax, five in Quebec and seven in Winnipeg. This course would not have taken place given the small number of participants at each location under present face-to-face training conditions. This implies also that training needs may be satisfied as they are required. It is also foreseen that the use of these systems for training activities may facilitate the integration of the split work/school day and thereby increase training effectiveness.

In summary, it is foreseen that the use of these interactive systems may bring about a new era in training and development services to public servants.

BIBLIOGRAPHY

Brilhart, John K., Effective Group Discussion, Wm. C. Brown Company,
        Publishers, Dubuque, Iowa, 1974.

Cyrs, Thomas E. (1975-1976). Design Techniques for Interactive
        Media, International Journal of Instructional Media,
        Vol. 3(4), Baywood Publishing Co. In.

Christie, B. and Martin Elton, (1975), Research on the Differences
        between Telecommunication and Face-to-Face Communication
        in Business and Government, CSG, London, England.

Frenette, Charlotte (1976), La Téléconférence comme Médium d'Ap-
        prentissage: Déroulement et Résultats de l'Expérience
        de Formation à DC II, Commission de la Fonction publique,
        Ottawa. Présenté au 43ième Congrès de l'ACFAS, Sherbrooke,
        Québec.

Institute for the Future (1976), The Camelia Report: A Study of
        Technical Alternatives and Social Choices in Telecon-
        ferencing.

Jull, G.W. and Nicole Mendenhall (1976), Prediction of the Accep-
        tance and Use of New Interpersonal Telecommunications
        Services, The Status of the Telephone in Education,
        Second Annual International Communications Conference,
        University of Wisconsin.

Knowles, Malcolm, The Adult Learner: A Neglected Species, Gulf
        Publishing Company, Houston, Texas, 1973.

Lortie, René (1975), Teleconference Financial Seminar, Public
        Service Commission, Ottawa, In-House Report.

Lortie, René (1976), Problèmes pratiques dans un domaine nouveau:
        Formation non-directive via satellite, Commission de la
        Fonction publique, Ottawa. Présenté au 43ième Congrès
        de l'ACFAS. Sherbrooke, Québec.

Lortie, René (1976), A Democratic Communication Model for Inter-
        active Tele-Communication Learning, Public Service
        Commission, Ottawa. Paper presented at the AMTEC Con-
        vention, St. John's, Newfoundland.

Mendenhall, Nicole and Michael Ryan (1975), L'Effet des communica-
        tions médiatisées: L'Affectivité sociale, la mélancolie,
        la fatigue, et le scepticisme de l'utilisateur, Communi-
        cations Research Center, Department of Communications,
        Ottawa, Rapport 1286 du CRC.

Mendenhall, Nicole and Michael Ryan (1977), A Survey of Teleconferencing Studies in the Public Service Commission, The Telephone in Education Conference, Wisconsin.

Ryan, Michael and James C. Craig (1975), Intergroup Telecommunication: The Influence of Communications Medium and Role Induced Status Level on Mood, and Attitudes Towards the Medium and Discussion, Department of Communications, Ottawa. Unpublished Report.

Ryan, Michael and Nicole Mendenhall (1976), Interaction: A Canadian Theme in Education by Satellite, Presented to the World Education Conference, Honolulu, Hawaii.

Ryan, Michael G. (1976), The Influence of Teleconferencing Medium and Status on Participants Perception of the Aestheticism, Evaluation, Privacy, Potency, and Activity of the Medium, Human Communication Research, Vol. 2, No. 3, Spring.

Stapley, Barry (1974), A Comparison of Field Trials of Teleconferencing Equipment, Communications Studies Group, London, England.

Tough, Allen, The Adult's Learning Projects, Research in Education Series No. 1, The Ontario Institute for Studies in Education, 1971.

Treurniet, W., Speech Patterns Analysis, Department of Communications, Ottawa. Report forthcoming.

Weston, J.R., C. Kristen and S. O'Connor (1975), Teleconferencing: A Comparison of Group Performance Profiles in Mediated and Face-to-Face Interaction, The Social Policy and Programs Branch, Department of Communications, Ottawa, Ontario.

Wigren, Harold E., Teacher-to-Teacher Communications, Report of the NEA's Pan-Pacific Satellite Experiment 1976, National Education Association, Washington, 1976.

OPEN CHOICE - NEW COMMUNICATION SYSTEMS AND APPLICATIONS AT THE

BRITISH OPEN UNIVERSITY

Dr. Peter Zorkoczy

Open University

Milton Keynes MK7 6AA, U.K.

1. INTRODUCTION

There is a growing interest worldwide in the extension of educational opportunities for adults. Continuing education, re-training, evening classes, part-time courses, extension studies and professional up-dating all indicate a shifting emphasis in educational provision - for people in jobs or at home. The developments in this field receive support from the International Labour Organization, which in 1974 enacted a convention on paid educational leave, and from national governments which devote resources and provide opportunities for intending students.

The British Open University, and its counterparts in the U.S.A., Germany, Israel, Iran, Pakistan and so on have been set up to deliver educational material to students at home by means of various telecommunication channels. That this is a successful and acceptable way of learning is shown by the steady increase in the number of students enrolled with the Open University (see Table).

There is an evident attraction for adult learners in the possibility of studying at home, organizing their own time and effort, obtaining qualifications at their own pace. These advantages are achieved, to a large extent, by the judicious use of a range of

Table - Annual total of undergraduates, Open University

Year	1971	1972	1973	1974	1975	1976	1977
Number	19,581	31,902	38,424	42,636	49,358	50,994	54,488

communication media - printed word, broadcasting, face-to-face contact, telephone and computer networks. But the use of such a variety of channels between learners and teachers brings its own problems. The foremost of these problems is cost: the establishment and running costs of these services exceed £20 million per annum. It is clearly in everyone's interest to make the best possible use of these expensive resources. There is, in the Open University, a strong motivation towards the study and development of the existing channels of communication and the creation of new ones.

This paper aims to review the advances made by the University in the field of interpersonal telecommunication, with the hope of sharing our experiences and results, and stimulating co-operation with all those interested, whether or not in the educational field. To place the more recent developments in context, it is useful to summarize, first, the features and problems of the existing communication system of the Open University, and then, to outline the planning considerations for future systems.

2. THE CURRENT COMMUNICATION NETWORK AND ITS PROBLEMS

The University's operations cover the entire United Kingdom since students come from all parts of the country. The University is organized on a regional basis, with headquarters in the new town of Milton Keynes, in Buckinghamshire. In each region there are study centres, catering typically for 200 students. The main vehicle for study materials is the printed text, sent by mail to students' homes. For most courses there are supportive television and radio programmes broadcast nationwide by the BBC. Face-to-face contact between a student and his personal tutor is rather restricted (about 20 hours a year), but they are in more frequent correspondence contact. For many courses a week's intensive summer school offers additional opportunities for face-to-face tuition, but at an extra cost. Assessment of students' progress is by written tests and examinations, and by computer-marked, multiple-choice assignments.

This basic correspondence broadcast and direct contact system is supplemented by a second network based on the telephone system. It consists of a distributed, time-sharing computer network with on-line terminals in study centres, using the public switched telephone network, and a private-wire network. This latter links the regional offices with the headquarters and is used mainly to support the administrative work of the University - for individual contact between members of staff, telephone conferences and for access to the University's data-base management computer from the Regions.

The overall system is functioning quite effectively, within its limitations. It has a remarkable record of reliability, linked as it is, to vital public services. From the University's point of view, the system is also cost-effective - it takes about half the cost of a conventional university graduate to produce an Open University graduate. But the cost savings are primarily due to the scale of operations. The University is, effectively, a mass-communication system.

The limitations of the system become apparent when considered from the students' point of view. Some of the more evident problems are the following:

1. the printed texts and broadcasts are aimed at the 'average student', with only a limited opportunity to cater for individual needs;

2. the communication is mainly one-way, from teacher to student;

3. the student has to attend a study centre or summer school to gain access to rapid two-way communication (face-to-face contact and computer terminal), the turn-round time of postal communication is rather slow for effective learning;

4. the principles governing the optimum use of the various communication channels for educational purposes are not yet known well enough.

The solutions to these and other problems of teaching and learning at a distance are under continuous review and study. They are strongly influenced, however, by the difficulty of forward planning in a period of financial uncertainty. Any new development must be able to show a very clear educational and economic advantage if it is to become an accepted part of the system. The systems I am about to describe have reached different stages of recognition within the overall communication network of the University: the use of the telephone to help with individual problems of students is now widely accepted; the use of the computer network for the diagnosis of learning problems has been implemented for several courses; the use of home-based terminals for guidance and advice is currently under study.

3.   TELEPHONE TUTORIALS

Work in this area has been motivated by the intention to make the contact between student and tutor more immediate and more

convenient than correspondence or travel to a study centre.   The
new system takes several forms:

1.   <u>individual, 1-to-1 contact</u> by telephone between tutor and
     student. This requires nothing more than a telephone call,
     but efficient use of time must be made.

2.   <u>telephone contact</u> between a group of students at a study
     centre and a tutor at a remote location. This is made
     necessary by the fact that more advanced and specialised
     courses have lower student numbers and, therefore, a
     proportionately lower number of tutors. Students and tutors,
     however, remain geographically dispersed and the area of
     coverage of each specialist subject tutor increases. As
     tutors are part-time employees of the University they cannot
     spend much time and money on travel and may have to hold
     tutorials remotely, with one group or even several groups of
     students simultaneously.

     Originally, the equipment included a loudspeaking telephone
     in the study centre, with a "roving microphone". Currently,
     experiments are being conducted on supplementing this with
     an "electronic blackboard" facility.

     This device, developed in the University by Chris Pinches and
     his co-workers, is based on the domestic TV receiver. The
     tutor 'draws' on the display tube with a light-pen. The
     diagram is digitized as a 256 x 256 dot array, and the digital
     data are transmitted in sound-modulated form over the
     switched telephone network. At the far end the picture is
     reconstructed and displayed on another standard receiver.
     The device can be operated in real-time using 2400 baud FSK
     coding. The data are sent along the same voice-line as the
     tutor is using for audio contact. Switching from voice to
     data, and back, is automatic when the light pen is being
     used. The tutor also has at his disposal an audio cassette
     recorder which stores, in a coded form, diagrams which he
     prepared in advance. He can then incorporate these in his
     presentation as required. The production cost of the equip-
     ment additional to the TV receivers is approximately £300.

3.   <u>telephone conferences</u>. These may take place with both tutor
     and students using their own domestic phones, and may also
     include groups using a loudspeaking telephone. The calls
     have to be booked in advance and are set up by Post Office
     operators. A recent development is a conference bridge
     located in a regional office which we can use to set up our
     own conference calls on the public network, and if necessary,
     link to the private network of the University.

3.1 Evaluation of Telephone Tutorials

The evaluation of these facilities is currently in progress. The University has set aside £9,000 for financing the cost of the calls in the first year and for informing students of the existence of the system. Initial results indicate particularly high utilization by housebound and geographically remote students. This is, of course, as expected, and it means that, for the first time, such students can receive the same support as other, more fortunately placed students.

Other findings point to the importance of preparatory work in advance of the telephone contact, to make the best possible use of time, and to the discouraging effect of line noise, distortion and fading on the effectiveness of the telephone tutorial. The importance of training of the tutors for this aspect of their job and of advising the students on the ways of getting the most out of the facility has also become apparent.

4. COMPUTER-BASED SYSTEMS

These developments are also aimed at giving home-based students personal support in a mass-communication system. In particular, they attempt to diagnose the student's problems with the teaching material and to suggest appropriate remedies.

The CICERO system involves the use of the 192 terminals of the University's time-sharing computer network. Students receive, in advance, a set of carefully designed multiple-choice questions on one part of the course which they study. The questions are structured so that the student's answer to them provides a guide to his understanding of that part of the course. At the terminal, the student types in his answers and then receives a printout of an-notated answers and advice on any further work needed in that area. Remedial assistance is also provided by the program.

A similar system is also being developed at the University to supplement the pre-programmed diagnosis routine with a live tutor. The system is based on a suggestion of Prof. W. Dorn of Colorado University who spent a year with us as a visiting scholar. When, in his dialogue with the computer, the student reaches a point where he considers the computer's response inadequate, he types in HELP! This message appears on the tutor's VDU terminal at one of three computer centres. The student and the tutor then both switch to voice contact on the same line as the student has been using to operate his remote terminal, and attempt to resolve the problem. The student then receives more on-line questions which test his ability to synthesize various topics or concepts including the one that has been causing the difficulty.

The tutor is also able to follow the progress of any one student on his VDU, without having been specifically asked to do so. He can, if he feels necessary, type a message to any terminal location and request the student there to get into audio contact. At the end of each session the student is given an assessment of his understanding of the course so far, and the implications of that for his study of subsequent parts of the course.

4.1 Evaluation of the Computer-Based Systems

These systems are in their first year of operation, so no final evaluation reports on them are as yet available. Initial indications of "consumer reaction" are quite encouraging. With the first system described, utilization has been of the order of 40%. This is higher than was expected in a course which involves no other computer work - a course in educational theory. The second system, tried out with a physics course, has attracted a somewhat lower response but those who have tried it - tutors and students alike - rated the system quite highly. One reason for this is that the advice given to students is not of the rather general type given in PLATO, CICERO and similar systems. Instead, the guidance is quite specific to the needs of an individual student, both in terms of the diagnosis and the remedy.

Some problems have become evident even at this early stage. The preparation of satisfactory educational software is very demanding, both in the expertise required and in the time taken. Also, the student is still required to travel to a study centre to gain access to a computer terminal - to some extent defeating the purpose of the system. The second problem is perhaps more easily surmounted than the first one. We are developing a low-cost computer terminal to loan to students who require it. It plugs into the domestic television set as a display and is equipped with a teletypewriter keyboard. It is acoustically coupled to the telephone line. The production cost of this device is about £150 in large quantities.

5. FUTURE SYSTEMS

The developments described so far were aimed mainly at improving two-way communication between tutor and student, for tutorial purposes. But these form only a part, albeit an important one, of the University's interests in interpersonal communication. Other concerns include the method of presentation and delivery of course materials, examination and assessment, assistance to people with sensory disabilities, career guidance and so on.

A system under consideration, which is likely to have a bearing on these concerns, is a Viewdata-type facility. The Viewdata service, as envisaged by the British Post Office, is based on transmitting text and graphical information from a vast computer database to subscribers with specially adapted domestic TV receivers, using the public switched telephone network. The user selects the data he wishes to consult by means of a small keyboard. The keyboard can also be used to send data to the computer for storage or processing. It is, therefore, an interactive system for information presentation with a limited processing capability. The cost to the user is the additional equipment to his television set (estimated at between £50 and £100) and the call and service charges.

This system, or a modified version, can serve, in the Open University context, as a means of presentation of course materials, although currently it lacks a hard-copy facility. The interactive feature offers the prospect of its use for assessment of progress and tutorial assistance. (Notice how such a facility incorporates the desirable features of systems described in the earlier sections, being home-based, interactive, with access to both telephone and computer information.) It can also carry a subtitle - textual information service for students with impaired hearing, a course-choice and career guidance service, and administrative information required by the student.

Another future system of a related type provides access to data stored in auditory form in a computer data-base. It does not require a display, only a standard telephone connection. Its main application, as currently envisaged, would be a service to students with visual impairment, such as a "talking book" facility. It would also enable students to record their own material, for later recovery either by themselves or other persons.

The introduction of these and other systems is conditional upon a number of planning considerations. These include:

1.   low unit cost. For general purpose use, the total cost of any new system is the product of the unit cost and the number of students in the University. The unit cost may be reduced somewhat if part of it is borne by the student himself. This could be the case, for example, with a Viewdata-based solution, as Viewdata is envisaged to offer a wide range of services, of which educational use is only one. Also, any new system will have to compete with existing or alternative methods of achieving the same objective. From a planning point of view, therefore, it is vital to keep cost trends under constant review, and phase the new developments appropriately. For example, the cost and penetration trends of

Viewdata are unlikely to make it a viable alternative to printing
and mail, as far as the University is concerned, until the mid-1980s.

2.   versatility of applications. In view of the high total cost
of any new system it must be able to serve a range of communication
requirements within the University, and offer a solution to several
problems listed in section 2.

3.   reliability. This is strongly related to the first two
considerations. A versatile system must have good availability,
yet reliability is often the first victim of any cost-cutting
exercise. The principle here is rather not to introduce an attrac-
tively priced new system if its quality and reliability are poor.

4.   ease of access. In the context of the University, systems to
which students have access from their homes are preferable to ones
requiring travel.

5.   simplicity of use. This should allow the student to
concentrate on the job in hand rather than be sidetracked by the
idiosyncracies of the system. Ease of error correction and
recovery are necessary considerations.

6.   educational relevance. Although mentioned last, this is
indeed the consideration which underlies all others. This implies
the thorough testing and evaluation of any new system before it is
introduced on a large scale. It must be able to carry with it the
conviction and commitment of all those who are involved in
operating it.

The Open University provides a unique test bed for the
development and evaluation of new ideas in interpersonal communication. There is ample room and good-will for co-operation and joint
progress with all those with similar objectives and interests.

ACKNOWLEDGEMENTS

The work reported here is being performed by members of the
University's Communication Research Group, as well as staff of the
Faculty of Technology, Student Computing Service and Regional
Tutorial Service.

SERIAL EXPERIMENTATION FOR THE MANAGEMENT AND EVALUATION OF COMMUNICATIONS SYSTEMS[*]

William A. Lucas and Suzanne S. Quick

The Rand Corporation

2100 M Street, N.W., Washington, D.C.   20037

When there is a decision to mount a demonstration of new communications systems in the field, controversy about the role of evaluation often results. Those who have operational responsibility for the system become its advocates, believing it to be sufficiently robust to succeed. They resent research that would divert resources from the central purpose of demonstrating success. In contrast, evaluators anticipate the chance of failure. They maintain that it is essential to construct systematic and rigorous research on a project so that future projects can learn from past experiences.

The conflict is likely to intensify when the demonstration is underway and something starts going wrong. Project personnel will quickly want to make changes, trying to correct the difficulty. Such action is clearly detrimental to the evaluator's efforts since the more rigorous their design, the more irrelevant the data usually will be if the program has been continually altered. But try, as an evaluator, to persuade a program manager that he or she should leave a failing field program unaltered so that you can rigorously establish the reason for failure in a final report.

One approach to this problem which serves the purposes of both program advocates and evaluators is to use a research design that consists of a series of short experiments that can mutate through successive generations. The manager is committed to holding each

[*]Views expressed in this paper are the authors' own and are not necessarily shared by Rand or its research sponsors.

discrete, experimental intervention constant, permitting rigorous
evaluation. The evaluator is committed to providing the results
of each experiment to the manager before the sequential experiment
is launched, and being prepared to design new instruments--however
imperfect--on the spot to evaluate this change. The manager receives
superior planning data in return for periodic inaction and concomitant anxiety.

To serve these purposes, a field project should have four
elements: serial experiments, a process measure, a robust outcome
measure, and a standard of comparison. The process measure is
essential to understanding how the system actually operates so that
the manager can be given guidance on how to direct change. The
outcome measure should be tied to a broad goal that will remain
relevant despite a variety of system changes so that one can
determine the relative value of each system generation. The standard of comparison is essential to a discussion of the value of
replicating the system and the generalizability of the results.

Having these factors in a project design has almost no disadvantages, and provides many opportunities to learn from demonstrations even when the starting assumptions are found to be in
error. There are, of course, many times when the single, grand
experiment is the only feasible route, but it seems likely that
many opportunities to establish serial experiments are missed.[1]
Here we report on the results of using this logic in the Spartanburg
two-way cable project.

THE SPARTANBURG INTERACTIVE CABLE EDUCATION PROGRAM

Since February 1975, a two-way cable project has been underway in Spartanburg, South Carolina, that compares the progress of
students who use simple home terminals to participate interactively
in televised instruction with the progress of students who receive
conventional classroom instruction. This effort is part of a
program of research supported by the Research Applied to National
Needs (RANN) program of the National Science Foundation on the
costs and benefits of two-way cable television systems, systems that
can be used both to send and receive signals from a home, agency, or
business. Research described here is one of several education and
training programs conducted in Spartanburg designed to test the
value of alternative forms of return communications on a cable
system.

Interactive cable television offers a technological opportunity
for education that lies somewhere between conventional educational

------

[1]Segmented time series data can often be used in lieu of serial
experiments, providing many of the same advantages.

television (ETV) and computer-assisted instruction (CAI). Like
ETV, students watch teachers on a television set and the programs
are designed to make heavy use of visual and workbook materials.
Like CAI, the students have data terminals that they use to record
answers to the teachers' questions and assigned exercises. With
interactive cable, however, the students' answers are immediately
corrected and reinforced by a teacher rather than by a computer.
This not only benefits the students by providing human as opposed
to machine feedback, but it also enables the teacher on cable television to have continuing information about the progress of each
student readily available as the class proceeds.

In this experiment, adult education at the high school level
was chosen as the substantive content to test the value of two-
way cable to the home. Using the two-way capacity, Spartanburg
Technical College has offered a series of home-based adult education
classes using an interactive cable television system in English
grammer, reading, and mathematics, the three subjects necessary to
prepare students for the General Educational Development (GED) examination for a high school equivalency degree. The text material has
been the Cambridge GED series and the content of the course is
similar to that of many GED programs around the country. In this
demonstration project, the progress of students in the conventional
classroom setting is compared with that of students in the "electronic
classroom" who view classes from their homes. Each of these students
has available an eight-button hand-held terminal to respond to
questions posed by the television teacher during a class period.

Students in the electronic classes can use their terminals for
three distinct kinds of student-teacher interactions. The most formal interaction is the quiz or question period. In this mode, the
teacher asks multiple choice questions from a workbook and the students punch in what they believe to be the correct answers. After
a brief pause, the teacher enters the correct answer through the
terminal and calls up a display on a CRT mounted in a lectern. She
can easily read a list of the student names, the answers they punch,
and the aggregate number of right and wrong answers. These results
are also recorded in the computer memory so that at the end of the
class the teacher has a hard copy of all student responses as well
as various summary statistics that aid in an assessment of individual
student progress.

In addition to questions posed in the question mode, the teacher
can initiate many informal questions, such as questions related to
procedures and student understanding of formal course content. Thus
the students can be asked to indicate whether they found an exercise
to be too difficult, or whether they have completed an in-class problem that has been assigned. A teacher wishing to receive this kind
of information from students can switch the system to a second,

informal mode. The switch commands the system logic to display student responses for the teacher but not to record these responses in the diagnostics being compiled in the computer memory.

The system also has been designed to allow for student as well as teacher-initiated communications. When the teacher is not asking questions, she can put the system into a third mode. Then when a student hits a button his or her name and an alphanumeric message appear on the teacher's CRT instead of a number for a multiple-choice response. In the current system, the student can send seven messages. These are "I understand" (indicating the student is ready to move on), "slow down," "give an example," "ask a question" (so I can see if I understand), "visuals are unclear," "I don't understand" (so please repeat and review), and "call me" (on the telephone). Each of these messages is printed on the student terminals by the appropriate button. Even though these messages are not a total substitution for the rich array of both verbal and nonverbal cues and messages that a student sends to a teacher in a conventional classroom, they do provide students with a means to communicate with the instructor with respect to how the material is being received. Like the other interactive capabilities, this capacity for student initiated signals was built into the system to make the experience in the electronic classroom as similar as possible to that in the regular classroom setting.

Verbal communication was also possible in later classes. The telephone was not used in the first class, but it was available in subsequent course offerings. A student could call the teacher-- and be heard over the system by the rest of the class should the teacher choose to lead a general discussion. Telephone use was occasional and supplemental in nature.

STANDARD OF COMPARISON

The demonstration has been structured so that the program can be evaluated relative to conventional classroom education. Each day, two teachers instruct the class, and their schedules are rotated so that both classes receive equivalent instruction. On a typical day, the math teacher, for example, instructs her class over cable from 8:30 to 10:00 a.m. and then drives to Spartanburg TEC where she teaches a regular class from 10:30 till noon. The reading teacher has the reverse schedule, instructing the conventional class at TEC until 10:00 a.m., and starting the same reading lesson over cable at 10:30 a.m. The curriculum is highly structured and is geared toward preparing students for the type of question that appears on the GED exam. A large portion of each class is devoted to students working on and discussing the answers to workbook problems and exercises that are similar to those on the

State GED examinations. Both classes were offered 4 days a week
for 15 weeks, a total of 180 hours of instruction. In sum, students
in both cable and tradition classes receive the same amount of instruction in the same subjects from the same teachers on the same
days. This arrangement enables us to compare the effectiveness of
interactive cable education in the home with conventional education
in the classroom while holding many factors constant.

OUTCOME MEASURE

The basic evaluation measure for this study was the Adult
Basic Learning Examination (ABLE), a standardized test of educational
achievement. The measure stood out because it had acceptable
reliability, and came in two versions for a pretest-posttest evaluation. Moreover, ABLE was the only technically sound examination
that had been normed on an adult population. This would enable us
to compare the scores of the Spartanburg cohorts entering the program with those of a national sample of adults with similar educational backgrounds. This was important because it helped us establish the extent to which the skill levels of students in our
experiment were similar to the skills of this type of student in
other parts of the nation. Being able to show that the Spartanburg
students are not atypical permits us to generalize the results of
our study to other adult populations.

NEED FOR PROCESS INFORMATION

But if one is to provide the manager some guidance on the
experiment, it is also necessary to provide insight into the ongoing
communications process. We therefore felt it was necessary and
important to supplement the outcome data with process information
that focused on the instructional dynamics of the cable and conventional classrooms. This process information could serve at least
three different roles in the demonstration.

First, the process information would help us precisely describe
the nature of the cable "classroom" and enable us to understand how
the two-way technology affects the pedagogical process. By comparing
systematic classroom observation data from both the conventional
and cable settings, we could establish how the move to cable had
altered such aspects of instruction as organization of time, the
nature of class activities, and patterns of classroom interaction.

Second, by providing a tool that would help us adapt the organization of content and teaching styles to the potential of the technology, the process information would be useful in project management.
We did not assume that the instructional dynamics we would observe

were necessarily those that had to be. In some cases the data we
collected might suggest that teachers were not taking full advantage of the technology. Our observations would enable us to offer
inservice training that was targeted at particular problem areas.
However, if such training failed to result in a realization of the
potential of the technology, we would have to entertain the notion
that teachers are not able to satisfactorily transfer their teaching
styles to the two-way cable setting. Such a finding would have
important implications for our assessment of the educational potential of the electronic classroom concept.

Third, the data from the observations could be used to help us
understand and explain differences in group outcomes if they occurred. In the event that student achievement in two-way cable class
was poor, we needed to know enough about the dynamics of each
teacher's style, level of interaction, and use of class time to
identify possible remedies. Alternatively, if the cable class
did well relative to the conventional group, it would be beneficial
to document the teaching style that had contributed to this outcome.

The Observation Instrument

These three uses of process information led to our decision
to develop a classroom observation instrument that would permit us
to compare the nature of the cable and conventional classes. The
underlying approach of the instrument was to focus on those areas
where one might expect to find important differences between the
cable and conventional classes and that would either indicate that
the teachers were not making full use of the interactive dimension
of the system or that the conventional and cable classrooms were
proceeding quite differently from one another.

The instrument was designed to focus on two distinct areas of
classroom pedagogy, the observation of the distribution of classroom
activities (the activity record) and observations of the frequency
of classroom interactions (the interaction record). Events falling
in either of the two areas are coded continuously in order to present a moving record of the instructional dynamics in the respective
classes.

The activity record is a system of categories used to code the
number of minutes devoted to different activities. The classroom
activity variables (Figure 1A) are simply descriptors of the types
of activities the students and teachers are engaged in at any given
time. The coder is responsible for tracking both the order in which
those activities take place and the amount of time devoted to each
activity. These variables were intentionally defined so they would
be grouped to distinguish along three different dimensions that we

believed to be of potential importance in comparing the instruction available to students in conventional and electronic settings.

The first dimension contrasts activities that involve subject matter development with activities not related to the instructional focus of the course. For instance, in a math class a period of time may be characterized by the teacher lecturing on how to solve quadratic equations or by students working at their desks on word problems. Both of these activities involve the "substance" of the course--the teaching of mathematics to students. In contrast, time spent on such activities as equipment adjustment or discussion of how the final grades will be determined (i.e., procedural information) is not considered relevant to the substantive purpose of the class.

A second way of comparing activities is by the extent to which they reflect an interaction between the teacher and students. Several of the activity variables have been included that indicate a high degree of student-teacher interaction. For instance, drill and substantive drill are characterized by the teacher repeatedly asking questions of students. Likewise, for classroom substantive discussion to be coded, there must be evidence of students' questions and comments being frequently interspersed during a teacher's presentation. In contrast, other activities such as individual work and teacher substantive presentation are only coded when the activity in the classroom is characterized by little or no interaction between the teacher and the students.

The third dimension for grouping the activity variables is concerned with whether or not the activity represented by the variables can appear in both the cable and conventional settings. Some of the activities such as drill, individual work period or teacher substantive presentation can be coded in either the cable or conventional classroom. Others such as classroom substantive discussion and student dominated activity can only take place in the traditional classroom. Similarly, equipment adjustment only makes sense in the context of the cable classroom. By including these variables, we can determine the extent to which teachers are prevented from translating aspects of their traditional teaching styles to the cable setting by virtue of the fact that the cable setting will not accommodate some activities. Similarly, we can determine how great a role the unpredictable demands of the technology play in contributing to the differing experience of students in the cable and conventional settings.

The second part of the classroom observation instrument is the interaction record. It deals with the nature and number of teacher and student initiated questions. The observer uses this portion of the instrument to keep an ongoing record of the types and frequencies

Teacher Substantive Presentation: The classroom is characterized by the teacher giving a presentation or explanation intended to convey subject matter related information.

Classroom Substantive Discussion: The classroom activity is characterized by verbal interactions between pupils or the teacher and pupils on the subject matter content being presented by the teacher.

Individual Work Period: The classroom activity is characterized by students working individually on assigned work.

Drill and Substantive Drill: The classroom activity is characterized by the teacher asking students narrow questions—that is, questions requiring one or two word replies; questions requiring specific responses. If, during the drill, the teacher either repeatedly asks students to explain how they arrived at their answers or the teacher interrupts the drill to substantially expand the answers, the activity is coded *Substantive Drill*.

Procedural Information: The classroom activity is characterized by the teacher giving and/or students eliciting instructions or information not directly related to subject matter content.

Student Dominated Activity: The classroom activity is characterized by an individual student making a presentation to the class.

Examination or Quiz: The classroom activity is characterized by students taking a quiz or examination on which they will receive a grade or other written evaluative feedback.

Equipment Adjustment: The classroom activity is characterized by efforts to adjust the cable equipment.

Teacher Works w/Subgroup: The classroom activity is characterized by the teacher intentionally separating out a subgroup of students with whom to work. (I.e., the teacher acknowledges that certain students need special help and therefore gives class members something to do so she can work intensively with those students needing special help.)

Non-designated Activity: The class is engaged in activities other than those designated in the other activity categories.

Figure 1A: BRIEF DESCRIPTION OF THE ACTIVITY RECORD CATEGORIES

The CLOSED Category: A closed question is coded when the teacher asks a question about the the subject under study for which the student(s) are to select the answer from a given list of alternatives.

The OPEN Category: An open question is coded when the teacher asks a question about a subject under study for which the student(s) must provide an answer without the benefit of having several response alternatives before them to choose from.

The UNDERSTAND Category: The understand category is used to code a) teacher questions aimed at determining whether students understand or are clear on subject matter related content that has been gone over in class and b) student initiated questions or comments that reflect the extent to which the student understands subject matter related material that has been covered in class.

The ROUTINE Category: The routine category is used to code teacher and student initiated questions that are procedural (as opposed to substantive) in nature.

The RHETORICAL Category: The rhetorical category is used to code questions that are not intended to elicit a response. Rhetorical questions are generally a reflection of a teacher's speaking style.

Figure 1B: BRIEF DESCRIPTIONS OF CATEGORIES IN THE INTERACTION RECORD

of questions asked in the classroom. The observer also notes whether the questions are directed at a specific individual or at the class as a whole. **Figure 1B** presents criteria for coding different interaction variables.

The interaction variables in this second category system have been defined with distant contrasts in mind that are similar to those previously described for the activity variables. On one level, the variables distinguish between those questions that relate to the subject matter content being studied (i.e., the open, closed, and understand questions) from those that deal with class procedures (i.e., the routine questions). On another level, the variables can be used to distinguish those interactions that can take place in either the cable or conventional classroom (e.g., the asking of closed-ended questions or questions of understanding) from those that can only occur in the conventional class (e.g., the asking of an open-ended question, or the occurrence of student-student interaction or teacher-student interaction during an individual work period). In addition, several variable categories require that the coder keep track of those instances where the teacher asks students to explain some aspect of their response. Such an explanation is clearly not possible over the cable system.

MEASURING THE IMPACT ON THE TEACHING PROCESS

The results of the first classroom observation records showed clearly that the pedagogical process over interactive cable differs dramatically from that of the traditional classroom. Based on a sample of over 100 ten-minute observation periods, both of the cable instruction and the conventional classroom, major differences were evident (Table 1). Note that teachers devoted 18.2 percent of the conventional classroom time to substantive classroom discussion. Of course, there was no opportunity for substantive classroom discussion over cable, and thus a central question was how the teachers sought to compensate for that loss of time. It was not by extending lectures, for the amount of time devoted to substantive presentation was relatively constant, 12.0 percent of classroom versus 13.1 percent over cable. The difference instead appeared in the amount of time devoted to individual student work. While the teachers had the students working individually on problems at their desks 29.7 percent of the time in the classroom, the proportion rose to an alarming 59.8 percent with the cable classes.

The interaction records also showed the impact of the technology. The absolute frequency of interactions in the cable classroom dropped off to one-third of the total interactions in the traditional setting. Norming the data to estimate the average number of interactions in a 90-minute class period, we found that the teacher addressed the

Table 1

DISTRIBUTION OF ACTIVITIES AND INTERACTIONS BY CONDITION

Distribution of Activity Time	Classroom	Cable
Teacher Substantive Presentation	12.0%	13.1%
Classroom Substantive Discussion	18.2	0.0
Individual Work	29.7	59.8
Drill	0.0	0.6
Substantive Drill	24.7	17.5
Procedural Information	9.3	7.2
Student-Dominated Activity	0.0	0.0
Exam	0.0	0.0
Equipment Adjustment	0.0	1.0
Teacher Works with Subgroup	1.6	0.0
Nondesignated Activity	4.5	0.8
Total	100.0%	100.0%

Frequency of Interactions per Class	Classroom	Cable
Closed-ended Questions	14.0	17.4
Open-ended Questions	35.2	0.0
Questions about Understanding	47.6	9.8
Procedural Questions	13.8	12.7
Rhetorical Questions	12.9	5.4
Student Dominated Activity	14.8	0.0
Total Frequencies	138.3	45.3

	Classroom	Cable
Total Minutes of Observation	519	494

classroom students with questions 138 times per class, compared to
45 times per cable class. The number of routine procedural questions was not very different, but there was a marked decline in the
number of questions about student comprehension of the material,
as well as a decline in rhetorical questions. While the cable class
was asked slightly more structured, closed-ended questions of the
type the data terminals facilitate, the rise failed to compensate for
the loss of the 47.6 open-ended questions the teachers asked the
average conventional class. And there were no student-initiated
questions or student-to-student interactions over the cable in the
first round.[2]

The shock was that only ten students had enrolled. In the City
of Spartanburg, 62 percent of the adults have not completed their
high school education. The evidence collected in the planning stage
of the project suggested that many adults dropped out because of
lack of transportation, or the need to take care of children at
home. Cable had been assumed to circumvent these barriers. The
first hypothesis was that a program may need time to win acceptance
and we assumed that enrollment would grow.

The first class size compromised our ability to reach conclusions
on the test data, but the results were encouraging. The two classes
had made roughly equivalent gains despite the differences in the
teaching process. Of the ten students that started the cable class,
all were sufficiently motivated to complete the 180 hours of instruction despite the lowered levels of interaction and the absence
of personal contact. Moreover, as measured by the Adult Basic
Learning Examination, their progress was comparable to that achieved
by the 13 of 25 students that completed the full series of conventional classes. There was, of course, the usual problem of interpretation because the students were unavoidably aware of their
uniqueness as participants in the first class of its kind, and
special attention effects may have also played a motivating role.

In this case, the decision was made to keep the second experiment relatively constant. As best we could tell from the ABLE, the
basic concept was effective. Now faced with the danger that our
student samples would be much too small, we decided to repeat the
program. Since the teachers, curriculum, technology, and goals were
the same, we planned to pool the student data if the student numbers

---

[2]Student use of data terminals to signal a request for a change
in the pace of the lecture or for a review were not coded since, among
other problems, it is not clear if that is equivalent to verbal or
non-verbal signals in the average discussion. In the second round
when a telephone was available the students could initiate verbal
questions.

continued to be small. Using the classroom observation data, we
would try to improve teacher performance using the technology, even
though this action would mean that there would be some differences
in the class dynamics between the first and subsequent experiments.

ADJUSTING TEACHER STYLE IN A NEW MEDIA ENVIRONMENT

When the teachers were shown the results of the first round
of observation data, they were surprised by the magnitude of the
differences in the pedagogical process in the cable and traditional
classrooms. In particular, they agreed that far too much of the
cable class time had been spent in individual work while too
little time had been devoted to substantive drill. As they reflected on the reasons for the disparity in their classroom behavior, it became evident that the absence of nonverbal cues from
students in the cable condition was leading them to prolong the work
periods. In both the classroom and cable class, students could
indicate when they had completed their assignment, but in the
classroom the students could then exert substantial, largely nonverbal pressure on the teacher to move on. Thus the teacher was
waiting longer for the slower students in the cable environment.
In addition, the teachers agreed to put more emphasis on drill. The
potential of the technology to involve all of a class in answering
questions was more evident, and the teachers felt that this strength
of the interactive system could be used more fully.

In the fall of 1976, the GED class was offered again along with
other interactive cable programs. The class was again small, and
the only technical change was that the cable students were permitted to use their home telephones to call in during class. Data
were collected for the GED cable class during eleven 90-minute observation periods over the 15-week period. The results showed that
the teachers did adjust their teaching styles to make better use of
the technology.

The data for the spring and fall classes for the math and language teachers illustrates the nature of the changes (see Table 2).
In the spring the mathematics teacher had spent almost two-thirds of
her time with individual work periods on the cable, whereas less
than a third of her classroom time had been devoted to individual
work. In the fall she gave half the cable class time to individual
work, and reallocated the time gained to drill. That shift, coupled
with increased requests that students signal their answers to exercises, doubled the average number of teacher-initiated interactions.
The math teacher asked her second cable class to respond to closed-
ended questions 32.5 times per class, or every three minutes.

Examining the findings for the language class, we see that in
the spring the teacher used classroom substantive discussion and

Table 2

DISTRIBUTION OF ACTIVITIES AND INTERACTIONS BY
TEACHER AND CONDITION

Distribution of Activity Time	Mathematics			English Language		
	Classroom	Cable		Classroom	Cable	
	Spring	Spring	Fall	Spring	Spring	Fall
Teacher Substantive Presentation	16.4%	18.5%	24.3%	0.0%	11.8%	20.9%
Classroom Substantive Discussion	22.7	0.0	0.0	26.5	0.0	0.0
Individual Work	30.7	65.9	49.0	32.3	54.2	42.3
Drill Activities	13.0	9.8	17.5	37.5	27.1	24.6
Procedural Information	6.7	5.8	6.5	3.7	6.9	5.3
Equipment Adjustment	0.0	0.0	1.6	0.0	0.0	0.6
Teacher Works w/ Subgroup	1.7	0.0	0.0	0.0	0.0	0.0
Nondesignated Activity	8.8	0.0	1.1	0.0	0.0	6.3
Total	100.0%	100.0%	100.0%	100.0%	100.0%	100.0%
Frequency of Interaction per Class						
Closed-ended Questions	2.1	5.7	32.5	32.4	35.6	41.6
Open-ended Questions	39.2	0.0	0.0	50.3	0.0	1.4
Questions about Understanding	34.1	19.8	9.8	92.6	3.1	6.8
Procedural Questions	14.5	8.8	24.0	4.6	21.9	25.1
Total Frequencies	89.9	34.3	66.3	179.9	60.6	74.9
Total Minutes of Observation	211	205	423	136	144	333

drill for presenting material in the classroom. This was also
captured in the interaction record, which showed that she was rapidly moving through an average of 180 queries per class. In the
cable class she had used relatively more time for lecture (11.8
percent) and individual work periods (54.2 percent as compared
with 32.3 percent) as well as devoting less time to drill over
the cable. In the spring class she too decreased the amount of time
devoted to individual study by being more self-conscious about
the absence of cues to move on. The time saved in her class was
invested in lecture, increasing that activity to 20.9 percent of
her class time.

Taken as a whole, these findings indicated that classroom
observation data can be useful in helping teachers become aware
of their teaching styles. In addition, the data indicate that
teachers, once aware of the pedagogical characteristics of their
class, are able to alter their teaching behaviors to better reflect
their own notions of how a class should proceed.

SEEKING EXPLANATIONS FOR STUDENT ACHIEVEMENT

The results of the second class again found that students
attending the electronic classroom made satisfactory gains. In
looking at the classroom dynamics from that perspective, several
explanations begin to emerge. Although there is a substantial
loss of class discussion and open-ended questions in the cable
class, the conventional classroom does not spend time in student
dominated discussion, rarely does the teacher walk around the class
to work with individual students, and the students initiate very
few questions. The observation data shows that the teachers in the
conventional class do not take full advantage of the range of
educational arrangements possible in the regular classroom environment, but not available on cable. Thus, the absence of significant
differences in educational achievement may reflect the fact that,
in actuality, the pedagogical differences between the two conditions
may not have been as great as might have been expected.

Another explanation comes from considering the different consequence of a closed-ended question in the two environments. Although these data seem to indicate that students in the cable classroom engage in much less interactive activity than those in the
conventional class, this difference may be more apparent than real.
When a teacher asks a question in a conventional classroom, only
one or two students need to respond. The other students are not
required to focus on the question, choose an answer, and receive
reinforcement. In the cable classroom with home data terminals,
every student in the class answers each question independently,
without guidance or cues from others. Particularly for passive

students, the nature and effect of this structuring of student attention and participation may be a key to the success of the electronic classroom.

RESULTS OF THE SERIAL DESIGN

When a serial design is used, the researcher can make the best of several worlds. It is a prudent strategy which can allow serious restructuring of the project without sacrificing evaluation. When the concept works as it did in Spartanburg, the data gives one confidence to leave well enough alone. But there are other advantages.

The first and more obvious opportunity is that one can pool the data. In all, we offered the GED course in Spartanburg three times. We had 10, 12, and 11 enroll in the course, and fewer took the ABLE posttest. If we had not had several experiments, kept them constant in terms of teachers, curriculum, and technology, and been able to pool the data, the conclusions would have been lame and tentative. The comparison group of students taught by the same teachers, using the same workloads and lectures, adds to our ability to reach a confident conclusion. The students in the electronic classroom learned as much as the students in the conventional class. Using mean percentile ranks on the ABLE as a criterion, we see on Table 3 that when the cable class began, they were weaker students. But they essentially held pace, and were just behind the conventional class on the posttest. Using percent gain as a criterion, they made similar progress. The dropout rate was slightly higher in the cable class, particularly in the last class when serious technical problems plagued the system. Thus there is the possibility that more weak students dropped from the cable class, artificially enhancing the average gains based on the posttest. Even so, that is not likely to be a strong enough factor to alter the basic conclusion: the proportionate gains of the cable were not statistically different from the conventional classroom.

The conclusions of the Spartanburg project support efforts to replicate the electronic classroom for home education. Since the basic communications functions of the system are outbound audio-video and return data, it seems clear that the approach could use broadcast television and telephone return and other technology mixes as alternatives to two-way cable. We know that the teachers can, with limited initial training, adjust their teaching styles to the electronic environment, and something about the problem that will arise when teachers are cut off from visual cues about their class. Such classes are probably equivalent to classroom instruction as it normally exists, even if they cannot provide instruction as good as that in the ideal class.

Table 3

STUDENT ACHIEVEMENT ON ADULT BASIC LEARNING EXAMINATION
(mean percentile ranks)

ABLE	Conventional	Cable Class
Pretest	(n=38)	(n=29)
Vocabulary	48.9	42.2
Reading	56.2	45.6
Computation	44.6	23.6
Problems	40.9	35.6
Spelling	53.5	49.8
Posttest	(n=32)	(n=21)
Vocabulary	55.2	62.0
Reading	68.0	69.5
Computation	71.8	54.1
Problems	60.4	50.6
Spelling	61.7	56.0
Gain (% of increase from pre to post)	(n=21)	(n=21)
Vocabulary	7.4	16.1
Reading	18.3	19.0
Computation	25.3	28.0
Problems	12.7	16.8
Spelling	10.4	5.3

More generally, we found that the use of serial design has
many advantages and would strongly recommend it for future field
work. Serial experiments, with both process and outcome measures,
and a standard of comparison served the combined purposes of the
management of innovation and research on its effects. The broad
conclusion is that the use of serial designs and a more creative
approach to communications system evaluation will increase the
value of research for the manager. That, in turn, creates incentives for more controlled management of the project, enabling
the evaluator to do better research.

THE DEVELOPMENT OF TWO-WAY CABLE TELEVISION:

APPLICATIONS FOR THE COMMUNITY*

Mitchell L. Moss

Graduate School of Public Administration

New York University

INTRODUCTION

Public uses of cable television have been one of the most widely-heralded applications of new communications technologies. Yet the promise of cable television, as a service delivery mechanism, has far exceeded its performance. In part, this is due to the regulatory environment which, until recently, has inhibited the growth of cable systems in the United States. In addition, the allocation of responsibility for developing public uses of cable television has not been considered as an explicit function of either public officials, citizens groups, or cable operators.

The rapid growth of cable television in the United States is now being widely projected as a result of the changing regulatory climate and the development of pay cable.[1] However, cable operators (unlike their counterparts in the telephone company or computer industry) rarely promote the application of broadband communications to the public sector.[2] Therefore, systematic

*This paper is based upon research supported by the National Science Foundation, Division of Advanced Productivity Research and Technology, under Grant No. APR 75-14311 A02. Opinions expressed in the paper are those of the author and do not represent the views or policies of the sponsoring agency. The author expresses his appreciation to Jody Brown, Barbara Felton, Martha Hirst and Robert Warren for their valuable suggestions.

attention must be given to those factors which influence
the development of public applications if the full potential of cable television is to be achieved. This
paper analyzes the process of developing public uses of
two-way cable television by examining the role of public agencies and their clientele in the formulation of
public service applications.

The task of introducing cable television to public
organizations is not a simple one. It reflects the
considerable difficulties which are generally inherent
in technological innovation within the public sector.
Unlike the private sector, where the marketplace rewards improvements in productivity which are achieved
through the application of new technologies, the public
sector has been characterized by an absence of such
direct incentives.[3] The lack of incentives is further
compounded by the existence of distinctive barriers to
the adoption of technological innovations in public
bureaucracies.[4]

The traditional approach to developing public
applications of cable television has been to identify
those services which can be most readily adapted to
broadband communications based upon the criteria of
technological and economic feasibility. The rising cost
of public services and the declining cost of telecommunications systems has enhanced the appeal of interactive cable telecommunications as a means of providing
public services more efficiently than traditional service
delivery mechanisms. Public services provided at the
local level are primarily characterized by their labor
intensive nature. The labor itself, such as a police
officer or teacher in the classroom, is often considered
to be the unit of service. Such labor intensive
services are especially resistant to increases in productivity where personnel expenses are regularly rising
and where the constraints of time and space limit the
capacity of the service unit to increase its output.
Consequently, the use of broadband communications has
been frequently proposed as a means of substituting
technology for labor in the delivery of public services.

Among the services which have been considered for
use in cable systems have been education, in-service
training, polling, information and referral, and monitoring of burglar and fire alarms. Despite the potential benefits to be gained from such applications, remarkably few public uses of cable technology have

actually been developed. Public sector agencies still
have little knowledge or understanding of how telecommunications can serve their goals. Even where a cable
channel is designated for the use of government agencies,
there is relatively little use of the channel for its
intended purpose.

Public bureaucracies, faced with the problems of
meeting federal requirements, managing budget cut-backs,
and putting out day-to-day brush fires, simply do not
perceive cable television as a potential vehicle for
fulfilling their needs. Consequently, there is little
willingness to invest scarce organizational resources
in a new technology which seems far-removed from the
daily rigors of administrative life.

The conceptual model underlying the development of
broadband communications has implicitly assumed that
public agencies can, or should, be the mechanism for
generating public uses of cable television. Given the
lack of information and incentives for technological
innovation and the presence of bureaucratic obstacles
to organizational change, an alternative approach is
useful to consider for the development of public uses
of cable television. Rather than asking how public
agencies can use cable television to provide services,
we might more wisely ask: how can citizens use cable
television to obtain public services?

Such an approach has formed the basis for an experiment in interactive cable television, which was
conducted in Reading, Pennsylvania over a thirty-month
period. The experiment was one of three projects
sponsored by the National Science Foundation in 1975 to
evaluate the costs and benefits of using two-way cable
television in the delivery of public services. Senior
citizens and public agencies were primarily responsible
for developing public service applications of two-way
cable television throughout the Reading experiment.

THE NYU-READING EXPERIMENT

Reading, Pennsylvania is an industrial city of
88,000, located 60 miles northwest of Philadelphia. The
interactive cable system was created for use by both
the senior citizens (who constitute 16% of Reading's
population) and the public agencies that serve them.
The two-way cable system consists of three neighborhood

communication centers which are interconnected. Local
government offices and high schools are also connected
to the system on a regular basis. The experimental
cable system was designed by New York University in
collaboration with the ATC-Berks TV Cable Company,
local service organizations, and senior citizens in
Reading.

From the outset, the design and implementation of
the project were the products of three criteria: the
needs of senior citizens, the technical configuration
of the cable system, and the requirements of evaluative
research. The basic methodology for assessing the
effect of the two-way cable television system employs
treatment and control groups which are surveyed on a
before-and-after basis. Control and experimental groups
have been designated for each of the neighborhood communication centers as well as the homes of approximately 125 senior citizens which were equipped with converters. The evaluative research permits the impact of
the cable system to be analyzed in three contexts: two-
way cable, one-way viewing with telephone call-in, and
those with no access to the system at all.

During the planning phase of the experiment, New
York University was responsible for soliciting the
participation of both senior citizens and service delivery organizations. At the outset of the experiment,
the city government and four public agencies had agreed
to participate in the project. Two agencies provided
space for the neighborhood communication centers. Other
than that, the only requirement for participation was
the willingness of an organization involved in serving
senior citizens to provide the time of its representatives for the two-way programming.

The neighborhood communication centers are equipped
with portable television cameras and monitors which
permit two-way communication among the three centers.
Initially, converters were installed in the private
homes of approximately 125 elderly citizens to allow
them to view the cable programming over their home television sets and to participate by telephone. The positive response by homeviewers to the interactive programming led to the subsequent decision to extend the
programming to the 35,000 local cable-subscribers in
Berks County.

Citizen Participation

A distinctive aspect of the Reading project is that senior citizens participate in virtually all aspects of the two-way cable system from planning to actual production. Programming consists of daily, interactive sessions which originate from the neighborhood communication centers as well as from various remote locations such as the City Hall, the local office of the Social Security Administration, the County Court House, and several high schools. The programs which are transmitted two hours a day, five days a week, are conceived and produced by senior citizens and representatives of local organizations.

Social service programming consists of information exchanges between senior citizens and representatives of social service agencies. The senior citizens, as producers of the programming, have used their control of the two-way system to make demands on other public agencies to participate in the interactive cable system. Rather than having the organizations assume responsibility for the production of programs, senior citizens, acting as both consumers of public services and as the clientele of specific agencies, are responsible for organizing public service programming.

In the Reading cable system, the focus of responsibility and incentives for developing public service programs has been shifted from the service delivery organizations to the senior citizens. By transferring the incentives for the utilization of the two-way cable system to the senior citizens, they can directly articulate their demands to service delivery organizations for specific types of cable programming. Organizations which might have no incentive to use the cable system independently are clearly in a substantially different position when they are responding to a request by an organized clientele group to participate in public service cable-programming.

The results of this process have been impressive. A diversity of public and quasi-public organizations use the interactive system to communicate with senior citizens. More than seventy agencies have participated in the programming. Twenty organizations are regular users of the two-way cable system and fifty have appeared on an occasional basis. Educational institutions account for 15% of the organizational programs, local governmental units comprise 21% of such programming, and

the social-service delivery agencies are responsible
for 49% of the programs.

Having senior citizens and service delivery organizations share responsibility for the delivery of
public services over cable television has influenced
both the scope and type of services provided through
the two-way cable system. Senior citizens have been
able to obtain the services they need from any source
which includes both the public and private sectors as
well as each other. For example, although a number of
seniors expressed a need for information on preparing
wills, the publicly supported legal services were not
statutorily permitted to provide such counsel. As a
result, local attorneys in private practice were sought
and donated their time for a program on the preparation
and execution of wills.

On other occasions, seniors produced programs in
which nursing home operators and the director of a
funeral home conveyed information about the cost and
nature of the services they provide, which, although
private, are still of vital importance to the elderly.
On numerous occasions, senior citizens used the interactive cable system for peer group counseling with such
personal problems as sexual activity, insomnia, and when
to stop driving a car. Often, the existing repertoire
of services provided by public agencies meets only a
portion of the information needs of citizens. The role
of senior citizens in determining the nature of the
interactive cable programming thus led to the provision
of services through cable television which otherwise
were not readily available to the elderly in Reading.

The involvement of the seniors in the planning and
production of the interactive cable programs has also
influenced both the nature of services provided and the
character of organizational participation. The initial
experimental design emphasized evaluation of the impact
of the two-way cable system on the utilization of federally-funded programs such as Medicaid and Food Stamps.
However, the subjects in this experiment also had a
strong role in defining the content of the experiment,
i.e., the programming. Because all senior citizens do
not necessarily believe in the value of such social
welfare services, many middle-class seniors were reluctant to have what came to be regarded as "their" cable
system used on behalf of such programs for the poor.
Like most population groups, senior citizens display
considerable heterogeneity, with regard to social class,

education, income, and preference for public goods and
services.

    In order to respond to the diversity of senior
citizens preferences, the interactive cable system
could not be solely oriented towards social welfare
programs but rather encompassed a broad array of services for the elderly. As a result, numerous local
organizations used the two-way cable system over the
course of the project and became familiar with both
cable television and the needs of the elderly. Public
and quasi-public agencies not primarily concerned with
the aged, participated in the programs on an ad hoc
basis when seniors sought specific information and
services from them.

    By relying on senior citizens to produce cable
programs, public agencies have been able to use the
cable system on an incremental basis, without investing
substantial amounts of staff time and resources in
planning and development activities. Programs can be
initiated on a trial basis and if the senior citizens
and participating agency were satisfied, then a regular
series of programs could be conducted. The typical
pattern of organizational innovation in which an entire
agency is required to adopt a new process or product in
toto was not characteristic of the Reading cable project.
The open-ended nature of the Reading cable system and
the fact that it depended not upon one agency but upon
an aggregation of public agencies encouraged innovation
and risk-taking by service delivery organizations.

                Impact on Public Sector Organizations

    There is no homogeneous set of public services that
can be simply or easily transferred to cable television.
The impact of the interactive cable programming varied
substantially among the different types of public orga-
nications. A structured questionnaire was administered
to all organizations which used the system on a regular
basis. The findings of this survey indicated that the
impact of the interactive cable television system varied
substantially among different types of public organizations.

    The most significant differences were found in the
goals that the programming served for elected versus
appointed officials. Five out of eight (62.5%) of the

elected officials reported that the major effect of the
system was "receiving input" and "allowing interaction,"
whereas only three out of fourteen (21.4%) of the non-
elected officials considered "receiving input" or
"interaction" as a product of the cable system. The
remaining appointed officials split about equally in
identifying the primary effect as an "additional medium,"
"providing publicity," "realizing the potential of
interactive cable television," and "don't know."

Differences between elected and non-elected officials were also found in their response to a question
concerning whether or not the organization's goals were
achieved by two-way cable programming. More than 75%
of the elected officials stated that their goals for
the interactive cable programs were reached while only
43% of the non-elected officials clearly felt that their
goals were achieved. The latter group was not explicitly dissatisfied but rather found it "hard to say"
whether or not their goals had been reached.

Such differences in the perceived functions which
the experimental cable system served reflects the
different roles of elected representatives and appointed
bureaucrats. For the elected politician, communication
with constituents is a basic element of the job; the
process of speaking through two-way television to citizens is, in itself, the product. It provides a means
of "staying in touch" with voters and of demonstrating
responsiveness to the constituency.

For the appointed official, the two-way cable
programming was designed to provide information and to
make referrals about specific social services. The
interactive cable system functioned as a means of
achieving increased service utilization by disseminating
information and responding to specific inquiries about
program requirements and regulations. The process of
communication between client and official was more
narrowly conceived and instrumental than the open-ended
citizen-government interaction where the process of
communicating was regarded as a valuable product in and
of itself.

Further, state and federal social services for the
elderly are characterized by a substantial body of administrative rules which limit the administrative discretion of public bureaucrats. The capacity of appointed
officials to respond to individual problems of such

programs is largely confined to explaining agency procedures and policies. Thus, there is relatively little
flexibility in bureaucratic responses to citizen concerns. Unlike elected officials, whose authority stems
from the individual citizen's vote and for whom the
receipt of citizen input is a recognized and accepted
aspect of the job; the authority of the social service
official is based upon their presumed professional expertise, a knowledge which is acquired independent of
their clientele. Professionals are socialized to perceive their training and skills as the source of their
knowledge and power and are accountable to fellow
bureaucrats, rather than to consumers.

     The impact of such service delivery programming is
primarily reflected in the increased knowledge and
awareness which senior citizens have with regard to
specific social services. Such information allows them
to be more informed and effective consumers but it does
not necessarily produce immediate demand for services.
For example, during the course of the Reading project,
there was extensive cable programming to encourage
senior citizens to participate in the federally-funded
Food Stamps Program, a program which the elderly have
traditionally resisted because of the stigma related to
such social welfare services.

     Although there have been no significant changes in
the utilization of food stamps since the project began,
there have been significant differences in the percentage of elderly who have considered using food stamps.
Such findings suggest that the use of service information obtained through the cable programming may require
a long period of time to emerge and even then may only
occur when particular needs arise. Furthermore, the
information about social services may simply represent
broadened options for the senior citizen which enhance
the individual's perception of the alternative "coping"
mechanisms available to him and act as a safety valve,
even if unused.

     For the social service administrator, the impact of
such cable programming is not immediately apparent since
any increases in service utilization are diffused and
difficult to identify. Thus, the provision of specific
social services provides far less feedback to the
appointed official. Further, it is less easily integrated into on-going organizational functions than the
citizen-government interaction and therefore requires a

a more substantial investment of the agency's resources.

By contrast, local politicians were able to use
the interactive cable system in a way that effectively
complemented their jobs. When asked about such neighborhood problems as street repair or the installation
of a stop sign, they either initiated steps that satisfied citizen demands or explained why the requested
action could not be taken. In matters where elected
officials were not familiar with specific issues, they
would attempt to get the necessary information and
convey the answer during their next program. When
complaints involved state or federal policy, local politicians were, not unexpectedly, more than willing to
identify another unit of government as the source of
the problem.

The teleconferences between senior citizens and
elected officials also served to heighten public officials' awareness of, and sensitivity to, the needs and
interests of the elderly. Information could be directly conveyed from citizens to the executives of municipal
departments without going through the filtering process
that normally occurs in bureaucratic settings. The
elected official could thus obtain accurate and regular
information on citizen concerns without leaving their
offices. Moreover, the two-way cable system personalized the interaction between the citizens and elected
official.

In the Reading cable system it was not uncommon for
citizens to address the mayor by his first name and for
the mayor to respond in kind. One homeviewer called in
so frequently from her home telephone that the mayor
came to recognize her by voice alone.

The difference in the perceived value of the two-
way cable system between the elected and appointed
officials suggests that direct accountability to the
electorate, as well as power to effect change, play an
important role in the impact of two-way cable programming. Thus, if interactive cable television is to be
used in the provision of social services, it is clear
that attention must be given to the character of the
relationship between the participating citizens and the
public officials, as well as to the nature of the service
function itself.

The findings of the Reading cable project indicate
that public agencies utilize interactive telecommunications for a variety of reasons. Surprisingly, production efficiency in the provision of goods and services
is rarely the basis for organizational participation
in the cable system. Public sector organizations
obtain a variety of benefits from participation in the
two-way cable system. Certain agencies regard the
interactive cable system as an innovative means of
providing outreach services which are otherwise conducted through staff visits to individuals and community
centers within the urban area. The municipal and county
governments view the cable as a mechanism for obtaining
citizen feedback on public policies and programs, while
other service agencies utilize the two-way programming
to disseminate information to clientele who are traditionally hard to reach. And, for some organizations,
the system serves as a tool for gaining exposure and
enhancing their status in the community.

Two-thirds of the public officials surveyed stated
that they participated in the cable programming because
it either provided a means of "giving information" or
"receiving input." The programs were frequently cable-
cast directly from the participating official's office
yet no mention was made of reduced travel or time costs.
Although such administrative economies may have been
achieved through the use of the system, they must not
have been considered to be sufficiently noteworthy to
the respondents.

Thus, the assumption that saving time or money
would enhance the appeal of new technologies for public
agencies was not found to be a relevant factor in the
development of public service applications on the
Reading system. Willingness to respond to the specific
demands of citizens, rather than to independently generate applications, was clearly the critical factor here.
This suggests that emphasis be placed on the role of
consumers rather than on the benefits to public agencies,
to facilitate the development of future applications.

Public choice theorists distinguish between supply
and demand considerations in the provision of public
services. The supply function involves the production
and management of services while the demand function
relates to the articulation of citizen preferences for
public goods and services.[5] Public administration has
traditionally emphasized techniques to improve "economy

and efficiency" in the management of public agencies
with relatively little attention being given to exploring ways of improving the expression of consumer
preferences for public goods and services. Such technological innovations as electronic data processing
and computer systems have commonly been designed to
enhance administrative control over large-scale public
organizations.

The experience in Reading indicates that interactive cable television can provide a powerful means
for citizens to articulate their demands to public
officials. Although citizen feedback is not generally
considered to be an element in the production of functional services, the elected officials clearly regard
this component of the cable programming to be useful
and valuable. In contrast to most technological innovations, the 'citizen as producer of public services'
as well as 'the public official as producer of public
services' can be effectively served by two-way cable
television.

The Community and the Cable System

The interactive cable system has generated a wide
array of effects on individuals as well as organizations
in Reading. Senior citizens are significant consumers
of public services. Moreover, they face substantial
problems of limited mobility and access to services.
The two-way cable system has emerged as a means for the
elderly to communicate with each other as well as with
public agencies.

A sample survey of the 19,000 subscribers in the
City of Reading was conducted in April 1977, six months
after the programming had been made available to the
entire community. Almost half of the cable-subscribing
households included at least one person age 60 or over.
Forty-five percent of these households with senior
citizens reported that someone in their household had
watched the interactive cable programming. Twenty-two
percent of the households including elderly residents
reported watching the programming regularly (two to
five times per week) and 8% reported watching the programming approximately once a week. Thus, almost 30%
of the cable-subscribing households with senior citizens
watch the interactive programming at least once a week.

The two-way cable system serves important social
and political functions by reducing isolation and providing a forum for the elderly to participate in local
governmental processes. Senior citizens can participate directly in local affairs without encountering
the time or travel costs of visiting city hall or the
institutional and psychological constraints of participating in formal public meetings. The abundance of
public service programs has created the equivalent of
a "one-stop service center" in which a broad range of
information and referral services are available over
interactive cable rather than in a conventional office
environment. What has clearly emerged is the development of a communications infrastructure that serves a
diversity of needs.

The two-way cable system in Reading was designed
to provide a specific set of goods and services. More
accurately, it has functioned as a community communications system. For senior citizens and public officials,
the primary value of the interactive cable system lies
in its ability to enhance local community communications.
Just as public officials valued the two-way cable system
for its communication rather than specific service
delivery functions, so did senior citizens find the
interactive capacity of the cable system to be its most
valuable attribute. Despite an abundance of social
service programs, senior citizens demonstrated a strong
preference for participant-oriented entertainment and
discussion programs by their attendance at the centers.

For a city such as Reading without its own broadcast television station, the interactive cable system
has emerged as a vital means of electronics communications between senior citizens and local government. In
an era when telecommunications has most frequently been
used to overcome territorial boundaries and thereby
transcend local values, the two-way cable system in
Reading demonstrates the potential for broadband communications to reinforce community consciousness and to
reflect the distinct preferences and priorities of an
age-based sub-group of the population. The Reading
cable system has fostered a sense of community at the
local level not by relying on elaborate technology
but by relying upon individuals and organizations
to design programs to meet their needs and interests.

Perhaps the most pervasive set of effects had been
on the social and psychological health of the elderly

and in their relationships with local officials and the
community-at-large. The two-way cable programming
serves as an important vehicle for social interaction
and participation in community affairs. Through their
involvement in the cable system, senior citizens have
developed a high degree of personal efficacy. Their
access to, and control over, the public service programming has enhanced their visibility in the community
and provided a forum for communicating in a collective
setting.

CONCLUSION

The experience of the Reading cable project highlights the importance of drawing upon citizens as well
as service delivery organizations in the development
of public service uses of cable television. Rather
than serving one single user, such as an educational
institution or hospital, senior citizens in Reading
have created a system that aggregates a mix of organizational users which use cable on behalf of the elderly.

The effectiveness of the Reading interaction cable
system is most visibly reflected in the support which
the community has given to Berks Community Television,
the local non-profit corporation which was created to
operate the system at the termination of the experimental phase. Berks Community Television has raised funds
from federal, state, and local sources and has expanded
the programming to include evening as well as daytime
sessions.

It has continued the institutional framework for
developing public applications in which senior citizens
are responsible for generating programs in collaboration
with service delivery organizations. The success of the
two-way cable system has led the city government to seek
out Berks Community Television's counsel and assistance
in designing new mechanisms for obtaining public partic-
icipation in municipal budget processes. The city and
other public agencies allocate funds to pay for their
use of the interactive cable system.

Rather than rely on a single governmental unit to
develop cable programs, numerous public and quasi-public
agencies use the two-way cable system. Although no one
agency requires an extensive amount of program time, the
combination of organizational participants makes

substantial use of the interactive cable system. Thus,
it is not necessary for any one entity to commit a
large portion of its activities to the cable system.
As a result, the traditional barriers to technological
innovation at the local level have not arisen in Reading.

The growth of the Reading cable system indicates
that if the full potential of broadband communications
is to be realized, public policies must recognize the
importance of citizens groups as a vehicle for developing public uses. It demonstrates the need to go beyond
reliance on public sector bureaucracies as the source
of public applications and to incorporate consumer
groups in the process of technological innovation.

REFERENCES

1. Les Brown, "Cable TV, After Lagging for Four Years,
    is on the Move Again in Area," New York Times,
    October 24, 1977; Douglas Davis, "Let's Hear It for
    the Cable," Newsweek, November 21, 1977.

2. Peg Kay, Social Services and Cable TV (Washington,
    D.C.: U.S. Government Printing Office, NSF/RA-
    760161, July, 1976).

3. Richard D. Bingham, The Adoption of Innovation by
    Local Government (Massachusetts: Lexington Books,
    D.C. Heath and Company, 1976); Irwin Feller,
    "Diffusion Milieus as a Focus of Research on Innova-
    tion in the Public Sector," Policy Sciences, vol. 8,
    no. 1, (March 1977); David J. Roessner, "Incentives
    to Innovate in Public and Private Organizations:
    Implications for Public Policy," in Administration
    and Society (Forthcoming, 1977); Everett M. Rogers,
    Diffusion of Innovations (New York: The Free Press,
    1962); and Lloyd A. Rowe and William B. Boise, "Or-
    ganizational Innovation: Current Research and
    Evolving Concepts," Public Administration Review,
    vol. 34, no. 3, (May/June, 1974).

4. George W. Downs, Jr. and Lawrence B. Mohr, "Concep-
    tual Issues in the Study of Innovation," Administra-
    tive Science Quarterly, vol. 21, no. 4, (1976); The
    Urban Institute, ed., The Struggle to Bring Techno-
    logy to Cities (Washington, D.C.: The Urban Insti-
    tute, 1971);   Gerald Zaltman, Robert Duncan, and
    Jonny Holbek, eds., Innovations and Organizations
    (New York: John Wiley & Sons, 1973; and

Robert K. Yin, Karen A. Heald, Mary E. Vogel, Patricia D. Fleischaver, and Bruce C. Viadeck, <u>A Review of Case Studies of Technological Innovations
in State and Local Services</u> (Santa Monica: California: The Rand Corporation, R-1870-NSF, February
1976).

5. Robert L. Bish and Robert Warren, "Scale and
    Monopoly Problems in Urban Government Services,"
    <u>Urban Affairs Quarterly</u>, vol. 8 (September 1972);
    and Robert L. Bish and Vincent Ostrom, <u>Understand-
    ing Urban Government</u> (Washington, D.C.: American
    Enterprise Institute for Public Policy Research,
    1973).

BEYOND STATISTICS

Red Burns

Alternate Media Center, School of the Arts,
New York University

"There are two principles inherent in the
very nature of things - the spirit of change
and the spirit of conservatism. There can
be nothing real without both - mere conservatism without change cannot conserve - mere
change without conservatism is a passage from
nothing to nothing."
- Alfred North Whitehead

INTRODUCTION

Senior citizens in Reading, Pennsylvania are programming,
operating and financing their own two-way TV system using the
local cable television operation. For two hours each weekday,
local people can turn to Channel Three to participate in the only
TV system of its kind in the world. Senior citizens are not only
responsible for operating the system, they are pioneering the exploration of a medium which is radically different from broadcast
television. The results that have been achieved and the interest
that the programs have created are in striking contrast to
society's stereotypes of older people.

Creating change is an art: measuring results is a science.
Much attention is rightly paid to evaluating the potential of the
new telecommunications technology by social scientists and planners. Such evaluation requires experimentation, but curiously
the process of implementing experiments receives little attention
in comparison to that given to research design and data analysis
in the United States. Our experience in Reading suggests that
attempts should be made to correct the imbalance. The responsi-

bility for concept development and implementation was distinct
from that for research design and evaluation in the NYU-Reading
Consortium project. This paper reports on some of the issues involved in the concept development and implementation effort which
we believe to have had an important impact on the form and function
of the experimental system.

BACKGROUND

In 1974 the National Science Foundation issued a request for
proposals; the subject of concern was experiments in the delivery
of social services using the two-way capability of urban cable
television systems. The Alternate Media Center, School of the
Arts, New York University had spent the three preceding years
working to develop cable TV systems with communities across the
country. This experience led us to believe that if tools, support,
time, and encouragement are provided people will design a workable
and humane system for themselves. We were excited. The NSF's
solicitation might give us the chance to test and demonstrate this
principle with a new medium: two way television. What was more,
we could be ambitious; since an experiment was called for, we had
freedom to fail. And, the opportunity to couple our activities
with rigorous evaluation was a welcome challenge.

Cable television had been the subject of many technological
and economic studies. Yet, none seemed to have seriously considered the possibility of software being developed by people who
were not media professionals. And, since the technology was developing rapidly we felt that it was critical for the people who
were to be served by a communication system to be involved in its
design and use if they were to realize its potential.

While cable television in the U.S. currently serves about 18%
of the 72 million TV homes, it has developed primarily as a distribution system for existing television stations. The promise
of multiple channels for local programming has been "stuck" in the
existing model of over-the-air television programming. Public
access cable television has had limited success when its objective
has been to make TV rather than to develop a community information
system. Therefore, we set out to investigate the structure necessary to create "locally-produced software".

The Alternate Media Center joined forces with the Graduate
School of Public Administration at New York University. We, at
the Alternate Media Center, took on the tasks of designing and implementing the application in collaboration with our research
colleagues who were responsible for the evaluation.

It was natural for the Alternate Media Center to choose to
work in Reading. We had established our first community video

workshop there in 1972. We already had relationships with community people, a number of public agencies, and the Reading cable system which has a strong commitment to its community. It was natural, too, to work with old people. The city has a higher than average proportion of senior citizens. Moreover, the NSF mandate was to explore the use of the medium for the delivery of social services and they are heavy consumers of social services.

The initial partnership, between the School of the Arts and the Graduate School of Public Administration at N.Y.U., was enlarged to include local partners: the City of Reading, the Reading Housing Authority, Berks Senior Citizens Council, and ATC-Berks Cable TV Co. This group as a whole comprised the NYU-Reading Consortium. During the proposal stage we coordinated the collaborative process of creating the basic design for a system which would be used by the two complementary client groups -- the old people and the agencies that serve them.

DESIGN

The plan was to set up three Neighborhood Communication Centers (NCC's) and several remote origination sites such as: City Hall, the Social Security office, the County Court House, schools, and other community facilities. The three NCC's were to be in use during all sessions. The remote sites would become part of the network as needed, generally once a week; they would use mobile terminal equipment which could be easily moved.

The locations for the NCC's were carefully chosen. We accepted the Reading Housing Authority's offer of space in a garden apartment complex -- Hensler Homes -- where elderly people made up 25% of the population and in a high-rise apartment exclusively for elderly people -- Kennedy Towers. The third site, Horizon Center, was space offered by Berks Senior Citizens Council in their new community building.

Comfort was regarded as important. The spaces we decided to use were in regular use for a variety of other purposes; they were not dedicated TV studios. There are trade-offs in this approach. If the space has been used for particular activities, there can be resistance to changing that space albeit for only a few hours a day. However, we believed that the important factor was that people be comfortable and familiar with the space. We wanted to encourage discussion and exchange, not TV production.

The technology design was intentionally simple and relied on inexpensive black and white equipment. This would make it fairly easy to use the existing cable facility and would allow us to meet the cost constraint imposed by NSF (i.e., no more than 10% of the budget should be spent on hardware). Equally important was the fact it would not mystify the users.

Each location would be able to originate and receive video and audio signals. There would be several microphones at each center, each of them live throughout. Everyone on the system could see and hear each other. Spontaneity was to be the rule rather than the exception. All participants at a NCC would share a single camera positioned directly beside the monitor. Since the system was conceived as a single entity, one of the outgoing pictures from a particular site was selected at the head end for return to all of the centers; alternately two of them could be combined for display on a split screen. The home viewers could respond by telephone and their voice was patched in at the head end so everyone could hear.

It was planned that the project would develop in five stages:

       8 months - start up, May through December 1975
       6 months - programming, January through June 1976
       2 months - summer hiatus to evaluate programming,
                   July through August 1976
       6 months - programming, September 1976 through
                   February 1977
       8 months - further data collection, analysis and
                   report writing, March through
                   October 1977

There were to be 12 months of actual programming. It was to be a closed-circuit system; approximately 125 homes would be able to see and hear the interactions between the three or more centers, and to join the discussions by telephone.

Two major changes were made subsequently. We did not interrupt programming during the summer to avoid losing momentum. The second change was more critical. Our experimental programming was made available to the full cable-subscribing community in late September 1976. Data from the first home-viewer survey had shown that many people were interested and participating in the programs. Also, the implementation team was receiving an increasing amount of feedback from the home viewers in the form of unsolicited phone calls and letters which showed enthusiasm, support, and interest in the system. Therefore, the decision was made, in collaboration with our colleagues, to revise the research timetable and extend the system.

Although these changes were unanticipated at the design stage, it was always clear that the processes of design and implementation would necessarily be intertwined. The initial design was intended

to provide a minimum structure for getting started. By the time
that the initial design was complete, our local partners shared our
interest and enthusiasm.

IMPLEMENTATION

When funding was approved, we arrived in Reading with high
hopes. We had worked there before, we had established relationships, we had a carefully planned design, we had considerable experience with telecommunications and somehow, we felt that by some
sort of magic the system would emerge. Yet, the responsibility was
a heavy one: we were about to create change in a community, but
the community had not asked for it. The NSF had, but that agency
was far away in Washington. Could we successfully translate our
rhetoric about community-based information systems into action?
How could all the pieces be fitted together with so many actors who
had different orientations, individual users, agencies, social
scientists, and implementers? What was inherent in the communications system that might organize people toward some common goal?

Technology by itself is an idiot. It takes people with spirit
and involvement to make it work. Clearly, one of the first things
to be done was to establish the value of the communications system
to all of its users once the overall design was agreed upon at the
broadest level. This was easier said than done, for at that stage
we could only hypothesize about the properties of the medium.

Two-way television, we rightly thought, would allow us to
escape from the paradigm of one-way television. For service delivery, the latter would have implied information distributed on a
one-to-many basis from a central source, the assumption being that
people would want this information and act on it. There is an old
conundrum about the tree falling in the forest -- if there is no
one to hear it, is there a sound? Two-way television would, on the
other hand, allow us to concentrate on the exchange of information.
It would also allow us to explore how people of very different
status might meet more equally in society. We were all conscious
of the forbidding environment of the public official's office or
of the TV studio. What philosophy of use might allow the technology to remove these barriers?

Economic restrictions had pointed toward bringing people together to use whatever the system would be, rather than toward
separating users in their individual homes. Even more important,
however, this would replace isolation with socialization. We deliberately set out to use the system as a socializing force.

From the outset, our belief was that the project should leave
something behind. If the people of Reading were willing to work
with us on an experimental research project we had a responsibility

to them which went beyond the research goals. The system should be valuable to them: our goal, as implementers, should be the creation of a system which would continue after we had left.

The people in Reading were no different from people anywhere else. They were accustomed to over-the-air TV: a slick medium. Agencies were used to giving information to people and TV viewers were used to their passive role. The models were firmly set in people's minds and there was no alternative metaphor. Working with both sets of clients to overcome the barriers imposed by these preconceptions absorbed a great deal of time.

But how? You can't walk into a community and say, "Here is a communications system. Use it creatively." People don't know what to do. Nor can you walk in and say, "Here is a communications system and this is what you are going to do with it." An approach was needed with sufficient flexibility that changes could easily be incorporated as people and agencies came to grips with identifying their needs and aspirations.

First and foremost, an atmosphere of trust between the implementers, researchers, and users had to be created. A great deal of time was spent listening to people. Staff had to be recruited and technical training programs set up. The entire staff was recruited from Reading; several were senior citizens. With a single exception, no one had been trained in video. The main qualities we looked for were openness, warmth, and interest in working with older people.

At the same time structures had to be created for program segments. We used this term to describe ways of using the medium for fixed periods of time. The contrast with conventional broadcast television programs is considerable. Although the current view holds that the problems are technological, economic, and concerned with software production, we believed that they lay primarily in creating new forms in which to use a new medium.

Since the approach was to allow senior citizens to develop their own talents, they were increasingly engaged in the decision making and program production. However, this was difficult to a-chieve. Many times the staff of locally trained, non-professional media people got in the way because they too wanted to "be creative". In addition, we had a commitment to the NSF to provide programming of a specific nature on a variety of social service subjects. What has emerged with trial and error is a system in which older people are the producers and staff people assist. It would have been expedient to get programming on the cable as efficiently and coherently as possible. But if the system was truly to reflect its users they had to be incorporated slowly but surely into the decision making and production.

The key to programming was interaction and exchange. Initially, however, people wanted to "make television". That was sexy. Exploring, exchanging, talking back and forth, and learning were not parts of their image of television.

We started with five hours a week of programming. Programs generally lasted 15 to 30 minutes. All were live; sometimes short videotape inserts, prepared under the producer's direction, were used. No single NCC was chosen as an origination point; different programs came from different centers. Half-inch video tape was successfully used for information spots. Seniors, working with staff, produced witty, serious, minute-long spots that gave specific social service information.

It was not long before the original five hours of programming a week were expanded to ten hours and our ability to accommodate all the ideas people offered became a major concern. Problems did not arise so much in production of "software", as in creating a system to deal with the creativity, energy, and imagination of the users. Small units for program decision-making were created. Each center had a committee open to everyone. These committees fed ideas to a monthly Program Committee. This overall committee consisted of representatives from each of these small committees, the board member responsible for programming, and members of the staff. One staff member, a senior from Reading who recently retired from the County Board of Public Assistance, was responsible for social service programming. The overall committee coordinated and scheduled programs.

Producing a program meant that an individual producer was required to contact guests, research the subject, set out the parameters of the discussion, and then act as host. Given the flexibility of the technological design, with many microphones available in each center and the opportunity to use the video to show someone who wished to interrupt, spontaneous interaction was a relatively easy matter. The host's job was to encourage discussion and to keep order.

For a quiz show, the producer had to line up the questions and answers as well as host the show. When videotapes were involved, there was additional pre-production planning. One of the programs on nursing homes used tape because the producer wanted people to "see" the insides of some of them. For "The Changing Face of Reading", one of the senior citizens who had been a volunteer fireman produced weekly tapes on what was happening in Reading (such a program had been requested by someone who was homebound). On one occasion extensive lighting was required and the producer went to his friends at the fire department to "borrow" ladders and lights.

Usually, tape was accompanied by live audio. Often partici-

pants commented along with the producer/host. For a program involving citizen-government interaction, the producer had to contact the elected official, confirm the time of the session, and
chair the discussion. The producer and participants have created
such a lively exchange that the rather stuffy category "citizen-
government interaction" provides Reading people with contact and
familiarity. Mixed with such programs were others focussing upon
the provision of social service information -- on Medicare, Medicaid, food stamps, and Social Security entitlements and regulations.

As styles developed people frequently disagreed about how to
"use the system". When folks waved back and forth to each other,
when they saw old friends for the first time in many years and used
the system to socialize, there were hard words from others that this
was trivializing the system. The disagreements were difficult,
sometimes discouraging, but in retrospect these initially controversial behaviors can be seen as the very strength of the system.

It was indeed true that if people were comfortable they would
get more out of their involvement, and it was also true that if
activity consisted only of people waving back and forth the system
might mot be used seriously enough to warrant the resources that
were being put into the project. We made a strong argument in
favor of the initial socializing: it helped people to feel comfortable and thus to lose their early inhibitions about the technology. Gradually, a time was set aside at the end of programming
each day for "Party Line" where such exchanges could take place,
while a discussion with the Mayor aobut city legislation required
the system to be used more seriously. Yet, without the easy
connection users had made to the system itself, the latter connection for purposeful business could not have been accomplished so
easily.

Throughout the implementation process we were keenly aware of
the need for there to be a continuing increase in user involvement.
It took time for enough people to become involved and provide the
support necessary to sustain the system. And as they became more
involved understanding grew, which made it easier to reach consensus. Initially, some people were reluctant to participate. They
stood to the side, but gradually, over time, they began to participate.

This approach seemed to require us to spend a very great deal
of time at meetings on small problems. Each person's concern was
regarded as important. We made many mistakes and tried not to make
the same one twice. We did not always succeed, but an intangible
spirit of cooperation surrounded us and came to our rescue. Indeed our acknowledged vulnerability was an ally; we needed help, we
asked for it and we received it.

SUBSEQUENT DEVELOPMENTS

The system grew organically throughout the implementation process. One example arose from a successful experiment with a microwave hook-up to a Kutztown high school about 20 miles away. The school arranged to have the local Kutztown cable operator provide them with origination capability to send a signal to the head end of the Kutztown College system so they could interconnect by microwave with the Reading group in the weekly program "Is There a Generation Gap?" which allows senior citizens and high school students to exchange views on controversial topics of current concern. This initial exchange led to a regular program in Kutztown between the seniors and the high school.

Working together with so many different people had taken up a great deal of time. We had been operating city-wide for only five months; yet, according to the schedule the implementation process should have been completed.

In January 1976, when actual programming began, a local Community Policy Board was formed to oversee but not to manage the system. Individual members were appointed by the chairman to take responsibility for specific jobs: finances (fund raising and operations), programming, legal matters, etc. While we were in Reading we were accountable to our sponsor, the NSF, and perforce were ultimately responsible for all decisions. By the time operations were turned over to the Community Policy Board at the end of the implementation period, we felt confident that there was sufficient interest, motivation, and ability in the community to manage the system independently.

Our concern now was whether the system had been in place long enough to be viable. The value of the system had to be expressed by the people in Reading. Money had to be raised to cover costs of $2,000 a week. For a small community that is a great deal of money. At the time of this writing, July 1977, the system is flourishing. The main financial support has come from within the community itself; small additional grants have been made by state agencies. The Board is now working to secure long-term funding and hopes for assistance from federal agencies such as the Administration on Ageing in Washington.

Our relationship to the Reading project has become that of an informal consultant and friend. The entire operation is locally run. It is continuing to grow. Since our departure, a six-point hook up was arranged for an energy teleconference. Community hospitals and local arts groups which were not part of the original design have become involved. Programming is replayed at night. Programming is being developed by more and more producers, all senior citizens. The initial organizational structure that in-

cluded programming committees at each of the NCC's and one overall
committee for scheduling has evolved further. Producers have regular meetings. They share ideas and resources. The Policy Board
meets monthly. Its members now include: program producers, senior
citizens, the former Mayor, a banker, a lawyer, the County Commiss-
ioner, the Mayor, representatives of educational organizations, the
cable company, and private business. Collectively they are responsible for what is now called Berks Community Television.

CONCLUSIONS

Currently the research activities that complemented our work
are drawing to a close and a final report is being prepared. It
would be inappropriate in this paper to draw conclusions about the
costs and benefits of such applications of two-way television.
Nevertheless, certain conclusions about the nature of the medium
and the process of implementing community-based projects are
suggested.

The broadcast TV paradigm of win big and lose big in an established market place is antithetical to the development of a system
which allows a community to identify itself. The two-way medium is
radically different from conventional television and must be treated
differently if advantage is to be taken of its distinctive properties. Also, it is quite different from one-way television enhanced
by digital response.

The use of interactive channels between separated groups of
people who are not media professionals can provide an exciting alternative to the usual TV software. By allowing the use of multi-
purpose spaces to which people are already accustomed, systems like
that in Reading may overcome barriers to communication imposed by
differences in status and forbidding official environments. And
their use for entertainment and socializing can reinforce the potential for a transfer of information.

Our experience shows that if full advantage is taken of the
interactive properties of the medium, then software production presents no serious problem. But care must be taken to develop an
organizational structure for group decision-making about future programs, their scheduling, and so on. This activity is expensive in
time. When the project started we understood much less about the
medium we were to work with. We needed to develop the system along
with our local partners so that we could all learn and adjust as we
went along.

Our objectives were distinct from, though consistent with,
those of our research colleagues and NSF. Our main concern was to
insure that by the time we left there would be a system in place

which the local community would want, and be able to keep going. Only on this basis could we justify asking the community for the energy and commitment we knew we would need if the potential of two-way cable television was to be truly tested. Our second concern was to explore the distinctive properties of a promising new medium.

Throughout the project we were confronted with conflicting responsibilities -- to different senior citizens, to local institutions, to members of the local staff, and to our research colleagues. Tension levels were frequently high and much time had to be spent on exploring alternatives and resolving conflicts. If we had not invested time and energy in building a strong base of trust and good will we would not have gotten far. The fact that this system, which was designed as an experiment, has become an integral part of the community suggests that we were not misguided in our commitment to community involvement in the implementation process.

In retrospect we were much too optimistic, limiting ourselves to such a short timespan; everything seemed to take several times longer than we had estimated. We were naturally interested in being successful, but had to guard against expedient methods which might give the illusion of quicker success. (A good example was to avoid having early programs prepared for, rather than by, the senior citizens.) We now believe that three years is the minimum amount of time for a major innovation to become rooted in a community.

Although it was obvious that one of the keys to success would be community involvement, when we arrived it could hardly be said that the old people in Reading formed an organized community. At the outset we invited the former Mayor, a senior himself, to set up the Community Policy Board. The Board was essential to the later total transfer of operations to the senior community, but it did little directly to form that community. This was due to the interactive television system itself.

One of the most challenging aspects of the project was to design a process by which an emerging community could itself design a system -- a system which would accommodate a very wide variety of different people. The system needed to allow people to enter easily at different points and to contriubte in ways consistent with their differing desires and abilities. Our most poignant conclusion is that as far as Berks Community Television is concerned, we are now redundant.

ACKNOWLEDGEMENTS

This paper draws on the work of the many people in Reading and colleagues at New York University who collaborated on the project.

It also draws freely on their experiences and ideas. These contributions are acknowledged with gratitude.

The project was supported by the National Science Foundation,
Division of Advanced Productivity Research and Technology.

The generous support of the John and Mary R. Markle Foundation
for the Alternate Media Center paved the way for the project.

DISCUSSION OF PAPERS BY DANIEL, MENDENHALL, ZORKOCZY, LUCAS AND MOSS

Discussant and moderator: Percy Tannenbaum

The Chairman opened the discussion with comments upon the importance of non-verbal influences upon learning. He then posed two questions:

1. Which public are we serving?
2. Is it desirable to encourage people to stay at home?

He pointed out that the situations described in the studies related to special populations. Shulman said that the special populations (the physically disabled) in question may have to stay at one location such as at home anyway. Wish thought that people did not want to stay at home all the time but that part-time working from home was a desirable compromise.

Tannenbaum was concerned for the lack of vicarious learning in remote situations. Lucas pointed out that skilled teachers can overcome this and that interaction can take many forms; in most cases it seems reasonable to assume that social good comes from increased interaction.

The discussion then moved to the general advantages of tele-communications. One speaker emphasized the view that telecommunications can provide a wider choice for students; another the value of long-range communications which permit one to draw widely scattered specialists into the traditional classroom. Zorkoczy felt that distance education systems must be dynamic and ready to experiment. They should allow people to choose the learning medium which suits them best.

The discussion then centerd on the motivation behind remote teaching experiments. Wish felt that the aim should be to look for need or social benefit and supply that need. Mandelbaum distinguished two strategies. One stems from a need or preference to pick off marginal cases, reducing the pressure for change on the establishment. The other locates settings where pressure on the system can be increased so as to generate change--as in the Open University. It depends whether the motivation is to be an agent of change.

Brownstein highlighted the difference between the Open University situation and the other studies which were experiments. The latter provided poor contexts for agents of change.

Burns felt that the Reading experiment has a proven socializing effect. Moss suggested that research can be conducted on both
the process of technological innovation and the impact of technology
on social processes. Tannenbaum said that often telecommunications
advocates were looking for a place to light instead of focusing on a
recognized need. Shinn argued that the people who have the need do
not have the knowledge to solve it.

The cost-effectiveness of telecommunications systems was then
discussed. Brownstein pointed out that while the cost of services
was increasing the cost of telecommunications was decreasing. It
was widely felt that decisions to use telecommunications were made
for political not economic reasons.

Mandelbaum maintained that sales were the driving force behind
technical development. This force might work in a direction
opposite to that suggested by the research. Bernemyr asked if there
was a danger of research on demand falling into the trap of saying
that new expensive innovations had no future. He cited the parallel
of colour T.V.

Tannenbaum thought too much emphasis in research was put on
the 'hindsight' of users, whose attitudes may often be biased. More
emphasis should be placed on need.

Section Four

INFORMATION SERVICES

The papers in this section describe work in the field of
Scientific and Technical Information (STI). First Carole Ganz and
Joel Goldhar review research findings relating to the behavior of
users of STI and draw conclusions of particular relevance to those
conducting research on new telecommunication services. They emphasize the need to be aware that changes in the use of one channel of
communication will lead to change in the use of other channels and
criticize those who treat new communication technologies as substitutes for existing services.

Gerhard Rahmstorf and David Penniman address the problem of
the individual scientist in the face of the rapid growth of the
literature. They present a roughly quantified model to describe
the current STI system; it is based on notions such as the average
reader and the average publication. They go on to consider some
proposals intended to make for more efficient interaction between
scientific and technical information and its users. Finally they
look forward to an electronic "universal text information system"
to which access will be possible through computer terminals.

EURONET is a data communications network which is being developed for the EEC. It will provide access via computer terminal
to about 100 STI data bases stored in about 30 host computers, and
is intended to become operational in early 1979. In their paper
Carl Vernimb and Garth Davies provide a brief description of the
system and of policies as to its use. They then comment upon
possible special features: an automatic referral service; standardization in user command sets; a search and retrieval algorithm
in which the user formulates a request by identifying a few relevant
documents; technical options and legal issues in document delivery;
and automatic translation.

Important questions will arise regarding the pricing of services
provided by EURONET. The paper by Tony Flowerdew, Christine
Whitehead and Jim Thomas draws on a study conducted for the EEC,
which sought to analyze how the demand for these services would be
affected by pricing structure and the level of prices. They identify

229

and briefly discuss various determinants of the demand for on-line
services, providing relevant results from a survey of 47 potential
users (organizations and individuals). The discussion is then
extended to future changes in demand and cost and to the relationship of demand with price over time. Time series data on the use
of the UK Medlars service are used to demonstrate the danger of
extrapolating demand data from a period when services are provided
free in order to forecast demand at more realistic long-run price
levels.

Bob Mason presents a second paper on the economics of STI
services, in this case Information Analysis Centers (IACs). After
providing background information about IACs and an introduction to
the issues involved in evaluating their costs and benefits, he
presents a model which is consistent with observations on the
demand for, and costs of, IAC services. Some numerical results,
obtained when the model's parameters were quantified, are presented.
The paper concludes with discussions of the present limitat s of
such research and of more general topics such as the impact of
technological developments and some international issues.

The final paper in this section, by Peter Davis and Ed Freeman,
deals with the assessment of telecommunications technologies within
the context of a future national STI system. As such it is, at
one level, concerned with the issues raised in the papers presented
by Ganz, Rahmstorf and Vernimb. At another level, it is concerned
with the methodology of technology assessment. The authors propose
that "technology assessment should become an integral part of an
ongoing planning activity which aims to take a more active stance
toward the creation of improved systems in the future." They
suggest a three-way typology for technology assessment: the
intentional system, the transactional environment and the contextual
environment.

Since the assessment of "technologies against the backdrop of
current conditions or extrapolated futures distorts the evaluation
process and perpetuates the errors of the past," Davis and Freeman
propose it be carried out in the context of an "idealized design."
A brief description of an idealized design of the US STI system
(the SCATT System) is presented early in the paper.

The review with which this section opens and the essay with
which it closes relate most closely to the broad range of current
issues arising in the evaluation of interactive telecommunications
systems. The other papers are more specific and, except for
Vernimb's contribution, adopt a mathematical approach.

THE IMPACT OF TELECOMMUNICATIONS TECHNOLOGIES ON INFORMAL COMMUNICATION IN SCIENCE AND ENGINEERING--RESEARCH NEEDS AND OPPORTUNITIES

C. Ganz and J. D. Goldhar

National Science Foundation

1800 G Street, N.W., Washington, DC 20550

INTRODUCTION

Over the last fifteen years, there has been increased interest in the storage, dissemination, and use of scientific and technical information. In looking for ways to increase the rate at which scientific knowledge accumulates and innovation progresses, science policy makers have supported an extensive body of research aimed at reducing the costs of information transfer and improving the timeliness of information available to the scientist and engineer. Technology-based advances in computer-controlled printing, machine-indexing, micro-reproduction, remote computer access, time-sharing, and cathode displays have been and are increasing the efficiency of the information-handling procedures used by scientific and technical professionals.

During the same time that these technology advances occurred, study after study showed that person-to-person exchanges of information among scientists and engineers were playing an important role. In his studies of the technical problem solving process, for example, Tom Allen has consistently found biases by engineers toward the use of personalized, informal, oral means of technical communication -- as opposed to more depersonalized, formal, and written channels.[1] These spontaneous, seemingly unplanned, and unmanaged, informal channels began to be seen as competitors to the formal, increasingly computer-based technical information retrieval and dissemination systems. Questions began to be asked: Why do we have to store information in elaborate computer systems when people are the most effective means for transmitting technical information to other people? Will the advances in information engineering make

informal communication obsolete? Or will the person-to-person
communication so indispensable to scientific progress limit the
improvements in efficiency being applied to the formal channels?
Few questions were asked, however, about the interactions, if any,
between these informal person-to-person networks and the formal
computer-based systems.

This conflict between policies and investments to improve
the efficiency of the formal channels of communication and the
increasing acknowledgement that scientific communication depends
on informal contacts and unplanned events bears directly on telecommunications research. A good deal of the current experimentation in the use of telecommunications technologies involves
attempts to stimulate progress in the efficiency of the person-to-
person interactions among researchers. Much current experimentation in the use of telecommunications technologies makes the
following assumptions: 1) improvement in the efficiency of
information transfer can improve scientist and scientific productivity, that is, increase the rate at which scientific knowledge
accumulates and innovation progresses; 2) the informal channels
of communication used by scientists and engineers do contribute to
scientific progress; and 3) these informal channels, as well as
the formal channels, can be planned, managed, made more technology-
intensive, in short, made more efficient.

One area of interest to telecommunications researchers has
been the informal communication channels used by scientific research
communities.(2) These researchers argue that letters, telephone,
visits, and professional meetings and conferences all involve delays
and inefficiencies that have grown in recent years. In addition,
rising costs and the increased numbers and needs of multi-
disciplinary research communities have made the existing informal
communication alternatives inadequate. Many highly-specialized
research communities, they further point out, are finding that not
enough of their members have interests in common to support a
professional meeting. The alternative offered by these researchers
is the general concept of utilizing computers and digital communication networks to facilitate direct written communication among the
group of researchers involved. The system is designed to provide
and manage regular and current informal communication channels for
a group as an alternative or a complement to their use of mail,
telephone, travel and group meetings.

Another area being researched currently is the informal
communication requirements of joint research efforts.(3) Unless
project teams are in the same location, satisfying the communication
requirements of cooperative projects and joint research is a
difficult and time consuming process. Yet it is becoming common to
find research efforts which involve interactions between a set of
laboratories, collaborators at university, and other research

centers, all geographically dispersed. In addition, certain types
of efforts like the implementation of models are almost impossible
to undertake unless members of the research team are located in the
same institution. But the availability of high-resolution video
displays could perhaps adequately substitute for physical proximity
in those scientific efforts heavily dependent upon graphics, equations, charts, etc.(4) Thus the National Science Foundation's RANN
Directorate is investigating the computer conference as a means of
allowing research project teams to confer even though they are
geographically distant. Other experiments are going on in NSF in
exploring the use of slow-scan T.V. receivers as multipurpose
communication terminals to be used for face-to-face communication
as well as for graphics, chart, and computer-stored information
exchange.

These attempts are at least ambitious. To varying degrees,
they involve planning and managing informal channels and the
ascertainment of the actual impacts of technology-assisted informal
interactions among researchers on the progress of science. Clearly
then these efforts do not adhere to the traditional view that only
formal channels of communication lend themselves to improvements
and utilization of technology. Contrary to the traditional view of
science information use, these efforts include the claim that
informal channels can be managed and planned, and can be factors
in improving the contribution of information-transfer activities
to scientific growth. The question then becomes: how realistic
are these assumptions in the light of our knowledge of information
use behavior? As an answer the following (1) details some of the
traditional research on information use patterns; (2) indicates
some more recent thought on scientific communication which shows
that formal and informal communication channels are interdependent
elements of a single system; (3) cites current research which
attempts to model how this mix of channels contributes to the
research process; and (4) proposes a set of researchable questions
which should be answered by future experiments in telecommunications-
assisted scientific communication.

TRADITIONAL RESEARCH ON INFORMATION-USE PATTERNS

Between 1955 and 1970, almost 1000 studies were undertaken
to analyze the information-seeking and use patterns of researchers.
Scientists' behavior in gathering information, the flow of information, needs for information and exchange of information were
looked at. Since its initial publication in 1967, the ANNUAL
REVIEW OF INFORMATION SCIENCE AND TECHNOLOGY (ARIST) has included
a chapter reviewing, summarizing and criticizing studies on the
information needs and uses of researchers. The various literature
reviews in ARIST, Paisley (1966), Moore (1972), Martyn (1974) agree

as to trends in the study of the use of scientific and technical information.

Paisley (1966) describe a twenty-year emphasis upon brute empiricism in the field of information behavior. Moore (1972) describes the early set of studies as "relatively simple tabulations of who uses what sources of information when." Martyn (1974) writes, "The first period of user study lasted until the mid-sixties." and "Most studies were carried out on relatively small numbers of subjects, drawn either from members of specific disciplines or from the scientific population as a whole, or less frequently from among the users of particular systems." (5, 6, 7)

Menzel in attempting to review the early period of user studies, cites Pelz and Maizell, among others, as having succeeded in correlating exposure to information channels with performance, as rated by colleagues and superiors, or as indicated by the amount of one's publications standardized by age. But as Menzel points out, correlations do not specify the particular way in which exposure to a channel contributes to research productivity. In addition, he says, there are serious problems in the measurement of performance and in the control of factors other than information exposure. (8)

Other studies during the early period attempted to evaluate the usefulness of the information disseminated. Herner tabulated the channels of information used in finding an answer or solution to a recent problem which had been identified by each interviewed scientist.(9) Scott tabulated the sources where scientists claimed to have gotten the idea for their most recent project.(10) Several pieces of research described the circumstances surrounding messages which arrived "too late," that is, which reached a scientist after their potential maximum usefulness to his work had passed.

The results of these early user studies were incorporated into efforts to improve the efficiency of the formal channels of science information transfer, that is, reduce the costs of journal articles, the library, abstracting and indexing services, and to improve the speed at which information could be made available. Computer-based information systems were developed to perform the functions of storing, retrieving, processing, and displaying information.

Many surveys of information use, however, documented strong reliance by researchers on informal channels of communication: face-to-face encounters; private correspondence; small group discussion; channels which allow two-way interaction between generators and users of information. In the late sixties, attempts began to be made to distinguish informal from formal channels of communication and show that such communication is not merely a practice of scientists and engineers, but a factor contributing to the advancement of scientific knowledge and technology. Typically, these studies used measures of productivity closely associated with academia. Individual scientists were evaluated according to their contribution to general technical or scientific knowledge in the field or their overall usefulness in helping the organization carry out its responsibilities, as measured by numbers of papers, patents, or citations in the literature.

Studies by Allen, Pelz and Andrews, Rosenbloom and Wolek, among others, discuss the relation between information flow and scientific productivity.(11,12,13) Allen reports on a series of studies of information use during R&D projects. A feature of his studies was the use of matched laboratories, two or more organizations working on the same problem under contract to the same government agency. He demonstrates that differences among research groups with regard to the proportion of time spent with various types of information channels, and also in the phasing of the use of these channels, were associated with the quality of their output or performance.

Pelz and Andrews drew data from 1311 scientists and engineers in 11 R&D laboratories. The performance of each respondent was evaluated by his co-workers and superiors on the basis of "contribution to general scientific and technical knowledge in the field" and "usefulness in helping the organization carry out its responsibilities." Productivity was also measured by counts of published papers and unpublished reports. They found that higher performance was related to a high level of communication with colleagues, that is, those who had relatively frequent contact with colleagues tended to perform at higher levels than those with less frequent contact. Frank Andrews further analyzed the data, concluding that the findings "tended to support the hypothesis that contact with colleagues could stimulate performance." Furthermore, they suggested that this was more likely to happen if the contacts were purposefully originated by people directly concerned -- the man himself or his colleagues -- than if they were unplanned or originated by some third party. Pelz and Andrews were also unable to find evidence of scientists for whom colleague contact was not useful in some measurable way. In a follow up study six years after the original study, Farris offers further support for the notion that informal contacts stimulate scientific performance.(14)

In an early study looking at technology-assisted interpersonal communication, Shilling, Bernard, and Tyson found that a policy of unrestricted long distance telephoning correlated highly with success in obtaining information but not with productivity. A policy of unrestricted travel however correlated highly both with productivity and with success in obtaining information.(15) Correlation and causation are however different.

Research on "invisible colleges" is also relevant here, since it attempts to track the effects of professional communication on the advancement of scientific knowledge. Using such techniques as surveys, and citation analysis, these studies demonstrate the importance of information-communication ties in scientific communications.(16,17) Crane attempts to test the hypothesis that science grows as a result of the diffusion of ideas. She compares the growth rate of research areas in which scientists were known to have been interacting with each other and with scientists who had not previously published in the area, with the growth rate of research areas in which researchers were known not to have been interacting in this manner. For fields in which the level of interpersonal communication was low, the intellectual development of the field was weakened, that is, the same problems were repeatedly selected by successive generations of authors and knowledge failed to accumulate. The fact that the same problems were repeatedly selected by successive generations of authors suggests that ideas diffuse more effectively when transmitted by individuals than by publications alone.

Menzel in 1967 summarized the importance of informal scientific communication by listing six major advantages it has over other communication channels:

1. Promptness; the ability of informal communication to bring new scientific developments to researchers with contacts in the field more quickly than their use of printed media.
2. Selective switching: the ability of informal communication networks to selectively match documents to the individual interest profile of each researcher better than formal current awareness services.
3. Screening, evaluation, and synthesis: information received from colleagues has already been screened from the large body of available literature and evaluated as being an important document to the receiver.
4. Extraction of action implications: when information is exchanged between colleagues, a judgment is often made about potential applications which adds an important component to the reporting of the research.
5. Transmitting the ineffable: information regarding specific problems in conducting research often, for many reasons, does not find its way to the literature. Colleagues who are close to the research can often provide this input.
6. Instantaneous feedback: because continuous messages are exchanged with instantaneous feedback, researchers are able to obtain useful criticisms more efficiently than through written media. (18)

Despite Menzel's laudatory comments, informal channels were in general represented as competitors to the formal channels for the time spent by scientists in accessing information. Their reliance on informal channels was often seen as a sign of deficiency in information resources, and the justification for improvement in formal systems. Informal channels furthermore were not thought of as amenable to improvements in efficiency and so not a potential area for planning and management attention.

Many policy makers expressed doubts about supporting informal communications. Informal communication is fortuitous, it was claimed, because while scientists working in related areas are likely to come into contact, there is no certainty this will occur. In addition, there is no assurance that subsequent action is based upon correct, complete, and the best available information. From this standpoint, an organization which attempted to manage informal communication might well wonder whether it wasn't encouraging communication which had not been subject to qualified evaluation, and thereby second rate or redundant work. These points were among the arguments leading to the withdrawal of government funding from one experiment which supported informal channels in the biological sciences.(19) Alongside this view that information transmitted through informal channels might be of uneven quality, economists of information, at this time, characterized informal channels of information as being costless and not amenable to investment, unlike the universe of formal channels.(20)

In addition, questions about which types of information are best transferred by informal channels or person-to-person contact, as compared with other means; and the interrelationships between informal and formal channels were not even asked.

In the late sixties, recognition of the importance of technological innovation in promoting economic development began to stimulate research into the nature of the innovation process. In particular considerable attention began to be paid in the innovation literature to the role of communication. Much of this empirical research strongly underlined the importance of good communications to the success of technological innovation and detailed studies were made of the sources of ideas contributing to the innovation process, the channels through which information is transmitted, and the characteristics of individuals who play an effective role as disseminators of information. As in the infor-mation-use literature, the most significant and certainly the most frequently cited finding was the relative importance of personal contact as compared with other modes of communication. But no real progress was made in understanding the explicit contribution of communication channels to the innovation process.(21)

    For the telecommunications researcher, then, traditional
studies of information use did show the importance of interpersonal
channels to scientific and technical work. It was clear from these
early studies that any effort to serve the information needs of the
research community must consider the reliance of the scientist or
technical user on interpersonal communication. However, because
interpersonal communication was generally seen to be an unplanned,
fortuitous event, investment and planning for informal communication might be futile.

              THE SYSTEMS VIEW OF SCIENTIFIC INFORMATION USE

    Research beginning in the 1970's, however, started to show
that the most typical instances of information access involved
successive interplays between formal and informal channels. The
results of one study indicated that although the users studied
could not rule out the possibility that formal sources would
completely satisfy their information needs, they did not expect
this to happen.(22)   From the initiation of each information
search, this study showed users felt that direct contact with a
competent person would be necessary. On the other hand, formal
sources were very important to the interpersonal exchange in
allowing receiver and source to share enough of a common background to allow mutual understanding. This began to introduce the
notion that use of formal channels and interpersonal communication
were interdependent.

    For telecommunications researchers, planners, and designers
of interpersonal communications systems, the important notion
coming out of this research is the balance that must be allowed
for between formal and informal information activities. That is,
efforts to support informal channels may be wasted unless for
example, journals important in the cataloguing of knowledge, are
also available. It may be true that people resort to face-to-face
communication even when good telecommunication is at their disposal
because, as has been suggested, face-to-face is richer than any
other mode. It has been suggested that in cases where there are
delicate transfers of information, or evaluation of information
which involves great risk of condensation of information in
problem-solving with great uncertainty, this occurs. It may also
be true as the results of some studies have shown, that even though
telephone traffic goes up when there are major problems to be
solved, these are followed by face-to-face contacts before the
problem is resolved.(23) But this is not an indictment of the
use of telecommunications technologies; it is an indication of our
lack of understanding of the complete conditions for assessment
and evaluation of sources of information in the problem-solving
process. Current efforts, then, to make informal channels more
efficient or less costly, through use of new technology, may be

wasted unless the formal channels, providing sources, subjected to qualified evaluation and review, are also supported, and available.

Recent innovation research indicates that the innovation process was most efficient when both personal contacts and formal sources were combined in a mutual supportive way. This occurred when personal contacts were used for locating relevant information in printed material and subsequently for translating this information into a form appropriate for the problem-solver. According to this research, this sequence is not merely additive. The use of the different channels, in tandem led to new and relevant information which would have been very difficult to obtain in any other way.(24)

In general, people need to be prepared for some communications. As Wolek has repeatedly pointed out,

> If the manager does decide to support some services for interpersonal communication, he must be careful to see these as elements in a larger system....If management has not provided literature sources which will allow participants (in a series of meetings) to share the same background and has not provided opportunities for participants to test their knowledge with friends beforehand, the expected change may not transpire....(24)

Thus, while the telecommunications researcher may be concerned primarily with improving the efficiency of informal channels, say, the meeting itself, it is clear that he must provide an adequate basis for these exchanges of information. He must ensure that the criteria used by the participants to judge relevance and validity of information are similar, and that factors such as common acquaintances and common memberships which affect the success of contacts, are allowed for. He must be involved in planning and managing the entire system of scientific communication available to information users in research groups or organizations. Morever, he must be prepared to consider that failure is due not to the inability of the system to replace face-to-face contact, but to the lack of planning of resources over which he did not indicate he required control. One of these points is that of providing the sources which users need to establish their background in a new area. That is, the researcher may have to make sure that the documents or collections of them, known as libraries, are available to users and prepare them adequately for interpersonal exchanges.

In addition to it becoming evident that mix and sequencing of
channels is necessary to bring about desired changes, it also
became apparent that seemingly voluntary contributions of information are based on prior knowledge of the receiver's interests.

The predominant use of local information sources in
operational circumstances is often explained as a manifestation of the simple economics of turning around
(or picking up the phone) and 'asking the next guy'.
The most important point is that there is a next guy
to talk to. The unifying element is provided by the
mission of the organization which relates task to task,
person to person, and which justifies the development of
a team with common interests. Organization, in the
sociological sense is the key element.(26)

This view is reinforced by recent work in economics of information which argues that attempts to improve information transfer
would be more successful if they were directed at the organizational arrangement influencing information use rather than the
channels themselves.(27) Some guidelines follow. If information
sources are to be easily accessible, the organization and its
professionals must have an established position in the profession.
The organization's professionals must also have sufficient
experience in that field so that they have the background to
interpret events correctly and to build on discoveries. Finally,
those in touch with professional developments outside the organization must have the internal contacts and experience necessary
to interpret developments for the internal staff. In general,
the information available to a scientist or engineer is going to
be influenced by the leadership roles in that field and the
relationships constituting collaboration in that field.

Studies such as Rosenbloom and Wolek deal with applied
research organizations. Other studies which consider scientists
as members of disciplines show a different set of institutional
factors influencing information channel use.(28) While the
basic researcher who establishes priority earns status, the
organization which establishes a lead over its competitors can
receive advantages in the marketplace. This discourages access
by researchers to information channels which allow leaks having
the potential of destroying chances for marketplace rewards.
Consideration of incentives to perform research, rewards for
priority of discovery, collaboration relationships between
researchers, as shown in these studies, implies a different mix
cf channels and different interactions between them, for the
basic, as distinct from, the applied research community.

Within the organization, the location of formal research groups, and management incentives for collaboration are going to affect information transfer between researchers and research groups. Furthermore, the informal groups which serve as nodes for informal communication networks within organizations typically arise within formal laboratory units, but are not co-extensive with them.(29) While the formal laboratory structure typically provides the focus and boundary conditions for such groups, it does not fully account for its emergence or composition. The fact is people build up relations over time irrespective of what their tasks or their formal positions are, and they are more willing to ask questions of those they know. Thus, the social systems of the organization such as the interpersonal relationships built up over time are going to influence access and use of information by researchers.

Studies about the utilization of research results consistently point out that the formal communication channels, effective within the disciplines, are parts of relatively self-contained social systems. Those who turn away from them by taking on applied interests and activities are frequently unable to develop formal communication channels which are effective. It has also been pointed out that there is a heavy degree of reliance on informal channels. But these differ radically from those used by the "invisible colleges" of disciplines, since they do not link outstanding experts, but individuals usually ill equipped to per- form information screening, evaluation, and synthesis.

In all these areas, a good understanding of the relationships among generators and users of research within the various scien- tific social systems is going to be needed to allow realistic assessments of technology-assisted interpersonal communication. The bottom-line is that communication and organization are inseparable. Much additional research is needed, especially in the areas of applied science and technology and utilization of science and technology, to make these suggestions operationally useful. We need to know the basis on which professionals enter into repeated informal relationships, the norms of behavior governing such contacts, and the management, institutional, and organizational policies influencing these contacts and relation- ships. Designers interested in improving and assessing the informal channels will not be successful unless they are aware of and can influence these social systems. They are going to have to realize that different sources of information available through different channels are necessary to bring about the desired increases in research effectiveness. Research and development embraces a broad spectrum of activities involving widely different social systems which will influence the require- ments and the uses of interpersonal channels of communication. Systems must be designed to take into account the user's position on the R&D spectrum.

RECENT STUDIES

Some recent studies in the innovation literature provide additional pointers to those interested in improving interpersonal communication. One variable which has been studied extensively is the researcher's professional field or discipline. Engineers do not read very much in comparison with scientists, and formal channels of informaiton therefore are not very frequently used. This fundamental difference between scientists and engineers regarding utilization of the formal channels is probably better documented and enjoys a wider community of agreement than any other characteristic of the information field. Furthermore, the engineer's primary mode of seeking information is through interpersonal communication, although, even here, the behavior of the engineer can be differentiated from that of the scientist. Scientists rely heavily on informal channels which connect them to sources outside their organizations. The engineer's interpersonal communication is more likely to be with colleagues within his organization or laboratory. In addition, engineers, who tend to be concentrated in the development end of the R&D spectrum are even less likely to seek out sources of information external to their organization. Not only are they more oriented towards interpersonal communication, but the R&D activities in which they tend to be engaged are operational matters which would encourage more extensive use of the interpersonal channels within organizations. In sum, engineers would seem a good target for innovations in the informal channels, which directed themselves towards intra-organizational flows of information.

Several other indicators have also been shown to influence use of interpersonal channels.(30) A group of researchers with the responsibility for a project through all its phases may not have the highly specialized knowledge or informal contacts of those who specialize in a single phase of research. Their research effort on a new project thus involves less informed and lengthier searches of formal sources. Their in-house contacts, are more likely to be in this same "early learning" pattern in this particular research effort and thus be of little help. For these reasons, a "project-dominant" group's dependence on formal systems would be much greater than that of groups organized around a single phase of research. If, however, the information needs of such a project-dominant group were tracked through the subsequent phases of R&D, shifts in their information needs from formal sources to in-house informal sources would occur. To manage and design the informal channels, then, we are going to have to track the interactions between formal and informal channels through the various R&D phases, with acknowledgement that interactions differ according to the structure of the research group.

Another variable influencing the use of informal channels is the rate of scientific and technical change in the field in which the research is going on. Thus, hot fields in which exciting developments are taking place at rapid rates will lead to increasing reliance on interpersonal channels to circumvent the delays in the formal network--if the field is characterized by free collaboration among researchers. Furthermore, when work in a field is accelerating, it may be a good time to introduce new communication services, especially ones which are inexpensive enough to be supported by small numbers of researchers.

A final variable affecting informal communication patterns does not depend on conditions in the research organization, or the rate of change of the field, but rather on the magnitude of the innovative effort itself, that is, whether it is incremental or discontinuous. The chief barrier to incremental innovations seems to be, not technical information, but the lack of information flow from the workers to management. When it comes to the information required for innovative efforts constituting a radical departure, the firm's current R&D personnel are likely to be inadequate, and new and appropriate specialized technical information will have to be brought in.

This list of factors affecting use of formal and informal channels is not meant to be exhaustive. It is only intended to suggest that research on information flows in the innovation process is maturing and that the current state of the art can provide useful guidance to telecommunications researchers.

CONCLUSIONS

It has been a long way from traditional information-use research which viewed informal and interpersonal communication as a threat to the technology-based formal systems, and as a set of lucky accidents which did not have the potential of being planned or managed. The results of more recent information studies present facts about researchers, research organizations, and the research process which should be encouraging to telecommunications researchers, and should improve the theoretical basis on which research in technology-assisted interpersonal communications can be conducted.

Some final advice is necessary to improve the conduct of current experiments. In general they have been stimulated by increasing demands to justify investments in new communication-information technologies in terms of increased efficiency. In general, they involve attempts to provide faster, less costly,

versions of what we now have, and produce results which involve
simple tradeoffs of these new technologies with existing services.
But the history of information-use studies reassures us that
improvements in the efficiency of any one channel of communication
will not automatically effect gains in scientist or scientific
output. Formal and informal communication is complementary.
Changes in one channel, in order to be effective, are going to
require support for changes in other channels. Even more
important, because organization and communication are interdependent, changes in the institutions affecting the relationships
between the users of this information are also going to be
required.

The implications are the following: First, the design of
services must not mean a faster or more comprehensive version
of what we have now. Increasingly in the information science
literature, there are demands for a theory of information-need
that is liberated from current information-gathering habits and
from doubts about the capacity of new technologies to deliver.
Radically new services are less likely to occur if design means
only a faster or cheaper way to serve current information needs.
The design question then is not whether technology can fill the
interpersonal communication needs of scientists as we now know
them, say, without the costly travel involved in attending scientific
meetings. The challenge is rather to describe the information
needs of professionals that will result from the successive inter-
play of meetings and say the use of telephone with video displays.
The contributions of such technologies to serving the information
needs of scientists will not be fully exploited if they are
designed and treated as substitutes, rather than additives, to
current information resources.

Secondly, the assessment of these technologies must involve
more than simple tradeoffs between existing services and their
technology-based substitutes. A particular assessment should
include the organizational innovations and institutional changes
entailed by these services. To evaluate improvements in the flow
of information, researchers must have an understanding of the
changes in social and organizational relationships which the
services require. We need to be able to judge these applications
of technology to communication processes as part of a bundle of
investments in improving science and technology.

With these criteria for design and assessment in mind, telecommunications researchers can begin to carry out experiments
aimed at overcoming some of the commonly cited deficiencies in
informal communication among researchers today. Some of these
are the following: First, it is difficult for young scientists

and technologists to gain entry into existing invisible colleges
or to form new ones. There is a possibility that technology-
assisted interpersonal communication can effectively enable a
larger body of people to have access to invisible colleges.
Secondly, scientific and technical meetings are not as effective
as they might be. There is a possibility that use of the telephone together with video display will be of great value, not as
a substitute for travel to meetings, but as a way of making
meetings more productive through previsit preparation and post
visit follow-up. Third, many research and development organizations are not managed to facilitate informal communication. There
is a need to experiment with new forms of organization in applied
science and technology which foster the conditions that ate known
to provide an effective basis for informal networks. Fourth,
individuals who act as valuable switching centers are frequently
overloaded. One experiment could involve evaluation of a variety
of managerial incentives which may motivate some less active
researchers to take on a more active key communicator role.(31)

These are only a few examples. The important point is that
a recommended improvement in the efficiency of one channel of
communication will not by itself increase scientific and technical
output. The research objective should rather be: the impact of
technology-assisted interpersonal communication on the efficiency
and effectiveness of the whole information system as measured by
the increased productivity of the user. Some subsidiary objectives
should be:

(1) the non-technological changes which have to occur in terms
     of organizational arrangements and individual behavior;
     and

(2) the new communications needs satisfied by the combination
     of technology and organizational changes.

Finally, we know that the channels through which scientists
communicate are not only affected by the various sets of social
and political mechanisms which are required and used to do science
and technology, but also can affect the different social systems.
In science, incentives to sustain research behavior at high levels,
mechanisms to authenticate priority of discovery, and to certify
quality of research influence the availability and use of both
formal and informal channels of communication. We know a good
deal about the social organization of science. But both current
knowledge of, and the quality of the relationships, between
scientific and technological communities and the society at large
which uses the end products of research, is poor. We know relatively little about how to manage and improve the relationships
between knowledge producers and users. Thus, one of the most

productive areas in which telecommunications researchers could
apply themselves would be to use technology-assisted interpersonal
communication to gain better understanding and more productive
relationships between the scientific communities and the public.
This area of research is also probably the most difficult. But it
may also have the highest potential for improving the contribution
of information transfer to the progress of science and technology.

REFERENCES

(1) Thomas J. Allen, Managing the Flow of Technology, MIT Press,
Cambridge, 1977.

(2) Murray Turoff, "An Intellectual On-Line Community," paper
prepared for AAAS Annual Meeting, February, 1977.

(3) Jacques F. Vallee, "Impact of a Computer-Based Communications
Network on the Working Patterns of Researchers: Design for Evalua-
tion of Effects Related to Productivity," American Sociological
Association, Annual Meeting, N.Y., August 1976.

(4) Stuart L. Meyer, Video Image Transfer Using Slow-Scan TV
Terminals, NSF-DSI 76-09479, National Science Foundation,
Washington D.C.

(5) William J. Paisley, "Information Needs and Uses," ARIST, 1968.

(6) E. Moore, "Information Needs and Uses," ARIST, 1972.

(7) John Martyn, "Information Needs and Uses," ARIST, 1974.

(8) H. Menzel, "The Flow of Information Among Scientists," Columbia
University, Bureau of Applied Social Research, New York, May 1958.

(9) Saul Herner, "The Information-Gathering Habits of American
Medical Scientists," Proceedings of the (1958) International
Conference on Scientific Information.

(10) Christopher Scott, "The Use of the Technical Literature by
Industrial Technologists," Proceedings of the (1958) International
Conference on Scientific Information.

(11) T. J. Allen, "Managing the Flow of Scientific and Technical
Information," unpublished doctoral dissertation, MIT, Sloan School
of Management, 1966.

(12) Donald C. Pelz and F. M. Andrews, Scientists in Organizations,
New York, John Wiley and Sons, 1966.

(13) R. S. Rosenbloom and F. W. Wolek, Technology and Information Transfer, (Boston, Harvard Business School, 1970).

(14) G. F. Farris, "Organizational Factors and Individual Performance, A Longitudinal Study," Journal of Applied Psychology, 1969, pp. 87-92.

(15) C. W. Shilling, J. Bernard, and J. W. Tyson, "Informal Communication Among Bioscientists," Washington, D.C., George Washington University, Biosciences Communication Project, 1964.

(16) E. Parker, W. Paisley, and Roger Garrett, "Bibliographic Citations on Unobstrusive Measures of Scientific Communications," Stanford Institute for Communication Research, Stanford, California, October, 1967.

(17) Diane Crane, Invisible Colleges, University of Chicago Press, 1972.

(18) H. Menzel, "Planning the Consequences of Unplanned Action in Scientific Communication," Communication in Science (Reuch and Knight, editors) London, Jand A. Churchill Ltd. 1967.

(19) Francis W. Wolek, "Policy and Informal Communications in Applied Science and Technology," Science Studies, 4, 1974, pp. 411-420.

(20) Geoffrey Newman, An Institutional Perspective on Information," International Social Science Journal, (UNESCO) Vol. XXVIII, No. 3, 1976.

(21) J. D. Goldhar, L. K. Bragaw, J. J. Schwartz, "Information Flows, Management Styles, and Technological Innovation," IEEE Transactions on Engineering Management, Vol. EM-23, No. 1, February 1976.

(22) Francis W. Wolek, "Preparation for Interpersonal Communication," JASIS, January-February 1972, Vol. 23 No. 1, pp. 3-10.

(23) Albert Shapero, "You Can't Do it All Long Distance," Fortune, February 1977.

(24) Ron Johnston and Michael Gibbons, "Characteristics of Information Use in Technological Innovation," IEEE Transactions on Engineering Management, Vol. EM-22, No. 1, February 1975.

(25) Wolek, "Preparation" op. cit.

(26) Wolek, "Preparation" op. cit.

(27) Newman, op. cit. p 472.

(28) William D. Garvey, et al., "Research Studies in Patterns of Scientific Communication," Information Storage and Retrieval, Volumes 8, 10, (1972).

(29) Allen, 1966, op. cit.

(30) Patrick Kelly, Melvin Kranzberg, The Flow of Information in the Innovation Process, Georgia Institute of Technology, 1977.

(31) Albert H. Rubinstein, et al., Field Experiments on Key Communicators, Northwestern University, 1977.

SCIENTIFIC COMMUNICATION AND KNOWLEDGE REPRESENTATION

Gerhard Rahmstorf and David Penniman

International Institute for Applied Systems Analysis

Schloss Laxenburg, Austria

1. INTRODUCTION

It is a well known and much discussed problem that the number of scientific and technical publications is rapidly increasing. The individual scientist or technical expert has to adapt his behavior to this trend. His adaptation may include:-

- Spending more time on searching and processing publications
- Ignoring a large share of new publications
- Restricting and specializing his field of work
- Improving the efficiency of his work

The capabilities of the individual to solve this overload problem are limited. He has a given time for scientific work and a given capacity to search and read literature. Spending more time on literature reduces the time left for creative and productive work.

It seems more appropriate to look for mechanisms to improve the system of scientific communication and science representation. Administrative mechanisms such as quality control are dangerous as they might restrict the diversity of contributions. We propose to assist the individual by use of a text information system for the primary information with controlled length and structure of contributions and improved text description.

This paper is an attempt to explain the need for such a change in scientific communication and science representation. To do this, we describe the current system and its impact on the behavior of the individual scientist. Then we compare the proposed text information system with alternative proposals.

The paper does not present a complete model of the current system, although attempts are made to identify characteristic quantitative indicators. These indicators must be discussed on the basis of more precise data, which are only in the process of being developed.

We talk in statistical terms about the average publication and the average scientist. We know that science depends very much on exceptionally qualified scientists, outstanding publications and we know the indicators may vary greatly in different disciplines.

2.   THE PRESENT COMMUNICATION AND REPRESENTATION SYSTEM

The system of scientific communication and knowledge representation consists of participants, i. e. authors and readers, and of all institutions and technical facilities, which contribute to the exchange of research results between scientists and experts (Fig. 1).

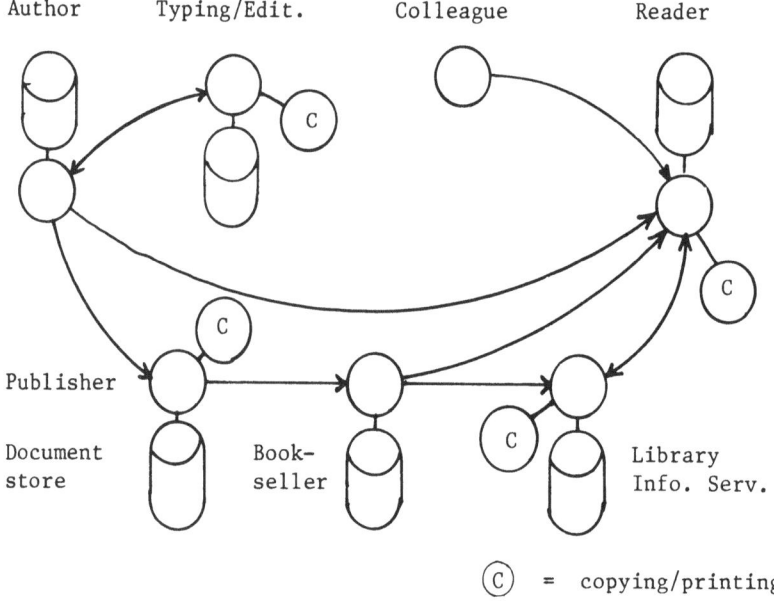

Fig. 1.   Document Flow

This system can be described topologically as a network whose nodes represent participants and institutions and whose arcs represent transmission channels connecting the participants and institutions.

2.1 Technical Facilities: Paper and Mail

Technical facilities of the communication and representation system are channels and storage media used to transmit and store scientific information.

The dominating medium for transmission and storage of scientific texts is still printed paper. Transmission is accomplished by mail. But the use of electronic communication technology and information storage by computers is growing.

2.2 Participants

Participants are the scientists who produce or read documents. All scientists read, most of them produce texts.

According to Anderla there are currently at least 10-12 million people involved in production and distribution of all kinds of scientific texts. Dobrov estimated the number of scientists as about 3 million (Dobrov 1974).

Each participant has his own information needs. This interest profile usually changes in time. Information needs are satisfied by reading physical documents transmitted from the author to the reader. The participant can either buy or borrow documents. Buying a document is more expensive for the individual participant, whereas borrowing requires the operation of libraries and has other disadvantages for the participant.

2.3 Institutions

Typing and editing services transform handwritten manuscripts into a document which can be used for copying, printing or distributing. Larger institutions work with a centralized typing department equipped with small text processing systems or with a larger central computer with text editing software directly available to the authors via terminals.

Proposals have been made as to how the production of journals can be improved by computerized editorial processing centers (Westat Inc. and Aspen System Corp 1974). But it is not believed that these centers will replace printed paper as the final medium for text.

The current system of printed publications is based on mediating institutions such as publishers, book stores or libraries, which support the authors by taking over editing, printing, copying, storing and distributing their texts. They also support the readers by searching for information and supplying them with documents.

King (1976, Page 63f) reports that journal articles used for citation were obtained most frequently through libraries (50.8 %), also by personal journal subscriptions (20.6 %), reprints (16.1 %) and colleagues (12.5 %).

2.4   Stock of Texts

The set of all text is called the "stock of texts". Copies of one text are counted as only one element of this set. The stock of texts is not increased by copying or printing more copies of a text, only by producing a new text, which is not identical with any existing text.

The size of the stock of text T is the sum of length of all texts m:

$$T = \sum_{i=1}^{m} T_i$$

It is estimated as 1000 million pages (50 · 10^6 texts of average length 20 pages).

Yearly production T_y is about 100 million pages.

These figures can be checked against figures about average productivity of a scientist, 32 pages per year. (see 3.6)

2.5   Stock of Documents

A text becomes a publication as soon as it is introduced into the communication system and is then potentially available for every participant. As we do not talk about private or non-public documents, we consider documents here as publications.

Two copies of the same text are considered as two documents. The set of all documents is called stock of documents. Most texts are reproduced in many copies. The average reproduction factor n is 10^3 to 10^4. Each new copy increases the document redundancy, which can be defined as:

$$r_d = \frac{D}{T}$$ with D = size of the stock of documents in pages.

The average length of a document is estimated as 50 pages. One document contains 2.5 texts on average.

2.6 Knowledge

The participants in the communication system are interested in the knowledge content of a document, rather than in the text or the physical document.

A model describing only the amount of texts transmitted and stored in the communication system does not give much insight into the efficiency of communication. Quantification of knowledge is a very difficult, but most important point. We discuss a possible approach.

Knowledge can only be transmitted by use of a language. The same knowledge can be expressed in English, German, or any other language and many subjects can even be described in a formal notation. After having decided on the language, the author is still free in the manner in which he uses this language. He can use special terminology or common language terms. He can write in a compact or a verbose style. He can arrange the order of his presentation according to different intentions and viewpoints. But for each text, one can determine whether these texts carry the same knowledge as another text or not.

Theoretically knowledge is a set of scientific propositions. One can roughly measure knowledge as the length of a minimal text in a given language to represent this knowledge. The amount of knowledge in the algorithmic language ALGOL is the length of the original text of the ALGOL 60 report. All other text books on ALGOL 60 do not contribute much new knowledge to this subject, although textbooks are absolutely necessary from a didactic point of view.

All knowledge contained in the stock of currently available text represents the stock of knowledge. This knowledge can be measured by the number of pages E needed to rewrite the information contents of the stock of texts without redundancy. To understand this concept, we propose the following thought experiment (Fig. 2). Suppose we could start again with publishing knowledge. The first text of length T_1 would then be completely new. A second one T_2 might refer to the first one, in part repeat the same knowledge as T_1 and in part add some new knowledge; likewise text T_3 and so on.

Having used the same unit (pages) for text and knowledge one can define the text redundancy of the stock of texts:

$$r_t = \frac{T}{E}$$
r_t is estimated as 10^1 to 10^2.

2.7 Relative Productivity of the Systems

If T_y is the yearly text production measured by the total number of pages of all texts and ΔE is the increase of knowledge reported in these papers measured by the number of pages needed to describe their new results without redundancy, then we could call

$$e = \frac{\Delta E}{T_y}$$

the relative productivity for this year.

It seems that relative productivity is decreasing over time. More and more texts are produced which intentionally and unintentionally report on subjects which are already described elsewhere.

The reasons for this increasing share of secondary literature and reinventions/rediscoveries is the overload and redundancy in the current systems. The ignorance which is a result of current overload feeds the production of more redundant texts. We will try to explain this statement by an analysis of scientific work.

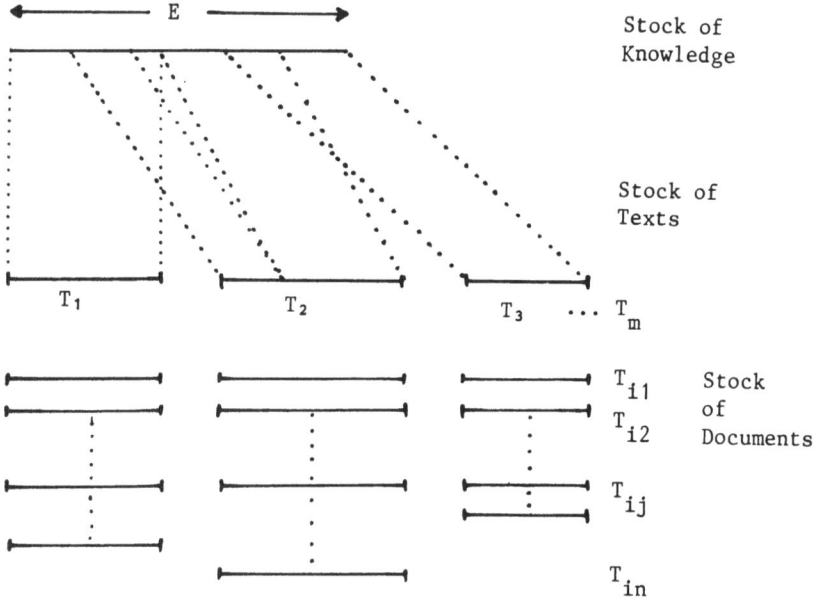

Figure 2: Relation between Knowledge, Texts and Documents

3. MODELLING SCIENTIFIC BEHAVIOR

3.1 The Pattern of Activities

A scientist participates actively and passively in the information flow of the network. He is receiving, processing, storing and producing texts and knowledge. To do his work he has 200 working days a year or 1600 working hours/year.

The daily work of a scientist can be characterized by the time spent in the following activites:

	Code	%	hours/year
Problem identification and Problem definition	PI	5	80
Literature identification and Literature ordering	LI	5	80
Browsing through literature	LB	10	160
Literature analysis	LA	10	160
Preparing personal documentation	DO	5	80
Creative work	CR	15	240
Producing own text	TX	15	240
Editing and reviewing text	ED	5	80
Oral communication, teaching, conference participation	CO	20	320
Reporting to management and administrative work, travelling	AD	10	160
		100 %	1600 h.

The figures given above are estimates based on our own observations and checked against other indicators which are discussed later.

How a scientist allocates his time depends on his working environment, his job and the subject of investigation.

We will discuss now how the average pattern of time slices is changed by growing knowledge and redundancy in the stock of publications. (Fig. 3).

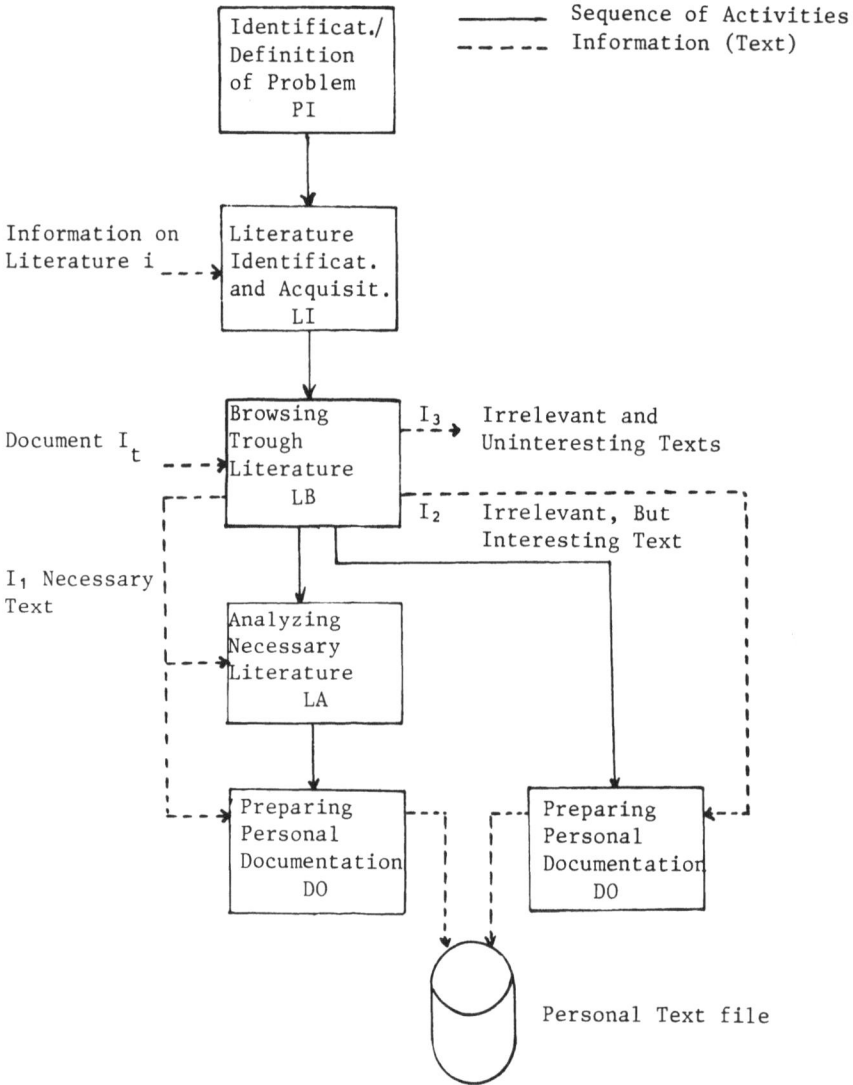

Figure 3:   Literature Related Activities

3.2 Literature Identification and Ordering

The quality of the incoming information on documents varies greatly, from title only to a detailed description and review.

The time needed for literature identification and ordering documents LI depends on the amount of information on literature i and on the quality of this information. Good bibliographic information must be well structured, precise, complete, up-to-date and should only refer to documents which are relevant to the reader. Even the best documentation services cannot satisfy these requirements.

The increasing number of publications which are sent into the system require more and more time for LI. However, improved documentation services might compensate for this trend partly.

3.3 Browsing Through Literature

A scientist has to check the literature which is sent to him for its relevance. This activity is called browsing through literature (LB). The time spent for this activity is proportional to the amount of incoming text (I_t).

$$I_t = a \cdot T_{LB}$$

a is the speed of evaluating the relevance of parts of texts. It is estimated as about 70 pages/hour.

The incoming literature I_t might be requested as a result of information which was available to the scientist, or it might be sent to him on the basis of a known profile of interest. Part of the literature is recognized more or less accidentally, for example, books or reports which were found in the library of a colleague.

The incoming texts I_t can be divided into three parts:

$$I_t = I_1 + I_2 + I_3 \quad \text{(pages)}$$

I_1 represents relevant parts of texts which will subsequently be analyzed carefully. Relevance depends on the interest profile of the scientist and his knowledge level.

I_2 represents parts of the incoming text which are not relevant for the current problem investigated by the scientist, but nevertheless interesting to him and worth keeping in his personal library or documentation for possible later use. If the document

is borrowed, copies of those parts of the document which might be
used later are desirable to avoid spending time in reordering and
other inconveniences.

I_3 represents completely irrelevant text and parts of text which
are of no interest. Complete documents can be irrelevant for the
scientist who requested the document because of insufficient or
misleading document descriptions. But an even more serious problem
is that many documents are only partly relevant and contain larger
parts with irrelevant text. Much time is spent in marking relevant
parts of text and identifying those parts which may be skipped.
Because printed documents cannot be split into relevant and irrelevant parts, one has to accept a document as a whole and, when
buying, one has to pay for the whole document.

The total input text I_t is estimated as about 11000 pages/year for
the average scientist:

> 2000 pages of journals and reports
> 2000 pages of books bought by the scientist
> 7000 pages of books borrowed from libraries.

The average number of book copies sold per scientist or engineer
has remained relatively constant over the years at about 7 copies
(King. p. 27). The time needed for browsing through this text is:

$$LB = \frac{I_t}{a} = \frac{11000 \text{ p/year}}{70 \text{ p/hour}} = 157 \text{ hours per year}$$

or about 10 % of the working time of a scientist. King (1976, p. 67)
assumes, on the basis of the SATCOM report, that the average time
spent by scientists for reading is only about 127 hours per year.
The following shares of I_1, I_2 and I_3 are estimates:

I_1 and I_2 each account for 2700 pages/year or each 25 % of the total
incoming text I_t. The individually irrelevant literature I_3 accounts
for 50 % or 5400 pages a year.

Scientific and technical literature is either bought by the
scientist or borrowed from libraries. The scientific and technical
library expenditures in USA in 1974 were estimated as 700 millions
of dollars or \$ 700 per scientist and student (King 1976, p. 55).
A major part of this money seems to be spent for transmitting texts
which were not used by the recipient.

3.4 Analyzing Literature

Those parts of the literature which are recognized as necessary
tests I_1 should be read carefully. These play usually and important
part in an article or book. The reader must obtain a sound under-

standing of the text in order to use the information for his own
investigation.

Usually a scientist will not have enough time to read all texts
which he has recognized as necessary. But we define I_1 as the
necessary information which is actually read.

The length of the text which is analyzed in a given time T_{LA}
depends on the speed of text analysis b

$$I_1 = b \cdot T_{LA} \quad \text{(pages)}$$

b is estimated as approximately 10 pages/hour.

The average daily input processing capacity of a scientist
depends on the time T_{LA} he has free for literature analysis. If
T_{LA} is 10 % of his 8 hours working time, then he can process the
following amount of text

$$I_1 = 10 \frac{\text{pages}}{\text{hour}} \cdot \frac{10}{100} \cdot 8 \text{ hours} = 8 \frac{\text{pages}}{\text{day}}$$

If 8 pages are analyzed in one day and if a year has 200 working
days, the total amount of literature thoroughly read by the average
scientist in a year is:

$$I_1 = 1600 \text{ pages/year}$$

If the reader understands and remembers all knowledge which is
contained in the analyzed information I_1 and if we suppose that all
text I_1 are new and non-redundant for the reader, then we can infer
that his personal knowledge level can be measured by the amount of
texts I_1 he reads per year.

The individual knowledge is a subset of the yearly increase of
the objective stock of knowledge E. The relation between them can
be regarded as the factor describing an individual's cognizance of
objective knowledge:

$$F = \frac{I_1}{\Delta E}$$

The objective stock of knowledge is growing. The proportion of
time available for literature analysis is decreasing. As a result
of both trends, every scientist knows less of the objective knowledge.

Suppose, that each text of the yearly produced 10^8 pages is
associated to only one working field and that the working fields
are defined such that the same number of text pages is produced in
each field. If we assume further, that each scientist reads I_1

pages a year and has a number of working fields q = 1 and that each
text is read by the same number of scientists, than one can conclude
the number of those hypothetical working fields w for 100 % individual's cognizanze of the produced texts of the working field:

$$w = \frac{m \cdot T_y}{n \cdot I_1} = \frac{100 \% \cdot 10^8 \text{ pages/year}}{1 \cdot 1600 \text{ pages/year}} = 62500$$

If we assume less working fields, the resulting cognizance
would be also smaller under the above assumptions. The number of
scientists per working field would be

$$s = \frac{10000000}{62500} = 160$$

3.5 Creative Work and Text Production

Scientific productivity depends on several factors such as
personal knowledge, intuition, phantasy stimulation by communication, motivation, organizational structure, instrumentation
(Bunge 1962). We restrict our discussion again to quantitative
aspects and statistical averages.

According to Hagstrom's (1970) analysis of the productivity of
scientists in six disciplines, a scientist produces about eight
articles of his own research, one review article, 0.3 textbooks
and 0.25 other books in five years.

Review articles and textbooks are considered as secondary
literature, because the knowledge mediated by them is usually reported in other research articles published earlier. We classify
this part of the total textproduction as redundant (TP_{R_2}) without
implying any pejorative connotation.

Those research articles or parts of the articles which are
intentionally written as original contributions to the stock of
knowledge, but are really not new, because the author was not aware
of other publications on the subject, must also be classified as
redundant (TP_{R_1}).

The last part of the total text production TP is the non-
redundant part of the text production, $TP_{\Delta E}$:

$$TP = TP_{\Delta E} + TP_{R_1} + TP_{R_2} \quad \text{(pages)}$$

If the average scientist produces 8 articles in 5 years and if
each article has 20 pages, the production rate would be 32 pages
per year or 0.16 pages per working day.

If 15 % of working hours are dedicated to producing texts (TX)
and 5 % to editing own texts (ED) and if half of this time is spent

for primary literature $(TP_{\Delta E} + TP_{R_1})$, the average scientist's production speed would be

$$c = \frac{32 \text{ pages/year}}{10 \ \% \ \cdot \ 1600 \text{ hours/year}} = 0.2 \text{ pages/hour}$$

The individual's originality can be defined as

$$o = \frac{TP_{\Delta E}}{TP_{R_2}}$$

If the individual's cognizance of the objective knowledge decreases, the production of non-intentional redundancy will increase.

It is an interesting question, whether one should ask the scientific community to spend more time in reading or in producing own contributions. It is a trade-off between lower productivity with a higher originality and a higher productivity with a lower originality. If the amount of new knowledge produced by the scientist is the same in both cases, one would prefer the first solution, because every scientist needs more time for literature identification T_{LI} and browsing through literature L_B, if the number of incoming text pages I_t is increased.

4. PROPOSALS FOR SCIENTIFIC REPRESENTATION

We have discussed some of the important activities of a scientist and indicated how individual behavior and performance depend on the information from the communication and representation system. The use of scientific texts and the interaction between scientists and the communication system would be more efficient if the texts were represented according to guidelines, which will be discussed now. But working in accordance with the proposed guidelines is limited by constraints on the economics of the system of printed publications.

4.1 Decomposition of Multi-text Documents

The smallest, physically non-decomposable unit of our current system of printed publications is the document. A document may contain several logically independent texts. A scientist is usually interested in text which represents only a part of a document. But printed documents cannot be decomposed. They are borrowed or sold as a whole. The system forces the reader to buy or borrow documents with texts in which he is not interested. Uninteresting texts are a burden on the communication system as well as on the private library and private documentation.

The system of printed publication also forces editors to publish text units of a larger size, because the relative price per page for a book of only a few pages would be too high. Finally, this economic constraint influences the authors to produce either journal articles which are put together with other texts or to produce books or reports of a larger size.

4.2 Decomposition of Multi-thematic Texts

Even if a document such as a reprint of an article or a monograph contained only one text, only some parts of the text would really be of interest to a scientist working on a specific problem. Much time would be necessary to identify relevant parts of the text and to extract this information from them.

The only way to reduce time for this work is by decomposing a text into several texts each of minimal length. Each self-contained part of a text should be produced as an independent document. Each text should have its own title describing the subject as precisely as possible.

But there are also psycho-linguistic limits to decomposition. Texts cannot be decomposed up to any size, as each sentence of a text is interpreted within the context of sentences around it. If an author introduces a new concept, he must use new terms or explain his own interpretation of a common term. To do this, he needs explanatory context in addition to the description of his new scientific information. Another factor which limits decomposition is the fact that any new proposition or scientific result is based on a framework of existing knowledge and existing theories. The more facts a text presupposes, the more difficult it is to understand this text in the way intended by the author.

The minimal length of a text depends on several factors. The average length of current journal articles still seems too high. Books are decomposable at least by a factor 1:10. The table of contents may indicate where the cutting lines should be.

4.3 Thematic Concentration

Another requirement following the same idea is to restrict a text to only one subject. Adjoining subjects and presuppositions of general importance should be discussed in separate text and referred to by citation. An author following this guideline would write in a style which is typical of handbooks.

Thematic concentration corresponds with decomposition: concentrating on one topic results in a non-decomposable text.

Thematic progression as understood in the linguistic school of discourse theory (Danes 1970) can take place, but only as a progression from the general to the more specific subject, not by moving from a subject to another subject at the same level. So for example one would allow in a text on the language Algol discussion of special parts such as declarative statements, but one should not include any description of another programming language or any discussions of applications of Algol.

4.4 Precise Description of Texts

A final guideline is directed to authors and to documentation services: every text should be described as precisely as possible. Many documents which are not well described by title or abstract are requested, transmitted and checked because the requester had some wrong expectation about the contents of the document.

Document description by keyword lists might be an appropriate method for the description of larger multi-thematic texts. A text retrieval system for short text containing the primary information as described in the next paragraph requires a more precise method such as the proposed description by phrases (Rahmstorf 1977).

More precise description is needed to avoid misinterpretation and to help identifying texts covering the same subject. It would help to make text redundancy more visible, to stimulate comparisons and encourage progress in the subject matter.

Everyone should be motivated to develop his own productivity. One should not punish production of redundant text, because several descriptions of the same subject written by different authors will help to understand the subject. Each text might contribute by another important viewpoint. The point is that it should be as easy as possible for readers to recognize which texts describe the same or closely related subjects.

5. SCENARIO OF FUTURE SCIENTIFIC COMMUNICATION

The representation of the primary knowledge by a handbook-like collection of small texts stored in a computer and accessible via a communication network is called here a universal text information system. We now compare it with the current system and with another approach based on formatted data bases (Fig. 4).

Documentation services currently supporting the scientific community do not use computer and communication technology to substitute printed paper as the main medium for the primary literature. Their data bases contain mainly bibliographic data, abstracts and other secondary information.

Properties	Bibliographic data bases supporting current system of printed documents	Universal text information system (handbook)	Universal fact retrieval system
Structure of primary information	Text	Text	Formatted data
Main transmission medium	Printed paper (document)	Electronic transmission	Electronic transmission
Unit of transmission	Document of any length	Text of minimal length	Single proposition or data record
Transmission technique/organization	Mail/book trade	Computer network/public	Computer network/public
Main medium of representation	Printed paper (document)	Electronic storage	Electronic storage
Organization of knowledge stock	Personal libraries and central public libraries	Public text data bases	Public formatted data bases
Personal knowledge file of scientists	Documents with any number of texts and any size	Text reproduced by terminals from text data base	Data base printout produced by terminal
Advantages	– no stylistic restrictions – no technical requirements	– efficient text updating – no time delays – reduced redundancy in transmissions, text storage and private files – Improved productivity	Same as 2
Deficiencies	– low speed, delayed distribution – redundant transmissions – redundant private libraries – high publicat. redundancy – unnecessary overhead load on individuals – document replacement, no updating	– scientists need access to computer network – decomposition of texts can result in difficulties – requires more control of individual style (handbook)	– scientists need access to computer network – limited to information, which can be formatted – data entry and retrieval require special knowledge

Fig. 4: Features of three Scenarios

The current system of printed publications is characterized by a high degree of freedom. Everyone is free to produce publications of any style and length. Any collection of texts can be published as one document.

But there are more deficiencies than advantages: production and transmission of printed documents are slow and result in time delays. Transmission costs of one page text of small publications are high. The production of larger multi-text publications and multi-thematic texts results in higher redundancy transmission, high publication redundancy, redundant private libraries, unnecesary overhead load on individuals and decreased originality.

Printed publications do not allow a permanent update. Updating is expensive, because a new edition of a printed text has to replace the printed document containing the previous version of the text.

More and more scientific and technical results are represented as formatted data in large computer files. Will formatted data bases substitute printed paper as the main medium for scientific information?

This seems to be unrealistic, because formatted data bases are limited to prestructured information. One can not convert the diversity of scientific discourse from free text into fixed data structures. Formatted data and numeric information is inserted according to a predefined classification scheme. Users entering or retrieving data must know and accept the predefined structures of the System.

Therefore we return to the proposed universal text information system. A scientist may use it for the following functions:

- typing and editing
- submitting final text to the system for public access
- receiving information on relevant texts of the store
- reading texts and generating copies

Information written by an author is immediately available for the whole community. Delays in production as well as distribution resulting from the medium of printed paper disappear. Text can easily be updated by the author without the expense, delays and redundancies involved in new printings of books. A reader can request copies of text or copies of part of the text. He need not buy or borrow complete volumes if he is only interested in selected parts of a text.

The system will only be used if a considerable part of actual primary texts are entered and available via terminal access for most scientists. The initial investment is high and requires governmental support. The system will partly replace functions of publishers, booksellers and librarians. Problems with copyright, privacy of unreleased texts and accounting should be solvable.

A universal text information system as proposed here is more than an "electronic journal" because it requires an almost complete coverage of the scientific discipline, a broader use by the scientific community and a special adaption in writing. Each text contribution should be as short as possible, non-decomposable and thematically concentrated. This handbook-like style reduces many problems which we have described already. As the average unit of text will be only a few pages, one can use a terminal printer to receive the primary literature.

More and more handbooks and encyclopedias of several disciplines show that knowledge can be described by independent and relatively short articles of a handbook. But such a handbook requires more control over individual contributions and more effort on the part of writers.

A universal text information system seems to be feasible using current computer technology. Online mass storage systems with a maximum capacity of 472,000 million bytes (472×10^9) are already installed in several large computing centers.

REFERENCES

Ackhoff, Russell L., et al. "Designing a National Scientific and Technological Communication System". The SCATT Report. University of Pennsylvania Press 1976.

Anderla, G.,"Information in 1985". OECD 1973.

Bunge, M., "Intuition and Science", Westport 1962.

Battelle, "Final Report on Development and Assessment of Scenarios for the Scientific and Technical Information Search System of the Future". Prepared for Access Improvement Program Division of Science Information of National Science Foundation, Battelle Columbus Laboratories, 505 King Avenue, Columbus, Ohio. Feb. 1977 (unpublished).

Danes, F., "Zur Linguistischen Analyse der Textstruktur", Folio Linguistica 4, Heft 1/2, 1970.

Dobrov, G., "Wissenschaft: ihre Analyse und Prognose", Stuttgart, 1973.

Greaser, C. U., "Alternatives to Traditional Forms of Scientific Communication", The Rand Corp. 1976.

Hagstrom, W., "Factors Related to the Use of Different Modes of Publishing Research in Four Scientific Fields". (In: Nelson, C. E. and Pollock, D. K. (Ed.) Communication Among Scientists and Engineers. Lexington USA 1970. P. 85-124.

King, D. W., et al. "Statistical Indicators of Scientific and
     Technical Communication 1960-1980." Vol. 1 A Summary Report
     for National Science Foundation, Div. of Science Information.
     King Research, Inc., Rockville, Md. 1976.
Nelson, C. E. and Pollock, D. K. (Ed.), "Communication Among
     Scientists and Engineers", Lexington, USA 1970, 346p.
Passman, S., "Scientific and Technological Communication", Oxford
     1969. 151p.
Rahmstorf, G., "Use of Semantic Networks for Information Retrieval".
     In: "Proceedings of the International Workshop on Natural
     Language for Interaction with Data Bases", January 1977.
     IIASA, Laxenburg, Austria, forthcoming.

COMMUNICATIONS ASPECTS OF EURONET

Carl O. Vernimb and Garth W.P. Davies

Commission of the European Communities

DG. XIII, Luxemburg

1. INTRODUCTION

On 24 June, 1971, the Council of Ministers of the European
Communities passed a Resolution, which defined basic community
policy for "coordinating the actions of the Member States regarding scientific and technical information and documentation".
In particular, the Council asked the Member States to coordinate
their actions in the following areas:

(a) the creation and rational development of systems for scientific and technical information and documentation, so that a
European Network could be established;

(b) the establishment of rules and procedures to ensure the co-
hesiveness of such a network;

(c) training of specialists and the education of users;

(d) technological progress in the science and processing of documentation.

The Resolution further specified that the creation of the
network is to be achieved by the "most modern methods" and to
provide services "to all persons needing to use such information,
under the most favourable conditions as regards speed and expense", thereby indicating the advantages of a European-wide network making full use of modern telecommunications and computer
technology.

Since the decision of the Council of Ministers, a complex process of the consultation has been undertaken between the Member States and Commission, the latter having been charged with the responsbility of implementing the Resolution. A special committe was established, known as the CIDST (Committee for Information and Documentation in Science and Technology), which is composed of persons responsible for drawing up policy on scientific and technical information in each of the Member States together with representatives of the Commission. This Committee's role has been, and still is, central to the formulation of the policies for implementing the Resolution and supervises the work of various task groups set up under its auspices.

As a result of this consultation process, a 3-year Action Plan (1) has been prepared, covering the period 1975-1977. The overall strategy of this Action Plan concentrates on three main areas:

1. Development and creation of systems in the various subject fields (e.g. agriculture).

2. Creation of a physical network for information handling.

3. Development of skills and tools in information technology.

The Action Plan has an overall budget of around 7 million units of account (approximately 8.8 million U.S. $). About half of this amount was dedicated to EURONET, the telecommunications network. It is being developed on behalf of the European Communities by a consortium of the PTTs of the nine Member States and will become operational in early 1979. Terminal access will be provided to about 30 host computers operating about 100 data bases.

After summarizing the present state of progress, this paper concentrates on recent research and policy issues regarding communications aspects of EURONET: referral services, a common command language, retrieval strategies, automatic translation and document delivery services.

2. STATE OF PROGRESS

2.1. Telecommunication Network

Recently, a consortium of contractors was selected by the PTTs to implement the telecommunication network for EURONET. The consortium selected is SESA-LOGICA, CARADATA, ITALSIEL, CHRISTIAN ROVSING and SAIT. The broad characteristics of the network are:

- packet-switching technology, based on an adaptation of the TRANSPAC network, which is in the course of development and which will form the French national public data network.

- 4 switching nodes, interconnected by 48 Kbits per second lines, located in Frankfurt, London, Paris and Rome.

- access facilities for terminals, located at the switching nodes and in Amsterdam, Brussels, Copenhagen, Dublin and Luxembourg.

- the standard interface defined in the CCITT recommendation 'X 25' will be used for connecting host computers (or any packet-mode terminal).

Implementation of the network is scheduled for completion by end-1978.

2.2 Data Base Services To Be Offered through EURONET

The primary type of service to be offered via EURONET will be dial-up terminal access for on-line retrospective searching of a wide variety of data bases. Although discussions are still under way on the exact mix of data base services to be offered in its initial stages, the broad spectrum of likely services is known. Over 100 different data bases have been offered for on-line access through EURONET on 27 host computers. The provisional details of these offerings are:

1. British Library (BLAISE - British Library Automated Information Service)	London area	UK Marc, US Marc, Medline, Sidline, Toxline, Chemline, Cancerline, other British Library data bases (including the B.L.Conference Index).
2. Info-Line	London area	CAS,INSPEC data bases, Derwent data bases, and other, mainly UK, data bases.
3. Computer Aided Design Centre	Cambridge	Engineering design.
4. National Computing Centre	Manchester	Computing Hardware, Computing Soft-ware, Computing Services, Computing Education, Computing Literature, Computer Installations.

5. Belgian Ministry   Brussels   INIS, EPIC.
   of Economic Affairs

6. DIMDI/FIZ 1   Cologne   Biosis,Cancerline, Poisons
                                        data bank, Excerpta Medica,
                                        IDIS (Social Medicine) CAB
                                        Index Veterinarius, Inter-
                                        national Pharmaceutical Ab-
                                        stracts, Hospital Affairs,
                                        Medline, Psychological Ab-
                                        stracts, Science Citation
                                        Index (Asca IV), Toxline
                                        Sport Science.

7. IDC/FIZ 3   Frankfurt Chemical Abstracts Conden-
                                        states (CAC),Chemical Ab-
                                        stracts Subject Index Alert
                                        (CASIA), Chemical Industry
                                        Notes (CIN), IDC-Literature
                                        documentation, IDC-Patent
                                        documentation, DECHEMA-lit-
                                        erature documentation,
                                        DECHEMA-Materials data doc-
                                        umentation , Literature doc-
                                        umentation on plastics, rub-
                                        ber and fibres, Literature
                                        documentation on progress
                                        engineering.

8. ZAED/FIZ 4   Karlsruhe Astronomy and Astrophysics
                                        abstracts, Compendex, Energy
                                        data base, NTIS, Nuclear re-
                                        search and technology data
                                        base (IKK), INIS, INSPEC-
                                        physics, Mathematics inform-
                                        ation system (NIS),NSA,
                                        Physics information system
                                        (PHIS), Physical data inform-
                                        ation system,High energy
                                        physics data base (HEP),
                                        Plasmaphysics Index, Plasma-
                                        physics Technology Index,
                                        Surface and Vacuum Physics
                                        Index, Cambridge crystallo-
                                        graphic data files, Evalu-
                                        ated nuclear structure data
                                        file(EBSDF),Conference calen-
                                        dar for energy,physics,math-
                                        ematics,Conference reports

		on aerospace, Data base on institutions.
9. DOMA, ZDE/FIZ 16	Frankfurt	DOMA mech.engineering literature data base,ZDE elec. engineering literature data base,DRE electrical engineering data base, Compendex, INSPEC-Electro-technology/ Computers and Control.
10. ZMD	Frankfurt	AGRIS, Agricultural sciences literature data base, Food Science and Technology Abstracts, SDIM, Deutsche Bibliographie.
11. Institut Textile	Paris	TITUS III.
12. Centre Inter-universitaire de Calcul	Grenoble	Cancernet Sabir, Pascal medical data bases.
13. Fédération Nationale du Bâtiment (CATED)	Paris	Ariane.
14. Necker Hospital	Paris	Drugs data bank (BIAM).
15. Thermodata	Grenoble	Thermodynamic data.
16. PLURIDATA	Paris	Chemical data banks (including crystallographic and mass spectrometry data, nuclear magnetic resonance data, DARC system).
17. INRA-Ministère de l'Agriculture	Paris	CAB, CAIN, Zoology, Bioclimatology, CDIUPA (food science technology).
18. French Host	France	CBAC, DARC System.
19. Institut de recherche des transports	Paris	IRRD.

20. SDS	Frascati	CAS, SCISEARCH, INSPEC-Physics, Electro-technology, Computers and Control, Compendex, NTIS, World Aluminium Abstracts, NASA, Electronic Components, Metadex,Environmental Science Index, Pascal Data Bases, Pollution Abstracts, Oceanic Abstracts, Biosis.
21. Commission of the European Communities	Luxembourg	Community data bases.
22. Datacentralen	Copenhagen	CAS, Food Science and Technology, Medline (through SCANNET), Compendex.
23. CNUCE	Pisa	Data bases on: ecology, geothermal science,iconography, oceanography, legal documentation, fine arts, citation index (Italian authors).
24. Interfaculty Computing Centre	Rome	US Marc, Marc Italy, SPIN.
25. Interfaculty Computing Centre	Naples	MTS.
26. CSATA	Bari	Meteorological data bank (concerning the south of Italy), agriculture.
27. Joint Research Centre	Ispra	Nuclear sciences, material properties and other data bases.

In terms of the product mix that is emerging, about one third of all the data bases on the above list are strongly related to the hard sciences, (chemistry, physics, metallurgy, electronics and so on). About a quarter are related to medicine and biology and the rest are distributed rather unevenly over various subjects, such as education, agriculture, law and the enviroment. An important point of concern is the identification of gaps. At present there seems to be a deficiency in the number of data bases useful to the non-research side of industry. Patents are one example, but other areas, product information, market research data and so on, seem to be inadequately represented at present.

Further analysis of the list indicates that, of the hundred or so data bases, 14 are offered more than once. Of those 14, eight are of US origin; the others are mainly European, although one or two are of an international nature. Two of the 14 data bases are offered three times, and four of them are offered four times. So there is some overlap in certain, mainly major, data bases in the offerings represented by the product mix of the hosts. In this regard, it is important to note a resolution which was agreed at the Community's CIDST Committee in July, 1976, which included the following statement: "Each Member State and the Commission intending to mount data bases on EURONET will make all necessary contacts with its partners, including the European Space Agency, in order to encourage avoidance of wasteful duplication". There is therefore the basic intention to cooperate with regard to the overall supply of services to EURONET.

2.3. Principles and Policies for EURONET Data Base Sharing

A broad consensus of opinion has been reached within CIDST and the Commission on many of the key principles and policies relating to the provision and use of EURONET services. These principles and policies specifically recognise the different interests of the many parties involved (users, PTTs, host computer operators, data base suppliers, equipment suppliers, the Commission, national authorities, etc.) and have the aim of presenting a balanced set of guidelines within which EURONET information services can develop. Among the important principles emerging are:

Equality of access. All community users should be offered ready, unimpeded and non-discriminatory access to and use of information services connected to EURONET.

Freedom of choice. Every user must be able to decide freely which information centre he will use.

Price. The price charged to Community users for access and use of information resources connected to EURONET should be as low as economically practicable, easily understandable to users and should be independent of the location of the user (excluding the local telephone charge for connection of users to national entry points to EURONET). It is recognised that the responsibility for fixing the tariff level for a given service rests with the service supplier, but it is hoped that EURONET service suppliers as a whole will gradually harmonize the structures of their tariffs.

Terminals. Users should be able to use a reasonable variety of terminal equipment, which may take the form of visual display units, teleprinters (TTY compatible) or line printers (200 lines per minute). The technical requirements placed upon user terminals should be kept to a minimum.

User convenience. Special care should be taken with regard to the convenience of users, as regards, for example, arrangements for using the network, referral and passage of users from one host computer to another, etc.

User evaluation of services. Appropriate means should be provided for assisting user evaluation of services, handling suggestions and complaints and generally consulting users on a regular basis.

User support. EURONET host computer operators must be prepared to make available sufficient user support to ensure adequate accessibility to their services in all Member countries. It is recognised that EURONET is mainly intended for improving on-line access to data bases. However, means should be studied. Furthermore it is planned that a referral facility should be set up to help users find data bases appropriate to their needs.

Standardisation. Standardisation should be introduced gradually and by voluntary cooperation among those concerned. It is expected that guidelines developed in the context of EURONET will have also an important influence in promoting cooperation in general among the Community's Member States in the field of scientific and technical information.

3. SPECIAL COMMUNICATION ASPECTS

EURONET is conceived as a device which should enable users to find appropriate information. The starting point in the information selection procedure is the user, demanding information. At the end of the procedure, there should be the same user, now using information. In between are familiar steps: selection of a network, selection of a host computer, selection of a bibliographic data base or hard data bank, and retrieval. In the case of bibliographic data bases documents, from which factual information will have to be extracted, have to be ordered and delivered. This scheme will be used to "locate" the special communication aspects discussed below. The graphic presentation illustrates that the closer the information selection event is to the bottom, the more relevant to the user is the selected information.

3.1. Referral Service

The purpose of the Euronet Referral Service is to make the user's task more convenient by suggesting short-cuts in the scheme of information selection procedures. It refers the user, demanding information, to the most appropriate host computers and data bases and banks and possibly even to qualified experts. In addition the referral service can provide information on tariffs, network characteristics, licensed terminals, etc.

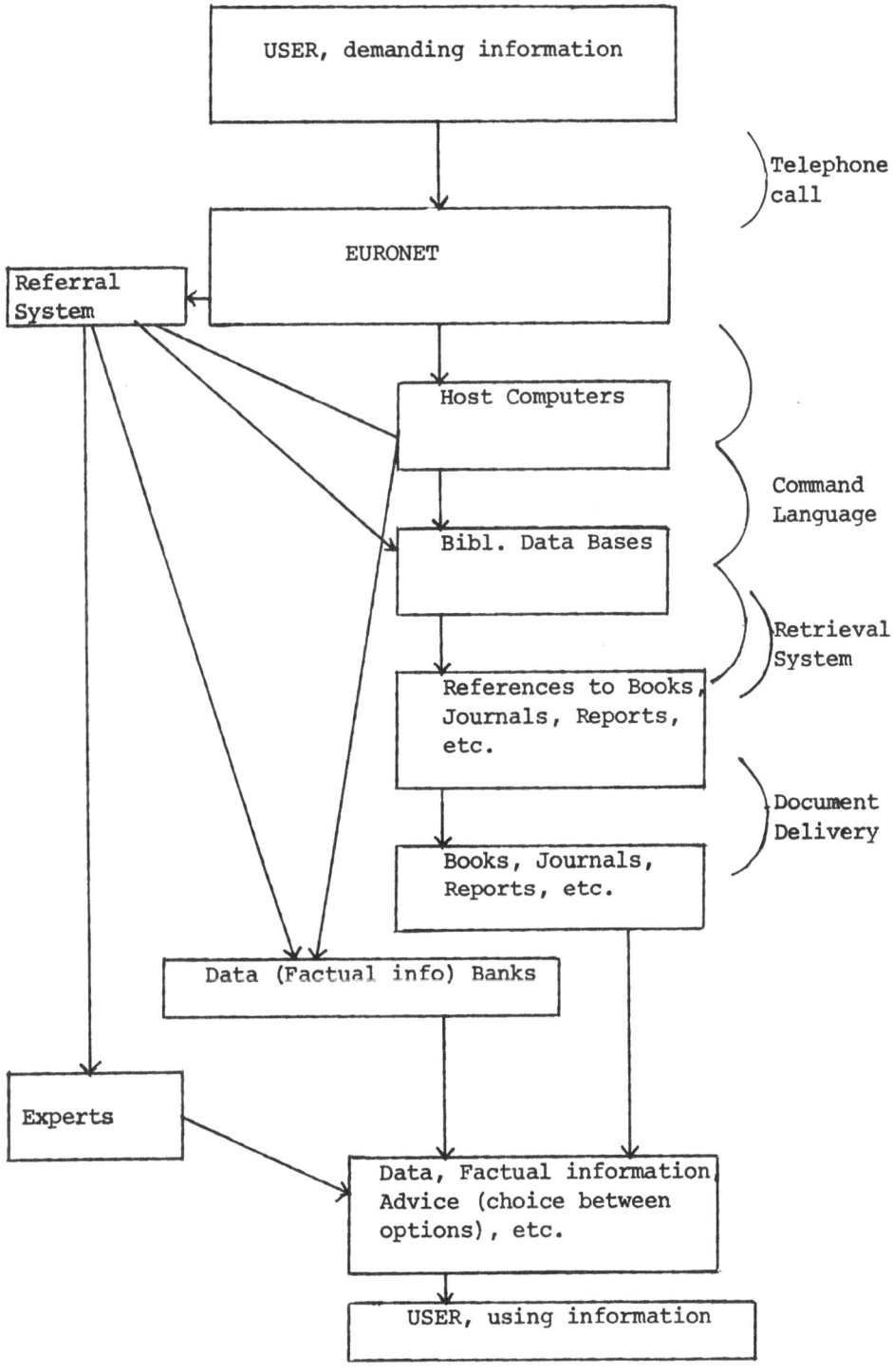

Information Selection Procedures in EURONET

As a first step towards the installation of a referral service ICSU Abstracting Board and EUSIDIC jointly will, by contract to the Commission, prepare a suitable standard description of data bases covering both their contents and the different treatments given to them by the host operators, e.g., whether it is possible to search by words in titles, by date, with right or even with left truncation, etc. The idea is of course, to make this standard description on-line accessible too.

An automatic referral system is technically feasible in which the query is entered in a free text form and in which the query terms are then compared with representative terms of the data bases made available through EURONET. A ranked list of suitable data bases would then be displayed or printed out. If and when such a system can be justified remains to be investigated.

In the scheme of information selection procedures there is an attractive path from EURONET via referral service and expert to useful information. I would imagine that subject specialists, e.g. in information analysis centres, would volunteer to give expert advice by the telephone free of charge, and, after having gained experience, would become professionals.

3.2. Standard Command Language

It is well known that users of interactive systems can have difficulty when changing from one system to another, particularly when most of their experience is with a single system. A way to minimize these difficulties is to introduce a degree of standardisation between systems, not by standardising on one particular system (as this could lead to stagnation by inhibiting development), but by developing a common definition of retrieval system functions and the commands that activate them. Therefore a study was launched to examine the feasibility of introducing a standard command set into EURONET. It was carried out by INSPEC.

The study (2) showed that it would be possible to search different retrieval systems using one set of commands, each of which would perform prescribed and therefore predictable operations. The views put forward in the study report gained such a wide acceptance that a follow-up (3) was launched aiming at the development of EURONET Guidelines. When a draft of these was presented to experts of the major hardware and software manufacturers, all commented positively on it; some are even considering implementing the standard command set soon. The Guidelines will also be submitted to ISO/TC46.

The following group of functions for which standard commands were suggested can be identified:

- general features (e.g., guidance or tuition for the user);

- entry and initialisation;

- database selection;

- query formulation;

- output;

- others (e.g., a transparent mode for switching to the original command set).

3.3 Retrieval Strategies

When interrogating a bibliographical information system through a terminal, the main obstacles to broader use are the constraints imposed by dedicated languages (e.g., thesaurus terms) and a dedicated logical algorithm (e.g., operations in accordance with Boolean algebra). As a consequence an intermediary between the end user of information and the information system is needed. This roughly doubles the costs of a literature search for the user.

The staff of the Commission designed and tested a system (4) for the automatic adjustment of queries addressed to information retrieval systems. It employs a structured thesaurus for the coordinate indexing of an average of at least five or six descriptors per document. Starting with at least two documents considered by the user as relevant to his inquiry, the system formulates different queries using the descriptors occuring in the relevant documents. Results are presented to the user for assessment of relevance; then the most efficient queries are automatically selected and loosened (broadened). The new documents retrieved are again checked for relevance by the user; and with the new set of relevant documents the loop starts again.

The automatic query adjustment procedure makes use of the fact that no one can decide better than the searcher which documents are relevant and which are irrrelevant, and nothing can determine better than the computer which descriptors were used for indexing relevant documents and for irrelevant ones. There are only a very few cases in which the documentalist using a traditional procedure cannot be replaced by the automatic procedure; they occur if at least two relevant documents cannot be found for initiating the automatic procedure. In all other cases, and this will be for more than 80% of searches, the automatic

procedure is superior to traditional searching procedures in
terms of both recall and precision. The main advantage is, of
course, that the user can, for a majority of queries dispense
with thesaurus look-up, Boolean strategy and descriptor weighting, and instead needs only to type in references for two known,
relevant papers. The economics of an online application of the
automatic procedure is being assessed under contract within the
framework of the EURONET development.

3.4. Document Delivery

The importance of a good document supply service for the
EURONET user is emphasied by the question: how can a user be
satisfied with EURONET if he gets references in a few minutes,
but has to wait weeks or months to receive the desired full documents?

The present situation will have to be improved considerably.
As a first step the Commission is launching a study of document
delivery which will comprise

- a detailed forecast of the expected document supply requests
  and their geographic distribution;

- an analysis of present document supply techniques, and an
  identification of technical and legal problems;

- a detailed description of options for solving the document
supply problem including performance (speed, reliability, user
acceptance, etc.) and costs, and recommendations on the options
to be chosen.

In my opinion there are two key problems to be solved, the
complex copyright situation and the choice between technical options. A uniform approach on copyright will have to be suggested which can be accepted by the Member States of the Community
and, hopefully, by other countries. Of the various technical
options those will have to be selected which would be heavily
supported. For example, should post or telex for transmittal of
requests be replaced by direct links between computer searches
in EURONET and computer transmission of requests? In other
words, should the user when screening displayed references be
given the opportunity to order a document by a simple command,
the order with the bibliographic data being transmitted directly
to a clearing house which would then provide the full document?

The present techniques of document transmission, such as loan, photocopy, and microform, are not fast enough. Should they be supplemented with, or replaced by, telefacsimile (through EURONET?), computer or cable television? Here also the question of optimal timing of recommendations and support arises. Within a year we expect to have some answers to these questions.

3.5. Automatic Translation

A special communication problem concerns natural languages.

The Commission of the European Communities (CEC) has launched a three-year "plan of action for the improvement of the transfer of information between European languages" (5).

The motivation behind this programme lies in the need to solve problems caused by the multiplicity of languages in the Community. One example of this is the volume of translation work generated by the Community Institutions, which requires the services of some 1300 people at present and is increasing at a rate of 10% per annum.

Consequently, the principal objectives of the plan are to create and implement multilingual tools which would improve not only the daily running of the Community institutions, but also the transfer of information from one language to another in the wider context of EURONET.

The programme will launch activities on several fronts, namely:

- automatic translation of texts in natural language;

- automatic translation of texts drafted in limited syntax;

- terminology banks;

- multilingual thesauri;

- technical infrastructure;

- assessment of applied research;

- encouragement of multilingualism.

We concentrate here on the two first items.

3.5.1. Automatic translation of texts in natural language. The programme has been planned in the light of two considerations:

(1) In the past the development of automatic translation systems for scientific texts has focused mainly on the Russian/English Language pair, which is of no particular interest to the Community.

(2) Disregarding the quality of the text supplied by the machine, no automatic translating system can operate properly in any sector without a substantial initial investment, equivalent to the establishment of a multilingual dictionary for the sector in question.

The Commission therefore intends to introduce an initial tried and tested system - SYSTRAN, developed by Toma (6) - into its departments. It intends to bring it first to the pilot-experiment stage and then to the operational stage for one pair of Community Languages and one specialised field. If expectations are fulfilled, the system will be extended to include other specialised fields and other Community languages.

Thus, the initial system will be gradually converted into a European instrument belonging to the Community and will be capable of:

- rapid draft translations in certain fields, allowing, for instance, an engineer to grasp the substance of a document without necessarily understanding all the nuances;

- translations of sufficient quality to pass directly to checking or revision, as is currently the procedure for drafts prepared by translators on the staff of the Commission.

3.5.2. Automatic translation of texts drafted in limited syntax. There are systems which provide satisfactory automatic translation of texts drafted according to limited grammatical rules. Such systems are particularly useful for the translation of summaries in several languages for information retrieval purposes.

It is intended that initially a tried and tested system of this type - TITUS, developed by Ducrot (7) - be applied to scientific and para-scientific texts, the drafting and translation of which are of interest to the Community; that the extension of this system to other specialised fields and to other Community languages be encouraged; that joint efforts be made gradually to lift most of the grammatical constraints which limit the use of the system; and if appropriate, that some staff be trained in the drafting of texts with reduced syntax.

4. OUTLOOK

A second 3-year action plan covering the period from 1978 to 1980 was submitted to the Council of the European Communities and to the European Parliament. One of its major goals is the provision for continuity of EURONET development and operation.

REFERENCES

(1) Decision of the Council of Ministers of the European Comunity adopting a first 3-year plan of action in the field of information in science and technology. Official Journal L100 of 21.4.1974 (Brussels).

(2) A.E.Negus, Study to Determine the Possibility of a Standard-ised Command Set for EURONET: Final report of a study carried out for the Commission of the European Communities, DG.XIII, Luxembourg, October 1976.

(3) A.E.Negus, Draft EURONET Guideline: Standard Commands for Retrieval Systems: Interim report of a study being carried out for the Commission of the European Communities, DG.XIII, Luxembourg, May 1977.

(4) C. Vernimb, Automatic Query Adjustment in Document Retrieval; Information Storage and Retrieval (in press).

(5) Plan of Action for the Improvement of the Transfer of Inform-ation between European languages, CEC Decision, CDM (76) 705 Final of 23 December 1976.

(6) P. Toma, Computer Translation: In its own Right, Kommunika-tionsforschung and Phonetik; IPK-Forschungsberichte,BD.50,1974.

(7) J.M. Ducrot, Le Systeme TITUS III. Report of the Institut Textile de France, Service Documentation, 1975.

(8) G. Anderla, Sharing Information Resources - a contribution to socio-economic developement - the European Effort, FID Symposium, 1975.

(9) R.K. Appleyard, The EURONET Project of the European Communi-ties, 7th Biennial IIS Conference, University of St. Andrews, Scotland, 1976.

(10) C. Vernimb, the European Network for Scientific, Technical, Economic and Social Information, Nachr. Dok. 28 (1977) No.1. p.11-18.

(11) H. Ungerer, EURONET: The EEC on-line Information Network, UNESCO Bull. Libr., Vol.XXXI,No. 3, May-June 1977,p. 128-133.

(12) G. Davies, Data Base Sharing in the EURONET Environment, NATO/AGARD Conference: The Impact of Future Developments in Communications, Information Technology and National Policies on the Work of the Aerospace Information Specialist, Oslo, 22-23 June 1977.

(13) W. Huber and P. Van Velze, Facilities and Access Arrangements for Data Collections on a Network, 2nd EASI (European Association for Software Access and Information Transfer) Workshop, Edinburgh, October 1976.

(14) W. Huber, EURONET, Contribution à la Réunion de travail "Banque de Donnée" des Chambres de Commerce et d'Industrie de la Communauté Economique Européenne, 14 septembre 1977, Bruxelles.

(15) H. Ungerer, EURONET, Contribution to Leergang Computer Communicatie, 31 Aug.-1 Sept. 1977, Technische Hogeschool, Eindhoven.

PROBLEMS IN FORECASTING THE PRICE AND DEMAND FOR ON-LINE

INFORMATION SERVICES

A.D.J. Flowerdew, J.J. Thomas, C.M.E. Whitehead

London School of Economics, Houghton Street,

London WC2A 2AE

INTRODUCTION

One of the most important recent changes in the provision of information for industry, government, research organisations and the general public has been the development of on-line information systems. Acceptance of these systems and therefore their growth and development has been much slower in Europe than it has in the United States. But many people believe that their use will increase enormously over the next decade both as a substitute for other information services and as an entirely new type of service especially as technology which exploits the inter- active nature of the system and understanding of how to use it become more developed. These advocates see on-line information systems providing not just the secondary source material that is their present main staple, but also details of product specifica- tions, up-to-date commercial and economic information, other kinds of numerical data and problem solving capability such as that required to assist medical practitioners in diagnosis.

A year or two ago we were asked by the European Economic Community to analyse the extent to which projected demand for on- line services would be affected by the pricing structure and level of prices. They are interested in this question because of the EURONET project, which aims to provide on-line information services throughout Europe via a network of linked computers. This task involved our first assessing the determinants of demand for on-line services as a whole, and then specificically analysing the effects of price as one of those determinants. The work was

carried out early in 1976.   On-line services had then only been
available in Europe for a short time and much of the use at that
time had been obtained free on an experimental basis.   So our
work can only be regarded as indicative.   However the results seem
to us to be relevant both for forecasting the potential growth of
this kind of service and for assessing its value to users.

In this paper, which expresses our own views and not necessarily those of our sponsors, we discuss first, the determinants of
demand, second the relation of demand to price and finally problems
in determining trends in price and demand.

DETERMINANTS OF DEMAND FOR ON-LINE SERVICES

Factors affecting the demand for any good include:

* the nature of the product
* its price
* the price and availability of substitute and complementary
  goods
* the closeness of these substitutes
* the level of income
* consumers' tastes
* the number of consumers.

The demand for information has a number of special features.

(a) In general the consumer does not desire information for its
own sake.   Rather he hopes that the information will help him to
obtain some other desired end.   Demand for information is thus
derived from the demand for other goods and services.

(b) It is very difficult to measure the benefits derived from
information, because:

  * of ignorance and uncertainty of the value of the information
    received,
  * there is a time lag between when the information is obtained
    and when it is used,
  * information is often collected for an unspecified future use,
    making it an investment with a very uncertain return,
  * the final consumer of information is often not the person
    who pays for that information.   The user may only be
    concerned about the time it takes him to obtain the informa-
    tion but the cost to the organisation that employs him may
    include as well as the cost of his time, the financial costs
    of obtaining the information, information service staff time
    and maybe other elements as well.   In some circumstances
    these costs may be passed on to user departments by a centra-

lised information service. If so, depending on the method used, the cost perceived by the user may be equal to or even greater than that incurred by the organisation.

* the unit of information is hard to define and the quantity paid for, such as contact hours, does not represent a quantity of information demanded, which could be the number of relevant references, or the verification or not of the existence of some required result.

(c) The production of information is often a by-product of the production of other goods. This is frequently the case with abstracts and bibliographies which may be by-products of the primary sources. It is also true for data which is collected for use as well as for subsequent reference. It is particularly important for on-line information. Data bases are generally compiled along with "hard copy" volumes, the computers used will often have been installed for other purposes, access to the data bases can be through general communication systems and the terminals themselves can also have other uses. Thus the availability of on-line information and its cost are jointly determined along with many other services.

(d) Almost all of the information that is available on-line can be obtained by some other means such as off-line searching of computerised data bases or searching of published "hard copy". The value of an on-line search differs from these other means in three ways:

    (i) the relative cost
   (ii) the time taken to obtain the information
  (iii) the quality of information which can be obtained using
        interactive searching may be different.

Because the nature of information is complex users of on-line information sometimes use a simplified means of quantifying output to assess the value of a service. In searching for secondary information two possible approaches are to count the number of titles obtained, or the number of useful titles obtained. In evaluating on-line services the second approach is preferable, since a major advantage of on-line searching is the reduction in irrelevant information which may be obtained by interactive use of the system.

(e) The cost to an organisation of using an on-line information system once this has been set up is made up of:

* the price paid
* the user cost of making the search, including the time
  of information specialists and researchers
* additional documentation costs resulting from the search.

When deciding whether or not to become a user the organisation must
also take into account:

* the cost of buying or leasing a terminal, or of adapting an
  available terminal with some loss of capacity for other work
* training of staff.

The cash price of a search may be a relatively minor part of the
total cost of using on-line services, but it is of greater importance in deciding whether or not to do a particular search.   This
price may include:

* royalty paid to the data base provider
* cost of access to the file
* price of transmitting the search, both on exclusive links
  and on open telephone lines
* cost of printing or displaying output.

(f) According to the theory of the firm, organisations will
purchase a good or service if the resulting increase in their
revenue exceeds the cost.   But no organisations actually make such
calculations in relation to every single purchase.   They are more
likely to make some more or less precise assessment of the likely
value of certain kinds of goods or services during a period and
then set a budget level in cash terms.   A budget may be no more
than a target or estimate of expenditure, but sometimes, especially
when finance is scarce, the budget may function as an upper limit
on expenditure, and this is particularly likely with large firms and
with goods and services not directly linked with short-run financial
goals.   Because on-line searching is a novelty, information departments in some firms may simply not have a budget for it, and
setting one up could involve lengthy studies and discussions.   Even
when budgetary provision for on-line services has been made, it may
not be at all flexible in the short run.   For instance if prices
increase halfway through a budgeting period, departments may have no
choice but to cut back use of the service, even though they believe
that it still represents good value for money.

Empirical Results

As noted above, there are not many users yet of on-line
information in Europe, so we could not usefully carry out a random
sample survey.   Instead we decided to concentrate on interviewing
major users, and also suppliers of both information services and
access systems, plus some organisations which catered for small users
and one or two organisations which were non-users although they
resembled in many ways organisations which did use on-line services.
As well as a number of personal interviews we sent a questionnaire
to organisations which we did not have the time or the funds to reach.

In total 47 organisations and individuals provided information.
The results of such a study cannot be regarded as definitive but
there were some significant patterns in the answers we received
which we believe are highly relevant to the future development of
on-line information services.

Determinants of Demand

What type of product is required?

The on-line searches being carried out at present are determined more by what is available than by what is desired.   The amount
of use that respondents expect themselves and others to make in the
future is strongly dependent on the type of material that will be
available on-line.   Most respondents were less interested in a
wider range of bibliographic data bases than in the provision of
commercial and numerical data.   This is illustrated in Table 1.

Respondent Type of informa- tion desired	Suppliers	Interme- diaries	Users including potential users	Information Specialists
Commercial and economic data	3	2	8	2
Product and research information	5	2	9	2
Numerical data	8	4	7	3
Problem solving capacity	8	4	6	3
No extension			1	
Total[1] number in each class	14	10	17	6

[1]Many regarded two or more types of material as potentially valuable.

Table 1

Desired Extension of Material Available On-Line

At what price?

The cost of using on-line information services varies a good deal depending on data base used, the supplier, the means of communication, and so on.   To give an idea of order of magnitude a figure of $20 per half hour search may be quoted as typical.   Most respondents believe that they would be relatively unaffected by small changes in price.   They think these could be absorbed without losses of useful output by improvements in the efficiency of the search methods used.

Respondents were asked how they would react to different real price increases.   The current pricing levels were taken as the base from which changes occurred.   Table 2 tabulates the answer of those who were willing to evaluate likely responsiveness to particular price ranges defined in this way.

These results suggest that at the present time most users and suppliers regard the product as worthwhile and that small changes in price would not result in very much modification of demand. Instead they would attempt to reduce the search time per query rather than, at this stage, to reduce the number of queries. Some respondents thought this could be done without loss of value from the output because they often did searches in greater depth than was necessary, because they enjoyed doing so.   Others felt they had not yet mastered the system and would concentrate on becoming more efficient by better preparation.   In other words many respondents felt that at current prices there was enough slack in their usage to absorb 10% cost increases resulting in the same value of output for the same budget.   Others felt that they could absorb up to 20% without difficulty and some said that they would not try to reduce their search costs unless prices rose by as much as 20%.   Thus with price increases of up to 20% the demand from

Increase in price	Expected reaction			Possibly stop using altogether	Number of answers	
	Zero	Small	Large			
5%	86%	14%	0%	0%	22	100%
10%	50%	50%	0%	0%	22	100%
20%	14%	41%	45%	0%	22	100%
50%	0%	9%	41%	50%	22	100%

Table 2

Reaction to Price Increases

existing users appear to be inelastic - the quantity of on-line searching would decline, other things being equal, but not so much as to decrease the revenue obtained.   Above this level a large effect on demand is expected, with 50% of users predicting that they would cease to use the service after a 50% price rise.   A second point is that not all elements of price affect demand in the same way.   For example, telephone costs are often not charged directly to the department deciding whether or not to use the on-line system.   The type of costs which are likely to change over the next few years are therefore very important not only because they will affect choice but also because if on-line systems become widely used cost allocation and budget procedures are likely to change.

Finally starting to use this service is a decision that is responsive to price.

The existence of a charge at all will be particularly important in organisations whose budgets are very specific.   Once a budget has been made available it seems not to be so difficult to obtain increased finance if consumers are satisfied with the product.   But proof of cost-effectiveness was clearly regarded as exceedingly important in both obtaining and extending the budget.

The problem of obtaining a budget at all is particularly important in relation to university and other academic usage.   In Europe most academic researchers are not expected to fund their own information provision.   It is expected that this will be available via a "well-found" library and is normally paid for out of a block grant to the institution.   If libraries or computer centres were to levy a charge it could be difficult for individual researchers to obtain funds for direct payment for information involved in on-line searching.

Alternatives

The existence of alternative means of obtaining the information and the relative price of such alternative means is also important in determining demand.   Although on-line searching is a new service it provides a product similar to that which was previously available in other ways and thus organisations can calculate its cost-effectiveness as a provider of information.   During the next few years most respondents expect staff costs to increase and therefore expect that on-line searching will become relatively cheaper.

The cross elasticity of demand between different on-line services appears to be high.   Most of those who could use more than one service had done so and had compared costs, usually per reference.

But by no means everyone automatically took the cheaper search.
Differences in the type of information and software available (free
text searching, provision of abstracts, coverage, etc.) were
regarded as important in that they affected both the value of the
result and related costs such as those of documentation and clerical
assistance.

FUTURE CHANGES IN DEMAND AND COST

Income Elasticity

Undoubtedly there has been a massive growth in demand for
information services over the last decades as the complexity of
production, sales methods and other activities have increased and
as firms have increased in size, diversified and extended their
markets.   This suggests an income elasticity very much greater
than unity for information as a whole.

In the public sector research budgets appear to be particularly
highly income elastic in the sense of being heavily dependent upon
the level of economic activity.   This also applies to government
department expenditure where the evidence of the last few years is
that public expenditure is growing faster than income in general but
that it is particularly subject to cuts in times of recession.

Thus we expect the growth in overall demand for on-line information to be heavily dependent upon the expansion of economic activity
in Europe but also that the growth in on-line services will be
more rapid than the expansion in overall output.

Expected Expansion in Demand at Current Price Levels

Respondents were asked to estimate the likely increase in their
own demand over the next ten years and also what level of general
development they expected.   Many felt that they could give no
estimate at all particularly because their usage was in the experimental stage.   This also means that percentage increases may look
very large but be reflecting really very small absolute usage.

	Own demand (both consumers and providers)	Total demand
Less than 100%	32%	11%
Between 100% and 1000%	54%)	89%
Above 1000%	14%)	
Total number of respondents	22	28

Table 3

Expected Increases in Demand over the Next Ten Years

For instance one intermediary who predicted between 300% and 400% increase in demand over the next ten years was estimating from a base of only 10 searches per month.   The typical view of users might be regarded as doubling use over the next couple of years but slower expansion thereafter.   The mean estimate for the ten year period would therefore probably be less than 5 fold with very few predicting more than 10 fold.

Two suppliers with considerable experience predicted growth rates of demand for their own products of as little as between 10% and 15% per annum and these together with other replies suggest that perhaps even the five fold estimate is somewhat of an over-estimate.   Consumers would probably be expected to estimate lower increase than suppliers because suppliers are taking into account new users of their systems while consumers may only be taking their own use into account.   For these reasons we would expect estimates of total growth in on-line demand to be greater than that for indi-vidual demand.

Cost Trends

However there could be off-setting cost effects.   Table 4 shows the expectations of cost changes found in our survey.

	Suppliers	Intermediaries	Users
Costs expected to go up			
Transmission	6	3	3
Provision	6	1	1
Costs expected to go down			
Transmission	0	0	0
Provision	2	4	3
Uncertain			
Transmission	4	4	3
Provision	3	1	2
Costs expected to stay same	1	1	4

Table 4

Expected Cost Changes in Real Terms

It is believed that some cost items will fall very considerably over the next decade as the cost of computer storage declines and access costs are reduced by the introduction of further nodes minimising the need for long-distance telephone communication. However there is evidence that at the present time charges do not reflect the full costs of production, although short-run marginal costs are sometimes covered. This is not a long-run equilibrium position and has most important implications especially for the type of information likely to be provided.

With respect to data bases it was suggested that on-line royalties were covering only a very small part of the total costs of the data base. Two estimates in relation to services where there was no government subsidy implied that the norm was usually between 5% and 10%. This is possible only because other sources of income - mainly sale of hard copy but also including SDI services - pay for the bulk of costs. However, it is generally agreed that these other sources of income were likely to become less and less important over the next few years. One information specialist to whom we talked thought that provision of hard copy secondary information sources would decline dramatically if on-line services really caught on and therefore the revenue from these could be drastically reduced leaving a shortfall presumably to be charged to on-line data base providers. The cost of producing a data base is expected to increase more rapidly than the rate of inflation. So total costs of data base provision are expected to increase substantially especially if on-line searching does become the main use of this type of information. Moreover if the types of information provided are extended as desired by consumers data base costs will rise drastically, not only because new bases must be started from scratch but also because of the need for up-dating.

Another element in the cost of data bases is the extent of subsidy. Many bases provided in the United States receive large subsidies or are provided free of charge from subsidised organisations. Many of the numerical data bases which might be made available on line are spin-offs from subsidised research. Not all these subsidies are likely to continue if it becomes obvious that the data bases are being used for a commercial service. Thus existing data bases are likely to become more expensive and will expect to draw more of their revenue from their on-line customers. Other potential data bases for which consumers have expressed a need would probably have to be financed completely from charges for searching at least at the outset. All of these factors suggest that the data base costs are currently unrealistically low and a ten-fold or greater increase would be by no means impossible. This could substantially offset the expected decline in computer and other costs. It is difficult to obtain evidence about the costs of hardware and software. Most suppliers expected there to be considerable new

investment in computer hardware specific to data base provision if
growth in demand is sufficient.   Few thought that there was
currently any massive excess capacity in terms of storage while most
agreed that increasing demand required new and extended data bases,
so that more capacity is going to be needed.   The computer hardware
required for numerical data bases and commercial and economic
material would probably be less costly, but would still require special provision.   A similar position is found with respect to
software.   More software will be required but the extra costs could
be offset by extra usage.   However if problem solving capacity is
seen as the basis for expanded demand, far more sophisticated programs entailing high costs will be necessary and these costs might
well not be offset by increased consumer demand.

     The possibility of higher, rather than the expected reduction
in costs as demand increases and costs can be spread over a
greater number of searches arises from the fact that development of
existing services has been a spin-off from other production processes.   On-line systems would probably not have been set up at all
if much of the technology had not been paid for from elsewhere,
many of the personnel had not anyway been working on similar
problems and even some of the hardware and software made available
from other sources.   But as on-line services grow they may have
to stand on their own feet and not gain from the fact that many
costs are being borne elsewhere.   Whether or not they will have to
do so depends considerably on the role that governments are
prepared to play in supporting research in information dissemination
and so allowing on-line services to benefit without paying the
full costs.   A number of respondents thought that governments should
be prepared to pay for basic research into software for problem
solving on-line services even when they believe that these services
should otherwise be expected to cover their costs.

     The fear that the relationship between costs and prices will not
remain as currently expressed is further strengthened by the belief
that the American firms are not finding on-line services profitable
at current price levels and that there is little excess capacity
available to allow costs to be spread over more users.   Moreover
with respect to production costs, very few of the hosts are
charging prices covering the full cost of the service to all users,
and most intermediaries are probably not covering all costs, but
on-charging any payments they make together with a proportion of
staff costs and allocated overheads.

     A final item in direct (as opposed to user) costs is that of
transmission.   Here there is great uncertainty reflecting difficulties in predicting the pricing policy of PTTs.   Some factors
however are relatively clear.   Transmission costs which relate to
distance will generally decline especially if EURONET is set up

because of the proliferation of nodes.   Speeds of transmission
are also expected to increase and therefore, excluding price changes,
transmission charges per search are predicted to decline.   Then
come the uncertainties.   Telephone charges have been increasing
faster than inflation over the last few years and this is expected
to continue, perhaps just about offsetting the gains from increased
speeds.   Finally, there is the cost of transmission from host to
node, where the position depends crucially upon the pricing
policies of the PTTs, and possibly upon technical changes such as
the use of satellite communication.

These factors together with the fact that up to now consumers
have often been accessing services at zero, or at least cut, price
experimental rates means that demand is likely to grow less fast
than that predicted by examining the expected demand of existing
and potential users at current price levels.

Overall, the major determinants of growth appear to be: the
rate of expansion of economic activity, the extent to which on-line
services are relatively cost-effective systems and most particularly
the speed at which new kinds of data bases can be made available
including far more advanced software for problem solving applications.

RELATIONSHIP OF DEMAND WITH PRICE OVER TIME

Among the factors expected to affect the demand for on-line
searching supplied through EURONET, the following appear to be the
most important.

1. The autonomous growth of demand, as more individuals and
organisations learn about the potentiality of these services.

2. Reaction to growth of the supply of new data bases and new
software systems.

3. Choices open to customers; especially whether or not
Lockheed, SDC or similar services will be available in addition to
EURONET.

4. Prices charged and user costs for services made available
through EURONET.

5. Prices and user costs of computing services.

6. Prices and user costs of alternative information services
such as off-line searching or manual searching.

7. Economic prosperity.

8. The nature and extent of the EEC/government control of the

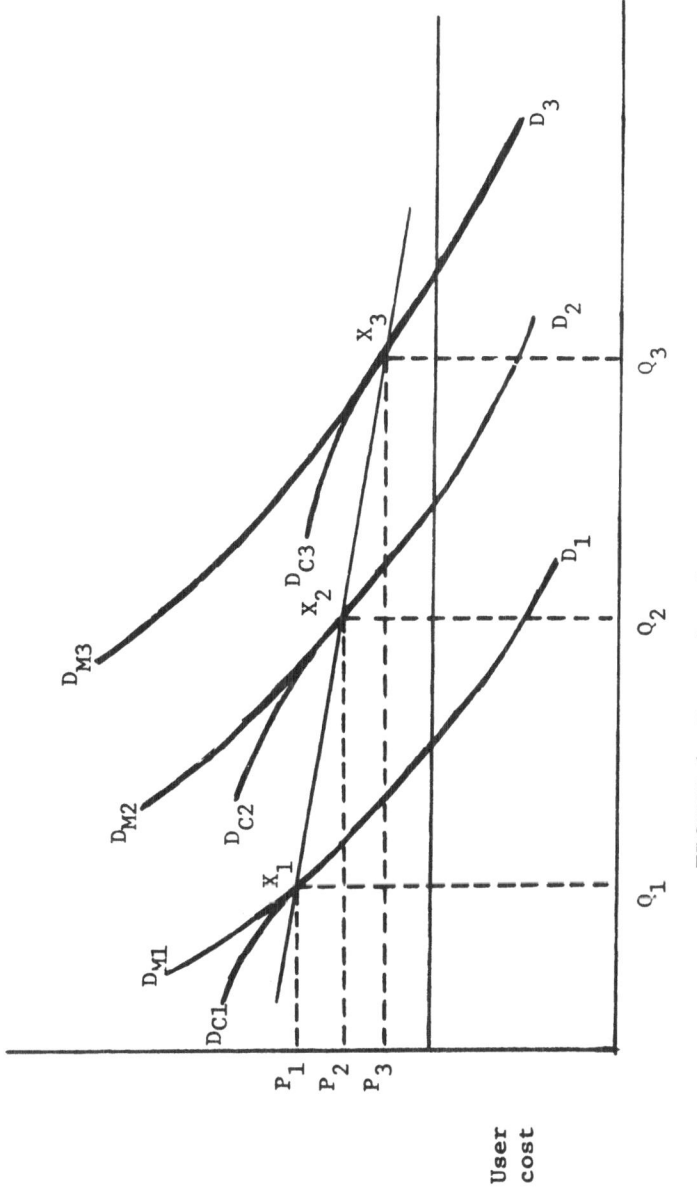

FIGURE 1 Demand relationships over time

facilities offered.

Figure 1 illustrates a simple model of demand taking into account only three factors: calendar time, price (and user cost) and market organisation.   This calendar time relationship will incorporate the effects of factors 1 and 2 above ; factor 3 is represented by market organisation.   The relationships assume implicitly certain expectations on factors 5-8.   Mathematically the model is an attempt to picture two relationships:

$$D_M = D_M(t, p + c)$$

$$D_C = D_C(t, p + c)$$

D_M is the expected European demand for on-line services assuming that they are all supplied through EURONET. D_C is the expected demand for on-line services through EURONET, given other computing services available as at present.   t represents a calendar year, p price and c user cost.

In figure 1 the curves $D_{M1} D_1$, $D_{M2} D_2$, $D_{M3} D_3$ represent demand curves in monopoly conditions for EURONET at 3 points in time, and $D_{C1} D_1$, $D_{C2} D_2$, $D_{C3} D_3$ represent demand curves in competitive conditions.   Because the competition would not be perfect the competitive curves are not shown as straight lines, that is we assume that some users will use some EURONET services even if prices are generally above those of other suppliers.   The figure shows price and user cost separately but on the same axis and indicates that even at zero price demand may not have reached its full potential as savings in user cost could increase demand still further.

As the demand curves shift to the right over time, our observations on price and demand at different times represent not points on a single demand curve but points on different demand curves, such as X_1, X_2 and X_3 in figure 1.   Clearly if we erroneously assume that the curve $X_1 X_2 X_3$ represents the demand curve over time, the flatness of this curve, relative to the slopes of the demand curves will cause us to over-estimate the price elasticity of demand. The graph shown as figure 2 which has been used to forecast demand corresponds to something like the curve $X_1 X_2 X_3$ which have tried in this way to relate historical changes in demand to changes in price.

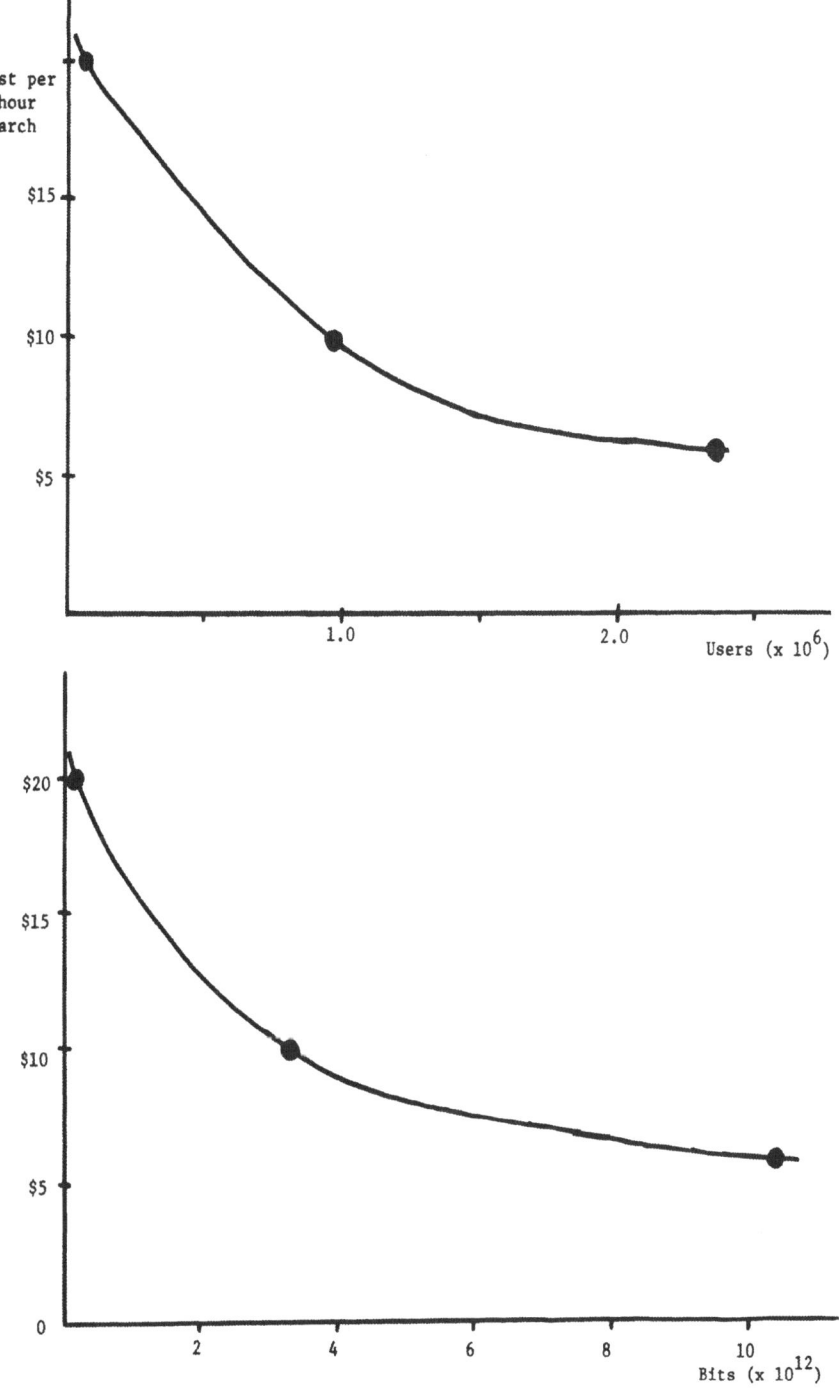

FIGURE 2 Hypothetical relationship between forecasts of
demand and costs

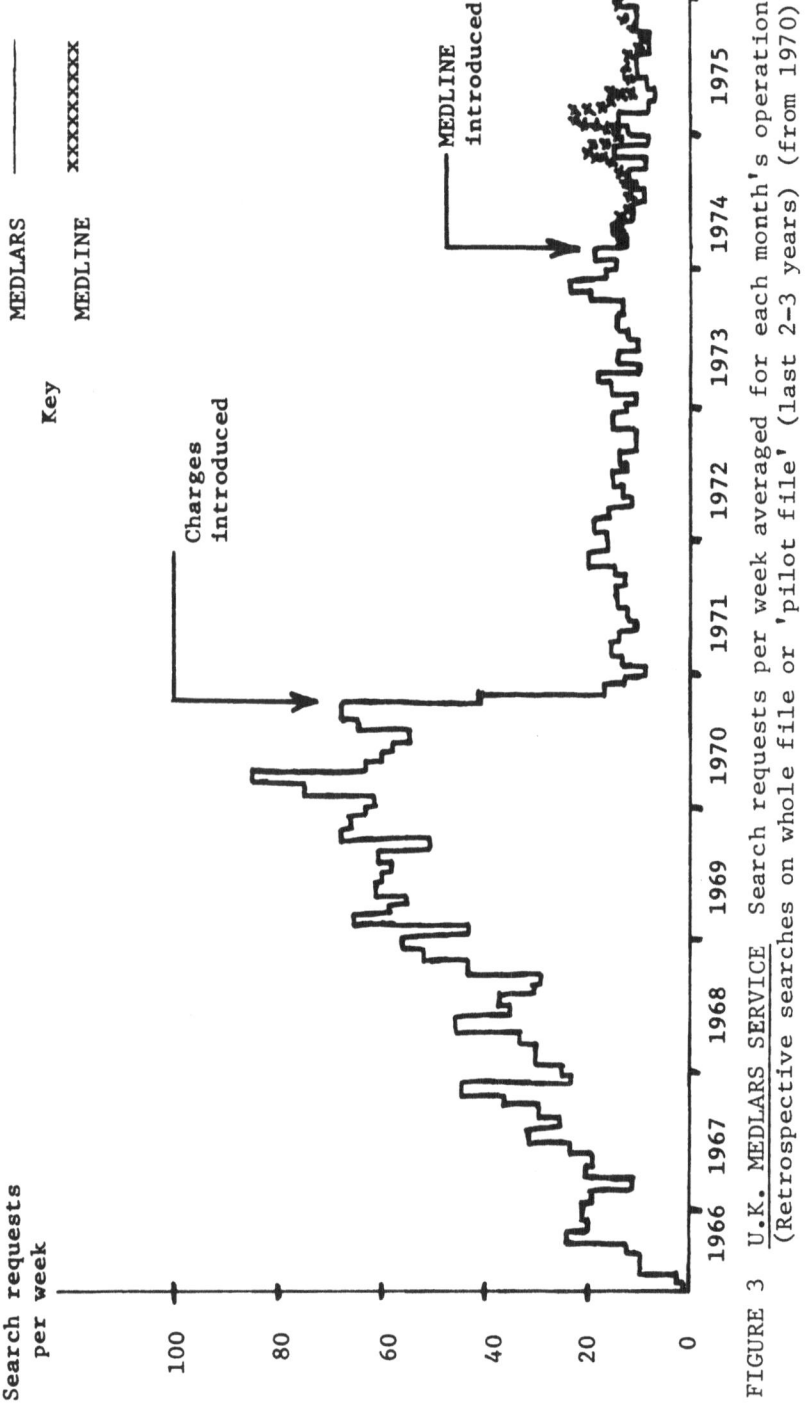

FIGURE 3 U.K. MEDLARS SERVICE Search requests per week averaged for each month's operation. (Retrospective searches on whole file or 'pilot file' (last 2-3 years) (from 1970)).

Figure 3, which shows the growth in search requests per week for the U.K. Medlar Service from 1966 to 1975, illustrates the potential errors in predicting future demand from data collected when the service is provided free: an extrapolation of the trend in the data from 1966 to 1970 would have been extremely misleading.

CONCLUSIONS

In this paper we have discussed the possible development of an on-line information service mainly catering to industry, government and the scientific/academic community. We have investigated the possible reaction of demand to future price changes and movements in prices as a result of various possible trends in costs. In evaluating the response of users to on-line services we have shown the dangers of attempting to extrapolate demand data collected when the services were generally provided free or at very low experimental prices to forecast future demand at more realistic long-run price levels. On the basis of our analysis of data collected in a sample survey we have drawn some conclusions regarding the way in which likely growth in future demand in Europe might occur as well as various possible developments in relation to costs.

These trends have important implications for the way in which information is retrieved within organisations and the trade-off between in-house provision and buying in information services. On-line information systems could revolutionise the extent to which information is cheaply and readily available and so affect business, government and industrial efficiency very greatly. But there are major difficulties, both technical and economic, ahead which themselves require considerable further research.

J.T.

G.W.

THE ECONOMICS AND COST BENEFIT OF ANALYSIS SERVICES - -

THE CASE OF INFORMATION ANALYSIS CENTERS

Robert M. Mason, Ph.D.

President, METRICS, INC.

290 Interstate North, Atlanta, Georgia 30339 USA

1. INTRODUCTION

Purpose and Overview

The purpose of this paper is to indicate current approaches and limitations to economic analyses of information services. The paper reviews relevant findings of recent research and discusses the economic and policy issues associated with the financing of information services. The paper examines the case of Information Analysis Centers (IAC's), specialized centers which provide information services characterized by data evaluation, analysis, and synthesis in a specialized field.

Making financial decisions and formulating financial policy for IAC's raise several issues: To what extent should IAC's and other analysis services be government supported (initially only, continually, or not at all)? What is the proper role of user charges for cost recovery? What are the anticipated economic impacts of changes in technology on IAC's and analysis services? What about international cooperation in financing IAC's? Knowledgeable resolution of these issues requires an understanding of the social and economic costs and benefits associated with providing the services, but current methodology does not permit adequate cost benefit analyses of such information services.

The remainder of this paper summarizes recent research findings and discusses the use and limitations of economic approaches. The following paragraphs present a brief background on IAC's, the services they offer, and their users.

Section 2 describes cost benefit approaches to examining the
economics of IAC's. This section discusses public and private
benefits, a conceptual framework and calculation model, and the results of sensitivity analyses using the model.

Section 3 discusses limitations to the current economic analysis methods and issues which are not easily amenable to economic
analysis, including conceptual and practical problems in performing
cost benefit analyses of information services. This section also
discusses the expected results of technological innovation on the
economics of IAC's.

Section 4 concludes with a brief discussion of research needs
and international issues.

Background

An Information Analysis Center (IAC) is a specialized information center. More specifically, it is

"...a formally structured organizational unit specifically
(but not necessarily exclusively) established for the purpose of acquiring, selecting, storing, retrieving, evaluating, analyzing, and synthesizing a body of information
and/or data in a clearly defined specialized field or pertaining to a specific mission with the intent of compiling,
digesting, repackaging, or otherwise organizing and presenting pertinent information and/or data in a form most author-
itative, timely, and useful to a society of peers and
management." [1]

There currently are more than 100 federally sponsored IAC's in
the United States [1]. Although these exhibit considerable diversity in size and technical discipline, they all share some commonality in activities and functions.

Figure 1 illustrates typical functions of an IAC. Critical
evaluation, or analysis, is the unique function that distinguishes
an IAC from other information centers or libraries which may provide
data collection, indexing, storage, retrieval, and repackaging services. Note that the evaluation feedback loop in Figure 1 indicates
that an IAC may conduct research to fill gaps in knowledge which
may be identified by the evaluation function. However, most IAC's
do not perform research beyond simple data extrapolations/inter-
polations or beyond investigations of data quality.

The observable outputs of an IAC may include both information
"products" and information "services". Table 1 lists eight basic

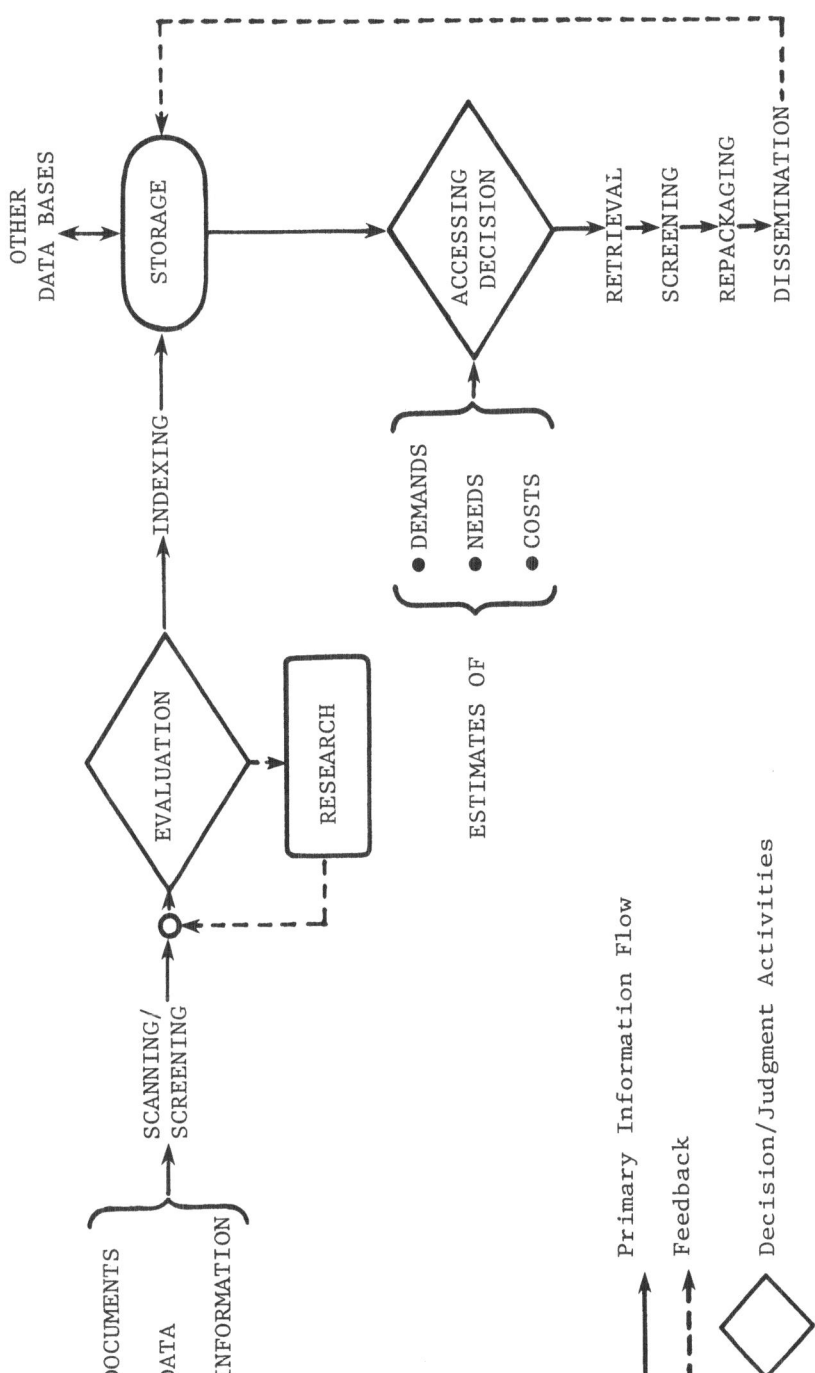

Figure 1. Diagram of IAC Functions

categories of IAC products and services, showing both customized services (such as a response to an inquiry, individualized for a particular group, organization, or person) and products provided for a large number of users (such as a handbook, or databook, aimed at a "mass market" of users with common needs).

Although the IAC concept in the United States is not new (as many as thirteen of what are now called IAC's were formed by the U.S. federal government before the beginning of this century [4]), 70% of the IAC's listed in the Directory of Federally Sponsored Information Analysis Centers [1] did not exist before 1960. The IAC concept became popular during the next few years, however, and the U.S. federal government established at least 62 IAC's between 1960 and 1969 [1].

The prominence and proliferation of IAC's evidenced in the United States during this period emerged from a growing recognition that existing discipline-oriented information systems were inadequate to meet the needs of users outside, or even within, the discipline. The publications from an expanding research and development community were producing an explosion of data, and this produced concomitant difficulties of separating relevant from unnecessary data and of distinguishing between valid data and data of questionable validity.

Another factor motivating the establishment of IAC's during the 1960's was the conviction that automated equipment would permit economies of scale and lower the costs of assuring the accessability of valid, relevant information. Because of high equipment costs and a disaggregated market, a sponsor (generally a federal agency, often the U.S. Department of Defense) was necessary in order to establish an IAC.

Table 1.   IAC Products and Services [4]

Handbooks/Databooks (Hard copy and magnetic tape)

State of the Art Reviews

Critical Reviews and Technology Assessments

Bibliographies (Hard copy and magnetic tape)

Responses to Inquiries

Current Awareness Services/Newsletters

Workshops/Seminars

Symposia Proceedings

Within the past decade, the rationale for establishing and maintaining IAC's with public funds has been questioned. In the early 1960's, the prevailing concept was that information should be freely available, but the late 1960's witnessed the emerging conviction that information is a commodity for which the user should pay. In response to this changing viewpoint, the U.S. Department of Defense, which originated and still sponsors many IAC's, instituted in 1968 a policy that called for centers to charge for their services and by 1972 to be recovering at least 50% of their contract [5]. The implementation of this policy emphasized the need for an understanding of the economic impacts and implications of policy and decision options relating to the providing and financing of IAC services.

2.   COST BENEFIT APPROACHES

Public and Private Benefits and Costs

Cost benefit analysis (CBA), as a procedure which provides economic information as an aid to decision making, aims at identifying and comparing the differences in economic costs and economic benefits to society resulting from choosing between alternative decisions or policies. In performing a CBA of information analysis centers to determine the net benefits of, for example, government support versus no government support, a critical issue is the distinction between "private" and "social" costs and benefits.

A private cost is what an individual (person, household, or firm) must give up in order to receive some good or service. A social cost is what society as a whole must give up in order for the individual to receive the good or service. For many goods and services, private and social costs coincide. For example, an individual who purchases a set of encyclopedias for $400 gives up $400 worth of other goods and services he could have purchased. Likewise, society as a whole (under fairly general assumptions) has given up $400 worth of other goods and services in order for the $400 set of encyclopedias to be made available. In other situations, private and social costs may diverge as, for example, in the case of attending a state university. An individual may pay $1000 for a group of courses at the university, but the cost to society of providing these courses likely exceeds $1000, since a state university typically is subsidized.

One can make similar statements for private and public benefits. While the measure of costs is what (resources) must be given up, the corresponding measure of benefits is "willingness-to-pay". Thus an individual may be willing to pay $1000 for the set of

university courses, but society as a whole may be willing, for
various reasons, to pay $2000 for the individual to have these
courses. The $2000 would include the $1000 that the individual is
willing to pay plus the sum of the many small amounts that other
persons would be willing to pay toward that individual's having
the added education. Others are willing to pay for the individual
to have more education because they personally feel better off.
They may feel, for example, that the increased education makes the
individual less likely to be a criminal and less likely to become
a welfare burden.

When private and social costs or benefits diverge, economists
say that externalities exist. An externality which exists and is
judged significant is often viewed as an indication that the private
sector of the economy is not producing that level of goods or services which, subject to overall resource availability, maximizes
the welfare of society. In such cases, direct or indirect government intervention may be undertaken in order to effect the desired
production level.

The costs associated with providing IAC services are the resources given up by society in order for the services to be offered.
These costs include the efforts of professional and nonprofessional
personnel, rent and utilities, materials and supplies, subscription
and other information service costs, etc. Most IAC's are managed
with careful cost controls, and, although accounting costs do not
necessarily coincide with societal costs, one can assign reasonable
values to cost parameters.

Table 2 indicates the private and societal benefits from IAC
services. Generally, the total social benefits would include both
the private benefits and the (additional, positive) social benefits
shown in the table.

User Factors Affecting Demand

Conceptually, benefits are measured by willingness-to-pay,
described by a demand curve, but there are no data on what an individual or society is willing to give up for each of the individual
benefits shown on the left side of Table 2. An individual user's
willingness-to-pay depends on several factors. Although the current
understanding of potential and actual IAC users is inadequate to
permit the construction of an adequate typology, most IAC users
typically are either scientists, engineers, or technicians [7]. A
meaningful typology for such users probably would include the
following dimensions [3].

Table 2. IAC Benefits [6]

_____

Private	Societal
- Time/Effort Saved (compared with alternative ways to obtain information)	- Codification of Knowledge
	- Second Order Benefits (e.g., arising from private benefits)
- Data Quality (e.g., precision, accuracy)	
- Data Reliability (e.g., low risk to use)	
- Data or Service Uniqueness	
- Intrinsic Value of the Information	

_____

Informational requirements/situational needs. Two cases may
be distinguished. In one case, the user desires background information that has no immediate utility. Information such as that
desired to maintain technical proficiency and current awareness has
been termed "nutritional information" [8]. In the other case, the
user is attempting to resolve an issue or to solve a problem; the
utility of the information is determined by how well it helps solve
the problem. The benefits of information services are likely to be
measured, or at least perceived, differently in the two cases. For
example, information used in solving a problem will have relatively
well-defined requirements of timeliness (urgency), reliability/con-
fidence, accuracy, and precision.

Informational resources. Potential users can have a range of
informational resources available other than the IAC. A potential
user with an effective in-house information service, ceteris paribus,
will be less likely to utilize IAC services than a potential user
without the availability of such a service. This dimension also
encompasses the nature of the information market in the technical
area of need. If the IAC effectively is the only source for the
needed information service (it is a monopoly), then the potential
user faces a different situation than if there are several optional
sources for the service.

Financial resources and procedures. Assuming an IAC charges a
fee for its services, a potential user who has no discretionary
funds or who has to justify expenditures for outside information
services is less likely to utilize an IAC service than a potential
user who does not have these financial constraints.

State of knowledge/awareness. A potential user who is technically up-to-date in the IAC's field of competence may be less
likely to utilize an IAC service than one who has greater information needs. However, a potential user who is unaware of the benefits of IAC services is not likely to become an actual user until
he recognizes the benefits.

User's perception regarding the nature of the expense. The
expense of utilizing an IAC service may be perceived either as a
problem-solving expense or as a capital expenditure. This perception might indicate which decision process a potential user employs
to choose whether or not to utilize the IAC service.

Service-Related Factors Affecting Demand

Several studies [7,9,10,11,12,13,14] indicate the information
service characteristics which may be important to the IAC user.
These studies were not all aimed directly toward IAC services, but
each focussed on how users select scientific and/or technical information. Table 3 lists and defines a set of characteristics synthesized from these studies. There have been no investigations
which indicate the relative value an IAC user places on each of
these characteristics, and neither has there been any attempt to
determine the relationships among these characteristics and the
IAC functions.

Model Criteria

The recent cost benefit study [3] generalized the findings
about the demand for IAC services in a set of model criteria or
"stylized facts", stipulating that many exceptions exist to any
generalization about IAC users. These criteria, although seemingly
contradictory in some cases, appear consistent with observations
and the limited research findings on the demand for IAC services and
for information services in general.

Low apparent willingness-to-pay for IAC services by potential
and actual IAC users. Subsequent to the initiation of fees for
services after a period of free services, the number of requests
for services drops substantially. (In some cases, the center experiences only a temporary drop in demand, which later grows at
approximately the same rate as before the initiation of fees. In
other cases, the demand has remained low.) Similarly, information
center managers have reported that a significant number of requests
for information are cancelled when the requestor is informed of the
charging/fee policy [15,16,17].

Table 3. Synthesized Set of Information Service Characteristics
          Perceived to be Important to an IAC User [3]

Cost	Dollar cost of using the delivery channel
Accuracy	Whether data is input and output without error
Currency	The "up-to-dateness" of information delivered
Response time	Elapsed time between a person's request for information and the arrival of the requested information
Ease of access	Availability of access when and where required
Ease of use	Ease of operating system/channel
Technical quality	Subjective evaluation by user of expected performance of delivery channel
Coverage of topic	Comprehensiveness of information delivered
Understandability	Ease of comprehending information delivered
Format	Physical arrangement in which information delivered
Media	Way in which the information appears (e.g., hardcopy/CRT display/microfiche)
Recall	Percentage of appropriate information in a file that is retrieved by system
Relevance	Percentage of information retrieved that is appropriate to user needs.

Apparent high elasticity of demand. Similar to the above observation, a seemingly small increase in price tends to produce a
relatively large drop in demand.

Small quantities of IAC services actually demanded. This observation is a judgment based on the size of the potential market
for IAC services. (It may reflect primarily a lack of awareness
of the availability of the services.)

Knowledgeable individuals who are informed about IAC services
but who appear to underutilize these services. As opposed to someone who is unaware of an IAC, this observation concerns potential
users who are well informed but do not utilize IAC's for other
reasons, such as the social factor mentioned above, or the lack of
appropriate financial resources or authorization.

Apparent "real" value of IAC services. Anecdotal evidence
[18,19] suggests that IAC services have high value to the user, in
contrast to the (apparently) low willingness-to-pay and low utilization of IAC services.

Interactions among service demands. This is a postulate that
has face validity; examples include (1) the reduction of demand for
an inquiry response service subsequent to the publication of a handbook, and (2) the stimulus for the publication of a state of the
art review, handbook, or bibliography arising from a high number of
inquiries regarding a particular subject or technical question.

The above observations suggest what forms IAC cost and benefit
models should take, and these models should assist a cost benefit
analyst in answering four questions: (1) What is the total benefit
from (cost of) providing a specified level of the particular services offered by the IAC? (2) What is the change in total benefits
(costs) which results from a change in the level of a particular
service? (3) What are the relationships among the benefits (costs)
of providing the various services, that is, (A) given that the IAC
offers a particular set of services at specified levels, what is
the change in total benefits (costs) resulting from an increase or
decrease in the level of one service, and (B) given that an IAC
provides a particular set of services, what is the change in total
benefits (costs) resulting from the addition or deletion of particular services from the set of offered services? (4) What are the
impacts on total benefits (costs) of implementing particular
innovations?

Cost Model

Based on the model objectives and the cost categories involved
in an IAC operation, a fixed plus variable cost model appears to be
the most appropriate general form [3]. At a given level of output,
the total cost (TC) may be expressed as the sum of the total fixed
costs (FC_T) and total variable costs (VC_T). By assumption, fixed
costs are independent of the level of output (over a broad range)
and variable costs are costs which vary with output level. IAC
fixed costs include: rent, telephone, furniture and equipment,
office supplies, subscriptions, advertising and marketing, copying
and reproduction, postage, travel, computer charges, and (certain)
salaries. Variable costs include: salaries associated with service
delivery, computer charges, reproduction and printing (e.g., of
reports), and supplies associated with service delivery.

For an IAC offering k services, the total cost can be expressed as

$$TC = FC_u + \sum_{j=1}^{k} FC_j + \sum_{j=1}^{k} VC_j n_j \, ,$$

where FC_u = unallocated fixed costs, FC_j = fixed costs associated with jth service, VC_j = cost per unit of output of jth service, and n_j = number of units of output of jth service. This assumes constant marginal costs associated with each service, and this should be a reasonable assumption for rather broad ranges of output levels.

Accounting for the time value of IAC resource expenditures is done by taking the net present value (NPV) of the total costs:

$$NPV_{TC} = \sum_{i=o}^{h} \frac{TC_i}{(1+r)^i} \, ,$$

where TC_i = total cost in ith year, h = time horizon, and r = discount rate.

Benefit Model

The observations of user behavior described above suggest a residual (or excess), risk-averse demand model framework to describe IAC service benefits [3]. The IAC user can, to some extent, supply himself with information similar to that which an IAC provides, and the demand for IAC services is thus the "residual" or "excess" demand. The amount of information which a user can supply himself (i.e., from non-IAC sources) may depend on the quality (and the utilization) of other IAC services - e.g., a good handbook may reduce the demand for the inquiry response service.

The model framework, postulating an IAC offering a handbook and an inquiry response service, is illustrated in Figures 2, 3, and 4. Figure 2 illustrates a potential user's supply and demand curves for information. The supply curve labelled "No HB" is the marginal cost of self supply without an IAC handbook and the other supply curve (labelled "HB") is that user's marginal cost of self supply with the IAC handbook. The latter is to the right of the former, indicating that possessing an IAC handbook reduces the cost of self supply.

Now assume the user has no IAC handbook, and the IAC charges a price of P^* per unit of information service. Under these conditions, the user will supply Q2 units himself (his supply curve

is beneath the IAC price for $Q < Q_2$) and will purchase $Q_3 - Q_2$ units
from the IAC. In general, the model indicates that the demand for
IAC information services is the difference between a user's demand
and supply curves below their intersection. Figure 3 illustrates
the demands for IAC services corresponding to Figure 2. Note that
the model, as described thus far, indicates the following: (1) the
highest price a user is willing to pay for IAC services is far more
elastic than the demand for all information services, and (2) the
user's possession of an IAC handbook reduces his demand for (other)
IAC services. (This latter statement is not surprising, since a
handbook may be, in a real sense, the product of experience in
other services.)

The model as constructed above is a deterministic model with
a rational, optimizing user, and it already accounts for several
of the criteria guiding its design. However, additional factors
must be added if the model is to account for the remaining observations. This is accomplished by developing a probabilistic model,
and the assumptions for such a model are summarized below.

First, assume that IAC services have a randomly distributed
value, and assume the user periodically must justify his use of an
IAC by demonstrating to an "auditor" that over the preceding period
the benefits from using the IAC have exceeded the costs. Assume
the user suffers a penalty (perhaps only psychic) if, at audit time,
benefits do not exceed costs; and, moreover, he enjoys only a small
reward if benefits do exceed costs. Under these circumstances, the
user will be risk-averse. He may be willing to accept a probability
of perhaps only 5% that at audit time benefits will not exceed costs.
It can be easily shown that even if the expected value of IAC use
is positive, the user may never find the risk acceptable, particularly if audits occur at frequent intervals and the opportunities
for IAC use are not frequent. This model shows it is rational for
some potential users not to use an IAC, even though its expected
long run value is positive. This result stems from the institutional justification process which IAC use often entails.

The model is a useful guide to a "first cut" cost benefit
analysis. Figure 4 illustrates the conceptual basis for a benefit
model which provides a framework within which to calculate lower
bound on benefits.

The total net benefits of an IAC are represented by the region
circumscribed by the points A B C H J G F . Similarly, the net
benefits from a handbook are represented by the region outlined by
A B C D E F, and the marginal net benefits of other services are
represented by the region G E D H J .

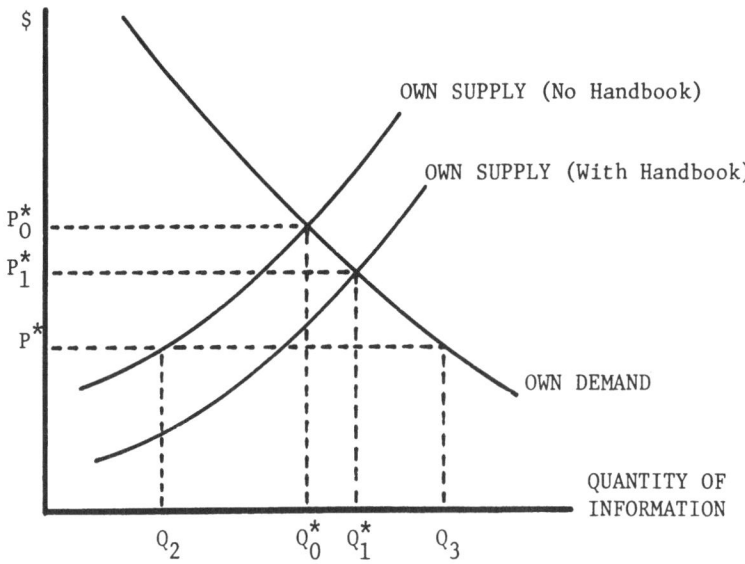

Figure 2.   User's Demand and Own Supply Curves For
Information (From Reference 3)

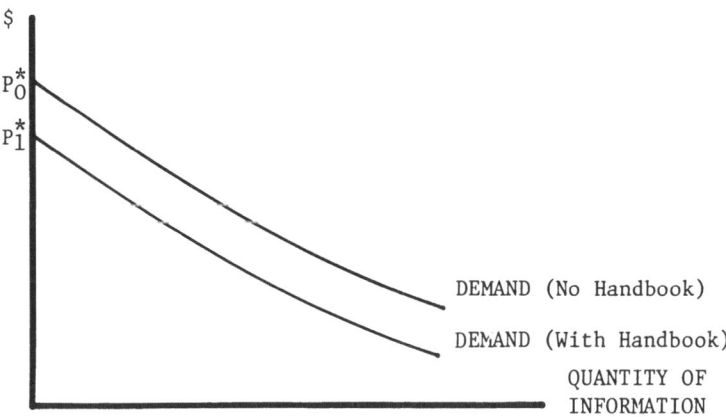

Figure 3.   User's Demand For IAC Service (From Reference 3)

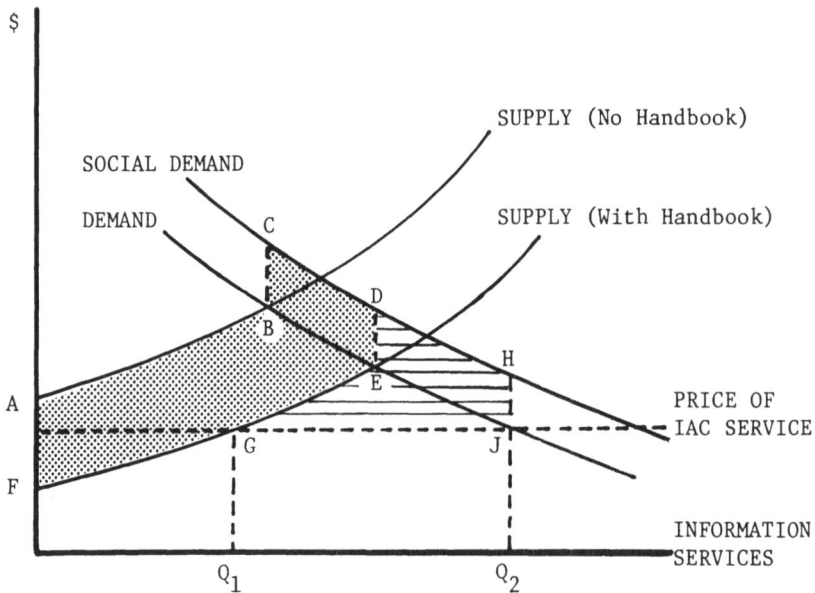

Figure 4.   IAC Economic Model, Handbook Example
(From Reference 3)

If $Q_2 - Q_1$, the demand for IAC services, is small, then handbook use accounts for the greatest portion of IAC benefits. The
value of handbook use is the area between the supply (marginal
cost) curves, and thus the value can be calculated from the product:
(cost savings per use) x (number of uses).

A similar approach can be used to estimate the value of an
IAC's inquiry response service: estimate the amount of time saved
per use and the value of this time and multiply the value per use
by the number of times the service is used each year. The product
is the annual benefit.

These approaches to calculating benefits provide only an estimate of the lower bound on benefits. Benefits other than time
saved (such as the other benefits listed in Table 2) would increase
the total annual benefit.

Results

The cost and benefit models were used with typical parameter values to calculate the costs and lower bound on benefits of the inquiry response service [3]. Using what is judged to be relatively conservative values, the net present value of benefits exceed the net present value of costs.

Perhaps more interesting are the results of a sensitivity analysis of the model to changes in parameter values. The sensitivity of an outcome to a specified parameter is the ratio of the percentage change in outcome to the percentage change in input. In other words, the sensitivity of the net present value to a particular parameter P is defined by

$$S_P = \frac{\Delta NPV}{NPV} \Big/ \frac{\Delta P}{P} \text{ ,}$$

where ΔP = change in value of the parameter of interest, P = baseline value of the parameter, ΔNPV = change in net present value (outcome) due to ΔP, and NPV = baseline value (value at P) of the net present value.

The sensitivity figure is useful for estimating output changes due to changes in parameter values. For example, if the sensitivity of the model to a parameter is 1.9, then a 10% change in the value of that parameter (from the baseline value) will change the output of the model by 19% (from the baseline value).

Table 4 presents the inquiry response parameter baseline values and sensitivities for the cost model; Table 5 presents the parameter values and sensitivities for the benefit model. Note that many individual elements are incorporated into the total costs, and thus the sensitivity to any one of the parameters is relatively low. The model is most sensitive to changes in time horizon, and this sensitivity is only 0.7. The benefit model shows greater sensitivity: 3.0 and -2.0 for annual benefit per user and for price, respectively.

3. DISCUSSION

Conceptual Issues

The model framework described in the previous section is based on fundamental economic principles and is consistent with observations about the demand for, and costs of, IAC services. However, the application of the benefit model thus far has been limited to

Table 4. Cost Model Parameters - Inquiry Response [3]

Parameter	Baseline Value	Expected Range of Values	Sensitivity
Fixed Costs			
Rent	$ 2000	$ 1500 - $ 3000	.03
Telephone	750	600 - 1200	.01
Furniture, Equip.	900	600 - 1800	.01
Office Supplies	450	300 - 1200	.01
Subscriptions	1250	750 - 2500	.02
Advertising & Marketing	2500	1250 - 6500	.03
Copying & Reproduction	600	300 - 1200	.01
Postage	450	300 - 900	.01
Travel	1750	1500 - 6000	.02
Computer Time	1500	900 - 4500	.02
Computer Terminal	450	300 - 750	.01
Salaries	18375	12580 - 25000	.25
Manager 7500			
Clerical 4000			
Professional 6875			
Fringe Benefits	2756	1875 - 3750	.04
Allocation Factor For Fixed Costs	.25	.1 - .4	.46
Variable Costs (Per Hour of Professional Effort)			
Salaries	$14.91	$10 - 20	.41
Professional $13.67			
Clerical $1.24			
Computer Time & Supplies	$5.00	$4 - 8	.14
Number of Professional Hours Expended Per Year on Providing Inquiry Response Service	2000	500 - 10,000	.54
Discount Rate (%)	6	3 - 12	-0.3
Time Horizon (Yr)	10	5 - 20	0.7

Annual Operating Cost $73,551

Annual Revenues (@ $20 per   40,000
    professional hour)

Adjusted Annual Operating Cost $33,551

$$NPV_C \text{ of Adjusted Annual Operating Cost} = \$246,935$$

Table 5. Benefit Model Parameters - Inquiry Response [3]

Parameter	Baseline Value	Expected Range of Values	Sensitivity
Time Saved Per Use (Hrs)	5	0 - 16	-
Number of Uses Per Year	1	0 - 10	-
Value Per Hour ($)	15	5 - 30	-
Annual Benefit ($)	75	10 - 1000	3.0
Price Charged by IAC ($)	40	0 - 1000	-2.0
Number of Users Per Year	1000	100 - 3000	1.0
Discount Rate (%)	6	3 - 12	-0.3
Time Horizon (Yrs)	10	5 - 20	0.7

$$\text{NPV}_B \text{ (Using Baseline Values)} = \$257,603$$

calculating a lower bound on private benefits by estimating the value of the time saved by an IAC service user. As shown in Table 2, "time saved" is only one aspect of the benefits from using an IAC. The data "quality" benefits, including the confidence with which a user can make decisions based on IAC data, are benefits derived from the functions uniquely associated with IAC's, namely, the functions of analysis, evaluation, and synthesis. These benefits should be the focus of IAC benefit calculations, but meaningful data which would permit such calculations do not exist.

One barrier to the collection of meaningful data on the benefits of any information service is the conceptual difficulty individuals often have in distinguishing between the value of the _information_ and the value of the _service_ that provides the information. As shown in Table 2, the intrinsic value of the information may be a valid measure of the benefit of an IAC service. However, this value is the value of the IAC service only if the IAC is a monopolist and provides unquestionably unique information.

Another difficulty in establishing the value of IAC services is that users perceive considerable uncertainty about the utility of the information they receive. Consequently, they do not know, a priori, what they are willing to pay for the information services.

These difficulties complicate data collection efforts that might further refine estimates of IAC service benefits and the procedures for calculating such benefits. There may be systematic

approaches which can help overcome these difficulties, but such
approaches have only been suggested for improving estimates of
private benefits. The social benefits, such as the codification
of knowledge, are believed to be real, but they present other
fundamental conceptual problems.

Impacts of Technology

The general area of electronic communications technology, as
it develops, is expected to have substantial impacts on IAC operations. The impacts are expected to be greatest in the time period
1980 to 1985. During this time period, four technological events
are judged to be significant [3].

A computer-linked network of IAC's could significantly affect
the currency, technical quality, coverage of the topic, and recall
capability. Such a network would also moderately affect response
time and ease of use.

The advent of the availability of natural English language
inquiry and updating is seen as being moderately significant during
this period. This capability would have a direct effect on the
inquiry response service and would affect accuracy, ease of access,
and ease of use.

The use of lasers for facsimile transmission (at a speed of
five million pages per second), particularly in conjunction with
image input and storage, could have a substantial impact during
this period. Such widespread use of highspeed data transmission
would affect the currency, technical quality, and topic coverage
of IAC services.

The final event which could substantially impact IAC operations is the common use of optical character recognition readers
with multifont capability. Such a capability would have a moderately significant impact on accuracy and currency of IAC operations.

The recent study of IAC economics examined three technical
scenarios for 1985. The scenarios included a "baseline", or sur-
prise-free, scenario, a scenario including a portable acceptable
microfiche reader, and a scenario including terminal-to-terminal
conferencing. The cost benefit model was applied to these three
scenarios and compared with the model results for 1975. Because
the model more carefully accounts for cost, the 1985 scenarios
exhibit more impacts on cost than they do on benefits. This indicates that the impact of technology on benefits will not be on the
lower bound of benefits, but is likely to be further reaching,
resulting in additional individuals having access to IAC services.

In general, the impacts of technology seem to be less on the distinguishing characteristics of an IAC (evaluation, data analysis, and synthesis) and on IAC management functions than on other functions. This suggests that IAC operations, although affected by technological change, will continue to have as their basis the key element of human judgment.

Other Considerations

Government intervention in the economics of the IAC service market can be exercised through its purchasing power, taxation or subsidation, regulation, or indirectly by influencing market participants. (Reference 20 presents a good discussion of the pros and cons regarding government intervention, as does Appendix A in Reference 3.)

In choosing to exercise some intervention mechanism, the government may be motivated by economic considerations such as optimal production and resource allocation or equity of information service distribution and availability. Other motivations can include considerations of national defense: there may be sufficient justification for maintaining an IAC to serve defense contractors even if the same information services normally would be available from outside the country. Similarly, defense and balance of payments considerations may justify an otherwise noneconomic decision to invest in an IAC to encourage the development of a strategic national technical capability.

4. CONCLUSIONS

Research Needs

Considerable effort is indicated before meaningful, comprehensive cost benefit analyses for IAC services can be performed. A useful next step would be the collection and analysis of data on how IAC users make decisions on the purchase and use of IAC services and data on the value IAC users place on the "quality" characteristics of IAC services. The "quality" service characteristics result from the application of human judgment in the IAC functions of data analysis, evaluation, and synthesis. The application of human judgment is an expensive cost element (the sensitivity studies indicated costs were more sensitive to personal services than other cost elements), and these aspects of IAC functions appear less likely to be affected by technological innovations than other aspects such as indexing, screening, storage, and repackaging.

Another important research need is the development of concepts and methods which would permit the measurement or estimation of social benefits of IAC services. The codification of knowledge is, prima facie, a valuable result of an IAC's operation, but there needs to be developed some approach that would indicate the value of this social benefit in a manner that will aid public decision making and policy formulation.

International Issues

The development and operation of an IAC, because it aims toward the collection, evaluation, analysis, and synthesis of data in a particular field, requires an investment "up front" and enjoys certain economies of scale. Consequently, it would appear economically desirable to have cooperation among several countries, creating and operating a single IAC to serve the needs of the scientists and engineers in each of the countries.

Two significant issues suggest the need for some international organization to oversee the development and operation of such international IAC's [21]: (1) Developing countries, being short of qualified scientists, technologists and information specialists, have particularly urgent needs for reliable, evaluated information in a usable form. (2) The projected growth of new scientific literature over the next 10-15 years is likely to increase interest in selective information systems as the nonselective systems can be expected to retrieve larger quantities of nonrelevant, along with the relevant, information. Quality is one good basis on which to screen and select from among the literature.

However, there seems to be little experience with international/multinational organizations of information analysis centers and remarkably little knowledge about the use of evaluated information beyond the country of origin. (This latter is in contrast to the traditional exchange of information and cooperation on data compilation [21].)

There seems to be little evidence that UNISIST or any other organization is taking the lead on promoting multinational IAC's. It may be that the problem of a diffuse, disaggregated market simply is compounded by the presence of national boundaries and organizational barriers, preventing the need for multinational IAC's from becoming an actual demand. If such is the case, then the best hope for establishing and maintaining a multinational IAC may well lie with scientists and engineers who see the need and are aware of the potential value of cooperation, rather than with information scientists or governmental/international organizations.

ACKNOWLEDGMENT

Portions of the research described in this paper were funded by the U.S. National Science Foundation, Division of Science Information, under Grant SIS 75-12741. The views, findings, and conclusions expressed are those of the author and do not necessarily reflect the views of the National Science Foundation. It is a pleasure to acknowledge the contribution of Dr. Peter G. Sassone in developing the economic concepts and the assistance of Mr. Brian D. Wright and Mr. G. William Spann in developing and implementing the cost and benefit models.

REFERENCES

1. National Referral Center, Directory of Federally Supported Information Analyses Centers; Washington, D.C.: National Technical Information Service, Third Edition, 1974.

2. Weisman, Herman M., Information Systems, Services and Centers; New York: Wiley-Becker-Hayes, 1972; p.139.

3. Mason, Robert M. (Principal Investigator), "Development of Cost Benefit Methodology for Scientific and Technical Information Communication and Example Application to Information Analysis Centers", Final Report on NSF Grant SIS 75-12741; Atlanta: METRICS, INC., February, 1977.

4. Lynch, J. F., Presentation at the DSA Meeting of IAC Managers, Johns Hopkins Applied Physics Laboratory, 10 September 1975.

5. Christensen, W. C., "Information Centers - DoD Policy on Cost Recovery", The Management of Information Analysis Centers Proceedings of a Forum; Washington, D.C.: COSATI Panel on Information Analysis Centers, 1972.

6. Suggested by Dr. Vladimir Slamecka and reported in Reference 3.

7. Corridore, Michael C., "Scientific and Technical Information Needs of Users or Potential Users of the DSA Administered, DoD Information Analyses Centers"; Cameron Station, Alexandria, Virginia: Defense Supply Agency; April, 1976.

8. Mick, Colin K. (Applied Communication Research), Personal Communication.

9. Auerbach Associates, Inc., "DDC 10 Year Requirements and Planning Study, Vol. II: Technical Discussion, Bibliography, and Glossary", June 13, 1976.

10. Clayton, Audrey, and Norman Nisenoff, "A Forecast of Technology for the Scientific and Technical Information Communities", Arlington, Va.: Forecasting International, Ltd., May, 1976.

11. Allen, Thomas J., "Managing the Flow of Technical Information", unpublished report, Cambridge, Ma.: MIT Research Program on the Management of Science and Technology.

12. Auerbach Associates, Inc., "DDC 10 Year Requirements and Planning Study: Interagency Survey Report", December 12, 1975.

13. Gerstberger, Peter G., and Thomas J. Allen, "Criteria Used by Research and Development Engineers in Selection of an Information Source", Journal of Applied Psychology, 52:4 (1968), p.274.

14. Rosenberg, Victor, "12 Factors Affecting the Preferences of Industrial Personnel for Information Gathering Methods", Information Storage and Retrieval 3:3 (July, 1967), p.99.

15. McCarn, Davis (National Library of Medicine), Personal Communication.

16. Summit, Roger (Lockheed), Personal Communication.

17. Gerstner, Helga (Toxicology Information Response Center), Personal Communication.

18. Kahles, John (Machinability Data Center), Personal Communication.

19. Hetrick, Samuel I., (Defense Logistics Agency), Personal Communication and written documentation from IAC users.

20. Flowerdew, A.D.J., "The Government's Role in Providing Information - A British Academic's View" in Robert M. Mason, et al. (Eds.), Information Centers and Services - The Economics, Management, and Technology, forthcoming.

21. Gray, John C., "Background Paper", prepared for the First Meeting of the UNISIST Working Group on Information Analysis Centers, UNESCO House (Paris) 3-5 November 1975.

TECHNOLOGY ASSESSMENT AND IDEALIZED DESIGN

An Application to Telecommunications

Peter Davis and Edward Freeman

Wharton Applied Research Center
University of Pennsylvania
Philadelphia, Pa. 19104

INTRODUCTION

This is an essay about the anticipatory assessment of telecommunications technologies for the production, dissemination and use of scientific and technological information (STI) in the United States. The focus is on the 1980's. The question posed is: What should be done, not only by government and industry, but also by the myriad interest groups with a stake in the information system, to generate and implement better alternatives in the next ten years?

We have two interrelated theses: (1) that current models of technology assessment (TA) have been inadequately applied in the evaluation of STI futures; and (2) that the potential of TA has been too narrowly conceived within the larger context of a planned change effort for a highly complex and pluralistic social system.

TECHNOLOGY ASSESSMENT AND THE STI SYSTEM

Technology assessment is an activity embedded in, and legitimized by, particular processes of social change. Originating with the development in the 1960's of populist interest groups concerned with a variety of environmental and ecological issues, TA has increasingly become an institutionalized accommodation to the movement which gave it birth. It has become, in large measure, an adjunct to the legislative process and it derives much of its legitimacy from the legislative branch.

Legislation is, of course, the traditional method of change
in western democracies. However, it is becoming increasingly clear
that in a number of vital areas legislative processes are extraordinarily ineffective. For example, actual reform of health care
and urban transportation in the United States lags far behind the
real capability for change. Similar conditions exist in the dissemination of, and access to, STI. Fifty years of haggling over
copyright laws is a testament to the ineffectiveness of STI legislation.

Yet while the legislative process remains stymied and ineffective, the recent history of STI has been unfolding in a way which
it to many disturbing. To some it is even alarming. Many observers believe that the explosion in the generation of information
coupled with haphazard and uncoordinated growth in editing, publishing, storing, retrieving and distributing capabilities is leading to system degradation. The lead time for the publication of an
article has increased significantly (it is not uncommonly 18 to 24
months for prestigious journals). At the same time the number of
journals has grown so much as to create widespread concern (according to the New York Times [1], as of 1975 there were 10,000 scholar-
ly and scientific journals published in the U.S.). The amount of
information produced each year has reached staggering proportions.
Chemical Abstracts alone will process 600,000 abstracts per year by
1980. Not only are users unable to affectively access, or deal with
much of this information, the key institutions which handle it are
faltering. Libraries are going through an acute financial crisis,
and even the largest ones are unable to keep up with acquisitions
expectations. Many publishers are struggling to stay afloat in an
era of rapid technological change. As a recent report to the President [2] pointed out, "the capital needed to modernize processes of
production (for publishers) is costly; publishing is becoming a
marginal industry." The effects are being experienced across the
board.

In cases such as these it becomes vital that we explore new
mechanisms for collaboration between the major actors in a complex
social field. We badly need applications of strategic planning
which emphasize the development of increased consensus about desirable directions of change, and which give these greater meaning
through the encouragement of improved appreciations of where the
systems as a whole ought to be heading.

Over the past two years we have been working with several hundred members of the loosely aggregated STI community in the United
States on such an application. The vehicle we have used in our
search for, and development of consensus is idealized design [3].

We believe that approaches of this type place new demands on, and open new opportunities for, technology assessment. Later in this paper we will explore the emergent possibilities.

IDEALIZED DESIGN AND TECHNOLOGY ASSESSMENT

An idealized design is a scenario describing what the system as a whole would be like if there were no constraints limiting its construction and development. It is almost a pure uncontaminated expression of the desired future. We use the qualifier "almost", because in practice it has been found that while such designs may be highly creative instruments, they may be so distant from current reality as to be ineffective in providing motivation for planned change. It is a well known finding of experimental psychology that aspiration levels which are too extreme do not effectively motivate behavior. For this reason we have imposed two practical constraints on an idealized design: (1) that it not involve any technology that is not known to be feasible, and (2) that the system designed must be operationally viable.

There are two interacting aspects of an idealized planning process -- design and evaluation. A given idealized design is a broadly conceived, holistic concept of a desired future. The technologies it utilizes are instrumental for the achievement of that future and can be evaluated in these terms. The impact of the design as a whole can be assessed and can be compared with the probable impacts of other futures. In this way technology assessment becomes the principal feedback loop in an iterative design process. At each state, the design is modified and features are dropped or added, until a better synthesis is achieved. Individual technologies are evaluated in a marginal sense. We can ask "what difference does it make if this technology is added to or dropped from the design?" Marginal costs and benefits can be compared.

From the point of view of the participant in this process, an idealized design inevitably represents, in the early design phase, a personal and somewhat ill-defined statement of preference. The level of commitment to the design may well be low if the underlying values are unclear. Often the very process of writing an idealized design is enormously helpful both for the clarification of ideas and values. However, the clarification process continues in a more rigorous mode in the next phase, which involves a comprehensive technology assessment. Such an assessment at this point serves two vital functions for the participant: (1) socialization, and (2) further value clarification. Internalized and obscured values are revealed. Other stakeholders react to the design, and subsequently become involved in its redesign. Impacts at the general societal level are investigated and the individual is forced to reassess his

position. The credibility of the technology assessment, at this
point, largely determines the commitment to the design.

There are two important aspects of this methodology which differ somewhat from more traditional applications of technology assessment.

First, we are dealing with complex, highly integrated technological and procedural configurations, not individual technologies
or relatively localized clusters of these technologies. This makes
sense for the field of information transfer because of the synergies involved. More than in most areas, the implementation of one
technology will strongly influence the advantages and disadvantages
of another. Thus, for example, the development and standardization
of terminal technology will strongly influence the development of
electronic mail and electronic editorial activities. We can anticipate the real effects of new technologies such as these only by
looking at the implementation of meaningful configurations which
subsume highly interactive alternatives. Furthermore, in the traditional process of impact analysis the "contextual field" within
which the technology is embedded is kept fixed. The technology is
then evaluated against the field. This may be satisfactory under
circumstances in which the value of the contextual field is considered adequate or even beyond question. But, it distorts evaluation when the performance of the contextual field is itself inadequate.

Second, we are dealing with a desirable and not a projected
future. Effectively in this process we assume that the future of
STI can be, in essence, what we want it to be. This undoubtedly
exaggerates the degree to which we can intervene and determine the
course of events. However, it provides a suitable redress to the
pessimism inherent in the "predict and prepare" philosophy underlying much of contemporary technology assessment. Technological
innovation is a creative act. In our view, technology assessment
must itself become a creative synthetic act, and not just a for-
malized analytic process. It must be concerned not only to anticipate futures but to be part of a continuing process of creating
them.

Technology assessments based on desirable futures help to
generate the creative input. In the remainder of this paper we
want to show how they may be applied and interpreted. We will first
describe an idealized design which has emerged out of our work with
the STI community. This design is called the Scientific Communication and Technology Transfer (SCATT) System [4].

THE SCATT DESIGN

The idealized design of the SCATT system contains over 120 features describing a comprehensive operating system for the production, dissemination, acquisition and use of scientific and technological information. Contributions to this design have been made by over 250 members of the STI community in the U.S. Further contributions and modifications have been made this year in twenty to thirty small group working sessions involving some of the more active constituents [5].

It is difficult to describe a system as complex as SCATT in a short space and yet give a feel for how the system would operate. All we can do is give the reader a very general outline of the basic ideas on which the system is built and refer him to a more complete description.

SCATT is, in spirit, a modern version of the futuristic designs for information transfer produced by Urquhart [6] and Kemeny [7] in the 1940's and 1960's respectively. The design is based on the technologies which we can anticipate will be available in the 1980's. It combines these with an organizational framework, and a set of operating policies and procedures to produce a coherent functional system to handle all aspect of information distribution, storage and retrieval. The operational focus of the system is the user. The bits and pieces of the system are put together so as to make sense from the user's point of view.

The organizational focus of the system is the community SCATT center. This is a local center, which is managed locally by the scientists and technologists of the area, and which provides the first and primary contact point for the user. The local center carries the principal burden of coordinating STI services for the user and providing the means of access. In addition to its role as the primary access point, the local center would store STI data of local interest (e.g., on local professional meetings) in its own data files and would manipulate these directly. For information which is not stored locally, the center is hooked up into a distributive computer network which ties it in with the national system.

We estimate that roughly 100 to 150 local SCATT centers would be required for the U.S. These local centers would, in turn, be associated with 10 to 15 regional centers. At the top of the pyramid would be the national SCATT center which would be the principal policy making unit in the system. The national SCATT center would be managed by members of the STI community and would report to the Congress.

The flow of communications through the network is shown in Figure 1.

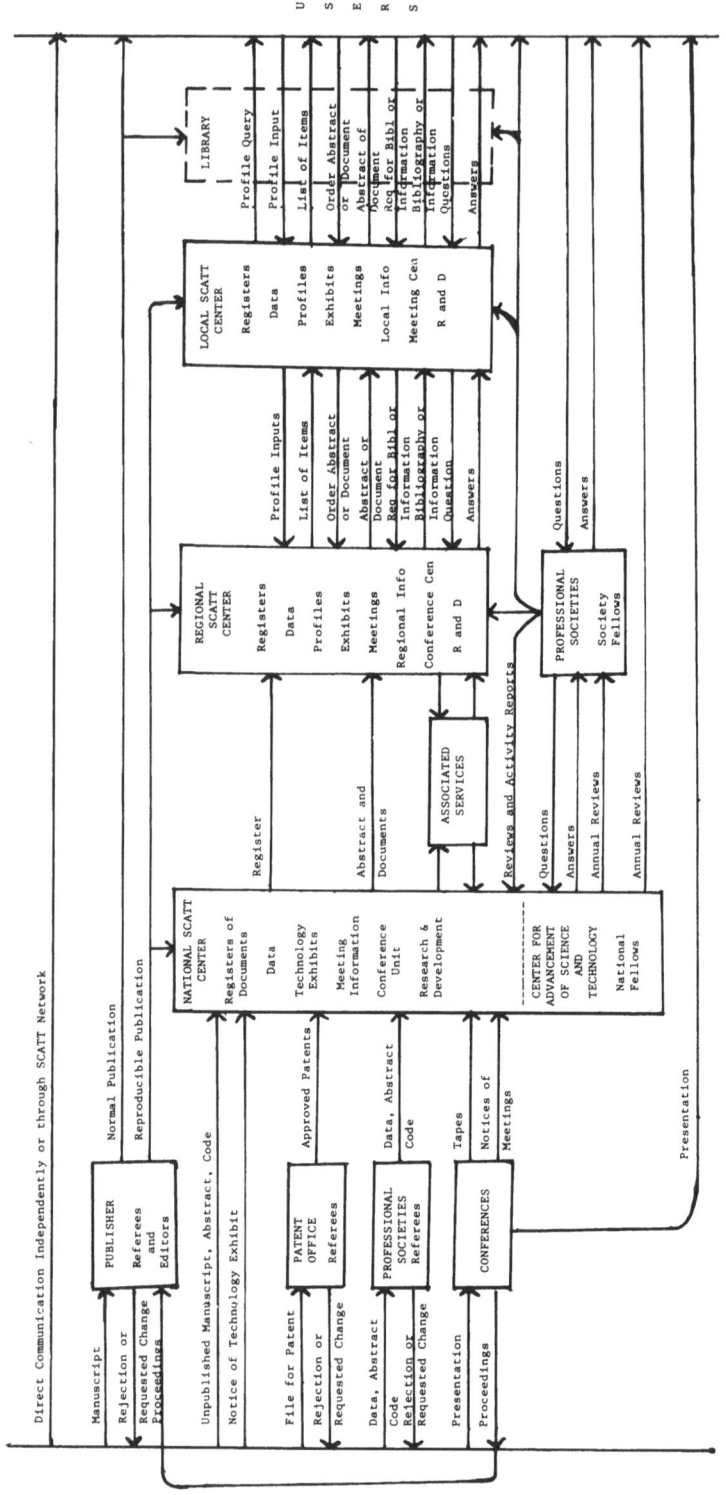

Figure 1

The SCATT system would emphasize, promote and facilitate informal, as well as formal, communications among scientists and
technologists. For example, the system would make facilities available locally for small meetings and regionally for large conferences and conventions. It would provide assistance in organizing
meetings and would conduct research to discover which conference
formats best serve different purposes.

The system would receive, and store electronically, manuscripts
submitted by authors. The foreseeable development of word processing technology would allow authors (or their secretaries), working at their desks, to produce these manuscripts directly in machine
readable form. The system would accept any manuscript provided that
it (1) was submitted in machine readable form, (2) was accompanied
by an abstract, (3) contained an appropriate coding indicating its
form and content, and (4) included an index.

The last three requirements would be used for coding and document storage. Special requests would, of course, be retrieved directly. Generalized searches would be carried out using similarity
index search procedures. The user could either request individual
items on his own initiative, or identify his needs and request his
material through a profile system operated by the local SCATT center. Users could request whole articles, or simply abstract and
reference material.

Publishers would continue to publish as they do now. Published
articles would not be included in the system but their associated
reference material would be (and hence would be accessible through
the profile system). Articles in the system which were subsequently
published could be withdrawn on publication. However, many would be
included in both hard copy and electronic journals. The possibility
for the author to "publish" his material through the SCATT system
would be a major incentive for the shortening of publication lead
times.

Technically the system would be a terminal-based operation. It
might be expected that by the mid-1980's the majority of users would
have their own terminals available. However, high-speed printers
and sophisticated graphical display devices would be installed in
special user sites. The SCATT system would be developed as part of
the computer based networks that are likely to become widely available in the near future. SCATT would be one of many such networks
linking professional workers in a variety of educational and research institutions. The networks, besides providing access to the
bibliographic data bases and information services of SCATT , would
also be the basis for additional services such as the "electronic
mailbox". Services of this type would allow the user to construct

"personalized networks" and would facilitate joint information handling by several colleagues in dispersed locations.

The software of the system would be developed to allow ready and easy access to SCATT, independent of the type of terminal hardware in use. In some cases the user would be allowed to write his own software to access the system in the way he saw fit. Additional advice to users on the use of SCATT services would be provided by the local centers, by professional societies, and by a group of national fellows located at the national centers.

Once set up, the system would be required to be financially self supporting. It could not accept operating subsidies. All services would be paid for by those using them. However, the users of SCATT could receive subsidies in the form of vouchers which could be used to pay for services provided by any of the local SCATT centers of any competing organizations. Such a system would allow the focusing of STI subsidies on end objectives (such as providing access to STI by graduate students through the subsidization of graduate students) rather than on the means of achieving those objectives.

A FRAMEWORK FOR EVALUATION

Technology assessment is a complex hierarchically-structured activity. Taken as a whole, it subsumes a vast compendium of levels of analysis, points of view and areas of concern. As an activity, especially when applied to complex systems of technologies, it is often so complex that only partial analyses can be attempted. Unfortunately, the scope and context of these partial analyses are seldom articulated. This we have found to be a major drawback in the development of methods of assessment for systems, such as SCATT, which could be expected to produce a wide range of impacts.

We can clarify some of the major issues and at the same time lay the framework for technology assessment of an idealized design through the development of a typology of assessment domains. We propose one such typology which groups impacts into three categories: the intentional system, the transactional environment, and the contextual environment. We will examine these categories and then show their application to the evaluation of SCATT.

The Intentional System

A technology is a set of instruments. Instruments are applied by users for the achievement of certain purposes. These purposes define what we might call the intentional system of the technology.

The intentional system of a given technology is simply all those effects which it was designed to achieve.

There are two major criteria for assessment of technology within the intentional system, efficiency and effectiveness. The efficiency of a technology is the degree to which it facilitates the achievement of a given goal. Its effectiveness concerns the extent to which the facilitated goal is an appropriate one. To what extent does the videophone improve the quality of joint decisions made by two business offices? This is a question about efficiency. To what extent does the videophone encourage joint decisions which would be better made separately? This concerns the effectiveness of the technology.

The Transactional Environment

While most of the TA literature is concerned with the intentional consequences of a technology, we are only beginning to tap the possibilities for evaluation of unintended or second-order impacts. New technologies invariably create a sequence of impacts flowing out of the intentional system in a chain of actions and reactions.

Impacts which can be linked directly to the implementation of new technology but which are not intended consequences (and therefore part of the intentional system) make up the transactional environment of that particular application.

Transactional impacts from a communications technology arise when behavior forming part of the intentional system of the technology disturbs some regulatory process which is, at least partially, external to the intentional system. For example, a reduction in the amount of travel with improved telecommunications might have a significant impact on office and family relationships. The increased "visibility" of staff in satellite office locations might lead to shifts in the corporate power structure.

The elements of the conceptual space in this analysis are behaviors, or more precisely, human acts. These acts may be clustered according to the different regulatory principles which govern them. The key cluster is clearly the intentional system. However, other clusters overlap with the focal cluster and give rise to a broader set of effects. The task of technology assessment is to discover the most important overlaps, to explicate the regulatory principles, and hence, to make predictions of specific impacts.

The Contextual Environment

The contextual environment contains a second, and more general, level of unintended consequences.

Technologies are embedded in natural and social systems which respond to change as systems. These responses include feedback mechanisms and synergistic interactions. Responses from one technological innovation may combine with those from another to produce wholly unanticipated effects. Thus, the impact of improvements in the technologies of warfare and telecommunications combined to produce radical shifts in the social order during the Vietnam era. Telephone and transportation technologies have combined to produce different patterns of family location and different expectations about the role of the extended family.

Transactional responses from the same innovation may also combine to produce effects which could not be predicted from analysis of the transactional processes taken separately. Thus, the combined impact of telecommunications on geographical dispersion and the autonomy of work groups may lead in Lawrence and Lorsch's terms [8] to a different level of differentiation or integration of the organization as a whole. This in turn might effect the organization's capability to deal with unpredictable changes, which might in turn lead to future reorganizations better capable of dealing with an uncertain environment.

Contextual impacts are analyzed at the level of the broad social field within which the technological change takes place. The relevant assessment principles are not individual regulatory processes, but properties of the field looked at as an integrated system. Contextual impacts are more remote and indirect. They are, therefore, that much harder to identify, and relate back to the initial technological changes. Remoteness of impact does not, however, imply "weakness" of impact. Because of the synergies involved, contextual impacts may be significant determinants of the quality of life, and it is correspondingly important that they be included in the assessment process.

ASSESSMENT AND THE INTENTIONAL SYSTEM

How far will the SCATT system go to meeting the needs of its users? The question must be phrased in two parts: (1) the efficiency question -- how far will the system go to meeting the user needs it is designed to meet?; and (2) the effectiveness question -- to what extent are the design needs the real needs?

In order to get a preliminary assessment of these issues we asked members of the STI community who attended our SCATT workshops

for their opinions. These workshops, which were generally conducted with five or six participants, involved a one-hour presentation of the design and a one to two-hour discussion of its features. Roughly three months after each workshop the participants
were sent a questionnaire. This questionnaire specified the basic
aims of the system and under each aim the specific operational and
management features designed to meet those aims. The respondents
were asked to specify the desirability of the aims and features.

A complete discussion of the results of this questionnaire is
reported elsewhere [9]. The details are not important here. However, they confirm the most important and the most obvious conclusion: Among experts in the field there is an enormous amount of
disagreement about both the general aims and specific system features. Even the old, and supposedly well worked over, controver-
sies such as on the use of profile systems, redundancy checks, re-
fereeing procedures, user comments on articles, etc., are still
hotly contested.

It is unlikely that many of these issues will be resolved without experimental demonstration. Yet over the years there have been
countless demonstration projects dealing with specific features contained in the SCATT design. The problem is that (1) few people are
aware of the experimentation that has taken place, and (2) there is
seldom an adequate basis for evaluating particular experiments or
comparing one experiment with another.

The design of valid experimental assessments of the intentional
system raises enormously complex issues. The ultimate concern must
be the value of the system to the user. The experimental design must
therefore take into account the various dimensions through which the
individual places value on the activity or activities of interest in
the demonstration. Brewster Smith [10] has suggested that there are
certain types of phenomenally objective requirements which give rise
to the experience of value. The three major types are:

        1. social requiredness -- what others require of me.
        2. personal requiredness -- what I require for myself
            (my own standards).
        3. objective appropriateness -- what the situation itself
            demands.

An experimental design will often address the third requirement
but neglect the other two. The consequence is often a sterile exercise with attention paid to the rules of the game but not to behavior deriving from a meaningful level of value. It is the neglect
of the personal and interpersonal levels of requiredness which has
led to the myth of the lazy or unmotivated user. Sterile experiments provide sterile responses.

Thirty years ago Bettelheim [11] developed the concept of a "treatment milieu" for the counseling of autistic children. What we need in the assessment of technologies is a better concept of the "experimental milieu". Further, the "experimental milieu" should not necessarily model the present but should offer a meaningful integrated experimental future. An idealized design provides, at least theoretically, the potential for the design of such futures.

In STI, experimental milieus are particularly difficult and expensive to set up because of the bibliographic data base requirements. Demonstrations require the establishment of the data that would be used in the actual system. To meaningfully experiment often requires setting up a significant portion of the actual system we want to test. The cost of doing this is prohibitive.

It is feasible to approximate adequate experimental milieus for groups of users with well prescribed and limited data or bibliographic needs. For example, certain highly specialized areas of natural science might have fewer than a hundred active participants publishing a few hundred articles a year. These groups might be increasingly set up as referents. There will still, however, remain the issue of whether the very tractability of the needs of these groups limits the generalizability of any conclusions drawn from such demonstration projects.

Ultimately we may be forced to accept the fact that there is no way to test, in a controlled and limited fashion, the salient features of those technologies and procedures which will lead to change in the whole culture of information transfer. The only adequate experimental milieu may be the real world.

If this is the case, then we must do three things: (1) adopt a universal design principal of adaptability. We must be able to adapt our procedures and technologies as we learn more about their performance in actual operation; (2) design improved feedback mechanisms so that valid information on the operation of the system is generated more rapidly and in a usable form; and (3) design more responsive management functions which accept the need for system modification and adaptation.

The actual operation of the system can, in this way, become an experiment and an on-going process of inquiry. This fusion of action and learning obviates the need for limited preliminary experimental tests.

ASSESSMENT AND THE TRANSACTIONAL ENVIRONMENT

With a system as complex and comprehensive as SCATT, the real problem at the outset is not so much the determination of the magnitude of the transactional impacts, but the <u>identification</u> of those impacts. This requires a base of experience. Obviously, with a hypothetical system of this kind we do not have direct knowledge from experience available. However, we can apply experientially-based knowledge in an imaginative sense.

Of course, this knowledge has to be developed and applied systematically. At the initial exploratory phase, we would suggest that the way to do this is through rigorous application of the participant observer methods of clinical research.

The appropriate model of explanation for participant observer methods is the pattern model. As Diesing [12] points out:

> "The linkages in a pattern model are not established statistically by observation of constant reoccurence. They are experientially causal, or functional, or involve the relation between a psychological state and its exemplification. Objectivity consists essentially in this, that the pattern can be indefinitely filled in and extended; as we obtain more and more knowledge it continues to fall into place in this pattern, and the pattern itself has a place in the larger whole."

Once an "objective pattern" has been established it can be analyzed for basic themes. The themes are used to establish the underlying regulatory processes. These processes provide the mechanism for assessing the transactional impacts.

The establishment of the initial patterns requires the involvement of individuals with an appropriate base of experience. The structure of this experience base is defined by means of what Ansoff [13] calls stakeholder analysis. The stakeholders in any organization have an immediate interest in both the inputs and the outputs required for the maintenance of that organization. The impacts in the transactional environment are experienced first by the stakeholders of the intentional system.

The first problem of analysis of the transactional environment is how to mobilize the experiences of stakeholders into a sustained effort at reflection on the impacts of the new technology on their areas of expertise. The second problem is how to integrate these into a consistent overall pattern of anticipations.

There are a number of artifacts which can be used to mobilize
the stakeholders to participate. Interesting, provocative designs
help, of course. It is also necessary that the designs be complete
enough that the stakeholder can begin to work on the material in his
imagination and project impacts in his own area of interest.

The stakeholder then can put together impact scenarios. These
scenarios can be analyzed for the patterns they contain. Subsequently underlying themes can be extracted. Patterns can be compared across different stakeholder groups, and checks can be made
for consistency in the interpretation of the impact felt by the different interest groups.

There are various interpersonal methods which can be used to
share information about impact projections between the groups [14].
In areas where there is a high level of conflict between the groups
these methods would not include face-to-face interactions. Planning
teams could serve the role of intermediary [15].

The point of all this is as follows: we all have natural, well-
developed procedures for dealing with, what we might call the familiar future. If we are told that someone will resign the office tomorrow, we have at least a partial map of the office situation which
will allow us to anticipate what the impact will be (or at least
bound the uncertainty about what that impact will be). This is the
kind of eventuality we know to expect, and if we are to be effective
we must learn to develop the observational facility and the methods
of analysis to assemble the cognitive maps necessary to deal with
the situation. Of course, some are good at doing this and some are
poor at it. However, the method is not entirely ad hoc and person-
alistic. It can be systematically developed. The most advanced
stage of development is in the participant observer methods of the
social sciences. Clinical and anthropological methods are cases in
point. Our argument then is that <u>we need to develop a clinical
method of anticipation of the transactional consequences of tech-
nology</u>. We need to use this method to guide observation in the
present and participate in discovery about the future. We need to
harness our experience not only for dealing with the present but
for appreciating the essential structure of our environment which
will determine the evaluation of the future.

We have only recently begun to systematically develop this
methodology for SCATT. Our design has, however, been widely cir-
culated among members of the STI community. As a consequence, we
have received unsolicited responses from a large number of readers.
Some of these responses are in embryonic form extraordinarily fine
examples of the kind of impact scenario discussed above. We are in
the process of carrying out a full analysis of these scenarios,
classifying responses into their appropriate environment contexts

and formulating follow-up inquiries.

Out of the many specific impacts a number of themes have so far emerged. One of the themes, for example, is that there are a vast number of relatively specialized agencies performing extremely valuable functions, particularly in the <u>transfer</u> of STI. These agencies are protected by aspects of the status quo which a system such as SCATT threatens to undermine. Thus, small publishers are protected as entities by the very technology of publishing. If we were to move to manipulation, storage and delivery of "electronic text", these companies believe they would lose their protection. It would only be a matter of time, they feel, before they were absorbed by the electronic data processing conglomerates or taken over by government. Similar concerns apply to the specialized abstracting, translating and current awareness services.

The regulatory principle can be explicated with further reference to economic and legal data on this economic system. The theme can also be responded to in the design itself. The implementation of a SCATT system would require substantial revision of the law. Among these revisions could be revised antitrust legislation designed to protect the small concern operating within the context of the new technology. Such legislation should certainly be contemplated, and would be the kind of measure that would greatly facilitate the search for a broader sense of consensus. Examples of other primary regulatory principles which have emerged include:

- the regulation of document authorship by academic tenure practices
- the regulation and development of informal collegial work groups

ASSESSMENT AND THE CONTEXTUAL ENVIRONMENT

Contextual evaluation is the most difficult and in many ways the most crucial for it is concerned with the effect of technology on the very fabric of our society. Unfortunately, the way contextual analyses are currently performed is both inaccurate and misleading. Contextual analyses are systems analyses. Contextual responses are systems responses and derive from systems regulating principles, not simple causal chains.

To carry out a contextual assessment requires a strong appreciation of the nature of the contextual field. Some individuals have either an innate sense or an acute awareness of significant movements in the social order. Many are able to express the essence of these changes in novels or sociological texts. There are, however, few examples of attempts to systematically <u>predict</u> changes

in the social order with the advent of new technological forms.
The great exceptions, such as Orwell's 1984 [16] stand out. But
they are few and far between. Not even the science fiction writers
have risen to the challenge. As Pournelli [17] has pointed out:

>"Given the obvious importance of social order to
>science fiction, it is amazing that out of all of
>the thousands of (science fiction) stories in print
>there are so few memorable social orders. It can
>be argued that one of science fiction's most impor-
>tant tasks -- other than entertainment -- is the
>explanation of human societies and examination of
>ways human institutions might respond to changes in
>technology, population, and even the passing of
>time. It is surprising, then, that so few science
>fiction writers set about creating social orders in
>any systematic way."

Insight and imagination don't seem to get us very far in this
task. The power of the "thought-experiment" seems to fade when
dealing with changes in the social condition.

As an alternative, some have turned to the use of historical
analogy. This approach attempts to use evidence from the impact of
technological change in the past to predict impacts of new technologies in the future. Preliminary analyses of these kinds have
been carried out in the case of medical technology [18] and the
space program [19].

The use of historical analogy may generate insights, but as a
generalized methodology it suffers substantial drawbacks. In any
particular case the positive analogy (or degree of similarity) encourages generalization but the negative analogy (or degree of difference) advises caution. It is always possible that while similar
in a great many respects, there are sufficient differences between
two cases to invalidate inference from one to the other.

A final and largely unexploited source of evidence comes from
social theory. Over the past twenty years there have been important discoveries in the theory of behavior of social groups. Bowen's
theory [20] of family process is an example, as is Lawrence and
Lorsch's theory [21] of organization.

These theories are typically holistic in nature and rather
weakly predictive of future events (the asymmetry between prediction and explanation is clearly evident in these cases). However,
the structural-functionalist theories of anthropology are also
holistic and weakly predictive and have been successful in a predictive mode in the past. (During the Second World War, for example, "anthropologists saved soldier's lives in the field by making

accurate predictions from highly qualitative data dealing with
South-Sea tribes" [22].

Such theories are generally applicable to technology assessments. We may not be able to predict consequences on undifferen-
tiated segments of society, but we should be able to draw conclusions about theoretically defined subgroups. For example, Bowen's
theory of family process talks about the differentiation of individuals and the distancing between pairs and coalitions in the
family system. If too much 'distance' develops, system regulatory
principles will generally throw one or more family members into
dysfunction of a specific kind (e.g., a behavior disorder). Ready
access to telecommunications (particularly television) could well
accentuate 'distancing' in these families and lead to more serious
consequences.

Such application of holistic theories requires interpretive
and clinical judgements from those familiar with these applications.
But by combining judgements about impacts at a variety of levels,
starting with the individual and moving through the small group and
organization to the societal level, it may be possible to develop
rich scenarios of contextual impacts. It may also be possible to
suggest intervention and design modifications necessary for the pre-
servation of vital interests (such as primary, meaningful work,
etc.).

There has been as much concern expressed about the impact of
SCATT on the social order as about any other issue. Even the technologists have expressed serious reservations. In many cases, unfortunately, the concerns are not articulated; but they do serve to
dampen the enthusiasm about participation in design activities.

While SCATT addresses the problem of design of the man-machine
and man-machine-man interaction in some detail, it says little of
the nature of the man-man interaction in the face of the new machine.
And this, in a way, is the crucial issue. How are we going to treat
each other in the context of these new communications possibilities?

In our interviews, concerns were expressed about some familiar
issues. Would the system lead to generalized information overload
and increasingly random behavior of those subjected to it? Would
the system lead to an increasingly repressive environment? (As one
of our critics put it, "SCATT would alter drastically the quality
of our lives and take us one step toward 1984 where Big Brother not
only pours his treasure into the scientific establishment, but then
owns it.") Would the system serve to reinforce a scientific elite?
Would the system generate an increased sense of personal loneliness?
(See, for example, C.W. Churchman's scenario [23] of the operation
of the scientific information system in "the Systems Approach".)

Most of the concerns are rather obvious and mundane. There appears to be very little original thinking being addressed to this issue in the STI community. In part, this is a consequence of of the lack of theoretical input to the discussion. In part, it results from a lack of systematic attention to the problem.

One of the advantages of using a design such as SCATT is that it is comprehensive enough to produce, in and of itself, major contextual changes. It is also concrete enough that systematic and directed attention to the problem of estimation of those changes is possible. The context of the problem is fixed and the issues to be addressed are relatively clear. The method of approach will need to be developed. There may be a need for 'search conferences' using relevant individuals from the various areas of social systems science. Speculative scenarios could be developed. Design responses could be suggested. Careful exploration may not provide many immediate answers, but it should open up a process of inquiry and surface vital issues which would not otherwise be raised. The "process" implications of contextual assessment may become the crucial ingredient in more thoughtful applications of technological change.

CONCLUSIONS

We have argued that technology assessment for telecommunications in general, and STI transfer in particular, should be carried out with a redefined sense of mission. Rather than confining itself to a rather narrow evaluative role, the TA function should become an integral part of an on-going planning activity which aims to take a more active stance toward the creation of improved systems in the future.

To assess technologies against the backdrop of current conditions or extrapolated futures distorts the evaluation process and perpetuates the errors of the past. Instead, we have proposed assessment in the context of an idealized design. Such assessment indicates the value of the new technology in a desired future and provides a clearer picture of, and motivational framework for, the planned change process.

Technology assessment, as an integrated part of a holistically conceived planning effort, provides two major benefits for those who participate in the effort. These are value clarification and socialization. It is our experience that even highly experienced individuals in the information transfer field have extraordinarily ill-defined values to guide their own assessments of the future. As a consequence, there is little sharing of concerns about (and hopes for) the future on anything but the most abstract levels. The major need is not for more technical information about technological impacts, but for better ways of conceptualizing, valuing, and developing a consensus over new alternatives.

We have gone on to argue that TA is a complex multi-faceted activity that should be analyzed in terms of its domains of application. We have suggested a three-way typology for this analysis. We have finally looked at the methods appropriate to each type of domain and have suggested applications in the area of STI transfer.

REFERENCES

1. New York Times, June 29th, 1975.

2. National Commission on Libraries and Information Science, National Information Policy, Report to the President of the United States (Washington, D. C.: Government Printing Office, 1976), p. 80.

3. For a detailed discussion of idealized design, see R.L. Ackoff Redesigning the Future (New York: Wiley, 1974).

4. See R.L. Ackoff, et. al, Designing a National Scientific and Technological Communication System (Philadelphia: University of Pennsylvania Press, 1976).

5. P. Davis and R.E. Freeman, SCATT: The Road to Implementation, a report to the National Science Foundation, June, 1977.

6. D.J. Urguhart, "The Distribution and Use of Scientific and Technical Information", Journal of Documentation, 3 (1948), 222-231.

7. John G. Kemeny, "A Library for 2000 A.D.", in Computers and the World of the Future, ed. by M. Greenberger (Cambridge, Mass.: The M.I.T. Press, 1962).

8. P.R. Lawrence and J.W. Lorsch, Organization and Environment (Cambridge, Mass.: Harvard University Graduate School of Business, 1967).

9. Davis and Freeman, Implementation of SCATT, pp. 57-65.

10. N. Brewster-Smith, "Personal Values in the Study of Lives", in The Study of Lives, ed. by R.W. White (Chicago: Aldine, 1963).

11. B. Bettelheim, Truants from Life (Glencoe: Free Press, 1955).

12. P. Diesing, Patterns of Discovery in the Social Sciences (Chicago: Aldine, 1971), p. 161.

13. H.I. Ansoff, Corporate Strategy (New York: McGraw-Hill, 1965).

14. I. Mitroff and M. Turoff, 'Technological Forecasting and Assessment: Science and/or Mythology?' Technological Forecasting and Social Change, 5, (1973) pp. 113-134.

15. Ibid., p. 125.

16. George Orwell, 1984 (Hermandsworth: Penguin Books, 1965).

17. J. Pournelli, "The Construction of Believable Societies", in The Craft of Science Fiction, ed. by R. Bretnor (New York: Harper and Row, 1976).

18. Barbara S. Marx, "Early Experiences with Hazards of Medical Use of X-rays, 1896-1906 -- A Technology Assessment Case Study", 1968 Program of Policy Studies, George Washington University.

19. Bruce Mazlick (ed.), The Railroad and the Space Program: An Exercise in Theoretical Analogy (Cambridge, Mass.: M.I.T. Press, 1965).

20. David Berensen, "An Interview with Murray Bowen", The Family, 3, 1976.

21. Paul R. Lawrence and Jay W. Lorsch, Developing Organizational Diagnosis and Action (Reading, Mass.: Addison-Wesley, 1969).

22. C. Argyis, Theory of Intervention (Reading, Mass.: Addison-Wesley, 1973), p. 115.

23. C.W. Churchman, The Systems Approach (New York: Dell, 1968), p. 116.

DISCUSSION OF PAPERS BY GANZ, THOMAS AND VERNIMB

Discussant and moderator: Gordon Wells

The main areas of discussion were as follows:

Need for Better Information

The reasons for the collection and review of scientific and technical information were discussed. Ganz said that one rationale was to improve the cost effectiveness of conducting research; there was evidence that improved information reduced duplication and increased research efficiency.

Goldstein suggested that people will communicate, come what may, and Thomas that people will collect and store information any way for a variety of purposes; this in itself was sufficient incentive to try to create more efficient systems. Gabbitas felt that it did not matter why people collected information. The key was that networks lead to a greater cross-fertilization of ideas and more contact across disciplines, leading to better research.

Causality

Ganz felt that, although more research is needed, evidence suggests that people who use information sources effectively produce better research. There is some debate, however, about cause and effect.

Charges

In answer to a question from Williamson, Thomas felt that few data are available on variation of the price-demand relationship over time although the change might be rapid. Goldstein remarked that present evidence is that the method of charging is important even when the service is a monopoly but, if there is competition, it is vital to base charges on costs (otherwise a competitor will spot the anomaly and "cream skim"). He and others pointed out that information services are very vulnerable to money saving measures-- telephone bills/budgets will be cut rather than sacking staff in hard times.

Hiltz took issue with the view that it is a form of "unfair competition" for government to use grants to provide free or heavily subsidized telecommunications services for research purposes. She

345

argued that experimental systems can never be competitive in the
early stages and that subsidies are essential. Even with zero charge,
the cost to guinea pigs is high from learning time, problems with the
technology, etc. Wells suggested that early subsidies were common
commercial practice by, for example, computer manufacturers.

Cowie warned of the pitfalls of asking clients how much they
would pay. Not only is it difficult for clients to assess the worth
of future services, but it is not in their interests to say how
much they would pay since this could influence the minimum price.
Often they are not the decisionmakers on the provision of the service, nor are they responsible for the budget, which pays for it.
Thomas agreed that such information (e.g., from in-depth interviews)
had to be collected and treated with caution.

Informal Aspects of Information Systems-Messages

Goldstein and Day pointed out that the ARPANET mailbox was an
afterthought--message services are a 'side-effect' that is very
widely used. Gabbitas said that message systems are already widely
used despite the fact that they are illegal under the policy of many
PTT's.

Vernimb said that Euronet could provide these services if PTT
policy allowed.

Back up Services

Rahmstorf asked how information was selected and edited in
Euronet. Vernimb said that more studies were being undertaken.
There are current deficiencies in speed of response and copyrights
are an important problem.

DISCUSSION OF PAPERS BY DAVIS, MASON AND RAHMSTORF

Discussant and moderator: Carole Ganz

The use of telecommunications as a mechanism for transferring scientific and technical information was addressed from two perspectives in the discussion: 1. the design of information transfer services and 2. the assessment of these services.

Design of Information Transfer Services

Both Davis and Rahmstorf addressed the design of information transfer systems. Davis focused on what an information system (SCATT) would look like if current organizational and economic constraints were not present and the system's design was based on information derived through the use of a technology assessment evaluation method. Rahmstorf chose to examine ways of improving the design of a current information system. Asked by Ganz to compare the design process of the two systems, Rahmstorf suggested that the systems are similar to the extent that they both represent large scale approaches to information transfer. The Text Information system is different from the SCATT system in its emphasis on structuring primary information in a special handbook-like style and in using non-Boolean text description.

Following this general discussion Rahmstorf suggested two problems in the design of current information systems which his study addressed. Current systems require mediators to assist with information inquiries. The precision of the output is not sufficient because of the semantic ambiguity of Boolean query languages. These problems, Rahmstorf stated, argue for an improved, more precise language for use by scientists which is closer to the natural language. These remarks led to a discussion by several participants on the effect of language on access to information systems. In this vein, Ohlman observed that some form of mid-range language is required for international use, and the international movement appears to be reviving. For example, the work of Charles Bliss on "Semantography" has at last found its application as a medium of communication for physically handicapped nonverbal children in Ontario and another pictogram-based language has been developed in Japan by Yukio Ota. Corresponding anti-Babel trends can be detected in recent developments in computer networks. Euronet, the European Community's planned network for scientific, technical, social, and economic information, has a multilingual program which will provide for automatic translation of scientific and technical texts drafted in natural languages.

Lucas suggested that a theoretically based language similar to that used by anthropologists might be a solution to the problem. However, as Ganz pointed out, basing systems on disciplinary languages implies that the disciplines of systems users are similar to those of systems generators. For example, interdisciplinary research users find it difficult to access systems based on discipline languages.

Another issue in systems design which also affects access is pricing policy. Mason pointed out that with respect to IAC's, the initiation of fees resulted in a relatively large drop in demand for these services. Subsequently, however, the demand appears to resume its prior growth at a rate near the prior rate. Anecdotal evidence indicates that charging for information services represents a problem for both the information supplier and the consumer. On the one hand, engineers and scientists may be embarrassed to request money for access to an external information system when they have been hired as experts. On the other hand, IAC managers tend to suffer from a "library syndrome" and do not want to charge for previously free services. Hard data on the impact of instituting fees is difficult to obtain since information services have traditionally been provided without charge.

It was pointed out that perhaps the real question with respect to access is how it is related to distribution. Lucas observed that access modes affect the distribution of information and suggested that if overhead rather than direct costs are used to pay for system access, distribution might be better. Ganz indicated that most arguments to date surrounding information systems are based on productivity increases, not on distribution issues. The question of distribution has only been looked at very recently. Hiltz and Ganz agreed that the relationship of information access to productivity is unknown. For example, we don't know whether less productive scientists can become more productive if they have access to better information.

Assessment of Information Systems

Assessing information systems was raised by Elton with respect to the state of the art of technology assessment methods. Davis indicated that the question which technology assessment was meant to address in his work was: how can the process of information retrieval be more productive and interactive? Since technology assessment literature tends to be very specialized, Davis et al., looked at levels of technology assessment in approaching the design of SCATT. These levels include assessments of the intentional purpose of the technology, the impacts which can be linked directly to the technology's implementation and the consequences of the technology which are unintended. In addition, Davis recommended the use of participant observers in evaluating information systems. He

suggested that the precision which this method lacks would be outweighed by the greater depth of understanding it would allow.

Mason also stressed that need to include assessments of factors
other than the immediate purpose of the technology. For example,
economic assessments should make it clear that economic factors are
only one element of evaluation. In the internation arena, he pointed out, national prestige or national defense needs may override
economic considerations. Lucas supported incorporating a range of
factors in evaluations, including any social and psychological costs
of the person using the system.

C. G.-B.

H.O.

Section Five

TELECONFERENCING AND COMPUTER CONFERENCING

This section comprises papers on three families of systems: audioconferencing systems which allow three or more individuals at two or more locations to talk with one another (sometimes they are supplemented with systems for the exchange of graphics or text); videoconferencing systems which provide two-way television connections in addition to the sound channels; and computer conferencing systems which allow dispersed individuals to use computer networks for real time or asynchronous keyboard communication with one another. Those unfamiliar with computer conferencing systems may find it helpful to read first the review on this subject in the next section. To avoid confusion it should be noted that some writers use the term teleconferencing to include computer conferencing; others do not.

The use of conferencing systems in health care and education has been covered in earlier sections. Here we are concerned primarily with their use in business and in government organizations.

George Jull reviews the results of surveys of users attitudes to four Canadian teleconferencing systems. Among other findings he reports that the acceptance of teleconferencing is strongly influnced (i) by the pressures of relocation coupled with the inconvenience of frequent travel and (ii) by its being found satisfactory as a substitute and a complement for some face-to-face meetings. The paper concludes by raising issues relating to the aggregation of services for delivery on common facilities. In this regard important traffic characteristics include bandwidth, traffic patterns, privacy, subscriber penetration and spatial distribution.

First in the UK, subsequently in a number of other European countries, surveys have been made of today's communications in business and government (in-person, by telephone, by mail and by Telex) with the idea that their characteristics could be used as the basis of projections of the extent to which new telecommunications services could substitute for them. The paper by Dormois, Fioux and Gensollen reports on a recent French study of this kind based on the use of communications diaries. They describe the methodology employed and some interim results.

351

In the next paper Bob Johansen and his colleagues show the
difficulties of making projections about the future use of teleconferencing and computer conferencing on the basis of current
understanding. The actual focus of their paper is the problem of
learning how to use these media. They discuss what can be learned
from social evaluations of conferencing systems, identifying a
number of key parameters. They then describe a "Teleconference
Tutorial" which they are developing to assist new users and to
serve as a research tool.

The paper by Dormois and his colleagues introduced the use
of data from communication surveys as a basis for projections on
the scope for teleconferencing as a substitute for established modes
of communication. (The reader will already have noted that some
papers have carried warnings that the substitution perspective may
be too limiting -- e.g., the papers presented by Ganz and by Jull.)
An important link in the necessary models is the function which
estimates substitutability from characteristics of particular
communications events -- especially in-person meetings. Such
functions have generally been constructed using one or a combination of (i) common sense and (ii) the results of laboratory experiments. These experiments have been conducted at the level of
individuals.

Art Shulman and Jerry Steinman point out in their paper that
the deployment and use of teleconferencing services depend upon
strategies adopted for coordination of communication among
organizational units. They review past substitution studies, in
particular the work of the Communications Studies Group (CSG) at
University College London, and discuss the limitations of the
approach. Then they introduce Galbraith's and Thompson's theories
of organizational communication. These are used to extend the CSG's
classification of meetings. Within this framework they discuss the
use of different conferencing media as substitutes for in-person
meetings.

Dave Conrath also approaches interpersonal communication from
the perspective of the organization. He reports on an ongoing study
to develop descriptive models of communications and organizational
structure. The paper describes how data were collected by several
methods from three different companies, two of them both before
and after the installation of a new telecommunications system.
Some results are discussed: who related to whom? what modes of
communication were used? for how long? and for what purpose.
Finally he shows statistically significant relationships between
choice of mode and (i) hierarchical rank (a surrogate for task),
(ii) department and (iii) whether "before" or "after". Communication content, however, was not found to be a satisfactory explanatory variable.

Mike Wish describes psychological research at Bell Laboratories on interpersonal communications behavior. This has focussed primarily upon the modality used, the context (or purpose) of the communications, and the relationship between the individuals concerned. Various issues have been investigated: the perceived effectiveness of different modes in different situations; perceptions of the ways in which people in different interpersonal relationships communicate with one another; and interpretations of videotaped interactions. Videotapes were also used to assess the relative importance of the audio and visual channels: for one group of observers sound was suppressed; for another group vision was suppressed.

The paper describes the measurment tools and analytical techniques that have been developed. The main results are presented and discussed.

Ric Irving presents a case study comparing the use, in a Canadian government department, of a computer message system with the use of the more sophisticated computer conferencing system which replaced it. Information is provided on usage statistics and answers to a questionnaire on attitudes. These have implications for the design of such systems and for the way they are introduced into organizations.

The section concludes with a statement by Craig Fields, calling for conferencing systems which are designed substantially to improve upon established forms of communication, rather than merely imitate them.

Taken as a whole these papers provide a comprehensive view of the current states of understanding and of methodology regarding the role of teleconferencing and computer conferencing in organizations. The papers by Jull and by Johansen et al provide useful non-quantitative introductions to some of the current issues in the field.

D.C.

B.J.

USE AND TRAFFIC CHARACTERISTICS OF TELECONFERENCING FOR BUSINESS

G. W. JULL

COMMUNICATIONS RESEARCH CENTRE
DEPARTMENT OF COMMUNICATIONS
OTTAWA, ONTARIO, CANADA

SUMMARY

This paper reviews results of some Canadian studies concerned with the use of teleconferencing for management and administration in business. These studies took the form of surveys to evaluate the attitudes of users of various audio and video teleconferencing systems in the federal government, the University of Quebec and Bell Canada.

It was found that the introduction and acceptance of the use of teleconferencing in various organizations has been strongly influenced by (i) the pressures imposed by decentralization or other constraints on frequent face-to-face meetings, (ii) teleconferencing being found to be satisfactory to substitute for, or complement, some face-to-face meetings, (iii) the users having favourable attitudes towards teleconferencing, and (iv) suitable technology being available.

A preliminary assessment of traffic characteristics has been made. These characteristics include bandwidth, traffic patterns, privacy, potential subscriber penetration and spatial distribution. It is concluded that more information on potential traffic characteristics of business and social service uses of teleconferencing is required. This information will be needed to define circumstances for which it would be sensible and economic to aggregate services for delivery on common facilities.

1. INTRODUCTION

The demand for business teleconferencing has developed very slowly, and systems for this purpose are currently used in only a few organizations in the Western world. In Canada, only about 12,000 meetings per year are conducted using the telephone conferencing service. Less than a dozen organizations are using dedicated audio and video teleconferencing facilities. Elsewhere, the use of closed-circuit television conference systems and video telephones has fallen short of the expectations of their promoters. This slow growth may appear surprising in view of the many cost and time-saving benefits which have been claimed and demonstrated for this tele-communications service. It immediately raises questions of whether teleconferencing is unsuited to many essential interpersonal communications tasks, or whether improvements in technology might lead to greater use.

Between 1972 and 1975, the Department of Communications (DOC) in Canada carried out studies on teleconferencing to determine its usefulness to assist federal government departments to meet their needs for improved communications between headquarters and outlying divisions. (Ref. 1) These studies, and those carried out in other countries, have led to a better understanding of the nature of teleconferencing, its benefits and uses.

In parallel with these studies of the use of teleconferencing for business, there has been a wide range of studies of the use of teleconferencing for medicine, education and community interaction. The technology used to deliver these social services is similar in many respects to that used for business applications. It is there-fore appropriate to consider circumstances in which it would be economic and sensible to aggregate business and social services, for delivery on common transmission facilities.

This paper first reviews factors which have been found to have an important influence on the introduction and acceptance of the use of teleconferencing for business. It then summarizes knowledge of some of the traffic characteristics of business uses which relate to the question of aggregation of business and social services on common facilities.

2. CANADIAN FIELD STUDIES

A number of Canadian organizations have installed a variety of audio and video teleconferencing systems. During the period 1972 to 1975, the reaction of many hundreds of users of these systems were evaluated as part of the DOC studies. The systems which were evaluated included:

(i)   The University of Quebec Audio Teleconferencing System:

This system uses dedicated lines, a specially-
developed conference switchboard and Western
Electric 50A teleconferencing sets, to provide
an intercity service between eight centres of
the University in various cities in Quebec.
(Evaluation was carried out by Communications
Studies Group (CSG), London, in collaboration
with the University of Quebec and DOC.) (Refs. 2 & 3).

(ii) Bell Canada Conference TV System:

This system was installed in 1972 to provide an
intercity video service between central locations
in Toronto, Montreal, Ottawa and Quebec City.
(Evaluation was carried out by CSG, London, in
collaboration with DOC and Bell Canada Marketing).
(Refs. 4 & 5).

(iii) DIAND Audio Teleconferencing System:

The system in the Department of Indian Affairs
and Northern Development (DIAND) uses the CN/CP
Telecommunications 'Broadband System' to provide
audio service between two locations in Yellowknife,
two locations in Whitehorse, (in the Canadian
North), and one location in Ottawa. Evaluation
was carried out by DOC, in collaboration with
officials of DIAND).

(iv) The DOC Audio Teleconferencing System:

This system also used the CN/CP Telecommunications
'Broadband System' to provide an audio service
between locations in Ottawa, Toronto, Montreal,
Moncton, Winnipeg, and Vancouver. (Evaluation was
carried out by DOC).

3.   FACTORS INFLUENCING INTRODUCTION AND USE

The results of the evaluation studies have been used to
identify factors which have had a strong influence on the introduction and use of teleconferencing. These factors include (i)
pressures imposed by decentralization or other major constraints on
face-to-face meetings, (ii) teleconferencing being found useful as
a substitute and complement to face-to-face meetings, (iii) the user

having favourable attitudes and (iv) suitable technology being available. These are discussed below. Other factors, such as potential cost-saving benefits will not be dealt with in this paper.

3.1   Constraints on Travel

The introduction and subsequent acceptance of the use of teleconferencing in the University of Quebec and DIAND have been strongly influenced by a number of somewhat unique constraints on business travel. In the case of the University of Quebec, structuring the university as a number of decentralized institutions across the province of Quebec, led to introduction of audio tele-conferencing and other telecommunications facilities. Shortly after, an opportunity arose to make extensive use of the telecon-ferencing facilities for urgent meetings, occasioned by student strikes. It is claimed that use of teleconferencing to satisfy this need convinced many participants that teleconferencing was useful. Since that time, use of the system has continued to grow, and it is now well-accepted as an important aid to increase overall contact between separated groups in this organization. (Refs. 2,3, & 6).

In the case of DIAND, with headquarters in Ottawa and outlying units in Yellowknife and Whitehorse in the Canadian North, pressures to introduce teleconferencing arose from the difficulties of travel between these locations. Coupled with these difficulties was the growing need for meetings between small and very large groups (of up to 60 participants) located in Ottawa, Yellowknife, and Whitehorse. Use of audio teleconferencing to meet this need has continued since the system was introduced four years ago.

Several other Canadian organizations, not subject to the same constraints on travel, have introduced teleconferencing but have made less use of their facilities. In the case of the Bell Canada system, two classes of users are distinguishable. First, interdependant working groups of Bell Canada employees located in Ottawa, Toronto, Montreal and Quebec have used video teleconferencing for frequent meetings over the past four years. On the other hand, groups outside the Bell organization have used these facilities only infrequently. This is in spite of the fact that the facilities have been available to them at no cost as a market-trial. In the case of the DOC, the constraints on travel and the need for meetings between Ottawa and regional offices in southern Canadian cities have not yet been sufficient to result in extensive use of their audio teleconferencing facilities.

Experience in other countries has also shown that unless travel is very inconvenient or time-consuming, teleconferencing is unlikely to be extensively used. In particular, a number of intra-city

teleconferencing systems have either been withdrawn or are seldom
used. Video systems used by the First National City Bank and the
Bankers Trust for meetings between locations in New York City were
used for some years and then withdrawn.

3.2 Teleconferencing as a Substitute and Complement to Face-to-Face Meetings

Another factor which has influenced the level of use of teleconferencing in some Canadian organizations has been the developing
awareness that teleconferencing can complement as well as substitute
for face-to-face meetings. For example, about 122 of 255 teleconferencing meetings in the University of Quebec were urgent or
'impromptu' meetings which might otherwise not have been held.
(Ref. 6). In the DOC, the use of teleconferencing generated about
four times the previous number of meetings between senior officials.
In addition, it provided other widely-separated individuals with an
opportunity to participate in meetings for the first time.

The characteristics of audio and video meetings held in
Canadian organizations illustrate the role of teleconferencing as
an aid in the on-going work of the organizations (Ref. 1):

(i)   The meetings were shorter than comparable
        face-to-face meetings, with an average
        duration of 100 minutes (for users of the
        University of Quebec and the DOC audio
        systems) and 133 minutes (for users of
        the Bell Canada video system).

(ii) For both the Bell Canada and the University
        of Quebec users, teleconferencing was used
        most frequently for meetings whose primary
        purpose was 'information-exchange' and
        'problem-solving' (over 50% of meetings).
        They did not use teleconferencing for
        meetings whose primary purpose was
        'bargaining' or 'getting-to-know someone'
        (less than 4% of the meetings).

(iii) Teleconferencing was used more frequently
        for urgent meetings in the University of
        Quebec than in Bell Canada (51% vs 29% of
        all meetings). This can be attributed to
        the greater convenience of using the University
        of Quebec system, in terms of interconnection
        time (less than 15 minutes), the average
        travel time to teleconferencing meetings

(less than five minutes compared with
about 20 minutes for the Bell Canada
system), and the availability of the
system at a large number of locations
of the University of Quebec compared to
the availability of the Bell Canada
system at only a few central locations
in Canada.

(iv)   Teleconferencing was not considered
suitable for confidential discussions.

(v)   In the University of Quebec, teleconferencing
was used to prepare for face-to-face meetings
and to follow up after meetings.

3.3   Attitudes Towards Teleconferencing

Business teleconferencing is an instrument for management
co-ordination and control.   When it is imposed on a complex
system of management responsibilities, the costs of failure could
be disruption of the system, or loss of control.   It is therefore
understandable that many organizations may be hesitant to introduce
teleconferencing.   Conditions must be favourable to ensure success
of its use; to a large extent, this requires favourable attitudes
and the co-operation of all those who must make it work.

In 1972, a survey of the needs of federal department officials
was carried out to determine attitudes towards the possible introduction of teleconferencing.   It was found that senior managers
perceived greater utility for teleconferencing and had more favourable
attitudes to it, than did middle managers or technical groups who were
sampled.   On the other hand, while relatively few were strongly in
favour, almost nobody opposed it. (Ref. 7).   Since then, several
federal departments, among those having the most positive attitudes,
have introduced teleconferencing.

With experience, users of audio teleconferencing at the
University of Quebec found it to be increasingly useful for a wide
range of meeting activities and were prepared to hold larger meetings.
On the other hand, first-time users of the Bell Canada Video System
liked it better than experienced users. (Ref. 1).

3.4   Availability of Suitable Technology

Our studies have found that a major source of dissatisfaction
with audio-only teleconferencing is the potential for misunderstanding

as to 'who-is-speaking' and 'who-wishes-to-speak'. This limitation
has been overcome in the Remote Meeting Table Audio System (Ref. 8)
used in UK government departments. With this system, signals are
transmitted which are used to identify speakers at their various
locations. Another source of dissatisfaction with both audio and
video systems arises from the use of voice-switching in the terminals.
At present, this is often necessary to overcome problems of acoustic
feedback in conference rooms. Other solutions, which minimize the
need for voice-switching, would be highly desirable.

   In spite of these unsatisfactory technical features of current
systems, for some groups the advantages of teleconferencing have
outweighed the irritations. Wider-scale acceptance of the use of
teleconferencing would be facilitated if systems permitted more
natural, free-flowing conversation.

4.   TRAFFIC CHARACTERISTICS OF INTERPERSONAL TELECOMMUNICATIONS

   Recent research on applications of interactive telecommunications
systems has found that the characteristics of telecommunications
facilities required for business teleconferencing are similar in
some respects to characteristics required for delivery of social
services.   In particular, business teleconferencing has used
channels with up to full video bandwidth, and these channels have
also been used for social service applications in medicine, education
and community interaction. In North America, business teleconferencing has been carried over terrestrial or satellite transmission
facilities provided by the common carriers. On the other hand,
while social services have been carried over facilities provided
by the common carrier, others have been carried on two-way cable
transmission facilities provided by operators of cable systems.
(Refs. 10 & 11). In these latter cases, cable-delivered social
services are additional to commercial television entertainment
services and may not be an economically-viable use of a relatively-
scarce resource which is principally dedicated to commercial
entertainment broadcasting and information services.

   Considering the foregoing, it is of considerable importance to
examine whether or not it would be economic and sensible to aggregate
some or all interpersonal business and social telecommunications
services with other commercial services for delivery on shared
facilities. If shared facilities were employed, economies of scale
could be realized and some of the proposed uses of telecommunications
facilities for delivery of social services could, for the first time,
become economically viable. On the other hand, the traffic characteristics of various services might be sufficiently dissimilar that
aggregation would lead to serious degradation in service to some

classes of subscribers. In this case, economies of scale could be
offset, and provision of specialized facilities would be more
sensible. In either case, in North America at least, policy
questions are raised with respect to the most satisfactory
institutional arrangements for delivery of these services.

To the best of our knowledge, the traffic characteristics of
various interpersonal business and social services are not yet
well-defined. Therefore it is premature to identify circumstances
in which it would be sensible to aggregate services for business,
education, medicine or community interaction. A start to this
identification can be made by comparing results obtained in studies
of various services. These results are presented for some important
traffic characteristics.

4.1   Bandwidth

Studies reported above have found that a significant fraction
of current business needs can be satisfied using channels of voice
bandwidth (4 KHz) or a few multiples thereof. Furthermore, it has
been found that some educational needs can be satisfied with voice
bandwidth channels (e.g. the intercity systems in use at the
University of Wisconsin-Extension (Ref. 9) and the University of
Quebec).

By way of comparison, some needs of business services, and to
a greater extent the needs of many social services can only be
satisfied using channels with bandwidths capable of supporting one
or two-way transmission of images. These bandwidths range from that
required for slow-scan TV (4 KHz) up to that for broadcast TV (6 MHz)
(e.g. the pilot two-way cable TV experiments being conducted at
Reading, Penn. (Ref. 10) and Spartanburg, S.C. (Ref. 11)).

4.2   Traffic Patterns

Studies of current business uses reveal at least two distinct
patterns of traffic. The first is the pattern arising when teleconferencing is used for urgent or 'impromptu' meetings. This need
is met by providing convenient access to terminals located within
the organizational setting, the capability for rapid interconnection
of participating locations, and adequate network capacity to minimize
waiting times. The second pattern arises when teleconferencing
meetings are scheduled some time in advance. In principle, this
need can be met by use of shared facilities, located in a convenient
central part of a city, such as the conference TV facilities of Bell
Canada, AT&T, and the BPO. (In practice, however, participants using
the Bell Canada facilities would travel no longer than about 20
minutes to get to these facilities.) For both patterns of traffic,

average circuit holding times ranged from about 100 minutes (audio)
to 133 minutes (video).

4.3   Privacy

A high proportion of users of Canadian systems has stated that
current facilities were not suitable for confidential discussions.
In our opinion, growth of the use of teleconferencing for business
is likely to depend on rectifying this situation, through provision
of adequate communication security facilities. Until improved
arrangements for security are provided, the potential for consoli-
dating business with other services appears limited.

5.   SUMMARY AND CONCLUSIONS

This review of Canadian experience has found that the introduction and acceptance of the use of teleconferencing has been
strongly influenced by (i) the pressures imposed by decentralization
or relocation of units of an organization coupled with the inconvenience of frequent travel, (ii) teleconferencing being found to be
satisfactory both to substitute for, and complement some face-to-face
meetings, (iii) the users having favourable attitudes towards teleconferencing and (iv) suitable technology being available.

The first and second of these factors appears to have been at
least as important as the promise of potential cost-savings as an
explanation of why some Canadian organizations and not others have
introduced teleconferencing. It suggests that favourable attitudes
and the co-operation of those who must make it work are assured after
experience confirms that teleconferencing is satisfactory for some
meetings and the effort required to use facilities is made very easy
relative to the effort required to travel.

These factors will influence some of the traffic characteristics
of business uses in the future. Important traffic characteristics
include bandwidth, potential traffic patterns, privacy, subscriber
penetration and spatial distribution. Further study is required to
define these characteristics, to consider whether or not degradation
of services might result if services were aggregated, and to consider
the economic feasibility of shared and specialized facilities.

6.   ACKNOWLEDGEMENTS

The author wishes to acknowledge major contributions to this
study by J.G. Craig, N.M. Mendenhall, and M.G. Ryan and other

members of the DOC staff who participated in teleconferencing
research. He also wishes to acknowledge the major contributions of
E. Williams, J. Short and H.G. Thomas of CSG (London) for research
analysis of some of the field studies conducted in Canada, the
contributions of researchers at Carleton University and the collaboration of Bell Canada, University of Quebec, Department of Indian
Affairs and Northern Development, and the Public Service Commission,
in field and laboratory studies.

7.   REFERENCES

1.   'Research Report on Teleconferencing' (Volume 1 and Volume 2),
      G.W. Jull, R.W. McCaughern, N.M. Mendenhall, J.R. Storey,
      A.W. Tassie, and A. Zalatan. CRC Report No. 1281, January 1976.

2.   'A Report on the Use of the Audio Conferencing Facility in
      University of Quebec', J. Short. CSG Report P/73161/SH,
      (sponsored by DOC), University College, London, 1973.

3.   'The University of Quebec Audio Conferencing System: An Analysis
      of Users' Atitudes', H.B. Thomas and E. Williams. CSG Report
      P/75190/TH, (sponsored by DOC), University College, London,
      July 1975.

4.   'The Bell Canada Conference Television System', E. Williams,
      CSG Report P/73137/WL, (sponsored by DOC), University College,
      London, 1973.

5.   'The Evaluation of Teleconferencing: Report of a Questionnaire
      Study of Users' Attitudes to the Bell Canada Conference Tele-
      vision System', E. Williams and Susan Holloway, CSG Report
      P/74247/WL, (sponsored by DOC), University College, London,1974.

6.   'La Telegestion a l'Universite du Quebec - bilan et perspectives'
      R. Barrette, J.C. Kilfoil, C. Horin, et P. Dumas. Rapport de
      recherche, Vice-presidence aux Communications, Universite du
      Quebec, 1975.

7.   'Canadian Civil Servants Perceptions of New Forms of Communica-
      tions Linkages', A.D. Cameron and J. Langlois (presentation at
      18th International Congress of Applied Psychology. Montreal,
      June 1974).

8.   'The RMT Teleconference System', CSG Report P/72024/RD,
      University College, London, January 1972.

9.   'Educational Telephone Network and Subsidiary Communications
      Authorization: Educational Media for Continuing Education in

Wisconsin', L.A. Parker, Educational Technology, February 1974.

10.   'Public Service Uses of Cable Television', M.L. Moss, (Presen-
       tation to the Second Symposium on Research Applied to National
       Needs, National Science Foundation, Washington, D.C., November
       7-9, 1976).

11.   'Moving from Two-Way Cable Technology to Educational Interaction
       W.A. Lucas. (Presentation at National Telecommunications
       Conference, Dallas, Texas, November 29 - December 1, 1976).

G.J.

N.G.

EVALUATION OF THE POTENTIAL MARKET FOR VARIOUS FUTURE COMMUNICATION MODES VIA AN ANALYSIS OF COMMUNICATION FLOW CHARACTERISTICS

M. DORMOIS, F. FIOUX and M. GENSOLLEN

DIRECTION GENEREALE DES TELECOMMUNICATIONS - SPAF -

TOUR MAINE MONTPARNASSE 75755 PARIS CEDEX 15

PREFACE

The estimation of the potential market for future communication services suggests an analysis of existing communications flows in terms of the characteristics of the new services. If we define a new communication mode, in terms of a distinctive type of offered service, an evaluation of the total traffic flow potentially transfered from the existing communication mode to the new one is then possible. For example, 7% of the total number of face-to-face meetings and 14% of the total number of telephone contacts could be transfered to the future mode "multiconference" (which permits several persons to confer by telephone simultaneously). The long term forecast of total communications flows also involves a model of the evolution of establishments from the view point of their communication systems organization.

The aim of this paper is to describe the methodology and preliminary results of an inquiry using diaries to register, during one week, the communications transmitted by the employees of different establishments via 4 main communication modes: Telex, Telephone, Mail and meetings. The analysis concerns 33000 communications, recorded by 2000 employees working in 60 establishments. Each communication is described by a set of variables concerning its function, its volume, the reasons for the choice of mode and whether or not an alternative mode could have been used.

The present systems of organization require transmission of a certain volume of information between business enterprises, production units, their headquarters and the services they require in order to operate properly. This information is transmitted by different communication modes, the main ones at present being mail, face-to-face meetings, telex and telephone. The choice of the mode depends mostly on the quality expected of the communication services proposed: speed of transmission, the possibility for several persons to confer simultaneously, the requirement of authentification of the correspondent's identity, respect of privacy ....

The services, on the other hand, depend on the role that the communication function assumes in the operation of the enterprise. If we wish to study in its entirety the future for communication modes, and to evaluate their potential market possibilities, we need to analyse the complete system of present communication flows. This should be done in terms of the type of service to be rendered, and related to the role of the type of communication in the operations of the said enterprise.

We have chosen this approach in preference to polling the potential users as to their attitudes vis-a-vis new communication modes, since this method, though often the only one possible, leads to results of questionable quality. Laboratory experimentation also provides a design approach for new facilities (for example the French visiophone), the aim being to adapt them as well as possible to the proposed utilization. However, there is an inherent risk that the potential market will be only imperfectly evaluated when one is dealing with new communication modes in the context of alternatives.

An analytical approach to present communication flows is initiated in this paper on the basis of the first sets of results from an ongoing inquiry. These preliminary results and the corresponding methodology are detailed below.

I. METHODOLOGY

I.1. Criteria for Choice of Sample

The choice of the sample was dictated by the two main preoccupations of the study. The first was to collect and store data on the communications of a number of establishments from which to radiate national projections. The second was more qualitative in nature, and concerned a knowledge of the operations of the establishments

from the viewpoint of their communication systems organization, both
internal and external. In order to accommodate these two aims, the
inquiry dealt with 60 establishments based on a factorial design.

The factorial design was defined by the interrelationship of
three variables which characterise establishments: size, industrial
classification and geographic location*. These three criteria are
sufficient to describe the telephone traffic of the establishments,
as has been demonstrated in previous studies. There is every reason
to believe they are equally applicable when it comes to describing
the behaviour of the traffic registered for the three other modes
under study: mail, face-to-face meetings, telex.

While this segmentation satisfied our desire to obtain quantitative results, it did not, on the other hand, enable us to reach a
wide variety of organizational systems. Since there has been no
prior work done in this field, our analysis is based on partial observations. In order to further detail the previously mentioned
segmentation, we introduced a supplementary criterion. The additional variable is the nature of the establishment (company headquarters, dependent or independent establishment). It is known that
this factor is important with respect to the choice of teletrans-
mission equipment (telex, facsimile). Therefore it can be considered
a criterion discriminating the types of system of organization likely
to be found in the establishment concerned.

In the service industries, we separated out "banks and insurance". Although this area employs only a relatively small number
of persons, the category proves of interest since it is innovation-
sensitive in communications. Thus we might expect to find the most
sophisticated forms of organizational communication configurations
in this sector.

I.2. The Sample

The number of establishments to be included was set at 60.
The factorial design mentioned contains one establishment per category, and 20 more were included to take account of the very large
number of establishments in some categories. The distribution of
the sample according to the fourth variable is also given (see
Table I).

---

* In fact this variable has only two possibilities: Paris or the
Provinces. The reason lies in the highly specific role played by
Paris in the French context.

These sixty establishments employed 17,850 employees. The inquiry was based on this set of establishments. By an equivalent group approach, described below, and by the existence of a high percentage of non-transmitting staff members, only 2,000 persons participated in the data input for the communication flows. A total of 33,000 itemised communications were analyzed. See Figure 1 for the overall scheme of the inquiry.

DISTRIBUTION OF ESTABLISHMENTS

INDUSTRY SECTOR \ EMPLOYMENT SIZE GROUP		50-99	100-199	200-499	+ 500	Total
CONSTRUCTION	PARIS	1	1	1	1	4
	PROVINCE	8	3	3	2	16
TRANSPORT AND TELECOMMUNICATIONS	PARIS	1	1	1	1	4
	PROVINCE	1	1	1	1	4
DISTRIBUTION AND SERVICES	PARIS	1	1	1	1	4
	PROVINCE	3	3	1	1	8
INSURANCE AND BANKING	PARIS	1	1	1	1	4
	PROVINCE	1	1	1	1	4
ADMINISTRATION	PARIS	1	1	1	1	4
	PROVINCE	3	3	1	1	8
TOTAL	FRANCE	21	16	12	11	60

Q U O T A S

LOCATION \ NATURE	COMPANY HEADQUARTERS	DEPENDENT ESTABLISHMENT	INDEPENDENT ESTABLISHMENT
PARIS	6	6	8
PROVINCE	8	16	16

TABLE I

SCHEMA OF THE INQUIRY

		RATE	SAMPLE
STEP I Choice of the esta-blishments (cf I.3)	CLASS I CLASS 2 CLASS 40 40 classes of establishments defined by the variables: geographical location, industrial classification, size, nature.	1/600	60 establish-ments (more than 50 em-ployees)
STEP II Choice of the transmit-ters (cf I.4)	SERVICE I • Transmitting employee * Non transmit-ting employee + Transmitting employee in sample ⊕ Equivalent group of transmitters SERVICE N In an equivalent group, only one employee records his communications.	1/9	2000 employees distributed in all services
STEP III Recording of the communi-cations transmit-ted by the employees in sample (cf I.5b)	Teleph. | Telex | Mail | Teleph. + recorded communication ⊕ communications with same structure. Only the communications + are recorded (with a weight of 2 in this example) N.B. Intra service communica-tions are not recorded	1/1,2	33 000 communi-cations Teleph.: 18 000 Telex: 1 500 Mail: 11 000 Meeting: 2 500

FIGURE 1

I.3. Representation of the Sample

We have restricted the scope of the study to those establishments employing more than 50 persons, in order to have only "efficient" establishments in the sample, and because the most interesting configurations require a minimal staff size. The proportion of the establishments studied to the total number possible is 1 to 600. The important thing, however, is the number of items of communication in our sample. The high number enables us to hope that the results can be taken as representative with respect to the volume of communication.

Concerning our aim to analyze the organizational operations of the establishments, it is certain that the most important types have been included in the study. A wide variety of services have been detailed. The representation of the total population in this aspect, therefore, would appear to be highly satisfactory.

I.4. Inquiry Parameters

Three conditions determined the methodology used to gather the data within any given establishment:

a) The data base was to contain the widest possible variety of information related to operational aspects of the establishment. Information on both internal and external communications, initiated by the staff of the establishment, was to be as detailed as possible.

b) The data base had to be capable of computerised processing.

c) The gathering of the data was to create as little perturbation as possible in the normal activities of the establishment.

The first of these conditions was related to the nature of our work-basic research. We knew of no precedents. Thus the widest possible variety had to be included, even though certain variables used to describe the items of communication may turn out to be irrelevant. Such a result needs to be proved by research rather than based on ill-founded a priori judgements.

The second condition reflected the quantity of data to be processed.

These two aspects together defined the tools to be used in the study. The third defined the way the inquiry was to be carried out. Whereas the gathering of the data concerning the establishment itself was easily managed through a questionnaire and an interview with the head of the establishment, the setting up of the data base for the communications was very delicate. If one is to have each employee note down every call placed over five consecutive days, plus giving a detailed description of each call (at least 30 seconds for each entry), this indeed perturbs the organization to an extent that few establishments would accept. Two flexibility points were allowed regarding the detail.

a)   In the case of a call with a structure similar to a previously placed one, there was no separate entry, but the weight of the previous call was to be incremented by one unit.

b)   A set of equivalent groups was set up for all transmitters (persons placing calls). These were transmitters with identical behavioural patterns and together they formed an "equivalent group." If it is possible to identify such groups before the inquiry begins, then the sample can be restricted to a single person in the group, taken as representative of his equivalent group.

I.5. Data Collection Instruments

A questionnaire and an interview were used to obtain data from the head of the establishment. A preliminary inquiry was conducted with respect to the employees, and a communication logbook proved necessary in order to note the volume of the traffic in terms of communications flow.

a)   The establishment questionnaire was arranged into two series of questions. The first series served to describe the establishment from a socio-economic point of view: turnover rate, industrial classification, physical plant. There were also variables which described operations: EDP management or not, decentralised management or not. The latter category is very important if one wishes to establish the relationship between the operations of the concern and the traffic generated in communication flow.

The second series of questions was aimed at describing the communications equipment available, the felt needs and perceived future requirements. We were also interested in knowing the rules laid down by the management regarding the utilization of the equipment. Some questions were included for the purpose of establishing cross-checks.

Lastly, the organization chart for the establishment was obtained. From the total of 60 charts thus provided, an archetypal chart was constructed, with a common nomenclature for all the services (220 basic services existed). In this way we developed a simple means to encode and to compare the various organization structures.

The interview with the head of the establishment proved useful
since it gave an idea as to how top management saw the existing communication system. Generally, such a reflection had not taken place
before, but a series of questions helped (questions concerning the
services rendered by the various modes available, services introduced with new modes, the evolutionary trends of the establishment's
equipment and the reasons underlying the changes). The second function of the interview was to throw light on certain phenomena specific to the establishment which were discovered during an analysis
of the early data.

     b) The preliminary inquiry lasted one full day. All members
of the staff working in the establishment were requested to log all
communications transmitted, as a function of the four modes, but
without describing them. With this data, it was possible to build
up equivalent groups, the principles for which were set out above.
Within a given service, two members of staff were considered to be
equivalent if they transmitted an equivalent number of calls in each
of the 4 modes, and if they had the same level of qualification (the
hypothesis that the calls transmitted would be of the same structure,
since they were needed to fulfil the same task, had been checked out
in other establishments prior to this inquiry). In the course of
this classification we rarely came across equivalent groups with more
than five members.

     c) The diary was used to register the calls transmitted in the
4 modes, their discription being given by a set of 13 variables common to all 4 modes, plus a dozen or so variables specific to each
mode. Each question required a single unique answer.

     The diary for each mode was unique, and was printed on one side
of two recto-verso log-files. A specific diary could be used to describe a maximum 10 communication items (calls). The one for telephone calls placed, as an example, can be found in the appendix.

     The group of objective variables breaks down as follows:

- Services rendered by the mode:

     Nature of the information transmitted: plans, photographs,
     drawings, objects, samples, figures, encoded data, visual
     representations of locations, equipment.

     Required transmission time (routing).

Message authentication.

Types of relationships: feedback of correspondent's reactions, constraint to reply, privacy, multiperson circulation of information, rapid decision context, deference, call to a mobile correspondent.

Special features: transmission reliability, guaranteed delivery, guaranteed speed, a posteriori classification.

- Function of the communication item:

Object of the message: information transmitted, information received, instruction transmitted, instruction received, negotiation, coordination.

Important contact.

Contact within or outside the enterprise.

Function of the correspondent.

Activity of the correspondent.

Nature of the mail item: contract, form, note.

- Volume of the information:

Duration of the contact (or length of message).

Number of persons participating.

Location of the correspondent.

Frequency of contact with the correspondent.

- Transfer (alternate modes):

Best alternative mode for transmission of the message other than the mode used.

II. PRELIMINARY RESULTS

II.1. Description of Traffic Transmitted via the Various Modes

A log entry is made only for communication between establishments and for communication between different services within the same establishment (calls within the same service are left out because they are very numerous and do not lend themselves to the detailed description used for our inquiry). Only the transmitted communications are recorded, not those received.

Estimated total contacts in France by mode

	Telephone	Telex	Mail	Face-to-face
N° of contacts per week (millions)	98	1.7	100	15.0 (6 0 ftf)
N° of contacts per year (47 weeks) – in billions (10^9)	4.6	0.08	4.7	0.7 (0.3 ftf)

TABLE 2

Table 2 is an extrapolation of the frequency of communications within France, based on our sample. Regarding the face-to-face meetings, in order to make this mode comparable with the others, the figure assumes that meetings bring together an average of 3.5 persons. This gives an average 2.5 contacts per meeting.

The total frequency of traffic is subdivided into internal and external traffic, and by distance (see Table 3).

Concerning the volume of information transmitted during each itemised contact, the following approximations are used:

- the length of the document sent as a mail item (30% of letters contain less than one page) or sent as telex messages (70% of telex messages are less than one page long).

- the length of the contact time-wise and the number of persons involved in the telephone calls and the meetings.

Distribution of internal-external traffic, by frequency of contacts

	Telephone	Telex	Mail	Face-to-face
Contact within the enterprise (intra)	43	30	22	38
Contact between enterprises (inter)	57	70	78	62
Contacts within same location	30	2	13	34
Urban (same town)	46	24	63	41
France (domestic)	20	34	21	24
Other (foreign)	4	40	3	1
TOTAL	100	100	100	100

TABLE 3

We note the volume of telephone calls in which more than 2 persons are involved (18% traffic see Table 4). This is an indication of the market potential open to technical innovations leading to "conference facilities".

Distribution of number of persons involved in a contact, mode

Mode

No. of persons:	Telephone	Face-to-face meetings
2	82	41
3	16	33
4 or 5	2	14
6 to 10	0	12
TOTAL	100	100

TABLE 4

Distribution of contacts in terms of duration

Duration of telephone call:		Duration of face-to-face meeting:	
- less than 2 min	25	- less than 10 min	11
- from 3 to 5 min	45	- from 10 to 30 min	21
- from 6 to 10 min	25	- from 30' to 1 hr	32
- from 10 to 20 min	4	- from 1hr to 2 hr	10
- more than 20 min	1	- from 2hr to 4 hr	13
		- more than 4 hr	12
TOTAL	100	TOTAL	100

TABLE 5

Face-to-face meetings are clearly longer than telephone, approximately by a factor 10 (see Table 5).

II.2. Description of the Alternate Mode Option

For each telephone call, telex message, mail item or FTF meeting, the transmitter noted the mode that he could have used as an alternative (as an example, for the telephone mode we asked "if you did not have the telephone, which of the following would you have used as an alternative: travel, telex, mail, facsimile, none?")

Distribution of alternate mode option, by mode used

mode alternative	telephone	telex	mail	ftf meeting
telephone	-	45	20	39
telex	14	-	5	1
mail	44	28	-	29
ftf meeting	37	11	22	-
facsimile	2	1	2	1
none	3	15	51	30
TOTAL	100	100	100	100

TABLE 6

Looking at Table 6 we find that there is a very low degree of projected transfer to facsimile. This a recent and little known mode, and this underscores the difficulties encountered when making an inquiry into attitudes about innovations with which one has little experience.

The non-transferred element enables us to judge the inherent advantages of the services rendered by each mode. For further detail, an in-depth analysis is needed for the types of service requested.

II.3. Description of the Types of Services
Requested and their Levels

The services requested under each mode heading correspond to the type of transmission of information involved. Thus, 50% of the travel-meeting context was aimed at "seeing goods, or equipment", 12% of all mail-items were accompanied by samples or objects. Like-wise, 60% of the mail items contained "plans, photographs, drawings" and approximately 45% of the meetings were aimed at seeing plans or photographs.

Concerning telephone and telex, i.e., alpha-numerical trans-mission, 40% of the telephone calls, and more than 60% of the telex calls were used to "transmit figures or encoded information".

Some of the services effectively provided by a given mode were:

- rapidity of making contact: 25% of all telephone calls took less than 5 minutes to complete and 44% took less than 1 hour (the res-pective figures were 4% and 28% for telex).

-authentication of the call, 38% of the telephone calls required a follow-up in writing, and 75% of the letters were sent out in order to "authenticate the contents of a prior call" (the equivalent for telex was 66%).

- feedback on the correspondent's reaction: this was of importance in 40% of the telephone calls and 70% of the meetings.

II.4. Description of Functions to be met
by Communication in Organizational Operations

The distinction made per type of service leads to an analysis of the function being met, i.e., the role played by the mode in an organization's operations. Such an analysis can be carried out from:

- the motivation for the contact: this may have been to give or receive information, instructions or orders, or again may have been to elicit cooperation or for negotiation (Table 7).

- the activity of the transmitter and receiver of the contact. The distribution of the activity of the correspondent shows a highly differentiated utilization of the various modes available (see Table 8.)

Distribution of contacts in terms of the object of the message

	Telephone	Telex	Mail	FTF meetings
Info. given	22	50	31	16
Info. Received	30	8	7	25
Instruc. given	23	22	52	19
Instruc. received	7	5	5	14
Cooperation	12	6	2	17
Negotiation	6	9	3	9
TOTAL	100	100	100	100

TABLE 7

Distribution of contacts in terms of the activity of the correspondent

	Telephone	Telex	Mail	FTF meetings
Administration	21	10	52	26
Production(design)	12	24	6	10
Production proper	30	10	11	19
Sales	16	15	8	28
Purchases	7	20	4	8
Other	14	21	19	9
TOTAL	100	100	100	100

TABLE 8

II.5. Model of the Total Traffic Generation Mechanisms

It would prove advantageous to study in detail the communication generation mechanisms, prior to a message being assigned to a transmission mode. In this case, we are led to reflect on the relationships that exist between a certain kind of operation of an enterprise (i.e., a certain type of organization), and the characteristics of information in terms of communication items needed in the exchange. The latter might be defined by content, by motivation and by the type and level of the service requested. An outline of a means to establish such a procedure can be seen in figure 2.

Of course the operations of an enterprise and the evolutionary trends of enterprises depend not only on the information transfer system that can be implemented, which is the only point dealt with in this paper, but on the processing, storage, interrogation and retrieval systems and means available.

Lastly it should be borne in mind that the study of the impact brought about by introducing a new mode is not merely limited to an evaluation of the potential market and the probable consequences of its utilization on organizational operations. To determine the real possible development, consideration should be given to certain variables specific to the offer itself, such as:

- the balance between the costs of supplying the service and the order of magnitude of the tariffs that might lead to the mode being adopted.

- the commercial policy envisaged that may help or hinder the penetration of the potential market (in particular, the policy applicable to the terminal equipment).

II.6. Evaluation of the Potential Market for Multiconference

To explain the overall purpose of our research let us give an example.

The "multiconference" option permits several persons to confer simultaneously by telephone. To evaluate this communication mode, we set up the proportion of communications transmitted by each existing mode which could be transmitted by this option. The sum of these figures indicates the total possible potential traffic at this point in time.

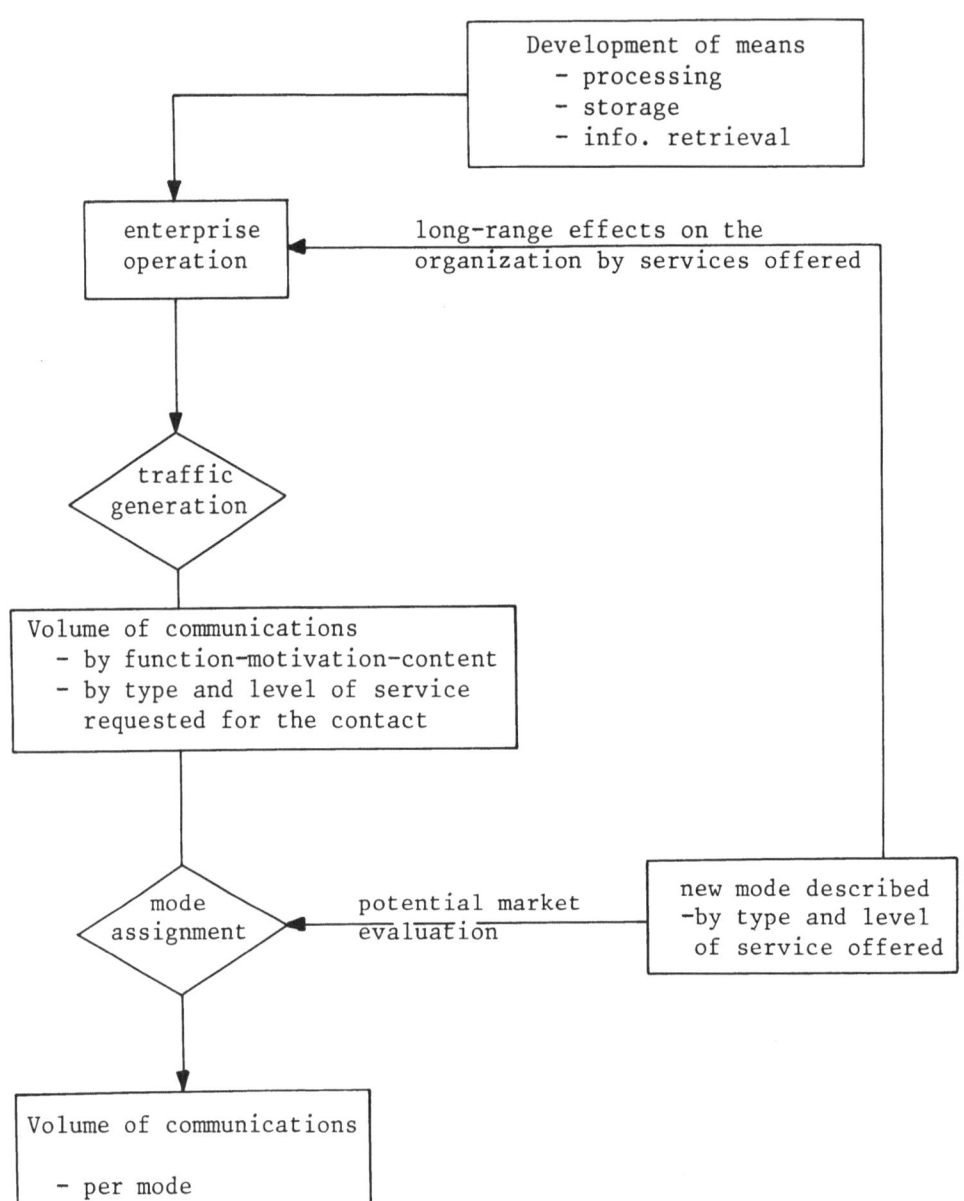

Figure 2

Transfer from "Face-to-face" to "multiconference". The communications to be transfered have to present the following features:

Meeting between more than 2 persons,

Meeting between different establishments,

Meeting without need to see plans or photos,

Meeting without need to see locations or equipment,

Contact shorter than two hours.

The number of meetings which have all of these features is 7.0% of the total number of meetings. Multiplying .07 times our estimate of the total number of face-to-face multiperson meetings in France each year gives us a first cut at an estimate of the potential market.

Transfer from "telephone" to "multiconference". Only two features suggest the use of the new option:

Contact between different establishments;

Contact between more than two people.

Of all of the telephone contacts 14% satisfy these criteria. This outcome would be treated in the same way as that for face-to-face meetings. Obviously these initial market estimates would have to be modified by as yet unobtained knowledge of user response to such a system as multiconference.

Note, this is just a superficial example. Clearly much remains to be done before we have an acceptable means for determining the potential market for new modes of communications. Nevertheless, we think our approach is headed in the right direction.

APPENDIX

The following page shows the form of the diary used to record data on telephone calls placed.

TELEPHONE CALLS

Hour of day (0 to 24 hr.)?	
Duration of call (in mins.)?	
Location of correspondent: On premises=0, same town=1, France=2, abroad=3.	
He is: in house=1, external=2.	
His function: 1/2/3 (see cover explanation).	
The contact is for the first time=1 Several times per day=2, per week=3, per month=4.	
His activity: 1/2/3/4/5/6 (see cover explanation).	
Is he especially mobile: YES=1, NO=2.	
Does he have telex equipment YES=1, NO=2, No idea=3.	
Object of message: 1/2/3/4/5/6 (see cover explanation).	
Call was extremely important=1, ave.=2, little=3.	
Was the message transmitted in French ? YES=1, NO=2.	
Transmission of figures, encoded info: YES=1, NO=2.	
This call had to be set up within 5 mins.=1, inside 1hr=2, in the day=3, in the week=4.	
How many persons took part in the call ?	
You set up contact to: establish personalised contact: YES=1, NO=2.	
have correspondent feedback: YES=1, NO=2.	
oblige correspondent response: YES=1, NO=2.	
Did the call require a written confirmation ? YES=1, NO=2.	
Circulate info. or gain advice from persons involved: YES=1, NO=2.	
Confidential nature=1, discrete=2, NO=3.	
If you cannot telephone and if you can use the following means, which would you use: travel=1, telex=2, mail=3, fac simile=4, None=5.	

Have you noted all your calls ? If not how many
have you missed?

Thank you

LEARNING THE LIMITS OF TELECONFERENCING: DESIGN OF A TELECONFERENCE TUTORIAL*

Robert Johansen[1], Jacques Vallee[1], and Kent Collins[2]

[1]Institute for the Future, Menlo Park, CA
[2]Charles F. Kettering Foundation, Dayton, OH

A growing number of optimistic articles on the promises of teleconferencing have piqued the public's interest in new electronic media for small group communication. Yet, there exists a considerable gap between teleconferencing research to date and practical application. Bridging this gap requires accelerated development of organizational techniques to facilitate real-world applications of teleconferencing systems. At present, even "trying out" teleconferencing is often difficult.

TRYING OUT TELECONFERENCING

One organization which is trying out the new teleconferencing technologies is the Charles F. Kettering Foundation of Dayton, Ohio, U.S.A. The Kettering Foundation is a nonprofit organization concerned with devising ways for diverse individuals and institutions to come to grips with several major social problems. Its activities are currently focused on program interests in elementary and secondary education, urban and international affairs, and scientific research into factors related to world food supply. Many of these program activities involve close collaboration with other groups in the United States and abroad and hence depend upon numerous conferences for development and on-going coordination.

---

*The authors of this paper are all involved in the "Intermedia Project" at the Institute for the Future, with support from the Charles F. Kettering Foundation. Other members of this research team include Kathleen Spangler and R. Garry Shirts.

Because of this orientation, the Kettering Foundation would seem to be a likely organization for practical tests of the theoretical promises of teleconferencing. So two years ago, enticed by articles on teleconferencing in publications such as *The Futurist,** the Kettering Foundation began to explore the new media to determine whether they could be applied to its program activities. This exploration resulted in a series of prototype computer-based teleconferences in 1976. The primary intent of these conferences was to give the Foundation's program staff direct exposure to one teleconferencing medium--to use it to meet established communications needs which were typically fulfilled through mail, telephone, and face-to-face meetings.

"THE BEAUTY AND THE BEAST"

Evaluation of these initial computer-based teleconferences reveals a number of paradoxical attitudes and perceptions about the medium. In a sense, the medium has been viewed as a "beauty-and-the-beast" technology.

For some users, for example, the comparatively narrow bandwidth of computer-based conferencing places a valuable limit on the range of communication--by accenting the cognitive and diminishing the affective. One individual interested in the Foundation's teleconferencing activities wrote:

> Computer-based teleconferencing is a highly cognitive medium that, in addition to providing technological advantages, promotes rationality by providing essential discipline and by filtering out affective components of communications. That is, computer-based teleconferencing acts as a filter, filtering out irrelvant and irrational interpersonal "noise," and enhances the communication of highly-informed "pure reason"-- a quest of philosophers since ancient times. In management, for instance, computer-based teleconferencing would filter out rhetorical malarky, and in politics, it would filter out demagoguery.

Such a view is typical of new users of computer-based teleconferencing, but it is probably an overly simple view. What one participant sees as "pure reason," may be seen by others as sterile. One participant in the Foundation's computer conference commented:

---

*Jacques Vallee, Robert Johansen, and Kathleen Spangler, "Computer Conferencing: An Altered State of Communication?", *The Futurist*, June 1975, pp. 116-121; and Murray Turoff, "The Future of Computer Conferencing," *The Futurist*, August 1975, pp. 182-195.

I missed not only the faces but the voices of people. Therefore, the ideas communicated really lacked a personal touch
for me.

As we discovered in surveying and summarizing the results of
social research on various teleconferencing systems,* the relative
advantage of each medium depends on the context in which it is
used. Hence, it is hazardous to speak confidently about the
"strengths and weaknesses" of any teleconferencing medium without
first considering carefully the environment in which it is applied.
A strength in one situation can be a weakness in another.

COMPARISONS WITH OTHER MEDIA: THEY CAN BE MISLEADING

To potential users, some attributes of various teleconferencing systems will resemble those of other communications systems
with which they have had prior experience. At times, this similarity can facilitate the introduction to a teleconferencing system.
But in other cases, it can be an obstacle.

For example, the Foundation has not progressed quickly in implementing prototype audio conferences. One reason is the antipathy
toward ordinary conference calls and toward the Speakerphone®. Also,
the telephone has historically been seen as (and marketed as) a
person-to-person communications medium--not as a group communications medium. Most people have neither the inclination nor the
skills necessary for effective group communication via audio teleconferencing. Coupled with the rather crude state of equipment
generally available for audio teleconferencing, it is simply difficult to encourage this type of communication. The Foundation's experience points out that potential users' attitudes toward teleconferencing will be shaped by prior experience with systems which are
superficially similar but which, in reality, are quite different.

A LEARNING PROBLEM

In its current stage of development and use, any form of teleconferencing represents an unknown to the vast majority of potential
users. It stands apart from their accustomed way of communicating
with one another. Even among those who initially express an eagerness to "try out the system," virtually all participants in the
Kettering Foundation's conferences have reported some anxiety during
their initial sessions.

----

*Robert Johansen, Jacques Vallee, and Kathleen Spangler, *The
Camelia Report: Technical Alternatives and Social Choices in Teleconferencing.* Institute for the Future, Report R-37, sponsored by
the Charles F. Kettering Foundation, 1977.

What emerges from the experience of the Kettering Foundation--
as well as comparable prototype application efforts--is the realization that teleconferencing truly does present an "altered state
of communication." The ability to capitalize upon this new form of
communication requires a new set of skills for conference leaders
and participants alike.

These skills do not appear to be easily mastered through vicarious learning. Rather, only direct and somewhat adventuresome experience leads participants to become increasingly facile in their
communication. While most can master the basic mechanics of teleconferencing within a few days of use, participants generally report that it takes them three to six weeks to feel totally comfortable with the medium. Only then do they begin to take the
medium for granted and concentrate fully on the substance of communicating with their fellow conferees. Just as some persons are
more effective face-to-face communicators than others, it seems
that there are special skills for effective communication via teleconference.

These observations suggest we should focus on ways to help new
users *learn* to communicate via teleconferencing media. Early anxi-
eties and misconceptions must be overcome. A willingness to experiment with new media should be encouraged. Finally, means should
be developed to allow new users to feel "at home" with the system
before they must rely on it for their actual communication needs.

LEARNING FROM SOCIAL EVALUATIONS OF TELECONFERENCING

Only tentative generalizations about learning can be drawn from
the social evaluations of teleconferencing to date. These evaluations have relied largely on laboratory experiments, often involving only two-person communication rather than small groups. Subjects typically have had very brief exposure (sometimes less than
one hour) to the teleconferencing media they were asked to use and
evaluate. While the tasks have usually been chosen to be representative of the "real world," they have necessarily been somewhat
contrived, geared toward comparison of results. There are relatively few examples of people who have used teleconferencing systems
over extended periods of time as a normal part of their work or
social life.

Such brief exposure to teleconferencing media under the artificial circumstances of the laboratory can provide only limited information about learning to use these media effectively on a long-
term basis. Laboratory studies, however, *have* served to open the
door on this important area of research and to suggest basic

variables which are important to track in long-term studies.* For
purposes of this paper, we are most concerned with those variables
which relate directly to learning to use teleconferencing media
effectively.

The beginning point for such a selection of key variables is
to identify basic elements of group communication via teleconferencing. In earlier Institute for the Future research, these elements have been grouped according to five categories: properties
of the medium, task, rules, person, and group.** These general
groupings provide a structure for examining the literature on social evaluation of teleconferencing media, from which the following
learning-related variables emerge:

- personal communication style (and history) of participants

- group task(s) to be performed

- group structure and leadership

- sense of social presence

- teleconference arrangements

PERSONAL STYLE: OLD TECHNIQUES IN A NEW ENVIRONMENT

Personal communication style (and history) refers to those
habits and experiences which individual participants bring to the
teleconference. Everyone has, to some degree, a tested set of communication techniques. When exposed to a new communications medium,
participants will generally carry some of these techniques into the
new environment. The ease of this transition process will depend
on both the specific communication styles of participants *and* the
characteristics of the teleconferencing medium being used. For
instance, participants can use computer-based teleconferencing
according to their own schedules: they "join" a teleconference,
see what has been said since they were last present, perhaps make
comments of their own, and leave. Such a usage pattern might seem
initially attractive to busy people, and, in fact, computer-based

---

*Our point here is not to disclaim the value of laboratory ex-
periments in this field, but simply to point out that research must
not be *limited* to laboratory experiments if longer range implications are to be understood.

**A taxonomy of specific elements under each of these headings
is given in Robert Johansen, Richard H. Miller, and Jacques Vallee,
"Group Communication through Electronic Media: Fundamental Choices
and Social Effects," *Educational Technology,* August 1974, pp. 7-20.

teleconferencing can work effectively in some situations where
scheduling meetings would be very difficult. However, this same
characteristic of computer-based teleconferencing--its potential
for asynchronous communication--can grate against the communication
styles and history of some potential participants. For example,
those managers who work best by responding to interruptions will
find that computer-based teleconferencing has no facility to "interrupt" them and call for their participation. There is no equivalent of a ringing telephone or a banging gavel. An "interrupt-
driven" manager will thus have to restructure his communication
habits if he is to make effective use of computer-based teleconferencing.

Each teleconferencing medium has its own characteristics which
might "mismatch" the communication styles of potential participants.
At a very subtle level, these mismatches might involve the way people form alliances in small groups or the way they bend protocol to
their advantage. Individual cognitive styles may vary, too, and
new media may place some stresses on styles developed for communication via other media. The point here is that personal communication styles are important to the ways in which people respond to
teleconferencing. While the specific characteristics of each medium
are only beginning to be understood, this interface with past experience will be crucial to the ways in which people learn to use teleconferencing.

GROUP TASKS: WHICH ARE MEDIUM SENSITIVE?

Once a communications process begins, *group tasks* are perhaps
the most visible factors which affect the learning process. And
the perceived importance of the task to the participants is basic:
without a high need to communicate, problems are likely to arise no
matter what medium is used. It therefore seems appropriate to
assume that learning will be faster if the task at hand is perceived
as important. Beyond this almost self-evident (but often avoided)
assertion, it also seems important to consider the complexity of
the tasks to be pursued. Connors, Lindsey, and Miller suggest the
general principle that the more complicated the task, the more the
visual channel is likely to be perceived as necessary.* Hammond
and Elton note that video is better than audio when reactions of
others must be carefully noted.** These findings, while still

---

*Mary M. Connors, George Lindsey, and Richard H. Miller, *The
NASA Teleconferencing System: An Evaluation,* Ames Research Center,
National Aeronautics and Space Administration, 1976.

**Sandy Hammond and Martin Elton, *Getting the Best Out of Tele-
conferencing,* Communications Studies Group, London, England,
P/76075/HM, 1975.

tentative, imply that the complexity of the task at hand will be an important factor in the success or failure of a teleconference--especially if new users are involved. The general wisdom seems to be: if it is important to form a complex image of other participants, then video or face-to-face is likely to be more desirable than audio or print (computer-based) teleconferencing. However, it may be possible to develop special strategies to allow the effective use of "narrowband" media for complex tasks.

Williams and Chapanis, in a survey of experiments involving teleconferencing media comparisons,* conclude that the following tasks show no media effects: "information transmission" (simple information transfer, such as the contents of a business letter), simple "problem solving," "generating ideas," and "interviewing." Williams and Chapanis also identify the following tasks which *do* seem to be affected by variations in communications media: "getting to know someone," "persuasion," "negotiation," "detection of lying," "coalition formation," "balance between cooperation and conflict," and "obedience." In the experiments on which these findings are based, the tasks were specific laboratory assignments made to subjects under experimental conditions.

The implication of these findings for learning to teleconference is that some tasks are likely to present more problems than others. (Indeed, tasks such as persuasion and negotiation often present difficult problems no matter *what* medium is used.) *The laboratory experiments have shown that media effects occur when subjects are exposed to teleconferencing for brief periods of time. They certainly have not proved that people cannot learn to use teleconferencing media effectively for tasks such as bargaining, negotiation, and other complex tasks.*

GROUP STRUCTURE: A STRONG LEADER HELPS

The problems of *group structure and leadership* may also be affected by the quality of training in the use of teleconferencing. However, it seems likely that some group structures will work better in teleconferencing than others. Also, whenever a group is involved in any new experience, strong leadership is likely to ease the learning process. What is not clear is how long this need for guidance continues after the initial learning process is over. Is there an inherent need within teleconferencing for stronger leadership than in face-to-face meetings, even when the conferees

*Ederyn Williams and Alphonse Chapanis, "A Review of Psychological Research Comparing Communications Media," in Lorne A. Parker and Betsy Riccomini, eds., *The Status of the Telephone in Education*, Madison: University of Wisconsin-Extension Press, 1976, pp. 164-168.

are experienced? The evidence at this point is mixed. One field
experiment found that the time spent in maintaining group organization is higher for both audio and video conferencing than for face-
to-face.* However, a series of laboratory experiments at Carleton
University concluded that video teleconferencing promotes a sort
of "unorganized informality" which does not require strong leadership.** In evaluations of computer-based teleconferencing, the
general feeling is that there *is* a need for strong leadership--
probably even more than in a comparable face-to-face meeting.***

SOCIAL PRESENCE: MEDIUM CHARACTERISTIC OR SOCIAL SKILL?

The *sense of social presence* provided within a teleconferencing medium will also have important effects on the new user. Several studies have noted an initial preference for face-to-face over
both video and audio teleconferencing,**** presumably because of
the greater familiarity and sense of social contact. Short,

---

*J. R. Weston, C. Kristen, and S. O'Connor, *Teleconferencing:
A Comparison of Group Performance Profiles in Mediated and Face-to-
Face Interaction,* The Social Policy and Programs Branch, Department
of Communications, Ottawa, Ontario, Canada, Report No. 3, Contract
OSU4-0072. In this field experiment, students used the teleconferencing system for a period of several weeks as part of a course
evaluation process.

**Donald A. George, D. C. Coll, L. H. Strickland, S. A.
Patterson, P. C. Guild, and J. M. McEown, *The Wired City Laboratory
and Educational Communication Project, 1974-75,* Carleton University,
Ottawa, Ontario, Canada.

***See, for instance, Jacques Vallee, Robert Johansen, Hubert
Lipinski, Kathleen Spangler, and Thaddeus Wilson, *Group Communica-
tions Through Computers, Volume 3: Pragmatics and Dynamics,*
Institute for the Future, Menlo Park, California, U.S.A., Report
R-35, 1975. This study emphasizes the importance of leadership
within computer-based teleconferences.

****See Brian Champness, *The Perceived Adequacy of Four Communica-
tions Systems for a Variety of Tasks,* Communications Studies Group,
London, England, E/72245/CH, 1972; and *Attitudes Towards Person-to-
Person Communications Media,* Communications Studies Group, London,
England, 1972; Bruce Christie and Martin Elton, *Research on the Dif-
ferences Between Telecommunications and Face-to-Face Communication
in Business and Government,* Communications Studies Group, London,
England, P/75180/CR, 1975; Michael G. Ryan and James G. Craig,
*Intergroup Telecommunication: The Influence of Communications Medi-
um and Role Induced Status Level on Mood, and Attitudes Toward the
Medium and Discussion,* Communications Research Centre, Department
of Communications, Ottawa, Ontario, Canada, 1975.

Williams, and Christie, in a major book on social evaluation of
teleconferencing, view social presence as inherently linked to each
particular medium:

> Although we would expect it to affect the way individuals per-
> ceive their discussions, and their relationships to the per-
> sons with whom they are communicating, *it is important to em-*
> *phasize that we are defining Social Presence as a quality of*
> *the medium itself.* We hypothesize that communications media
> vary in their degree of Social Presence, and that these varia-
> tions are important in determining the way individuals inter-
> act. We also hypothesize that the users of any given commun-
> ications medium are in some sense aware of the degree of Social
> Presence of the medium and tend to avoid using the medium for
> certain types of interactions; specifically, interactions re-
> quiring a higher degree of Social Presence than they perceive
> the medium to have. Thus, we believe that social presence is
> an important key to understanding person-to-person telecommun-
> ications.* (emphasis added)

Even if one assumes that social presence is inherently linked
to each particular medium, there is some question about the uni-
formity of perceptions of social presence. As indicated earlier,
a print-based medium may appear to lack social presence for some;
others, however, find in it a strong sense of interpersonal contact.** Perhaps the sense of social presence can be cultured by
teleconference leaders or encouraged by initial learning sessions.
At any rate, it seems doubtful that "social presence" is a static
variable welded unwaveringly to each teleconferencing medium.
Rather, we hypothesize that social presence can be influenced
directly by skilled teachers and leaders--though certainly basic
constraints will be associated with each teleconferencing medium.

TELECONFERENCING ARRANGEMENTS:
AS IMPORTANT AS MEETING HALL ARRANGEMENTS

As with meeting rooms for face-to-face conferences, the physical arrangements of the teleconference rooms are crucial to the

---

*John Short, Ederyn Williams, and Bruce Christie, *The Social
Psychology of Telecommunications,* London, England: John Wiley and
Sons, Ltd., 1976.

**This sense of social presence has been documented for several
conferences in Jacques Vallee and Robert Johansen, *Group Communica-
tion Through Computers, Volume 2,* Institute for the Future, Menlo
Park, California, U.S.A., Report R-33, 1974; also Philip Spelt,
"Evaluation of a Continuing Computer Conference on Simulation,"
Behavior Research Methods and Instrumentation, Spring 1977.

communication which evolves. The principle is simple: put people in a stiff, formal setting, and they are likely to feel (at least a little) stiff and formal. Beyond this effect of formal surroundings, the technical quality of various facilities can also influence communication. Teleconference facilities of varied quality can provide basic imbalances in the group communication process. In a recent Institute for the Future report, we speculated about the effects of a specially equipped video teleconference facility linked with lower quality temporary facilities:

> . . . when Clemmons spoke, he sounded like a stern parent allowing his squeaky-voiced children but a few minutes of his valuable time. His voice boomed through the system, seemingly sucking the decibels from the other timid microphones.*

This is a rather extreme example, but one which illustrates the need for a relative balance of the technical quality of the various rooms involved in a teleconference meeting.

For group-to-group teleconferencing, another important consideration involves the placement of group members at various locations. Several studies have pointed out the potential for "we" to "they" tendencies as communication varies from within-terminal to between-terminal patterns.** The group dynamics become complicated as subgroups are actually meeting face-to-face while communicating with other subgroups via teleconferencing. Often, uneven distribution of participation will result simply from the geographic distribution of the group. However, the effects of these imbalances need to be understood more fully, and techniques need to be developed to overcome the potential negative effects.

Teleconferencing arrangements also include such basic procedures as scheduling. Especially for novice users, it seems important to have some sort of regular teleconference schedule. Noll, in a study of the use of video teleconferencing by Bell Laboratory employees, emphasizes the importance of regularly scheduled meetings.*** The same seems true for audio teleconferencing and perhaps

---

*Robert Johansen, Jacques Vallee, and Kathleen Spangler, *op. cit.*, p. 42.

**Anna E. Casey-Stahmer and M. Dean Havron, "Planning Research in Teleconference Systems," McLean, Virginia, USA, Human Sciences Research, Inc., 1973; J. R. Weston, C. Kristen, and S. O'Connor, *op. cit.*; Ederyn Williams, "Coalition Formation over Telecommunications Media," *European Journal of Social Psychology*, Vol. 5, 1975.

***A. Michael Noll, *A Study of the Communications Activities Performed by Users of the Bell Labs Video Conferencing System,* American Telephone and Telegraph, Morristown, New Jersey, and Bell Telephone Laboratories, Inc., Murray Hill, New Jersey, USA, 1976.

even more so for computer-based teleconferencing. In the latter
case, participants are typically responsible for setting and maintaining their own schedules.

WHAT CAN BE LEARNED?

It is clear from social evaluations to date that much uncertainty remains with regard to the process of learning to use teleconferencing media effectively. What now appear to be "characteristics" of teleconferencing media may simply be results of poor
system design or factors which must be overcome (or used to advantage) by new users if they are to maximize the communication potential of teleconferencing. On the other hand, teleconferencing
media will undoubtedly place *some* constraints on even the most experienced user groups.

*A central research task, then, is to map the potentials and
limits of teleconferencing with a precision not heretofore accomplished.* However, these potentials and limits do not lend themselves well to study via conventional social science methodologies.
A new teleconferencing medium cannot be evaluated as an isolated
component in the communications process; the medium is difficult to
study as an "independent variable" in the classical sense. *Rather,
we propose a more active (and thus less controlled) research approach which focuses on (1) learning to use teleconferencing media
effectively and (2) exploring the potential impacts of widespread
use.*

A TELECONFERENCING TUTORIAL

Teleconferencing research is advanced enough to offer some advice for developing teleconferencing systems which are adapted to
varied group needs. The research also offers insight into the
dynamics of user groups to caution them regarding potential limitations of the tool. This knowledge, however, does not mean that
the full range of potential applications for teleconferencing can
be predicted; neither does it mean that trustworthy information
about its impact on the people who use it and on the society around
them is easily available. In order to explore these complex effects, we are developing an approach which combines some techniques
from small group research with those of simulation gaming to produce a "tutorial." This tutorial will be designed to give novice
users a feeling for the power of each medium as well as a realization of its weaknesses.

The tutorial is rooted in the social evaluations of teleconferencing media which have occurred to date. Special attention has
been given to the learning variables summarized earlier in this

paper. For instance, the tutorial is designed to provide participants with a direct exposure to a variety of *group tasks*--both
those for which teleconferencing appears well suited (e.g., information exchange) and those for which the use of teleconferencing
is questionable (e.g., bargaining or negotiation).

It is difficult to describe the tutorial to someone who has
not experienced it. The task of explanation is similar to explaining baseball to someone who has never heard of it. Thus, it may
be most helpful to imagine that you are a participant in the tutorial. Here is what would happen:

1. You would be given basic tasks to accomplish during an
    upcoming meeting (actually the tutorial exercise). The
    purpose of the meeting is to decide which medium--audio,
    video, or computer-based teleconferencing--best serves
    the needs of a hypothetical group. Five participants are
    assigned media preferences based on the communication
    needs of different divisions of the group. One person is
    a leader who attempts to facilitate a unanimous decision,
    and two others are "experts." (The experts have research
    data from social evaluations of teleconferencing.)*

2. The meeting (i.e., the tutorial) begins; it is held over
    audio, video, or computer-based teleconferencing, or face-
    to-face. A typical meeting might last 1-1/2 to 3 hours,
    although an asynchronous computer conference might stretch
    out over several days, with each participant moving in and
    out of the conference.

3. If the group cannot agree on a unanimous choice among
    media (in fact, it is very difficult, given the tasks
    assigned to the various participants), some participants
    may decide to form a new group of like-minded persons.
    The "rules" of the tutorial place limits to how easily
    such a spinoff can be formed, but three or more partici-
    pants can decide on this strategy.

4. The tutorial ends when a unanimous decision is reached, a
    spinoff group has been formed, or a stalemate has occurred.

5. Each participant fills out brief posttutorial reporting
    forms, and a "debriefing" discussion follows.

---

*It is important to note that the tutorial does *not* involve
role-playing or simulation in the sense of modeling a specific
social environment.

The tutorial has two basic goals: (1) to allow participants
to experience the strengths and weaknesses of communication via
different teleconferencing media; and (2) to introduce participants
to the results of social evaluation studies of teleconferencing.
It can be used more than once with the same people, but with different media; thus, participants can "try out" different leadership
styles or styles of self-presentation. In this way, the tutorial
provides a context for practicing the more subtle skills of small
group communication via teleconference.

RESEARCH USES OF THE TELECONFERENCE TUTORIAL

In the teleconference tutorial, we hope to provide rapid
learning as well as an entertaining and absorbing experience. Our
interests, however, go beyond the learning of novice teleconference
users. We wish to exhibit clearly the limitations of current media
in order to help teleconference system designers. We expect to
create an "environment" in which we can investigate group behavior
via teleconference.

The tutorial can be "instrumented," allowing us to gather information about such issues as strategies for use of teleconferencing media, barriers to successful use, interpersonal or cross-
cultural communication problems, and variations among teleconferencing media. Exploration of the latter issue will be most appropriately pursued in experimental laboratory situations where the
tutorial represents an assigned task. Several teleconference research laboratories have already expressed an interest in using the
tutorial for such cross-media comparisons.

Even when the tutorial is used for training novice teleconferencing users, however, posttutorial reporting forms will allow for
research-oriented information gathering. These forms, together
with tapes or transcripts of actual uses of the tutorial, should
provide the opportunity for making comparisons across widely varied
groups. While lack of control over group selection and other key
variables could limit the type of comparisons which might be made,
these uses of the tutorial should provide rich opportunities for
learning about the possibilities and limitations of teleconferencing media. For instance, barriers to the acceptance of the media
could be documented. If the sample groups were from different cul-
tures, hypotheses about cultural variations might be developed;
these could then be tested under more controlled conditions. Also,
varied strategies of teleconference usage--at both personal and
group levels--could be documented so that new users would have a
better sense of the range of possibilities. General findings such
as these would certainly expand current knowledge about possible
uses and misuses of teleconferencing.

CONCLUSION

The teleconferencing tutorial will be publicly available in the fall of 1977. It is one attempt to develop a flexible approach to evaluating and learning to use varied forms of teleconferencing. Many such approaches will be necessary if we are to plan creatively for the communication potentials and pitfalls of teleconferencing. With new technologies, such a planning and learning process ordinarily takes years; many people must experience--and be influenced by--the technology in operation. Our tutorial is one attempt to compress this waiting period and explore future effects before they occur.

INTERPERSONAL TELECONFERENCING IN AN ORGANIZATIONAL CONTEXT

Arthur D. Shulman and Jerome I. Steinman[1]

Washington University

St. Louis, Missouri 63130

Communication in organizations functions primarily to reduce uncertainty (Galbraith, 1973). Because of increasingly turbulent environments (Emery and Trist, 1965; Terreberry, 1968), many organizations today, be they health service delivery, legal, educational, or manufacturing, are confronted with the necessity of dealing with increased uncertainty and thus with growth in the volume and complexity of communications. Employing new communications technologies is one means of coping with this problem. This paper is concerned with identifying the appropriateness of employing interpersonal teleconferencing modes--audio, video, or computer--for that purpose.

Decisions to deploy and use interpersonal teleconferencing systems are made at three organizational levels: 1) the level at which coordination strategies are formulated, 2) the work unit level, and 3) the individual level. Decisions on selections and placement of interpersonal telecommunications systems are made at the organizational strategy level, while decisions about actual use of telecommunication systems are made and controlled at the work unit and individual levels. The large body of empirical research on use of telecommunications systems has been conducted at the work unit and individual levels. This paper attempts to relate that research to organizational strategies for handling coordination among work units, a task which has not been attempted so far. The major tenet of this paper is that the deployment and use of teleconferencing systems as substitutes for face-to-face meetings depends upon contextual effects and, in particular, upon the strategy adopted to coordinate communication among organizational units.

THE SUBSTITUTION PERSPECTIVE: AN OVERVIEW

Past empirical work evidences two major thrusts: 1) a concern
with establishing conditions for cost-effective substitution of telecommunications systems for existent face-to-face communication; and
2) a concern with introducing new patterns of coordination among
organizational units. Most research on teleconferencing has adopted
the substitution perspective. Most substitution studies employ
highly controlled laboratory or field simulations using contrived
tasks and ad hoc groups (i.e., participants usually do not know one
another beforehand, nor do they have any further contact on completion of the imposed task).

Substitution studies have focused on task outcomes rather than
processes. Results have been rather consistent. Much, but not all,
of the information exchanged at the work unit level can be communicated equally well over a wide variety of telecommunication systems[2]
which vary in their capacities to provide audio-visual messages.
Furthermore, equivalent or better task outcomes are achieved with
the use of relatively inexpensive computer or audio systems than with
systems which contain a visual component. Perhaps equally important, differences in picture resolution are not as important as clarity of sound. These research efforts have been well summarized by
Williams (1974), Chapanis (1975), and more recently by Hough (1976).

THE CSG STUDIES

The Communications Studies Group (CSG), University College,
London, has been active in research employing the substitution perspective. CSG has developed the most widely used procedures and
instruments for evaluation of the substitutability of telecommunications. CSG's own work has culminated in a type allocation model.

Their type allocation model identified face-to-face business
meetings effectively accommodated via electronic media. That model
was developed from data gathered by CSG in three large surveys and
over 25 laboratory and field studies sponsored by the British Post
Office and Civil Service Department. The surveys identify the
nature of business meetings while the 25 plus laboratory studies
determine the effect on task outcomes of audio, audio-video, and
recently, interpersonal computer communication devices.

The CSG type allocation model consists of the following stages:

1) Definition of the different types of face-to-face meetings
   based upon field surveys conducted in Great Britain (Pye,
   et al., 1973; Connell, 1974). This will be referred to as
   the DACOM (Description and Classification of Meetings)

project.[3] Each DACOM task cluster is designated by its
most important variable (e.g., <u>delegation of work</u>, <u>form-
ing impressions of others</u>, <u>conflict</u>, etc.).

2) Allocation of each type of meeting to the least costly medium (usually audio) which effectively accommodates it.

3) Measurement of the frequency of each type of meeting.

4) Calculation, from 2 and 3, of the proportion of face-to-
face meetings transferable to various telecommunications
media.

Recent summaries of CSG's work (Williams, 1974; Short, Williams
and Christie, 1975; Hough, 1976) suggest that 40% of the surveyed
business meetings, <u>when considered as isolated business meetings</u>,
can employ audio systems without alterations in task outcome. This
substitution to audio can take place for problem solving, information seeking, and decision making. Substitution appears to cause
alterations in task outcome in meetings in which conflict, negotiation, opinion change or disciplinary action is involved, or in situations in which people who are not familiar with one another are
interacting.

SOME LIMITATIONS OF THE SUBSTITUTION PERSPECTIVE

Typical substitution studies employ highly controlled laboratory and field simulations using contrived tasks and ad hoc groups;
conditions which at least limit, if not render it impossible to
examine how cost-effective substitution depends upon the contextual
features of organizational strategies and communications environments. The effects of these limitations are apparent in the results
of studies not using the paradigm.[4] As Shulman has summarized these
results elsewhere (1975) they will only be briefly presented here.

Champness (1972), Chapanis and Overby (1974), and Strickland
(1974, personal communication) have studied groups using more than
one mode of communication over a period of time. Usually they find
sequence effects: the strategy developed during the face-to-face
and/or audio exchange was carried over to subsequent exchanges or
novel teleconferencing systems.[5] That is, when given a familiar mode
to converse on, groups quickly develop standard operating procedures
which carry over to other modalities. This suggests that once work
units have established a pattern of coordination in a familiar mode
such as face-to-face, switching to other new conferencing systems is
not difficult. Furthermore, meetings do not normally involve just
one task, nor is a task necessarily completed during one meeting and
over one modality. For instance, Conrath, in a study of one organ-

ization (personal communication, May 1975), found that individuals tended to sequentially use a mix of communication systems within their communications environment. Commonly, individuals tended to use face-to-face meetings for information exchange and sizing up one another. Actual decision making was conducted over the telephone, with confirmation of the decision provided in writing. These findings suggest the necessity to consider both the changing nature of task and the total communication environment.

The generalizability of substitution study results is questioned in studies investigating the use of teleconferencing systems when alternative means of conveying information were available. Conrath has conducted the majority of these studies. Conrath (1973) found that face-to-face meetings were associated with both the authority and the task structures of a manufacturing and sales organization, whereas telephone use was associated only with task structure and with distances over 60 meters. This strongly suggests that telecommunications are used differently than face-to-face communications.

A second field study replicates these findings. In that study, Conrath and Blair (1974) utilized two units of a computer software development firm and compared the communication flows among face-to-face, telephone and computer interface (Augmented Knowledge Workshop) modes. The new interpersonal mode of communication (in this case the Augmented Knowledge Workshop) did not replace written communication, but lowered the relative amount of face-to-face and telephone communication--thus supporting a substitution model. More importantly, the Augmented Knowledge Workshop caused an upward flow of communication.

A series of surveys asking individuals when and for what reasons they would use different media also suggests caution in generalizing results from ad hoc group studies. Individuals who had not used a new system usually suggested that it would have a limited use, while persons who had taken part in a demonstration suggested more uses. Overall, users and nonusers responded that audio-only systems-- and to a lesser extent, video systems--would not be used for the tasks involving strangers, nor for tasks which involve conflicts, negotiations or disciplinary actions (Champness, 1973; Christie, 1973, 1974; and Christie and Holloway, 1975). Christie (1974) also reports that the above social and organizational considerations assumed greater importance for respondents than did the cost of the devices. Further analysis (Christie and Holloway, 1975) revealed that many respondents cannot visualize using audio or video systems for any purpose; a finding also reported by Kollen and Vallee (1974) in the Bell Canada Communication/Transportation Survey. This phenomenon is perhaps best understood when individuals are conceived of as decision making organisms possessing limited capabilities. Simon (1976) has argued that human beings are not completely rational decision makers because of inherent limitations. Individuals are incapable of con-

ceiving of all possible alternatives. They can pay attention only
to a limited number of events. They tend to search for solutions
near other solutions or near the area where the problem occurs.

From this perspective it is argued that merely providing individuals in organizational or simulated organizational settings with
devices possessing radically different capabilities is not, in itself, a sufficient condition for full exploitation of the potential
of those new devices. It is, however, a _necessary_ condition. When
confronted with the availability of a teleconferencing system, people will tend to search near previous problems and solutions, that
is, they will tend to use the device in a way similar to the way in
which they previously used the most similar device in their repertoire. Only over a period of time will they gradually discover new
methods of employment.

There is support in the above studies for the position that even
the _use_ of telecommunication systems as substitutes for face-to-face
meetings depends upon contextual organizational effects. These effects have been studied only at the work-unit level.[6] However, the
major tenet of this paper is that deployment and use of teleconferencing systems as substitutes for face-to-face meetings depends upon
contextual effects and, in particular, upon the strategy adopted to
coordinate communication among organizational units. Note that all
tasks in which substitution results in a different outcome involve
situations in which there is uncertainty. Organizational strategies
and coordination modes employed to deal with uncertainty are discussed below; they rely on Galbraith's (1973) and Thompson's (1967)
conceptualizations.

ORGANIZATIONAL STRATEGIES AND COORDINATION MODES

The crucial property of Galbraith's (1973) theory is its focus
on information processing as a mechanism for uncertainty reduction.
Different organizational designs are appropriate to cope with differing degrees of uncertainty. Uncertainty is "...the difference
between the amount of information required to perform the task (of
the organization) and the amount of information already possessed
by the organization." (Galbraith, 1973, p. 5) Furthermore, as
uncertainty increases, more information is lacking, and more information must be processed by decision makers in the course of task
execution.

Galbraith proposes a hierarchy of seven organizational design
strategies, enabling organizations to cope successfully with increasing levels of uncertainty and, hence, with increasing demands on
their information processing systems. The three least complex,
least costly design strategies are: coordination through rules and
plans, hierarchical coordination, and coordination through setting

targets and goals.[7] Their capacities for information processing are limited. Since telecommunications systems are designed to facilitate communication, the focus here is on the four design strategies that attempt to cope with greater uncertainty. Use of any of these four also implies the presence of the first three strategies. Two of the four additional strategies reduce the need for information processing through: 1) the creation of slack resources, or 2) the creation of self-contained task-units.

By creating slack resources (e.g., slipping schedules, creating inventories) coordination and hence communication needs are reduced. This strategy tends to create what Thompson (1967) calls polled interdependence, i.e., units are interdependent in the sense that contributions from each unit are discrete, and the units do not directly support each other. Yet, if one unit fails to perform, the performance of the whole organization would be degraded. Pooled interdependence exists in all organizations.

The self-contained task strategy reduces information processing needs by providing each unit with its own resources thus reducing coordination of shared resources. The reduced coordination needs imply that there will exist at most, sequential interdependence. In sequential interdependence (Thompson, 1967) one work-unit's output is the input for a second unit; therefore, the second is serially dependent on the first work-unit. Since both units are part of the same organization, sequential interdependence implies the existence of pooled interdependence.

Creation of slack resources and creation of self-contained tasks both reduce the need for information processing. Although organizations using these strategies may employ teleconferencing devices, the next two strategies—vertical information systems and lateral relations—are more germane for telecommunications planners since, in contrast, they are designed to enable processing of all the information during task execution. The first strategy, creation of vertical information systems, involves developing mechanisms for processing the information during task execution without overloading the vertical channels. The capacity of vertical channels is augmented to process the information into decision relevant form, rather than overload the decision maker and channels with raw information. The cost here is the investment in the creation of the information system, e.g., computers, programmers, software, etc. Vertical systems tend to coordinate subunits through the hierarchy. Subunits so coordinated are at most sequentially interdependent.

The second enabling strategy, creation of lateral relations, selectively employs "lateral decision processes which cut across lines of authority. This strategy moves the level of decision making down to the level where the information exists rather than bring-

ing it up to the points of decision. It decentralizes decisions but
without creating self-contained groups." (Galbraith, 1973, p. 18)
Lateral systems contain either (1) reciprocal or (2) team interdependence or both.

Reciprocal interdependence refers "...to the situation in which
the outputs of each become inputs for the others....Under conditions
of reciprocal interdependence, each unit involved is penetrated by
the other. There is, of course, a pooled aspect to this, and there
is also a serial aspect....But the distinguishing aspect is the reci-
procity of the interdependence, with each unit posing a contingency
for the other." (Thompson, 1967, p. 55)

Van De Ven, et al. (1976) add a team mode of coordination. It
involves work undertaken jointly by personnel from different units
where inputs and outputs cannot be assigned to individuals since
"...the work is acted upon jointly by unit personnel at the same
point in time. Examples of team work flow in organizational units
include group therapy sessions in mental health units, a sports team
playing a game and a group of research colleagues designing a study
as a 'think tank.'" (Van De Ven, et al., p. 325)

Thompson (1967) proposes that his three modes of coordination
--pooled, sequential, and reciprocal--or types of interdependence,
are successively more difficult to coordinate, and that the cost of
coordination also goes up with each mode. This holds true when the
team mode of coordination is appended--it is probably even more costly than the reciprocal mode.

Therefore, lateral systems, involving reciprocal and team coordination modes, are the most difficult and the most costly types of
systems to coordinate. However, in some cases the cost of a vertical
information system may also be substantial and may approach the cost
of lateral systems.

TASK UNCERTAINTY

Organizational design strategy, which has just been shown to
be related to mode of coordination, does not determine the nature of
appropirate interpersonal telecommunications but rather interacts
with the degree of task uncertainty for each DACOM cluster of meetings. In this section we use Perrow's (1970) task uncertainty dimensions to classify the DACOM tasks as a function of the strategies and
patterns of interdependence that organizations use for handling uncertainty.

Perrow (1970) proposed that two dimensions serve to classify
the kinds of work, i.e. tasks, performed by organizational units and

Table 1

SUBSTITUTION MATRIX

DACOM Clusters of Tasks
(from Communications Studies Group)

		Non-Emotional			
		Delegation of work No forming impressions of others	Keeping people in the picture Report giving	Problem solving Discussion of ideas Information seeking Keeping people in the picture	Policy decision making No information exchange Some problem solving
Strategies (and major modes of coordination)	Perceived Nature of Raw Material				
Vertical (sequential)	Analyzability of search	Analyzable	Analyzable	Analyzable	Analyzable
	Variability	Stable	Stable	Stable	Stable
	Appropriate Conferencing System[1]	1) asynchronous computer/mail 2) 1-way audio	1) synchronous or asynchronous computer 2) telephone for exceptions	1) computer 2) audio	1) computer 2) audio
Lateral (reciprocal and team)	Variability	Unstable	Stable	Stable	Stable
	Appropriate Conferencing System[1]	1) audio 2) computer	1) audio 2) computer 3) mail	1) audio	1) computer 2) audio (?)

		Negotiation	Interpersonal Evaluation			
		Negotiation Bargaining No conflict	Conflict Negotiation Forming impressions of others	Forming impressions of others Information seeking	Forming impressions of others Conflict	Disciplinary interview Delegation of work
Vertical (sequential)	Analyzability of search	Non-analyzable	Non-analyzable	Non-analyzable	Non-analyzable	Non-analyzable
	Variability	Stable	Unstable	Stable	Unstable	Stable
	Appropriate Conferencing System	1) computer	1) face-to-face 2) video	1) audio 2) video	1) face-to-face 2) video	1) face-to-face 2) audio
Lateral (reciprocal and team)	Variability	Unstable	Unstable	Unstable	Unstable	Unstable
	Appropriate Conferencing System	1) audio 2) computer	1) face-to-face 2) video 3) audio	1) video 2) audio	1) face-to-face 2) video	1) face-to-face

[1]See text for rationale for order of conferencing systems

that appropriate organizational structures are contingent on the resulting task classification. A task is defined as the actions taken by individual unit members to transform the raw material in a desired output. The raw material may be symbolic, human or material. The first task dimension is analyzability of search, i.e. whether "...there are well established techniques which are sure to work" (Perrow, 1970, p. 75) to appropriately transform the raw materials. If there are such techniques then search is analyzable (well understood), otherwise it is unanalyzable (not well understood). The second task dimension is the perceived variability of the raw material and refers to"...the variety of problems which may lead to search behavior. Sometimes the variety is great and every task seems to be a new one demanding the institution of search behavior of some magnitude (whether analyzable or unanalyzable). Sometimes stimuli are not very varied and the individual is presented chiefly with familiar situations and few novel ones." (Perrow, 1970, p. 77) Perceived variability of the raw material is dichotomized into few exceptions (stable) and many exceptions (unstable). The two dichotomized variables yield four distinct categories of tasks.

In the substitution matrix, described below, Perrow's (1970) task classification scheme is employed in conjunction with meeting type (i.e., DACOM cluster) and predominant coordination mode to allocate types of interpersonal communication systems. The following section details how this is accomplished.

AN OVERVIEW OF THE SUBSTITUTION MATRIX

The substitution matrix for decision makers is designed for use as a tentative guideline by researchers and individuals responsible for decisions concerning investment in, and placement of, teleconferencing systems. Each of the tasks that are characteristically performed in meetings is discussed to illuminate the gains derived by considering organizational strategies as a major determinant of appropriate telecommunication system usage. The DACOM clusters are used to represent the tasks. The DACOM clusters vary in frequency of occurrence and more importantly in terms of the degree to which task variability and perceived analyzability facilitates connecting the strategies with the DACOM clusters. Table 1 summarizes how a classification scheme, dependent on the task and the organizational strategy, determines the prevailing patterns of communication.[8]

The matrix is rudimentary, reflecting limitations inherent in the existent data base. Other delimiters operating within the organization's external environment (e.g., legislation, terminal compatability of suppliers and clients) are acknowledged to influence decision makers' deployment of devices, as do the number and location of its work units but are not as yet fully considered in the

matrix. Additional factors operating at the work unit and individual levels are also not fully incorporated in the matrix. These
factors (norms for sharing information, power to choose mode, past
history with other teleconferencing systems, geographic separation,
the size and homogeneity of skills of its members, ability of work
unit membership, individual preferences and habits, and external
fringe benefits associated with travel), however, are presumed to
have more effect on the actual usage of teleconferencing systems,
once they are installed.

The tasks are divided into three groups: (1) those concerned
with non-emotional procedures usually present in most organizations;
(2) those involving negotiation with and without conflict; and (3)
those involving a large interpersonal evaluation component. Tasks
within each of the three groups often occur in sequential order.

For each DACOM task cluster, the rationale for any substitution
of teleconferencing systems for face-to-face meetings is described
for given patterns of coordination and task uncertainty. Table 1
shows the major modes of coordination associated with vertical and
lateral strategies, and the type of perceived uncertainty involved,
for each of the DACOM clusters. Cost and the perceived ability of
the teleconferencing system to convey personal interaction are then
considered in the assignment of teleconferencing system to each clus-
ter/strategy combination. Video, audio, and computer conferencing
systems are listed here in estimated, descending rank order of cost.
(The reader, however, should be aware of the problems in assigning
specific costs for each network; see Hough (1976).)

The systems, in descending order of their ability to convey a
sense of personal interaction, are: (1) video, (2) audio, (3) synchronous computer, and (4) asynchronous computer. This ordering is
based on Hough's (1976, p. 268) integration of studies by the Communications Study Group and the Institute for the Future. Two other
aspects of the task environment have consistently affected the ordering in the listing of alternative systems. These are urgency of
task completion, and perceived confidentiality requirements of the
task--specific effects will become clear in the discussions of the
clusters.

DISCUSSION OF INDIVIDUAL CLUSTERS

Delegation of Work through Policy Decision Making

The four DACOM tasks in the first set are usually connected in
a sequential but repetitive manner; individuals are assigned a
task(s) by an acquaintance and they keep each other in the picture,

often through oral reports. Problems or exceptions lead to problem
solving and information seeking, followed by policy decisions, which
then affect the delegation of work, etc. All four tasks are analyzable, i.e. the rules are well understood. Each task is discussed
below.

Delegation of Work. The major activity in this cluster is dele-
gating work among individuals who know one another. That they know
each other is inferred from the absence of forming impressions of
others. Secondary analyses of the DACOM survey data indicate that
this cluster of meetings is characterized by: a duration of five to
ten minutes, the presence of a boss, and the presence of individuals
from the same work unit or from different work units in the same organization. The original DACOM analyses indicate that delegation of
work occurs in about 12% of all meetings logged.

Delegation of work is straightforward and analyzable. In vertical systems, it is standard procedure accomplished through use of
formal authority. The instances in which delegation will not adhere
to formal authority are few; hence the task is stable. Communication
is often one way, from the top down. Consequently, in vertical systems, this activity is allocated to the two cheapest modalities: (1)
an asynchronous computer, or (2) a one-way audio system--including
telephones with an answering service.

Since task assignments tend to be negotiated in lateral systems,
delegation is presumed to occur infrequently; it is not standard procedure and is classified as a task having many exceptions. Some form
of two-way communication, with capacity for immediate feedback is
required because of the relative novelty of the task. Audio systems
appear to be the cheapest, most appropriate mode for this task and
are not excluded by other considerations (e.g., need for privacy).
Synchronous computer systems might be used, but potential users suggest that the personal sensitivity necessary in an unstable task environment is not provided. Since meetings involving delegation are
short, i.e., five to ten minutes, a telephone call or telephone conference will probably suffice.

Giving information to keep people in the picture. Report giving,
maintaining friendly relations, and maintaining morale are characteristic of this cluster and furthermore, individuals in the group know
each other. Past laboratory and field studies have not been conducted under these conditions, nor have they studied the activities in
this cluster. They have separated them into two subclusters: (1)
giving information to keep people in the picture and report giving,
and (2) maintaining friendly relations and morale. The following
discussion is restricted to subcluster (1) above.

In vertical systems, rules for giving information to keep people
in the picture and report giving are well understood. The task is

analyzable. Like delegation, it is part of the standard operating
procedure. Information is routinely passed upward, often through
non-personal means. The task is stable and analyzable for boss-
subordinate pairs. Allocation is to inexpensive asynchronous computer
or synchronous computer systems, but occasional exceptions may occur
because of inadequacies in the existing vertical information gathering system. Telephone connections might suffice for reporting exceptions but for the circumstance that passing information about exceptions up the hierarchy also involves passing information to an
individual who also evaluates performance. If one or more hierarchical levels is bypassed, a further problem intrudes, i.e., going over
the boss's head. Telecommunication systems preserving confidentiality
and some form of anonymity are one resolution of this dilemma. Albert
Seyler (personal communication, 1975) at Telecom, Australia has reported the use of an asymmetrical audio system for increasing the
upward flow of information. Conrath and Blair (1974) report similar
results using an Augmented Knowledge Workshop.

     Reciprocal interdependence, the predominant coordination mode
in lateral systems, leads to a multiplicity of horizontal as well
as vertical communication channels. Furthermore, the horizontal channels are likely to shift frequently, causing uncertainty about which
of the many connections among the units to use. Under these conditions, this task is initially classified as an unstable task. However, research by Dewhirst (1974) has pointed out that in successful
lateral systems (unlike vertical systems) norms and rules for sharing information quickly develop. The development of these norms reduces the uncertainty concerning who should be kept informed about
what, and given these norms for sharing information, the task is
stable. Therefore, relatively cheap teleconferencing devices can
be efficiently used for this purpose.

     Problem solving and general discussion of ideas. The major
activities here are problem solving involving information seeking,
general discussion of ideas, and giving information to keep people
in the picture. They capture the intermediate stages of the problem solving process--generating alternatives and their consequences.
Reanalysis of the DACOM data shows no relation to anything, except
when problem solving and information giving to keep people in the
picture are involved, then the meeting usually involves bosses and
subordinates, and individuals from other buildings. Approximately
19% of the meetings of the original DACOM survey include this cluster of activities.

     The lack of emotional content indicates problem solving is rou-
tine with well understood methods of arriving at conclusions and little disagreement over the goals. It is analyzable. Since routine
problem solving with a group usually develops over a series of meetings, individuals know one another and there is no "forming impres-

sions of others" activity present. Also, the absence of impression
formation and emotional overtones indicates that problems do not
reflect on competence or evaluation of attendees.

In a vertical system these meetings could be among information
consumers attempting to verify commonality in interpreting the information, or that the system was functioning properly, or to accomplish
some other non-controversial routine coordination not provided by the
system. The task is stable. As in the previous cluster, computer
conferencing or audio conferencing are promising substitutes. In
lateral systems these meetings would probably be the latest in a
series of meetings which, although originally non-routine have,
through experience, now become routine. The task is stable. Such a
meeting is exemplified by weekly and/or daily meetings for routine but
timely financial discussions within a bank. The feedback is short-
term and has no major long run effects. There is no change in the
scope of anyone's operations, nor is any major resource allocation
going on. The success of the Bank America Audio conferencing links
between San Francisco and Los Angeles supports allocation of audio
links here.

Policy decision making. The last cluster in this set concerns
policy decision making with some problem discussion but no collection
of new information. 8% of the DACOM meetings include policy decision
making. DACOM reanalysis indicates meetings of more than two individuals average 10-15 minutes with participants from another building.
Two-person meetings average one or two hours and occur mostly in the
same department between a boss and a subordinate. Teleconferencing
is examined only for non-dyadic meetings; participants in dyadic
meetings are usually already in close physical proximity.

Short average meeting duration, the presence of people from other
buildings, lack of collection of new information, and some problem
discussion are consistent with two types of meetings in a vertical
system. DACOM data is not adequate to distinguish the two. For both,
however, the task procedures are analyzable and there are few exceptions. One meeting is a formality; it serves to formalize a policy
decision already agreed upon. The other meeting, hypothesized to be
a more likely occurrence, involves a type of policy decision for which
the system is designed; the information system provides all the required information--hence no collection of information is needed--
and methods for arriving at the decision are well specified. In
both meetings, therefore, the rules are well understood, and the procedures are analyzable.

Consensus on the decision is presumably important--the communication modality should provide an opportunity for some sort of interaction, i.e., problem discussion--otherwise the decision could
probably be made by an individual. An audio or computer conferen-

cing system should be sufficient to provide this type of interaction, given there is no high emotional content.

This type of meeting--policy decision making--is hypothesized to occur less frequently in lateral systems than in vertical systems. When it occurs in a lateral system, it is likely to include lower level individuals (probably specialists, who have already made the decision) and their superiors, who have to authorize the decision. The superiors already know the decision, and are prepared to approve it, since they have probably been kept informed during the decision making process. To satisfy their responsibility to understand, they ask a few proforma questions generating minimal problem solving and little conflict given their previous inclusion.

In lateral systems, policy decision making is a final step. Interestingly, Conrath (1975, personal communication) has noted policy decision making may be by telephone after face-to-face conferencing, resulting in information exchange and sizing one another up. Therefore, in a lateral system, policy decision making is a formality-- again, there are few exceptions and analyzable search. The meeting could be held using a less costly medium such as the Planet computer system or simply a conferencing telephone.

Negotiation With and Without Conflict

The two DACOM task clusters in the next set both involve negotiation; one also includes conflict.

Negotiation without conflict occurred most often among bosses and subordinates and in groups of more than five individuals, some of whom traveled to the meeting from another building or location. Meetings often last two hours or more. These activities took place in approximately 6% of the DACOM meetings. Negotiation is not a process having well defined and understood rules and procedures; therefore, the task has been classified as unanalyzable. The necessity for negotiation implies differences among the parties present, while the lack of conflict would seem to imply that the differences are not important and/or are easily resolved.

In vertical systems most information follows a standard path from subordinate to superior. In this context the task is classified as stable with information primarily flowing upward. Major foci for bargaining and negotiation in vertical systems are the quality and quantity of information to be entered into the system. More and higher quality information transmitted upwards leads to more hierarchical control. Addition of telecommunication systems increases the quantity of information and so increases hierarchical control. Information quality may not increase simultaneously because of negative reactions to additional hierarchical control per-

ceived to be associated with use of the communication device. Use
of audio, or visual devices rather than computer systems reveals to
supervisors information concerned with one's social presentation
that the participants would normally like to control, (also known
as impression management by Social Psychologists). This itself may
influence the quality of so-called static information. Formal asynchronous and synchronous interpersonal computer conferencing systems
may cut down the impression management.

Bargaining and negotiation in lateral systems can occur across
numerous links making it a task with many exceptions. Also, the
involvement of individuals who have been working with other groups
will lead to extensive interaction among, as well as within, groups.
To facilitate intra-group discussion, this necessitates a communication system with some time-out qualities. An asymmetric computer
system might be ideal here except that the social, emotional issues
involved may demand more personal modes, e.g. face-to-face. Whether
this demand must be met depends upon whether, in previous interactions, the bargaining agents have established a rapport or routine
in face-to-face contact. If so, the task has fewer exceptions, and
in view of results on sequence effects, previously described, a
cheaper mode, such as computer conferencing, may suffice. However,
high security is usually desired in negotiation and bargaining since
confidentiality and limited access to outsiders is perceived necessary. Early users of computer conferencing often were concerned
about lack of security (Hough, 1976). Therefore, audio conferencing is the preferred mode.

Negotiation with Conflict. DACOM reanalysis characterizes this
cluster of meetings by the presence of both conflict and negotiation.
It loads positively on forming impressions of other, often involves
six to ten people, from different status levels, some of whom traveled to the meeting place. These meetings often lasted more than
two hours and occurred in approximately 7% of the DACOM meetings.
As with negotiation without conflict, the procedures are ill-defined
and the task has been classified as unanalyzable.

The standard operation procedures characteristic of the vertical information systems are probably not adequate here because of the
presence of conflict and forming impressions of others. Impression
formation leads to situations being perceived differently from one
instance to the next and thus to many exceptions. Impression management is also a major concern for reasons articulated in the previous
cluster. It follows that teleconferencing modes providing for a
low degree of feeling of interpersonal contact are unwarranted. This
task is allocated to face-to-face meetings. Video teleconferencing
is a second alternative most likely to be used when the participants
have had some prior experience with the system, when the system is
secure, and when the task is urgent.

The discussion of negotiation with conflict in vertical systems generally is germane for lateral systems; i.e., there are many exceptions and the task is unanalyzable. So, again this cluster is allocated to face-to-face with the proviso that conferencing with limited access and adequate security can be used. Where team coordination is practiced, however, an audio teleconferencing device could be used for this cluster. Because everyone gets to know one another quite quickly in the team mode of coordination, the need to form impressions of others should lessen. The participants would have had many interactions, many of which would undoubtedly have been held face-to-face. The patterns of coordination set up through face-to-face experience would carry over to the audio system, particularly when sandwiched between face-to-face meetings.

Interpersonal Evaluation

This last set of clusters involves activities with high degrees of interpersonal evaluation. Past substitution task allocation attempts have assigned this set to face-to-face meetings because of the perceived need for both visual and audio channels by novice users of teleconferencing systems. It will be shown that, although this is often desirable, it is not always necessary.

Forming impressions of other/with information seeking. DACOM reanalysis indicates these two activities occurred most often when five or more people were present from different departments and in longer meetings. This cluster occurs in 10% of the DACOM meetings. The frequency of occurrence of forming impressions may vary, but regardless of frequency it incorporates poorly understood procedures; it is unanalyzable. It can, however, be a routine operation (it is stable). In organizational structures where the coordination mode involves limited contact with the same individuals time after time, as in vertical systems, it is a routine (stable) activity. Therefore, when conflict is not involved, audio conferencing systems should suffice. However, individuals believe, albeit unjustly, that forming impressions of others necessitates a visually mediated system-- video or face-to-face. Krauss, et al. (1976) provides evidence that it is often easier to detect false information with just an audio transmission. Given these beliefs, acceptance of non-visually mediated systems is in doubt. If meetings involve unfamiliar participants, the task classification changes to unstable, i.e., many exceptions, and unanalyzable. This case is frequent in lateral systems because there is contact with a variety of people from different subunits. Indeed when relationships between individuals are defined by more than authority--often the case in lateral systems--both impression formation and its impression management counterpart become more complex; then video conferencing substitutes for face-to-face meetings. In addition audio systems are appropriate when participants, regard-

less of status, have a history of working together; such a group
would be found in the team coordination mode.

Conflict and impression formation of others characterize this
cluster. Giving information to keep people in the picture, information seeking, discussion of ideas and presentation of report were
also present. The activities occur most often in dyads or groups
made up of five to twelve individuals, in meetings of an hour or
more, with some attendees traveling from their office. About half
of the meetings had participants from other organizations and about
a third had participants from other departments. The cluster occurred in approximately 5% of the DACOM meetings.

The presence of impression formation suggests that either the
participants are strangers or that differences in attitudes and values lead them to search for possible alterations in previous opinions
of others. In either case, changes and readjustments occur during
the meeting. The absence of impression formation rules implies that
the process is poorly understood--it is unanalyzable. Because the
processes of forming impressions of others change over the course
of meetings, the task is unstable. The foregoing is true in organizations using either lateral or vertical strategies. Thus, this is
one cluster where knowledge of organizational structure is not useful for predicting appropriate teleconferencing devices. Consequently, allocation is grounded in research at the work unit level, for,
as suggested by Pye, et al. (1973), media differences do exist at
this level.

> The evidence from our psychological experiments is that
> greater opinion change and more favorable description
> of the other are obtained for a 'conflict of opinion'
> task via an audio medium than face to face (intermed-
> iate results were obtained for a video condition)...
> If opinion change and the formation of favorable im-
> pressions of other is held to be a desirable outcome
> of conflict rather than, for example, regarding con-
> flict as a useful way of releasing tension..., we would
> allocate this cluster to a narrow-band system. If
> opinion change and formation of favorable opinions
> are not necessarily desirable outcomes of conflict
> then allocation must be to video or face to face;
> since findings support the view that the presence of
> the other or his image increases perception of him as
> a person, video or face to face conditions seem likely
> to increase the 'opinion' rather than the 'objective'
> aspects of conflict. In the absence of any informa-
> tion about the value or usefulness of 'conflict of
> opinion' rather than 'conflict of objectives' this
> cluster is conservatively allocated to face to face.
> (pp. 53-54)

Disciplinary interview. Conflict and delegation of work characterize this cluster. DACOM analyses show that attendees are always
from the same department. In the booster sample, where department
was not identified, 40% of the participants are from different buildings, and 25% travel to attend. The meetings most frequently involve
two people. Three is the next most common size. Less than two percent of the DACOM meetings contain this cluster. Inability to specify the processes involved in conflict, an important part of this
cluster results in this cluster being classified as not very well
understood; it is unanalyzable.

In vertical systems, where control is a hierarchical task, the
disciplinary meeting is a regular part of the manager's job. The
when, where, and who of disciplinary interviews in vertical systems
are probably well specified; as to who the participants will be there
are few exceptions. Given that the few people are involved in this
task are frequently located near one another, most of these meetings
will be held face-to-face. Because the meeting is held along lines
of authority and the task is routine, audio conferencing facilities
can also be employed, particularly when alternating with face-to-
face interactions.

This cluster is unlikely to occur in lateral systems where work
is not delegated, or when delegated, the delegation is usually to an
entire group or team of individuals. Within lateral relations task
stability depends upon the complexity of overlap of authority and responsibility between subunits. Since complexity tends to be the rule
here, the task is non-routine with many exceptions. The extent to
which this task is non-routine and unanalyzable makes it highly unlikely that teleconferencing systems are a viable alternative. Following the rationale provided under forming impressions of others with
conflict, this cluster is allocated to the face-to-face mode.

Maintaining morale and friendly relations within the context of
giving information to keep people in the picture. These activities
are a subset of the second cluster of the first set (giving information to keep people in the picture). The rationale for separating
the two was discussed there. This cluster is being discussed last
because of difficulties in classification.

The original DACOM analysis, and the reanalyses, provide few
cues about the context for these activities. Maintaining morale
and friendly relations are suspected to occur most often as a corol-
lary activity in meetings convened for other reasons. There is just
not sufficient data to support any speculation about allocation of
modalities for this cluster, especially in light of evidence indicating that there are a multiplicity of means by which individuals can
maintain morale and friendly relations over any communication mode.

SUMMARY OF MATRIX RESULTS

The above discussion illustrates the necessity to consider task meetings within their organizational strategy context and the dangers inherent in doing otherwise. Cautiously, given the adequacy of the existent data base, a summary will be offered of some general phenomena evidenced, ex post, in the matrix.

Only three of the four possible tasks in Perrow's (1970) typology appear in the cells of the matrix: (1) analyzable search, few exceptions-routine, (2) unanalyzable search, many exceptions--nonroutine, and (3) unanalyzable search, few exceptions--craft or skill. The missing cell is (4) analyzable search, many exceptions--engineering. In routine meetings, what occurs is relatively predictable, and there are existing procedures to handle the few exceptions. Meetings exist which require developed skills (Perrow's craft type); where participants become individually skilled in handling the small range of possible situations, very similar to the way in which a carpenter learns his craft. Finally, there are non-routine meetings, where there are neither the skills present, nor the procedures available, to deal with the many novel situations--procedures and skills must be created anew in each instance. There is no equivalent to an engineering type technology available for meetings, i.e., there exists no set of procedures prescribing consistent and effective action for meetings involving many different situations.

There are some consistent allocation patterns in the matrix: (1) asynchronous and synchronous computer conferencing systems for routine type meetings, (2) audio systems for skill type meetings, (3) face-to-face for non-routine meetings, and (4) video was second choice for non-routine meetings and second choice to audio for skill type meetings. Since more clusters were classified routine in vertical than in lateral systems, asynchronous and synchronous computers were more often assigned to tasks in the vertical system. More skill clusters were identified in lateral systems; therefore, audio conferencing systems were assigned more often.

It would seem that design strategies and the corresponding patterns of coordination adopted by organizations--be they social service delivery, education, government or private business organizations--lead to some task-meetings being more routine than others. The more routine a task-meeting, the more narrow band teleconferencing systems can be employed. Even non-routine meetings can be conducted over audio channels, particularly when the meeting is embedded in a sequence of meetings, some of which are held face-toface. The preceding suggestions are perhaps a bit premature given the weaknesses in the data upon which they are based. The conclusion that organizational context is crucial is not premature. At the beginning of this paper it was observed that almost all studies

on teleconferencing have been devoid of an organizational context.
Hopefully, this will not be the case in the future.

USE OF THE SUBSTITUTION MATRIX

The matrix is designed for use as a tentative guideline by researchers and individuals responsible for decisions about investment and placement of telecommunication systems. It is used as
follows:

1) Identify the distribution of the prevailing patterns of
coordination in which face-to-face meetings are used among the units
of an organization (see Van De Ven, et al., 1976 for scales). In
some cases, this may require a communication audit.

2) Eliminate from consideration all coordination links between
units which are physically located within approximately 65 meters of
one another. As Conrath (1973) has illustrated, face-to-face communication will be used most often within this distance. This is not
surprising. Thompson (1967) has pointed out that most organizations
locate reciprocally interdependent units next to one another.

3) Identify the distribution of DACOM tasks in meetings in
which teleconferencing systems might be substituted.

4) Using the perceived variability and analyzability of the
task, examine the predicted substitutability ratings for each task
cluster.

5) Take into consideration the mediating external environmental
factors, work-unit factors, equipment, and individual factors listed
previously and make a decision to retain the face-to-face format or
to substitute the teleconferencing system indicated in the matrix.
Before using a second or third choice system, read the discussion
section for that cluster.

BEYOND SUBSTITUTION

It should now be apparent that it is both feasible and desirable
to maintain the basic substitution perspective within an organizational context. There is, however, still another major problem; the substitution perspective does not allow for innovations in communication
usage sparked by the introduction of new teleconferencing devices.
Goldmark (1973), in his New Rural Society project, Etzioni (1975) in
the Minerva project, CSG members (Elton, Williams, Pye, Christie),
and others, have speculated about innovation, but evidence for innovation is sparse, consisting of small demonstration studies.[9]

The innovation perspective is difficult to research because, unless there is a crisis, users initially tend to substitute; that is, to use new modes in the same way as old modes. Further, barring a crisis, only over a period of time will they tend to discover new and innovative uses of new teleconferencing devices. Thus little is likely to be learned about specifics of innovation without conduc- ting some rather expensive, and extensive, long range research. It is, however, possible to speculate about the general direction of long term interactions between new teleconferencing systems and or- ganizations in which they are placed.

Organizations are most often designed to eliminate the most difficult (and costly) coordination problems by containing interde- pendencies within the smallest unit (Thompson, 1967). If all recip- rocal interdependencies cannot be so contained, an attempt is made to locate all reciprocally dependent units in the same department. The same logic can be extended to larger units if all reciprocal interdependencies are still not contained by departmentalization. Organizations adopt the above strategies because coordination by reciprocal interdependence is costly, and these costs rapidly es- calate as geographic organizational distance increases.

Assuming that as new electronic communications systems are devel- oped, as the costs of teleconferencing systems decrease, and as more is learned about how they can best be employed (these are, of course, interdependent accomplishments), there will then be a large increase in the ability to meet the coordination needs occasioned by recipro- cal interdependencies and a concurrent decrease in the costs.

This stream of developments will then have two effects: (1) the grouping of organizational units will become far less of a constraint and other organizing imperatives (as market, product, population served, etc.) will be better satisfied, and (2) more opportunities to create reciprocal interdependencies will be recognized and taken advantage of. As more reciprocal interdependencies are realized, small increases in use of impersonal coordination mechanisms, mod- erate increases in use of personal coordination mechanisms, and large increases in use of group coordination mechanisms have been shown to occur (Van De Ven, et al., 1976). In this case, both the number of scheduled and unscheduled meetings increase at a rapid rate as work flow moves from sequential to reciprocal to team modes of coor- dination. These findings have major implications for the accessi- bility characteristics of telecommunication devices to be installed in organizations. In the cases where work flow interdependence is moving towards becoming reciprocal or team, the number of unsched- uled meetings in both cases increases dramatically, creating a need for teleconferencing systems that are immediately accessible.

One caution before concluding. Much of the speculation on both

the substitution and the innovation perspectives does not emphasize
understanding the interaction of organizational strategy and teleconferencing systems over time. With minor exceptions, past evaluations of teleconferencing have been short lived, with no follow-up
to ascertain if, when, and how much the systems are utilized for different purposes. The cost effectiveness computed prior to introducing a particular system or immediately afterwards will probably
differ considerably from a similar evaluation after usage has become routine.

It was previously concluded under the substitution perspective
that narrow-band systems can be used effectively when the tasks have
become stable and analyzable (i.e., routine). The same principle
takes a slight twist under the innovation perspective. With the
introduction of new teleconferencing systems, individuals may sud-
denly realize unforeseen possibilities for sharing information. The
new associations may in fact continue without teleconferencing. That
is, teleconferencing devices may serve as catalysts and then, over
time, would no longer be needed for effective organization and communication flow.

Obviously, before long term investments in teleconferencing
facilities are made by organizations, research is needed that addresses itself to understanding the organizational conditions under which
these results would occur and how long they would remain stable.

FOOTNOTES

[1]There is no senior author here; the authors share equal responsibility for both the faults and merits of this paper.

[2]There is much redundant communication between individuals despite large individual differences in preferences for sending and receiving information on specific channels. Berman, et al. (1976) and the Shulman and Mutschler (1976) review support this conclusion.

[3]The DACOM studies include an initial study identifying the clusters, and the secondary Post Office and Booster Sample surveys. Data on frequency of occurrence, status make-up of individuals, and department affiliation are from the original DACOM survey. Booster Sample reanalysis yielded information on group size, meeting length, and participant travel to meetings. The Communications Studies Group was kind enough to provide these sets of data for our reanalyses. Roger Pye of CSG was most helpful in this matter. A detailed description of the secondary analyses is found in Konen and Shulman, 1976.

[4]There is another reason to question the generalizability of the conclusions. Secondary analyses of the British Post Office survey data suggest the laboratory analogue used has not been representative of the make-up of groups who currently use face-to-face meetings. Analysis of the data also indicates that many activities in meetings are not associated with the size or structure of the group --the activities include information exchange, problem solving, delegation of work and policy decision making. Surely, each of us has spent much time in meetings characterized by these activities--and wonder why we had the meeting in the first place. Interestingly, it is these activities which can be conducted over any modality.

[5]At least half a dozen other studies have had units communicate over more than one communication mode. Unfortunately, that data has not yet been analyzed for sequence effects.

[6]In general, what is said here is not new. The problems in studying communication patterns in work-units devoid of their organizational context were documented over nine years ago by Cohen, et al. (1969). The application to teleconferencing planning is new.

[7]These three design strategies are described in Galbraith (1973).

[8]Penley (1977) has found a strong empirical relationship between Perrow's (1970) task classification scheme and intra-unit and inter-unit communication patterns.

[9]Examples in first category are: use of teleconferencing in open universities, and work on telemedical diagnosis. An example of the second is the two-way CATV experiments in Reading, Pennsylvania.

REFERENCES

Berman, H. J., Shulman, A. D., and Marwit, S. J. Comparison of mul-
    tidimensional decoding of affect from audio, video and audio-
    video recordings. Sociometry, March, 1976, 39, 83-89.

Champness, B. The perceived adequacy of four communications systems
    for a variety of tasks. Technical report. Communications
    Studies Group, University College, London, 1972. (E/72245/CH)

Champness, B. The assessment of user reactions to confravision to
    analysis and conclusions. Technical report. Communications
    Studies Group, University College, London, 1973 (E/732250/CH)

Chapanis, A. Interactive Human Communication. Scientific American,
    March, 1975.

Chapanis, A. and Overby, C. M. Studies in interactive communication.
    II: Effects of similar and dissimilar communications channels
    and two interchange options on team problem solving. Percep-
    tual and Motor Skills, 1974, 38, 343-374.

Christie, B. An evaluation of the audio video conference system
    installed in the department of the environment. Technical
    report. Communications Studies Group, University College,
    London, 1973. (W/73360/CR).

Christie, B. Perceived usefulness of person-person telecommunica-
    tions media as a function of the intended application. Euro-
    pean Journal of Social Psychology, 1974, 4, 366-368.

Christie, B., and Holloway, S. Factors affecting the use of tele-
    communications by management. Journal of Occupational Psycho-
    logy, 1975, 48, 3-9.

Cohen, A., Robinson, E., and Edwards, J. Experiments in organiza-
    tional embeddedness. Administrative Science Quarterly, 1969,
    14, 208-221.

Connell, S. The 1973 office communications survey. Technical re-
    port. Communications Studies Group, University College, London,
    1974. (P/14067/CN).

Conrath, D. Communications environment and its relationship to or-
    ganizational structure. Management Science, 1973, 20, 586-603.

Conrath, D. and Blair, J. H. The computer as an interpersonal com-
    munication device: A study of augmented technology and its
    apparent impact on organizational communication. In Proceed-

ings of the Second International Conference on Computer Communication, Stockholm, Sweden, August, 1974.

Dewhirst, H. D. Influence of perceived information-sharing norms on communication channel utilization. Journal of the Academy of Management, 1974, 14, 305-315.

Emery, F. E., and Trist, E. L. The causal texture of organizational environments. Human Relations, 1965, 18, 21-31.

Etzioni, A. An engineer-social science team at work. Technology Review, 1975, 27-31.

Galbraith, J. Designing complex organizations. Reading, Mass.: Addison-Wesley, 1973.

Goldmark, P. C. The new rural society. Papers in communication, No. 5. Department of Communication Arts, Cornell University, Ithaca, New York, 1973.

Hough, R. W. Teleconferencing systems: A state of the art survey and preliminary analysis. Final report prepared for the National Science Foundation, May, 1976. Grant No. SSH74-22611.

Kraus, R. M., Geller, V., and Olson, C. Modalities and cues in the detection of deception. Paper given at the meetings of the American Psychological Association, Washington, D.C., September 2, 1976.

Kollen, J. H., and Vallee, J., eds. Travel/Communication Relationships, Bell, Canada, July, 1974.

Konen, P., and Shulman, A. D. Who, how and why of group meetings: A functional analysis. Paper presented at Midwestern Psychological Association meetings, Chicago, Illinois, May, 1976.

Penley, L. E. Organizational communication: Its relationship to the structure of work groups. In Readings in interpersonal and organizational communication by R. C. Huseman, C. M. Logue, and D. L. Freshley, eds. Boston, Mass.: Holbrook Press, Inc., 1977.

Perrow, C. Organizational analysis: A sociological view. Belmont, Calif: Brooks/Cole, 1970.

Pye, R., Champness, B., Collins, H. and Connell, S. The description and classification of meetings. Technical report. Communications Studies Group, University College, London, 1973. (P/73160/PY).

Short, J., Williams, E. and Christie, B. The Social Psychology of
    Telecommunications. London: John Wiley and Sons, 1976.

Shulman, A. D. Mode of communication and communication flow: A
    review of telecommunications research. Paper presented at Hu-
    man Factors meetings, October, 1975, Dallas, Texas.

Shulman, A. D., and Mutschler, E. Individual variation in verbal
    and nonverbal cue utilization: decoding of affect. Presented
    at Psychonomic meetings, November, 1976, St. Louis, Missouri.

Simon, H. A. Administrative behavior. (3rd ed.). New York:
    Macmillan, 1976.

Terreberry, S. The evolution of organizational environments. Ad-
    ministrative Science Quarterly, 1968, 12, 590-613.

Thompson, J. D. Organizations in action. New York: McGraw-Hill,
    1967.

Tyler, M., Cartwright, B., and Collins, H. Interaction between
    telecommunications and face-to-face contact: Prospects for
    teleconference systems. Long Range Intelligence Division (TSS6),
    British Post Office, August, 1975.

Van De Ven, A. H., Delbecq, A. L., and Koenig, R., Jr. Determinants
    of coordination modes within organizations. American Sociolo-
    gical Review, 1976, 41, 322-338.

Williams, E. A summary of the present state of knowledge regarding
    the effectiveness of the substitution of face-to-face meeting
    by telecommunicated meetings. Type allocation revisited.
    Technical report. Communications Studies Group, University
    College, London, 1974. (P/74294/WL).

ORGANIZATIONAL COMMUNICATION BEHAVIOR: DESCRIPTION AND

PREDICTION

David W. Conrath

Department of Management Sciences
Univ. of Waterloo, Waterloo, Ontario, CANADA
&
Institut d'Administration des Entreprises
Univ. d'Aix-Marseille, Aix-en-Provence, FRANCE

INTRODUCTION

For the past several years I have been contending
that the essence of the concept "organization" is to be
found in the interpersonal communication networks which
exist within any defined organizational boundaries
(e.g. Conrath, 1973). This holds for both theory and
practice. While the specifics of the contention appear
to be new, the importance of communication to an organization has been argued in the literature on organizational analysis for many years (e.g. Barnard, 1938;
Guetzkow, 1965; Hage, Aiken & Marrett, 1971). Furthermore, at the level of small groups interpersonal communication has been used as the basis for measures of
organizational structure (e.g. Bavelas, 1950; Leavitt,
1951; Mackenzie, 1966). Unfortunately, there is little
if any evidence that these measures have been applied
to, or are relevant for, organizations past the size of
small groups. What is lacking are measures of structure
based on communication behavior that are applicable to
large and complex organizations.

The argument supporting the contention that interpersonal communication is the basis of organization is
rather straightforward. On the one hand an organizational
relationship, to be exercised, requires communication.
If two individuals have no means by which they can relate
to each other, or if in fact no means are exercised, then

the relationship between the two exists only in an
abstract sense. There is no vehicle for interdependent
activity; yet the concept "organization" has value only
to the extent that there is value in interdependent
activity over and above that which can be accorded to
independent activity.

On the other hand interpersonal communication
provides the opportunity for the coordination, the
organization of multi-person activity. Whether a network has been established on normative grounds, or by
happenstance, or through social relations, it is the
means through which interdependent activity can be
recognized and can be accomplished. The network becomes
the de facto organization - the state or manner in which
people are arranged so that their interdependent activity
can have value with respect to some joint purpose.

To give an example, let us suppose that A is placed
in direct authority over B. If A cannot or does not
communicate with B there is no means by which A can
exercise his authority, i.e. can direct and/or control
the behavior of B. Let us further assume that C can and
does communicate with B, and that C has convinced B that
C is in a position to reward and punish B. In such a
case C has de facto authority over B whether or not the
normative structure provides such a relationship between
the two. The lack of communication in the first instance
indicates that at least part of the formal authority
structure exists only in an abstract sense. In the
second instance an organizational relationship exists in
fact, whether or not such a relation is recognized in
any formal or normative sense.

Data on organizational communication patterns are
essential if we are to build an accurate descriptive
model of an organization's structure. This conclusion,
however, is insufficient by itself. At a minimum,
answers to three interdependent questions must be sought:
What data on interpersonal communication behavior are
relevant? How might these data be processed to provide
a comprehensible model of an organization's structure?
To what uses can we put these descriptive models?

These three questions provide the focus of an
extensive research program in process at the University
of Waterloo and also at the Institut d'Administration
des Entreprises, Université d'Aix-Marseille. This paper
will take only a cursory look at answers to the first
two questions. The primary objective will be to consider
several answers to the third.

A DATA BASE

The questions posed regarding data relevance, their use for measures of structure and the applications of these measures, require answers that can be provided only by empirical research. Unfortunately, there is little existing theory and/or empirical work in the literature upon which one could base such an effort. Furthermore, in addition to the interest in the structural implications of organizational communications behavior, we were interested in the technological implications, especially as they might affect the design of new telecommunications systems. Thus it was necessary to collect as robust a data base as possible, including everything that we had reason to believe might be relevant, recognizing the limitations of subject time and effort inherent in field research.

A detailed description of the data base and the methods used to obtain it would overwhelm this paper. Nevertheless, a summary is clearly in order. Further detail can be obtained directly from the author and hopefully will also appear in future publications.

Three distinctly different methods were used to gather data: one-shot questionnaires, interviews and personal diaries of communication behavior. All of the methods were applied to each of three organizations. For two of them the data were collected both before and after the installation of a new computer-controlled telecommunications technology.

The Sample

The first organization studied was a small manufacturer (annual sales in the 5 to 10 million dollar range) of plastic products. All of the management and staff personnel were included in the sample (59 subjects), while none of the first-line operatives were. The communication behavior of the latter was restricted by the machinery with which they worked. As the company did not receive the new communications equipment, data were collected just once.

The other two organizations were studied both before and after the installation of a new telecommunications system. One was a branch of an international insurance brokerage company. Everyone but the messenger boys, several salesmen and the chairman of the board was included in the sample, approximately 90 persons for

each study. The other organization was a branch of a
large public utility. Here all of the managers and
staff personnel were included with the exception that a
few of the large staff of engineers were not. This
sample contained 124 people for the "before" study, 90
for the "after" one.

Questionnaires

The questionnaires had a number of facets. The
first part contained over 100 statements reflecting
attitudes towards various communication technologies
and ways of handling specific communication situations,
and perceptions of one's communication needs, of the
physical environment in which one must communicate and
of the psychological environment (organizational climate).
Responses were made along a five point Likert scale which
ranged from "strongly agree" to "strongly disagree".
The statements were based on the presumption that such
attitudes and perceptions were likely to influence
communication behavior.

The second and third parts of the questionnaire
differed between the "before" and "after" data collection
sessions. The "before" session asked for an expression
of sentiments toward each of up to 10 persons with whom
one must communicate relatively frequently. Interpersonal
sentiments were measured using the semantic differential
technique, the selection of the ten sets of bipolar
adjectives based in part on the results of Wish (1973).
The critical incident technique was also employed. Each
respondent was asked to describe one or more recent
situations which were examples of particularly bad or
especially good communications. The data on interpersonal sentiments were gathered because we suspected that
how one felt about someone else would affect how the
two interacted. Critical incidents were obtained to gain
insights on the effects of certain communication situations. Several members of the International Communication
Association had indicated success using such a technique.

The second and third parts of the "after" questionnaire focused on telecommunications technology. One
part contained a number of descriptions of communication
situations which could be handled by means of several
different technological alternatives. The respondents
were to indicate how satisfactory they thought each of

the alternatives would be. The other part asked for
information specific to the value of the various
features offered by the new communication system.

Interaction Analysis

A separate questionnaire, which we labelled the
Interaction Analysis, was designed to gather data about
the communication aspects of one's job. Despite the
fact that the research of Burns (1954) and others since
him have found that managers spend between 70 and 90
percent of their time communicating, none of the task
or job description taxonomies focuses on communication
requirements. Rather, they indicate what one does on
one's own and for what one is responsible, without
suggesting how one's own actions relate to the work of
others nor how one's responsibilities are to be executed
with respect to others. The Interaction Analysis, in an
effort to correct this oversight, attempted to obtain a
description of one's job in terms of its communications
content. Each person was to indicate with whom he or
she related, the processes involved (informing & advising,
directing & controlling, discussing & problem solving,
persuading & negotiating, asking & answering), the mode
of communication used, an estimate of the percentage of
one's time at work spent on each person-process-mode
combination, and certain requirements for communication
that have technological implications. Because of the
complexity of the data requested, the questionnaire was
divided into distinct steps. Furthermore, it was given
to each subject only once, the assumption being that the
communication content of most jobs would not change in
the interval between the two collections of data.

Interviews

Approximately one week after each subject had
completed the questionnaire and the Interaction Analysis
he or she was interviewed. The purpose of the interview
was to gain additional insight about particular, especially extreme, responses to the two forms already
completed, and to get people to describe the kinds of
communication problems they have and to think of ways
that technology might help to alleviate them. The
interviews also allowed our respondents to express their
concerns about our study and to ask us about our
motivations and expectations.

Communication Diary

Each subject was asked to keep a diary of 100 consecutive interpersonal interactions commencing on a specific date and time. The form used required no more than a series of check marks, except for the name of the other party or parties. The data recorded included: the identification of the other party, the initiator of the interaction, the mode used, the elapsed time of the event (one of four categories), the success of failure of one's attempt to reach another person (and if a failure what was done about it), the process or processes involved, and several aspects related to technology.

DATA: RELEVANCE AND RELIABILITY

Our data on attitudes provided little of interest. Very few attitude responses correlated with observed behavior (the communication diary data), the number being no more than one would expect by chance. As one finds so frequently in the literature, the only things which correlated significantly with attitudes were other attitudes.

While the above result did not surprise us, we were surprised to find that interpersonal sentiments appeared to be unrelated to communication behavior, especially such obvious aspects as the choice of mode.

The critical incidents which were described, without exception, concerned problems of interpersonal relations and not technology. It is the behavioral aspects of communication which dominate one's perception of what goes on during an interaction, but this may partly reflect the fact that most people take communications technology as given. As the variety of communications technology increases and becomes more commonplace, people may become more sensitive to variations in the effect that technology may have during interpersonal communication.

The responses to the Interaction Analysis were disappointing. It was clear that the respondents were not used to thinking about their jobs in terms of communication. For example, they were not aware of how much time they spent communicating with specific others. The cross-check reliability calculations made on these data indicated a rate of disagreement which averaged

about 70 percent. Interestingly, the aggregate responses
to questions having technological implications were
more reliable. However, these data could be collected
much more easily using other procedures.

Interviews were very useful, and we would make use
of them again in any future study even though the man-
hour cost is substantial. We found that they served
three purposes. They enabled us to estimate data
reliability and validity. They provided insights that
one could not obtain without the kind of interaction
which takes place during an interview. And of equal if
not greater importance, interviews gave us the opportunity to explain what we were doing and why, enabling
us to gain and maintain subject cooperation. Without
this kind of personal rapport we doubt that one could
obtain reliable data from such an extensive effort as
that which we undertook.

The communication diary data are at the core of our
analyses of organizational communication behavior. The
data yield reasonable cross-check reliability - an error
rate of less than 50 percent - in terms of sender-receiver
agreement. This is consistent with the results obtained
by the few competent uses of the diary technique that
appear in the literature. Furthermore, based on the
substantial monitoring conducted by the research team
during this phase of our study, we feel that the only
way in which one is likely to increase reliability is
to institute automatic recording devices. Telephone
calls, for example, could be recorded via a switcher.
But then this approach might lead to the spectre of
"big brother is watching you."

REPRESENTATIONS OF STRUCTURE

The search for measures of structure, especially
those which might give us some new insights about organizational structure per se, is a major current research
activity. Unfortunately, despite considerable time and
effort, as yet it has yielded little in the way of
tangible results. Most of the structural analysis done
to date has relied on network models, descriptive and
normative matrix representations of who relates to whom.
The fundamental link involves a dyad, and the measures
of the relationships which have been used are either 0,1
or cardinal, such as the measurement of the frequency
of interaction which takes place along any given channel.

While models based on the above measures are very
useful, we are not convinced that they get at the essence
of "organization". What we would like to be able to do
is to group people together who have a high need (descriptive or normative) to relate within the group relative to their need to relate without it. The intra-
group relations should be dense relative to the intergroup relations, and the intra-group network should be
essentially an all-channel one. Furthermore, we want to
identify those persons who are members of two or more
such groups, those who are not members but may link two
or more groups, and those who do not fit within any group
structure. These would then become the basic units of an
organization, and the process which enables us to form
such groups would be used again to form groupings of the
groups. This would be done repeatedly until everyone
who is a member of the formal organization being analyzed
is eventually formed into the super-group.

We have used the term "relational clustering" for
the above procedure, even though it is not yet perfected.
While we have written heuristic programs which can do
the grouping, what we lack is a measure by which we can
evaluate such programs. We need an algorithm, which can
be justified in terms of organizational behavior, that
can tell us when we have reached the point of "optimal"
clustering - when we have discovered the basic units
which make up the organization. Any ideas that one might
have along this line would be most welcomed, even
including a proof that such an algorithm cannot exist.

ORGANIZATIONAL COMMUNICATION BEHAVIOR

Who Relates to Whom?

This is an obvious first question, but it can be
answered only in the context of some concept of organizational structure. Since the most widely recognized
concept is the authority structure we have used this as
our basis of comparison.

The data gathered from our five samples are shown
in Table 1. This shows the number of recorded communications initiated by our respondents to other members
of their organization. It also shows the distribution
by percentage within each sample. The recipients have

Table 1: Number and Percentage of Initiated Interactions by Authority Relationship of Recipient

Recipient	Firm X -- # %	Firm Y before # %	Firm Y after # %	Firm Z before # %	Firm Z after # %
Superior Direct	218 13.7	310 14.3	196 10.0	292 12.9	174 10.1
Peer Immediate	163 10.3	213 9.8	220 11.2	468 20.6	326 18.9
Subordinate Direct	165 10.4	155 7.2	96 4.9	915 40.3	738 42.7
Superior Diagonal	328 20.6	641 29.7	588 30.0	149 6.6	114 6.6
Peer Indirect	287 18.1	519 24.0	520 26.5	333 14.7	246 14.2
Subordinate Diagonal	429 27.0	323 14.9	342 17.4	112 4.9	131 7.6

been divided into six categories according to their authority relation with the initiator: direct superior - senior in rank and in the direct chain of command, immediate peer - same rank and reporting to the same immediate superior, direct subordinate - junior in rank and in the direct chain of command, diagonal superior - senior in rank but not in the direct chain of command, indirect peer - same rank but reporting to a superior other than the respondent's, and diagonal subordinate - junior in rank but not in the direct chain of command. Since our respondents recorded both initiated and received interactions, and since roughly half of all interactions were with persons outside of the organization being studied, the entries in Table 1 comprise about one-fourth of the total sample.

Before commencing our observations we should mention that a higher proportion of respondents from organization Y had no subordinates than was the case for the other two firms. Thus it is not surprising that there are fewer interactions with subordinates within Y. What is surprising, however, is the rough similarity of patterns between X and Y, and the great divergence between those

two and Z. Less than 35 percent of the initiated interactions within X and Y were to direct superiors, peers
and subordinates. For firm Z more than 70 percent of
the traffic stayed within the authority and immediate
peer network. Even though one could argue that our
observations are not strictly independent, any adjustment
for independence would still lead to a difference which
is statistically very significant (would occur by chance
less than one in a thousand times).

In searching through the rest of the data to seek
a possible explanation for the difference, one thing
stood out. An examination of the various organizational
climate indices indicated that organization Z was
relatively low in its emphasis on human relations and
relatively high in its emphasis on following rules and
regulations and on risk aversion, when compared to X and
Y. All three were roughly comparable in their stress on
productivity. Considering the aspects of organizational
climate in which we found differences, one might label
organization Z as bureaucratic, especially in comparison
to the other two. It stressed those things typically
associated with bureaucracy, and placed relatively
little stress on human relations.

What this suggests is that communication within a
bureaucratic organization is much less likely to deviate
from the network provided by the formal organization.
This seems reasonable since it is consistent with the
normative models of bureaucracy. Perhaps all we have
really discovered is that people recognize a bureaucracy
when they are in one. In this light it is interesting
to note that several senior members of Z kept stressing
their emphasis on human relations during discussions
with members of the research team, whereas the subject
was scarcely mentioned by senior members of the other
two organizations.

By What Mode?

The choice of mode to communicate within the firm
was relatively similar across the three organizations
studied. The data in Table 2 show that the telephone is
used more heavily for indirect than for direct organizational relations, and that company Z used the phone more
than X, which in turn used it more than Y. Though it is
not obvious based on the data presented, these results
all reflect the dominate influence in mode choice -
distance. People are more likely to be located near

Table 2: Number and Percentage of Initiated Interactions
          by Mode Used

Firm	Relation	Face-to-face #	Face-to-face %	Telephone #	Telephone %	Written #	Written %
X	Direct	445	81.5	86	15.8	15	2.7
X	Indirect	734	70.3	260	24.9	50	4.8
Y	Direct	1107	93.0	62	5.2	21	1.8
Y	Indirect	2429	82.8	331	11.3	173	5.9
Z	Direct	2375	81.5	279	9.6	259	8.9
Z	Indirect	627	57.8	400	36.9	58	5.3

their subordinates, peers and superiors than they are
near others. Company Z personnel were scattered in
several buildings located throughout a reasonably large
city (300,000+). Company X respondents were in three
separate locations, though the majority were in one
building. Everyone sampled in firm Y worked on the same
floor of a large office building.

A previous study of the relationship between mode
choice and distance indicated that the use of face-to-
face contacts dropped off dramatically when people were
located further than 150 feet from one another (Conrath,
1973). Other analysis of the data base described herein
confirmed this finding. In very few instances were
subjects located more than 150 feet from each other if
they were on the same floor of a building. Thus, we
were not surprised to find that the proportion of interactions between two people on the same floor which were
face-to-face was approximately 80 percent. Yet, as soon
as two different floors were involved, even if adjacent
in the same building, the proportion of communications
that were face-to-face dropped to roughly one-third.
The difference, of course, was made up for by use of the
telephone. Clearly telecommunications provide just that
- the ability to communicate over a distance.

Table 2 also provides some additional evidence of the relative bureaucratization of organization Z. A substantially higher proportion of communication within the formal structure was in writing for Z in comparison with X and Y.

For How Long?

As the data in Table 3 indicate, there is not a great deal of interest to be found in the study of the length of interactions. Short ones, those requiring less than 3 minutes to complete, predominated. This was especially the case for firm Y, which appears to reflect the nature of the insurance business as much as anything else. One can also note that except for company Z contacts which were along the lines of the authority structure lasted longer.

Table 3: Number and Percentage of Initiated Interactions by Elapsed Time

Firm	Relation	Minutes			
		< 3	3 - 15	15 - 60	> 60
		# %	# %	# %	# %
X	Direct	297 54.4	198 36.3	40 7.3	11 2.0
	Indirect	683 65.4	317 30.4	36 3.4	8 0.8
Y	Direct	849 71.3	290 24.4	38 3.2	13 1.1
	Indirect	2306 78.6	533 18.2	77 2.6	17 0.6
Z	Direct	1721 59.1	916 31.4	201 6.9	75 2.6
	Indirect	635 58.5	382 35.2	47 4.3	21 1.9

The only result of interest which was lost by the summarization of the data is that persons from Z indicated that they spent relatively more time on interactions with superiors than with other members of their organization. This was not the case for the respondents from the other two firms.

Using What Processes?

Table 4 indicates the percentage of interactions in which a given process was perceived to have taken place. Since more than one process might have occurred during an interaction the sum of percentages exceeds 100. The groupings of organizational relations are based on relative rank, whether or not the relation was direct or indirect. This is because the greatest differences in the process patterns were yielded by the relative rank of the "other party" rather than the specific organization or whether or not the communication was along the lines of the authority structure.

Table 4: Number and Percentage of Initiated Interactions by Process and Rank of Recipient

| Process | Rank of Recipient | | | | | |
| | Superior | | Peer | | Subordinate | |
	#	%	#	%	#	%
Informing, Advising	1538	51.1	1515	46.0	1780	52.3
Directing, Controlling	93	3.1	140	4.2	547	16.1
Persuading, Negotiating	60	2.0	98	3.0	77	2.3
Discussing, Problem Solving	518	17.2	674	20.5	530	15.6
Asking, Answering	1320	43.9	1471	44.6	1246	36.6

Informing and/or advising appears to have taken place in about half of all the interactions recorded. The proportion is not much less (just over 40 percent) for the asking and answering of questions. At the other extreme our respondents felt that they did relatively little persuading or negotiating or selling type of activities. The low response might also reflect a perception that this process does not appear to be quite

as legitimate as the others. One is more apt to assume
that he is informing rather than persuading someone
else. This is the kind of difficulty that one faces
anytime one attempts to obtain communication content
data. The combination of semantic problems and the
hesitation that people have recording activities that
might be seen by others as of questionable value are
likely to lead to responses that are difficult to
interpret.

Looking at the processes used to initiate interactions up, sideways and down, we find little to distinguish those directed to superiors. Interactions with
peers are somewhat less likely to be of an informing,
advising nature, and more likely to involve discussion
and problem solving. This is what we would expect among
peer relations. Communications initiated to those junior
in rank, as seems very reasonable, contain a relatively
high proportion of directing and controlling processes.
Furthermore, one finds less discussing and problem
solving, and asking and answering of questions.

While the differences among the three organizations
were not great, two warrant mention. Respondents from
firm X made relatively heavy use of discussing and
problem solving processes. This may be one reason why
the company is perceived to be human relations oriented.
Persons from Z indicated a relatively high use of
persuading and negotiating, especially when initiating
contacts with peers and superiors. Perhaps the formalism
of the bureaucratic structure forced one to do more
"selling" of ideas and the like.

MODELING MODE CHOICE

The pattern of communications varied according to
the mode used. Earlier research has shown this (Conrath,
1973), and so has the analysis of the present data base
(e.g. see Table 2). We expect, therefore, that new modes,
such as interpersonal computer communication or an all
pervasive video network, may create their own particular
networks, or in reality their own organizations. Thus,
we would like to be in a position to predict what the
organizational effects might be with the introduction of
new forms of communication.

A first step toward this goal is a better understanding of mode choice under existing circumstances.
We have already noted that distance is a major factor.

The relative location of people clearly influences the mode chosen to interact with another. There are other factors, however, which are also likely to affect mode choice. Certainly there is good reason to think that the content of an interaction or the tasks which one must accomplish on one's job would influence mode choice.

Our data permit us to consider communication content in two ways. One is the interaction process or processes which took place. The other is the use of communication aids - e.g.: printed matter, existing images, physical objects, filing systems - during the interaction. We considered these two aspects both separately and collectively, with disappointing results. At the very best we were never able to explain more than 15 percent of the variance in mode choice. Furthermore, the findings were not consistent across organizations. About all that one can conclude is that the telephone is used relatively heavily for asking/answering and persuading/negotiating processes.

We next turned to look at the effect of tasks on mode choice, but given the above results and given that we are primarily interested in individual communication profiles rather than specific event behavior, we searched for a way to measure one's entire job. Unfortunately, nowhere were job descriptions adequate for coding so that we could use them as one or several variables. Furthermore, but with one exception, the job descriptions completely ignored the communications aspect of one's job. Thus, we had to accept a surrogate.

The surrogate chosen was one's rank in the organization's hierarchy and one's departmental affiliation. Since each of the companies studied was organized along functional lines, jobs at the same rank within a given department were quite similar. As we were also interested in the impact of the new telecommunications system, we wished to examine the "before and after" effect on mode choice as well. Thus, we ran a three-way analysis of variance (rank, department, study), using the proportion of one's interactions that were made by telephone as the dependent variable. The results for organizations Y and Z can be found in Tables 5 and 6 respectively.

All three factors are statistically significantly associated with the extent to which one uses the telephone. Together they explain approximately one-third of the variance across individual behavior for each of the two firms. In both cases one's departmental affiliation

Table 5: Analysis of Variance of Percent of Interactions
         by Telephone, Firm Y

Source of Variation	Deg. of Freedom	Mean Square	F	Stat. Signif.
Main Effects:	10	2786.3	9.08	0.001
Department	6	3473.3	11.32	0.001
Rank	3	2150.4	7.01	0.001
Study	1	2597.1	8.47	0.004
2-Way Interactions:	22	181.0	0.59	0.999
Dept./Rank	13	187.8	0.61	0.999
Dept./Study	6	188.0	0.61	0.999
Rank/Study	3	129.3	0.42	0.999
3-Way Interaction	10	362.7	1.18	0.307
Explained	42	844.6	2.75	0.001
Residual	141	306.7		
Total	183	430.1		

Table 6: Analysis of Variance of Percent of Interactions
         by Telephone, Firm Z

Source of Variation	Deg. of Freedom	Mean Square	F	Stat. Signif.
Main Effects:	13	1777.2	8.14	0.001
Department	9	1542.8	7.07	0.001
Rank	3	3765.2	17.25	0.001
Study	1	1278.9	5.86	0.016
2-Way Interactions:	31	467.0	2.14	0.001
Dept./Rank	19	652.5	2.99	0.001
Dept./Study	9	82.3	0.38	0.999
Rank/Study	3	25.8	0.12	0.999
3-Way Interaction	10	115.9	0.53	0.999
Explained	54	717.4	3.29	0.001
Residual	154	218.2		
Total	208	347.8		

explained the greatest amount of variance, though one's rank was not far behind. While the "before and after" effect was statistically significant, it accounted for only about three percent of the variance in telephone usage. It is also interesting to note that only one of the interaction effects was significant, the department/rank effect for firm Z.

Our next step will be to look at the effect that distance had on these results. To what extent was the variation in telephone usage among departments and among hierarchical ranks related to the relative distances of the people with whom they communicated?

CONCLUSIONS

There is more than adequate evidence to suggest that organizational structure and organizational communication have a great deal to say about each other. Unfortunately, very little has been done looking at the relationship between the two, especially from an empirical point of view. What we have done is only the first few steps of many which must be taken to correct this oversight.

The value of such research is substantial. Not only will it tell us much about the phenomenon called organizational structure, but we can better understand the impacts of other phenomena on structure as well. In particular, we need to be in a position to estimate the impact of new communications technology on an organization's structure - the network of interpersonal relations - before we implement that technology. To do otherwise is as foolish as to suggest the implementation of a new industrial technology without attempting to gauge its effect on the environment. Certainly the human environment is as important to our well-being and the quality of life as the physical environment. And communications has a tremendous impact on the former.

REFERENCES

Barnard, C.I. (1938): The Functions of the Executive, Cambridge, Mass.: Harvard University Press.
Bavelas, A. (1950): "Communication Patterns in Task Oriented Groups," Journal of the Accoustical Society of America, 22, 725-730.

Burns, T. (1954): "The Directions of Activity and
    Communication in a Departmental Executive Group,"
    Human Relations, 7, 73-97.
Conrath, D.W. (1973): "Communications Environment and
    Its Relationship to Organizational Structure,"
    Management Science, 20, 586-603.
Guetzkow, H. (1965): "Communications in Organizations,"
    in March, J.G. (ed.), Handbook of Organizations,
    Chicago: Rand McNally, 534-573.
Hage, J., Aiken, M. and Marrett, C.B. (1971): "Organ-
    ization Structure and Communications," American
    Sociological Review, 36, 860-871.
Leavitt, H.J. (1951): "Some Effects of Certain Communi-
    cation Patterns on Group Performance," Journal of
    Abnormal and Social Psychology, 46, 38-50.
Mackenzie, K.D. (1966): "Structural Centrality in
    Communication Networks," Psychometrika, 31, 17-25.
Wish, M. (1973): "Individual Differences in Perceptions
    of Dyadic Relations," paper presented at the 81st
    Annual Convention of the American Psychological
    Association, Montreal.

Partial financial support of this research by the Canada Council
is most gratefully acknowledged.

MEASURING THE DIMENSIONS OF INTERPERSONAL COMMUNICATION

Myron Wish

Bell Laboratories

Murray Hill, New Jersey   07974, U.S.A.

INTRODUCTION

One of the most recent lines of psychological research initiated at Bell Laboratories deals with the influence which various
factors have on the way people communicate and interact with each
other and with the development of better procedures for measuring
verbal and nonverbal aspects of communication. An important long-
range goal of this research on the structure and dynamics of interpersonal communication is to create a knowledge base that will be
useful in planning, designing, and evaluating telecommunications
systems. In this regard, knowledge about the factors influencing
the process and outcome of interpersonal communication may clarify
the communications requirements for different purposes and for
different segments of the business and residence markets. Although
it is a long road from basic research in a new area to practical
applications and planning, we feel that the development of a sound
foundation about the nature of dyadic and group communication is
likely to be a wise investment.

Our studies have focussed primarily on three factors that
influence the way people communicate and interact with each other:
(1) the modality over which the communication occurs (e.g., telephone, face-to-face, visual telecommunications); (2) the situational
context or purpose of the communication (e.g., to work out a
compromise); and (3) the relationship between the communicators
(e.g., supervisor and employee). Undoubtedly there are many other
factors, such as the organizational structure and climate in which
the communication occurs, that are also of great importance. However, we have had to limit our scope at this stage of the endeavor.

The first investigation in this research program dealt with perceptions and opinions of PICTUREPHONE® Service, as compared to telephone and face-to-face communication (Wish, 1975a). This was an interview study, conducted in 1972, using two groups of respondents. One sample included 64 executives in the Bell System Corporate PICTUREPHONE Network, while the other was comprised of 173 Chicago businessmen who had had no prior experience with PICTURE-PHONE Service. The latter group did, however, receive a thorough demonstration of the capabilities of the service, and one third were actually interviewed over PICTUREPHONE.

Both the Chicago and Bell respondents perceived very large differences in the effectiveness of the three modalities (PICTUREPHONE Service being rated as more effective than telephone, but less effective than face-to-face communication) for making a good impression or "sizing up" another person. Both groups also felt that PICTUREPHONE Service was not very promising for very important situations such as "working out the major issues of a business deal."

There were some substantial differences in the kinds of business situations which the Chicago and Bell interviewees perceived as being most and least effective for PICTUREPHONE Service. For example, while the Chicago sample judged PICTUREPHONE to be more effective (relative to the other modalities) for inter-company than for intra-company communication, the Bell executives judged it to be as good for internal as for external communication. The Bell respondents' ratings of PICTUREPHONE Service for relatively unimportant communications were somewhat higher than those of the Chicago businessmen, which suggests that appreciation of PICTURE-PHONE Service for ordinary, day-to-day communication may develop as one adapts to the medium. This is compatible with the Bell group's view of PICTUREPHONE communication as more relaxed, "low-keyed," and natural (than the Chicago group viewed it). Still other data indicated that the importance of various features of a telecommunications service (e.g., graphical capabilities and conferencing applications) depend on the communication habits and the levels of the people in the network as well as on technical capabilities.

The study, which revealed many interesting perspectives on communications via alternative modalities, served to motivate further investigations to discover the most important dimensions on which communication should be measured and modalities compared. This was a clear example of applied problems stimulating the need for basic research! Knowledge about these dimensions would simplify other studies of communication modalities, and would facilitate the selection of a more comprehensive set of business or social situations. This research on the dimensions of interpersonal communication involved the use of statistical procedures called

multidimensional scaling, or MDS for short. In fact, MDS was
largely developed at Bell Laboratories (Shepard, 1962; Kruskal,
1964; Carroll and Chang, 1970), although important developments
have been made elsewhere.

Before describing these studies on the dimensions of interpersonal communication, however, we shall give a brief description of
what MDS does. This may help to clarify why particular methods
were used and how the dimensions were extracted from subjective
ratings. Greater detail about MDS is provided in the above references as well as in a monograph by Kruskal and Wish (1978).

MULTIDIMENSIONAL SCALING

Multidimensional scaling is a name applied to certain
statistical techniques that are used to obtain spatial representations of the underlying or latent structure in a matrix of
"distance-like" numbers. Thus, if the numbers in the data matrix
(or table) indicated the geographical distances between various
European cities, an MDS analysis would provide a reconstruction of
the map. In most MDS applications, however, there is no a priori
information about what the "map" or multidimensional space looks
like, nor even the number of dimensions that it has. Moreover, the
data values generally correspond to "subjective distances" between
stimulus objects based on one or more rating methods. Nevertheless,
these powerful techniques can determine how many dimensions are
needed to explain or reproduce the data, what the dimensions are,
and where the objects are located on each dimension. A newer procedure called INDSCAL (Carroll & Chang, 1970) also provides information about the salience, or weights, of the various dimensions
to each of the individuals who made direct or indirect judgments
about inter-object distances.

It is important to point out that the computer locates the
objects on the map, but the researcher has the task of labelling,
or interpreting, the dimensions. Interpretation of a dimension
involves distinguishing how stimuli at one extreme differ from
those at the other. This can be done statistically by multiple
regression, or by more intuitive examination of the multidimensional
space. Although most MDS procedures require rotation of the axes
of the "map" in order to interpret the dimensions, this difficult
step is generally unnecessary when INDSCAL is used.

DIMENSIONS OF INTERPERSONAL RELATIONS AND HYPOTHETICAL COMMUNICATION

One of the factors that is likely to affect the way people
communicate and interact is their relationship to each other. Thus,

the typical communication between supervisor and employee is likely
to be quite different from that between husband and wife. In one
of our questionnaire studies (Wish, 1976a) we systematically investigated this relational component of interpersonal communication.
In the study under consideration 87 college students made several
kinds of judgments about the way people in various interpersonal
relations typically communicate and interact with each other. One
of the tasks was to rate the degree of similarity between all possible pairs derived from a list of 25 interpersonal relations.
For example, if a subject thought that the typical communication
between a supervisor and employee was very similar to that between
parent and teenager, he would circle a high number on a scale from
$\underline{1}$ to $\underline{9}$. Lower numbers were used to indicate that the communication
in the two designated relations was very different.

A matrix of similarities among interpersonal relations (the
stimulus objects) was obtained for each person participating in the
study. The numbers in such a table were assumed to reflect the
subjective distances between the interpersonal relations; that is,
the greater the rated similarity, the smaller the subjective distance. These data were analyzed by means of the INDSCAL computer
program. (Unlike other MDS procedures, INDSCAL analyzes several
matrices simultaneously.)

Four dimensions emerged from the multidimensional analysis.
The first distinguished interpersonal relations in which the communication is typically competitive and hostile from those which are
generally more cooperative and friendly. Dimension 2 was based on
the power or dominance structure of the dyad, ranging from relations
in which power is equally divided to those in which one person
exerts considerable control over the other. The third dimension
contrasted relations in which the communication tends to be task
oriented and formal with those involving family or other personal
interaction, while the fourth distinguished intense relations from
those involving infrequent and superficial contact.

There was strong external statistical evidence for each of
these dimensional interpretations. This support came from another
judgmental task in which the students rated one interpersonal relation at a time on numerous bipolar scales. For example, on one
scale subjects circled a high number if communication between
members of the specified dyad was thought to be very cooperative
and a low number if it was viewed as rather competitive. There was
a correlation of .97 between mean ratings (based on the entire
group of subjects) of the 25 interpersonal relations on the
"cooperative vs. competitive" scale and the stimulus projections,
or coordinates, on the first dimension. Similar statistical
evidence validated the interpretations of the other dimensions.

Almost identical results (Wish, 1976; Wish, Deutsch, and Kaplan, 1976) arose from analyses of data from the other rating tasks used in the study. Moreover, these dimensions were also obtained in a later study (Wish, 1975b; Wish and Kaplan, 1977) in which subjects rated hypothetical communication episodes, which were constructed by factorially combining a subset of the interpersonal relations with a variety of different situational contexts. For example, one stimulus was "supervisor and employee pooling their knowledge and skills to solve a difficult problem," and another was "business partners attempting to work out a compromise when their goals are strongly opposed." In this latter study, however, there was one dimension interpreted as "task oriented vs. nontask oriented" and another labelled "formal vs. informal;" these two dimensions merged in the study only involving the interpersonal relations.

DIMENSIONS BASED ON PERCEPTIONS OF VIDEOTAPED COMMUNICATION

In the previous studies subjects made judgments about hypothetical rather than real communication that they could observe. The results are important in their own right since expectations about communication may serve as a baseline for evaluating actual communication and interaction. However, there is no assurance that what is in the subjects' heads reflects the structure of real communication behavior.

More recently (Wish, 1978) we conducted a study in which the stimuli were videotaped segments of interpersonal communication. The study as a whole involved many different methods of data collection, and had a major aim of integrating the results from several approaches to arrive at better procedures for measuring and coding interpersonal communication.

The stimuli were 20 scenes of about one and a half minutes each, all of which involved communication and interaction between a pair of individuals (not the same pair in each scene). Seventeen scenes were excerpted from the twelve one-hour broadcasts of the TV series, "An American Family," while the other three were taken from a study by Bricker (1975) in which unacquainted subjects discussed social issues on which their opinions differed. The "American Family" series was intended to provide a candid view of the naturally occurring communication which members of one family (the Louds) had with each other and with their friends and associates. We attempted to select scenes representing a wide variety of interpersonal relations and situational contexts so that the domain would be roughly as broad as those from the studies of hypothetical communication.

VERY COOPERATIVE : 1 2 3 4 5 6 7 8 9 : VERY COMPETITIVE

VERY FORMAL : 1 2 3 4 5 6 7 8 9 : VERY INFORMAL

VERY UNEMOTIONAL : 1 2 3 4 5 6 7 8 9 : VERY EMOTIONAL

VERY RATIONAL : 1 2 3 4 5 6 7 8 9 : VERY IRRATIONAL

VERY GLOOMY : 1 2 3 4 5 6 7 8 9 : VERY CHEERFUL

FIGURE 1: Portion of Questionnaire Form for Rating Communication
and Interaction in a Scene

In one part of the study 51 subjects viewed one scene at a
time, and then rated the communication between the two individuals
on 24 bipolar scales. A portion of the questionnaire form for
rating the communication and interaction in a scene is illustrated
in Figure 1. Thus, for each scale there were ratings of the 20
scenes by each subject.

In order to analyze these data by MDS methods we converted the
original data tables to matrices of distances between scenes. By
using a profile distance formula (explained in Wish and Kaplan
1977), a separate matrix of inter-scene distances was derived for
each bipolar scale. This set of matrices was used as input to the
INDSCAL computer program. There were 24 matrices (each 20×20)
in all, one for each bipolar scale on which subjects rated the
scenes.

Five dimensions provided a good representation of these
"American Family" data. The interpretations of dimensions were
based on dimension weights for bipolar scales obtained directly
from the INDSCAL analysis; the higher the weight for a bipolar
scale on a dimension, the more relevant it is to that dimension's
interpretation. This is very similar to using factor loadings
for various tests for interpreting results from factor analysis.

Figure 2 shows the scales having the highest weights on each
dimension. As indicated by the interpretations, "Cooperative vs.
Competitive," "Intense vs. Superficial," "Task Oriented vs. Nontask
Oriented," "Dominance vs. Equality," and "Impersonal and Formal vs.
Personal and Informal," these dimensions almost exactly replicate
those based on the hypothetical communication episodes. For con-
venience, these dimensions will be referred to as Cooperativeness,
Intensity, Task Orientation, Dominance, and Formality, respectively.

DIM. 1: COOPERATIVE VS. COMPETITIVE

VERY COOPERATIVE VS. VERY COMPETITIVE
INDIVIDUALS HAD VERY SIMILAR VS. VERY DIFFERENT VIEWS OR GOALS
NO ATTEMPTS AT PERSUASION VS. FREQUENT ATTEMPTS AT PERSUASION
VERY FRIENDLY VS. VERY HOSTILE

DIM. 2: INTENSE VS. SUPERFICIAL

VERY INTENSE VS. VERY SUPERFICIAL
VERY EMOTIONAL VS. VERY UNEMOTIONAL
VERY GLOOMY VS. VERY CHEERFUL
INDIVIDUALS DEEPLY ENGROSSED VS. UNINTERESTED AND UNINVOLVED

DIM. 3: TASK VS. NONTASK ORIENTED

ENTIRELY TASK ORIENTED VS. NOT AT ALL TASK ORIENTED
VERY PRODUCTIVE VS. NOT AT ALL PRODUCTIVE
VERY CLEAR GOALS VS. NO CLEAR GOALS OF THE INTERACTION
VERY RATIONAL VS. VERY IRRATIONAL

DIM. 4: DOMINANCE VS. EQUALITY

ONE PERSON TALKED MUCH MORE THAN THE OTHER VS. NEITHER PERSON
                TALKED MORE THAN THE OTHER
ONE PERSON TOTALLY DOMINATED THE OTHER VS. NEITHER PERSON
                MORE DOMINANT THAN THE OTHER

DIM. 5: IMPERSONAL AND FORMAL VS. PERSONAL AND INFORMAL

VERY IMPERSONAL VS. VERY PERSONAL
VERY FORMAL VS. VERY INFORMAL
VERY RESERVED AND CAUTIOUS VS. VERY FRANK AND OPEN

FIGURE 2: Bipolar Scales with Highest Weights on Each Dimension

Having found essentially the same dimensions for real as for
hypothetical communication, there is now much stronger evidence
that these represent basic aspects of communication that should be
measured in future laboratory and field studies. Two important
questions that might be asked are "what is the relative importance
of the audio and visual channel for each dimension?" and "what are
the verbal and nonverbal behaviors that are associated with the
respective dimensions?". The first of these questions is addressed
at this point.

RELATIVE IMPORTANCE OF AUDIO AND VISUAL CHANNELS

In order to assess the relative importance of the audio and
visual channels, we presented the American Family scenes
to other subjects in different ways. One group listened to the
videotape without sound (video-only condition), while another heard
an audiotape of the conversations (audio-only condition). Ratings
were made on the same bipolar scales after presentation of each
scene. (Although it will not be mentioned in the ensuing discussion, there was another subject group who made their ratings after
reading the transcripts of the various scenes.)

The ratings on a single scale by subjects in one experimental
condition were reduced by computing mean ratings (averaged over
subjects) for each scene. There were, therefore, 24 sets of means
for each modality of stimulus presentation. For each bipolar scale
correlations were computed between mean ratings of scenes based on
the audio-visual subject group and those based on each of the other
experimental groups.

If the audio-visual ratings on a scale correlate more highly
with the audio-only than with the video-only ratings, then one
could conclude that the audio channel was weighted more highly
than the visual channel. In other words, the impressions of communication formed by subjects who could both see and hear what was
going on would be more predictable from ratings by subjects who
could only hear than by those who could only see the interactions.

Overall, the correlational analyses showed that subjects'
ratings in the audio-visual condition were determined much more
by what they heard than by what they saw. Moreover, for the scales
having high weights on the Cooperativeness, Task Orientation, and
Formality dimensions, the visual channel added almost nothing; that
is, nothing that could not be predicted by the audio-only ratings
by themselves.

The visual channel did, however, make a substantial contribution to the Intensity and Dominance dimensions. In fact, the

audio-visual ratings were correlated almost as highly with the video-only as with the audio-only ratings for all scales having high weights on the Intensity dimension. The same is true for the scale, "One person totally dominated the other vs. neither person more dominant than the other."

These results clearly demonstrate that the visual channel conveys more nonredundant information about some dimensions of communication than about others. It appears to be most important when emotional or affective considerations are of concern. Thus, a person listening to an audiotape of a conversation would be able to form an accurate appraisal of the cooperativeness, task orientation, and formality of the communication, but would get a distorted impression of its dominance and intensity levels. The same would be likely to apply if a person in one audio-teleconferencing location were trying to form an impression of communication between people at another location.

RESEARCH DIRECTIONS

These findings have potential importance for the assessment of situations for which visual telecommunications might be useful since they point out the dimensions on which the visual channel makes the greatest contributions. It is important to recognize, however, that the audio-only and video-only ratings are based on observer judgments when one channel is inaccessible; all of the original conversations were conducted face to face. Thus, although there may be a close correspondence between what we have reported here and the kinds of differences associated with communicating via alternative modalities, these results do not directly tell how important it is for communicators to be able to see each other during a conversation. Our current and future laboratory studies are aimed, however, at exploring the importance various communication channels have on the process and outcome of dyadic and group communication. One example is a recent study by a member of our group (Geller, 1977).

Geller's study had an interview format, with one subject in each dyad being randomly assigned the role of interviewer, and the other being the interviewee. Half of the interviews were conducted face to face; the others were over an audio intercom system between two rooms. Interviewees in one experimental group were led by the experimenter to believe that it was in their interest to come across as dominant individuals, while those in another group were given a motivational set to appear submissive. At the completion of the interview, the interviewer (who was unaware that the other person had been given motivational instructions to appear dominant or submissive) rated how dominant or submissive a person the interviewee was.

Interviewees were more effective in creating their desired impression in the face-to-face than in the intercom condition; that is, the interviewer's ratings of how dominant or submissive they were agreed more with the impression the interviewees wished to convey. Evidently, the visual channel transmits information that facilitates the impression management process. In this regard, Geller is analyzing videotapes of these interactions to discover the verbal and nonverbal cues associated with effective presentation as a dominant or submissive person.

Geller's results have something in common with those found in an earlier investigation in which she collaborated with R. M. Krauss and C. Olson. (R. M. Krauss and C. Olson are both at Columbia University; Krauss also serves as a consultant to Bell Laboratories.) This study also involved face-to-face and intercom interviews in which subjects attempted to manage the way they presented themselves. The interviewees were instructed to tell the truth on half the questions, and to lie when other questions were asked. The interviewers rated the truthfulness of each answer on a 7-point scale. The face-to-face mode was advantageous to the interviewees; that is, they were more effective at not getting caught when they lied. Looked at from the other perspective, however, interviewers were more effective at detecting deception in the audio intercom than in the face-to-face condition.

In these studies, as in most interaction, "visual access" means that a person can both see the other and be seen during the communication. It is, therefore, impossible to tell how much of the difference between the face-to-face and intercom condition is due to the ability to see the other person, and how much is attributable to being seen. On the one hand, being able to visually monitor the reactions of the other person may serve an important feedback function; that is, knowing how one is coming across may allow for more effective control of self presentation. On the other hand, it is advantageous to be seen if information in the visual channel enhances the impression you wish to convey. Accordingly, Valerie Geller, Peter Bricker, and Vivien Tartter are contemplating studies in which visual access is asymmetric - audio-visual transmission in one direction and audio-only in the other.

Our current research also involves the development of procedures for measuring communication at a more microscopic level. One approach (in collaboration with Professor Roy D'Andrade of the University of California, San Diego) is a speech act coding scheme for measuring certain linguistic aspects of a conversation. This work has shown strong correspondences between various speech act variables and the dimensions discussed in this report. We are also using a computerized system for tracking the moment-to-moment flow of communication. By moving a slide bar, or "joy stick", back and forth along a 9-point scale, an observer can record how cooperative, how dominant, how intense, etc. a communicator appears to be as the conversation goes along. The position of the "joy stick" is sampled periodically, analog-to-digital converted, and stored in the computer memory. The computer then prints out graphs showing the time trends for each dimension and for each person being evaluated. We will be carefully examining these plots to find out what is going on, verbally and nonverbally, when large changes occur in such a plot - for example, when a person changes from being perceived as very dominant to somewhat submissive. This will help to clarify the verbal and nonverbal correlates of each dimension.

In conclusion, the measurement tools we are developing are likely to play a vital role in our future studies. If one cannot measure what is going on during an interaction, it is extremely difficult to assess the role which the visual and other channels play in the interpersonal communication process. Although our current methodology has a heavy emphasis on microscopic analysis, we hope to develop simplified measurement procedures that can be efficiently applied in field studies as well as laboratory experiments. For example, we will use these techniques in our anticipated teleconferencing studies to determine whether there are differences in the kinds of verbal and nonverbal behaviors between people at the same location as opposed to those occurring across the teleconferencing link. Such analyses may provide insights into group processes (e.g., leadership, coalition formation, coordination) associated with different teleconferencing arrangements.

REFERENCES

Bricker, P. D. Modality and process: A look at overlooked looking behavior. Paper presented at Human Factors Society Meetings, Dallas, October 1975.

Carroll, J. D. and Chang, J.-J. Analysis of individual differences in multidimensional scaling via an N-way generalization of Eckart-Young decomposition. Psychometrika, 1970, 35, 283-319.

Geller, V. J. The role of visual access in impression management and impression formation, unpublished manuscript, Bell Laboratories, 1977.

Krauss, R. M., Geller, V. J., and Olson, C. Modalities and cues in the detection of deception. Paper presented at American Psychological Association Convention, Washington, D. C., September 1976.

Kruskal, J. B. Multidimensional scaling by optimizing goodness of fit to a nonmetric hypothesis. Psychometrika, 1964, 29, 1-27.

Kruskal, J. B. and Wish, M. Multidimensional Scaling. Sage Publications, Beverly Hills, 1978, in press.

Shepard, R. N. Analysis of proximities: Multidimensional scaling with an unknown distance function. Psychometrika, 1962, 27, 125-140, 219-246.

Wish, M. User and non-user conceptions of PICTUREPHONE Service. In Proceedings of the 19th Annual Convention of the Human Factors Society. 1975 (a).

Wish, M. Subjects' expectations about their own interpersonal communication: A multidimensional approach. Personality and Social Psychology Bulletin, 1975, 1, 501-504 (b).

Wish, M. Comparisons among multidimensional structures of interpersonal relations. Multivariate Behavioral Research, 1976, 11, 297-327.

Wish, M. Dimensions of dyadic communication. In S. Weitz (Ed.), Nonverbal Communication, 2nd. edition, New York: Oxford University Press, 1978.

Wish, M., Deutsch, M., and Kaplan, S. J. Perceived dimensions of interpersonal relations. Journal of Personality and Social Psychology, 1976, 33, 409-420.

Wish, M., and Kaplan, S. J. Toward an implicit theory of interpersonal communication. Sociometry, 1977, 40, 234-246.

COMPUTER ASSISTED COMMUNICATION IN A DIRECTORATE OF

THE CANADIAN FEDERAL GOVERNMENT - A PILOT STUDY

R. H. IRVING

UNIVERSITY OF WATERLOO

WATERLOO, ONTARIO, CANADA

INTRODUCTION

Computer Assisted Communications (CAC) has received considerable attention over the last few years. The main thrust of this attention has been studies of Computer Conferencing in a variety of environments, and studies of CAC in what Machlup[1] has termed the "knowledge industry". Some of those who have contributed to the literature are: The Institute for the Future[2], which has studied the use of its own FORUM and PLANET systems in some detail; Murray Turoff, who developed the first large scale conferencing system (EMMISARI) and who is currently developing further applications of CAC[3,4,5]; D. W. Conrath and J. H. Bair, who have studied applications of CAC to the "knowledge worker"[6,7]; and the Bell Canada Business Planning Group which has used and studied the NLS system (On-Line System) developed by the Stanford Research Institute[8]

A common denominator of many of these studies has been a case history approach, generally with an emphasis on computer conferencing. While this work has been of value in supplying evidence of how computer conferencing operates and what its possible applications may be, little progress has been made towards a conceptual understanding of such systems or towards studies of applications with other than relatively sophisticated users. Since we feel that these areas deserve attention the pilot study presented here is a slight departure from the mainstream since it focuses on messaging systems rather than conferencing systems and since the application described was one of day-to-day communication between novice computer users.

This study was carried out as a starting point for further

455

work aimed at developing a generalized model of mode choice. Data
were collected to provide insight into actual communication behaviour and to be of interest to system designers, since to date
there has been little guidance for the design of these systems.
Additionally a variety of attitudinal data were collected so that
correlations could be established between expressed attitudes and
behaviour as a first step in the development of a framework on
which to hang our model.

Since this research will be data-based we will first present
details of the application, the systems studied, the methodology
of the study and the results. These results will serve as a basis
for the first links in our framework.

The Host Organization

The study of CAC took place in the Non-Medical Use of Drugs
Directorate of the Department of Health and Welfare of the Canadian
Federal Government.

NMUD, as it is known familiarly, was formed in April, 1971 in
response to a growing concern in Canada over the non-medical use of
drugs and other substances. It is one of eight directorates within
the Health Protection Branch of the Department of National Health
and Welfare.

NMUD consists of an Ottawa headquarters and five regional
offices located in Halifax, Montreal, Toronto, Winnipeg and Vancouver. Each of the regional offices is responsible for liaison
with provincial agencies and for the monitoring of regional programs
within its own area. The headquarters office is responsible for
administration of nationwide programs and for coordination of the
activities of the regional offices.

The whole group is dedicated to social change as a method of
dealing with drug abuse and as a result has a strong humanistic
bias. While it is perhaps surprising that a group with such a
strong "people" orientation became one of the first Canadian organizations to use CAC for day to day communication, this is a reflection
of the innovative approach to problems that characterises the group.
A further reason for their choice of CAC was a feeling on the part
of upper management that this communication system would provide
improved communication between the regions as well as between the
regions and the headquarters office.

At the time of writing it is not known if communication was
actually improved. While the reaction of the users was mixed the

general consensus appears to have been positive. Apparently the
systems were used for a variety of purposes ranging from social
facilitation to the discussion of policy questions. In a future
study considerable attention will be devoted to collecting data in
these areas.

The Timing of the Study

     The study was begun in an interval between the discontinuation
of one system and the inauguration of another. The old system (MFT)
was not used after August 1975. The new system (MINT) was introduced on May 1, 1976. During the period from September 1, 1975 to
April 30, 1976 data on the past use of MFT were collected. In
addition an attitude questionnaire was administered to those who
had used MFT and to those who were slated to use MINT. At the same
time a statistical package was designed to collect data on the use
of MINT.

Description of MFT and MINT

     While both systems may be classified as CAC systems, they were
rather different in design and in purpose.

     MFT was essentially a conferencing system that also contained
a message capability and a retrieval feature. The conference
facility allowed several conferences to take place at once and had
the option of allowing analysis of the opinions of the participants
through use of the Delphi polling technique. The system was implemented on a privately-owned computer network and was accessed by
means of acoustically coupled terminals.

     MINT was designed mainly as a message discussion system and as
such lacked sophisticated conferencing features. The message feature
allowed any user to send a private message to any other user. MINT
notified the sender when the message was picked up and notified the
recipient that a message was waiting when he signed onto the system.
As soon as the message was received it was removed from the system's
memory. Provision was also made for a user to determine if anyone
else was online since the system designer felt that this might promote more simultaneous interaction.

     The discussion feature was fairly simple in that any number of
users could form a discussion group on the topic of their choice by
implementing one command (CREATE). There was no provision for a
chairman or for complex conference procedures, though one could
review old conference statements by means of a simple retrieval
system. Since the feature did not allow control of the interaction

by formal means the term "discussion" was used to distinguish this feature from the more regulated approach of a conference facility. The term also emphasised the informality of the approach.

Both MINT and MFT had similar message facilities, hence the description of the former is adequate to explain the latter. The more sophisticated questionnaire features of the latter were not directly relevant to the study reported here, and thus they will not be detailed. For example MFT had several sophisticated statistical routines for the development and analysis of questionnaires. These were not used to any great extent and thus were not of importance to the average user.

METHODOLOGY

Two different sets of data were collected. The first was on the actual use of the systems. The second was a questionnaire that invited responses to a set of statements related to communication in general and to CAC in particular.

The data on the use of MFT were collected by analysis of the bills for the computer time that was used during the period that MFT was extant. These bills indicated, for each account, the amount of CPU and connect time used, as well as the number of characters transmitted each month.

The data on the use of MINT were much more extensive than for MFT since a package was designed which collected the data which were felt to be of interest. These data were:

1) The number of messages between each user.
2) The number of characters in each message.
3) The number of CPU seconds used for each message.
4) The number of connect minutes used for each message.
5) The elapsed time between transmission and reception of each message.
6) The frequency of usage of each available command for each user.
7) The number of errors made on each command for each user.

These data permit not only calculation of message characteristics, but also analysis of the developing communication networks. In addition the data base provided the opportunity to investigate correlations between use of the system and expressed attitudes as indicated by the attitude questionnaire.

The attitude questionnaire consisted of 95 statements or questions to which the subjects responded via a five-point Likert scale. The questions related to the following seven areas:

1) Relations with Others.
2) Input/Output Requirements.
3) Implied Requirements for CAC.
4) Attitudes to Computers and Requirements for Computer Access.
5) Attitudes to Various Communication Modes.
6) Job Environment.
7) Requirements for Various Telephone Features.

The last category, seven, was included so that data comparable to those from another study could be collected. Categories one and six were included to gain insight into the organizational climate and to characterise the organization in terms of general descriptors.

The questionnaire was administered to 54 subjects who either had used MFT and/or who were slated to use MINT. Where possible the questionnaires were administered in person. In some cases, such as the more remote regions, the questionnaires were sent by mail and the participants were informed of the study either by mail or by telephone.

Of the 54 questionnaires distributed 40 were available for analysis in this paper. Of this number only 16 persons used MFT, completed the questionnaire, and used MINT. This relatively small number was partially caused by attrition and transfer.

RESULTS

Use of MFT

The basic results of the data collected are presented in Figure 1 and Table 1. Aggregate data across all users by month are presented. Data on individual use is not presented as it is too voluminous.

The most striking characteristic from the graph in Figure 1 is the strong relation between connect time, CPU time and the number of characters transmitted. It appears that any one of these variables is as good a measure of use as the others since the form of the three graphs is identical.

One of the problems with the use data for MFT lies in the fact that characters transmitted cannot be separated from characters received. Additionally one can not distinguish between private messages and statements in conferences. These problems substantially reduce the value of the data in describing behaviour on this system.

Despite the aforementioned problems, it is possible to estimate the number of characters that could have been transmitted on MFT by

using the average typing rate for MINT (this is calculated in the
following section). During the 3,090 hours of connect time (Table 1)
that were used it was possible to send 15,759,000 characters. One
also assumes that all connect time was spent in typing. This is a
rather liberal assumption since some time must have been used in
receiving and in reading messages and statements. It is justified
here since the calculations are only used for purposes of illustration.

Since there were roughly six times as many characters charged
to the average account as were likely to have been sent in the time
available, the difference can be, in part, attributed to the reception of messages from others and to the reading of statements in
conferences. This is a striking example of the multiplicitive
effect of a system with multiple addressing capability. With a CAC
system one can send the same message to a large number of addresses
with the same cost as sending one message.

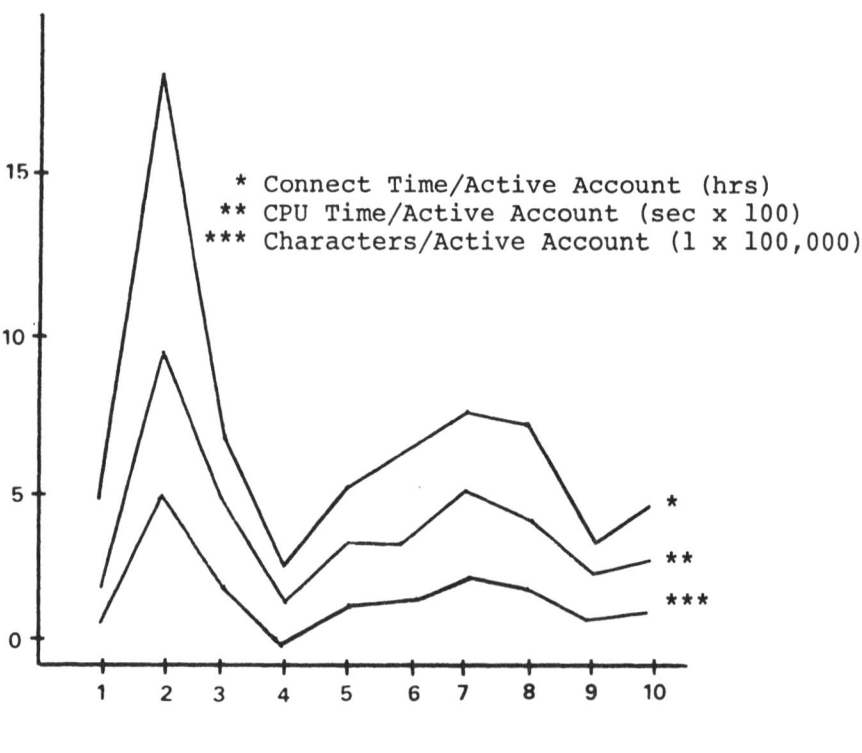

Figure 1. Use of MFT System by Month

TABLE 1   Use of MFT System

Month	Number of Accounts	Total Connect Time (Hours)	Total CPU Time (Seconds)	Total Number of Characters
1	28	136.6	6,038	3,051,907
2	40	715.9	38,549	20,886,861
3	40	278.0	19,946	9,079,282
4	50	144.8	9,726	3,648,644
5	37	197.2	13,425	6,418,184
6	51	316.4	17,936	9,917,144
7	56	442.4	30,150	13,601,964
8	44	309.4	19,229	9,181,583
9	65	231.0	16,918	7,424,620
10	67	318.6	20,340	10,216,734
Total Usage -		3,090.3	192,320	93,427,923

Average Number of Active Accounts - 48

The MFT data were also used to establish a profile of system use by rank ordering participants in terms of the amount of connect time they used. Since the graphs of both CPU time and number of characters transmitted were of the same form as the graph for connect time, either of these variables could have been used with similar results. The results of the rank ordering will be presented in conjunction with the attitude data.

When the MFT system was introduced to NMUD the senior management issued a directive requesting all users to spend roughly twenty minutes per day on the system. It was interesting to note that the average number of connect minutes per working day for the ten months was 19.3.

Use of MINT

Results drawn from the data collected on the use of the MINT system are presented in Figure 2 and Tables 2 & 3. From the graph in

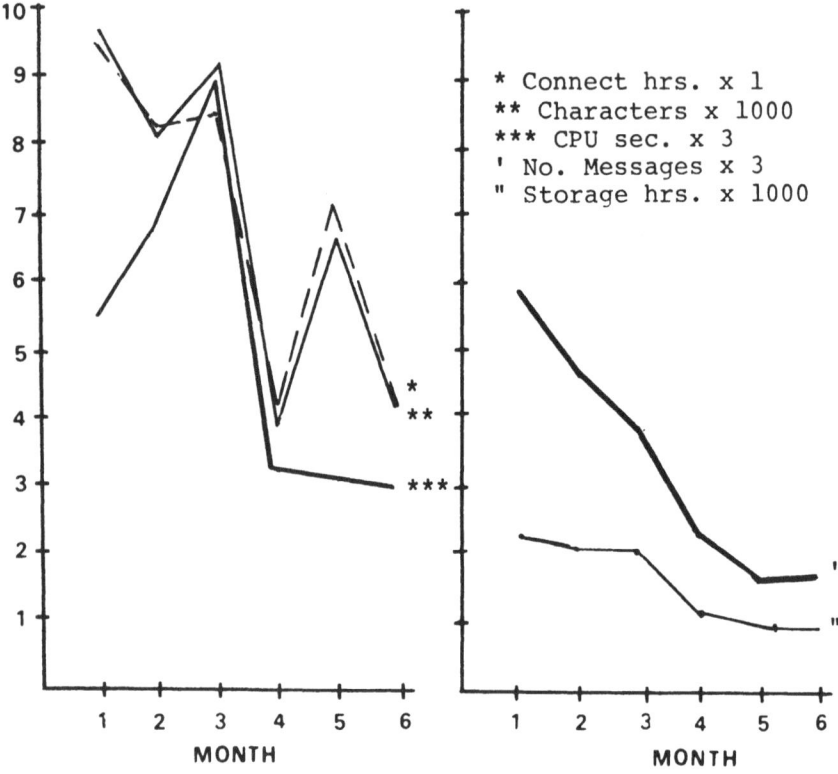

Figure 2. Use of MINT System by Month

Figure 2 we see that the correspondence between connect time and
the number of characters transmitted is as close as for the MFT
system. However, the relation between CPU and these two variables
is not nearly so close. We cannot as yet explain this difference
but it may be due to design differences such as a lack of conference facilities in MINT.

It is of interest to note that the graphs for both MINT and
MFT show the same tendency to decrease; though in the case of MINT
this decrease is more dramatic. This may be partially due to the
fact that the latter period of data collection took place over the
summer months when many people were on vacation.

The average number of accounts active on MINT was 25; this
indicates a much lower level of use than for MFT. One explanation
for this lower level is the fact that the use of MINT was voluntary

TABLE 2   Use of MINT

Month	Number of Active Accounts	Number of Messages	Total Connect Time (Hours)	Total CPU Time (Seconds)	Total Number of Characters	Total Storage Time (Hours)
1	38	664	37.4	642	181,664	51,500
2	34	481	28.5	704	144,450	43,000
3	25	285	21.7	674	117,578	29,500
4	20	166	8.8	198	37,836	10,300
5	19	125	13.9	175	66,248	9,500
6	16	109	7.1	144	34,304	7,200
		1,830	117.4	2,535	582,089	151,000

Average Number of Active Accounts/Month – 25

TABLE 3   Average Use of MINT

MINT

Average No. Accounts Active/Month	25
Average No. Messages/Account/Month	12
Average No. Characters/Message	318
Average CPU Seconds/Message	1.39
Average Connect Hours/Message	64 (3.84 minutes)
Average Storage Time (Hours)/Message	82.5 (3.43 days)

--there was no directive from management encouraging use of the
system.

Since the average message was about 320 characters long and
took an average of 3.84 minutes to compose, the average input rate
(at five characters/word) was 17 words per minute. These messages
are somewhat longer than those reported by other researchers. The
Institute for the Future[9] reported average message lengths of 380
characters for conference statements and 178 characters for private
messages. The greater length for private messages reported here may
be due to a variety of factors such as the nature of the work, the
style of use and the organizational environment in which the study
was conducted.

The average storage time is of interest since this, in conjunction with the average message length and frequency of use, determines the amount of disc space required to store the messages. In
our case the storage time of 3.43 days was coupled with relatively
low use; hence the demands on storage were not high.

Comparison of MINT and MFT

Table 4 presents the use data for both systems in a form that
allows comparisons between the two systems to be made easily. Based
on our previous analysis of the number of characters that could have
been sent on MFT we now know that the figure of 194,161 characters/
active account/month is artificially high. Using our previous estimate as a basis the new figure becomes 32,831 characters/active
account/month. While almost ten times higher than the corresponding

TABLE 4   Comparative Use of MFT and MINT

MFT*

Average No. Connect Hours/Month = 6.44 (19.3 minutes/working day)

Average No. CPU Seconds/Month = 401

Average No. Characters/Month = 194,641

MINT*

Average No. Connect Hours/Month = 0.77 (2.4 minutes/working day)

Average No. CPU Seconds/Month = 16.7

Average No. Characters/Month = 3,829

* All averages are per average number of active accounts.

figure for MINT, this figure is more in line with the differences in the average connect times for the two systems. The average connect time for MFT is eight times higher than the average for MINT.

The use of CPU time for MFT is roughly 25 times higher than for MINT. This is understandable when one considers the fact that heavier use was made of retrieval facilities, for example to retrieve statements in conferences.

Finally it is interesting to note that the average number of connect minutes per message (3.84) is greater than the average daily use of connect time for MINT. This means in effect that the time taken to compose and send a message was longer than the average time that each user spent on the system each day. This leads us to the conclusion that the use of the system must have been rather sporadic, a conclusion that is supported by the relatively long storage time per message.

The Attitude Questionnaire

The responses to the attitude questionnaire were correlated with the rank order use of connect time for MFT and for MINT. The only attitudes that had a significant correlation with use were the attitudes toward the computer as expressed by questions 31 and 81.

Q 31.   I feel that computers are rather tedious to work with.
Q 81.   Computers are creating a less personal work environment by
        reducing the amount of human interaction.

        The responses to questions 31 and 81 were combined to produce
an overall rank order of attitude. This 'generalised attitude to
computers' had a correlation of -.60 with the use of MFT and -.50
with the use of MINT. Interestingly enough the use of MFT had a
correlation of +.67 with the use of MINT. All correlations were
significant at the 0.99 level.

        Since no attitude measures were taken prior to the use of MFT
one cannot determine the degree to which the experience with MFT
contributed to the formation of the attitude to computers. Even
though the process of attitude formation is uncertain, the persistence of the attitude once it has been formed is rather remarkable.
While there was an eight month period when no CAC system was available, a significant correlation between attitude and use persisted
when the data for MINT were gathered. If we can assume that prior
experience played a major role in the formation of this attitude
then it seems that this example illustrated well the importance of
ensuring that novice users have a positive first experience with a
new communication system.

        Despite the fact that a high correlation was found between use
and attitudes the best predictor of use seems to be past use. In
fact, this seems to be a good bench-mark for comparison with any
model that pretends to be predictive in nature.

                        DISCUSSION AND CONCLUSIONS

        On the basis of the results obtained we can state several
tentative conclusions regarding the design and use of CAC systems.

        First, the differences in the use of a conferencing system
versus a messaging system are quite striking. The use of a conferencing system in terms of connect time, CPU time, number of
characters transmitted and storage time for these characters was
much higher than the use of a messaging system with a similar
number of subscribers. While in the pilot study the conferencing
system used roughly ten times the connect time of the message
system, some of this difference must be attributed to the management
request for participation. Such support may be desirable for the
introduction of any new system to an organization since it lends an
air of importance to the effort and some individuals who otherwise
may have not participated may be encouraged to do so.

Second, we conclude that the storage of messages for a system
similar to MINT will not be a problem. Given that there are on the
average 25 accounts active each month and that each account sends
an average of 12 messages of 318 characters during the 20 working
days we find that for each day roughly 4,800 characters will be
entered into the system. Since the average storage time for a
message is approximately 3.5 days the system must have capacity
sufficient to store at least 16,800 characters at one time. Even
if we assume that high variance could double or triple requirements
the capacity of a disc is still far in excess of likely demand for
storage. Thus for a message system like MINT, where messages are
removed from system memory once they have been received, storage
will not present any problems. This illustrates one further difference between message systems and conferencing systems. In a
conferencing system the conference statements are generally stored
until the end of the conference. If several conferences are active
on the system at the same time the demands on storage capacity could
increase dramatically.

Another type of capacity requirement of interest to designers
is the core requirements of the system program. In its APL version
MINT required 100,000 bytes of core. The Fortran version required
some 180,000 bytes. Perhaps these requirements can be reduced somewhat as the programming of these systems becomes more sophisticated.
No comparable data are available for the MFT system.

Third, we conclude that much of the interaction that took place
on MINT was asynchronous since the storage time for messages was
relatively long and since the average number of connect minutes per
user per day was small. Since CAC systems are touted because of
this ability to provide asynchronous interaction it is pleasing to
note that this facility was of apparent value to the users of MINT.

Fourth, on the basis of just the volume of use of the MINT
system we conclude that a system capable of supporting two or three
terminals in simultaneous operation would have been sufficient. In
practice, the number of terminals required is likely to be based on
the location of the users rather than on the volume of traffic.
Since many more than three terminals are likely to be required it is
probably wise planning to assume that as many as five or six may be
'up' at the same time.

Another interesting aspect of use is the relation between use
and individual attitudes to the technology employed. On the basis
of findings that both attitudes to the technology and prior use
were factors in the use of MINT, we conclude that the generalised
attitudes to the technology and experience in the use of that technology contribute to individual levels of use.

SOME IMPLICATIONS

Since messaging systems are used differently from conferencing systems it appears that they should be marketed as separate items. Furthermore the potential markets may be quite different. For example, since messaging systems make fairly light demands on computer time, and since they are relatively simple to program, many organizations which have a computer may decide to build their own system.

Conferencing systems tend to demand much more computer time and are much more complex to program. Because of these factors it is unlikely that many private organizations will choose to develop their own systems unless they have expertise in this area. Rather it appears that the development of conferencing systems will be in the hands of individuals or groups who wish to provide this service to the general public.

While the systems have many differences they are related by patterns of individual use and attitudes. In other words if an individual dislikes computers in general he is not likely to make much use of either system. If he has had unsatisfactory experiences on one system he is less likely to use another, and perhaps quite different system.

There are many approaches to solving this problem. For instance a carefully prepared training program would certainly help ensure that a user's first experience with CAC was a positive one. Undoubtedly the final approach must be to make the services supplied by CAC relevant to the needs of potential customers.

We would argue that the only way to find out what the needs of the potential user is by conducting controlled field experiments with people who are willing to use CAC to conduct their normal affairs. Such experiments are now in the planning stages. In a few months we hope to be able to report our findings.

REFERENCES

[1] Machlup, Fritz; 'Production and Distribution of Knowledge in the United States', Princeton University Press, 1962.

[2] The Institute for the Future; 'Group Communication Through Computers', 3 vol. report, 1975.

[3] Turoff, Murray; 'Conferencing via Networks', Computer Decisions, Jan. 1973.

[4] _____; 'Computerized Conferencing and Real Time Delphis: Unique Communication Forms', Proceedings of the Second International Conference on Computer Communication, Stockholm, Sweden, 1974.

[5] _____; 'The State of the Art: Computerized Conferencing' Views from ICCC 74, 1975.

[6] Conrath, D. W. and Bair, J. H.; 'The Computer as an Interpersonal Communication Device: A study of Augmentation Technology and Its Apparent Impact on Organizational Communication', Proceedings of the Second International Conference on Computer Communication, Stockholm, Sweden, 1974.

[7] Bair, J. H.; 'Experiences with an Augmented Human Intelligence System: Computer Mediated Communication', Proceedings of the S.I.D., vol. 14/2, second quarter, 1973.

[8] Day, Larry; 'Computer Conferencing: An Overview', Views from ICCC 74, 1975.

[9] Institute for the Future; 'Group Communication Through Computers', vol. 3, p. 121.

EXPLOITING THE TELE- IN TELECONFERENCING

Craig Fields

Defense Advanced Research Projects Agency

1400 Wilson Boulevard, Arlington, VA 22209

Until recently we have been involved in the development of a "hi-fi" teleconferencing system. The system is meant to assist geographically distributed groups in holding teleconferences that are as effective as if all of the members were holding a more conventional meeting in the same room. The approach has been to utilize relatively sophisticated communications and computer technology to provide adequate bandwidth among the conference participants, good automated control procedures, powerful tools for recording the conference, and automated speaker aids. Problems of choosing the bandwidth for teleconferencing systems are roughly of two types. First, there are cost-effectiveness issues associated with the use of voice, pictures, video images of the faces of participants, and so on. Second, there are effectiveness issues regarding the importance of small body movements, precise tone of voice, and other subtle communication clues. Control procedures that might be built into the system include provisions for conference roles, such as a chairman, an electronic agenda with automatic speaker recognition and scheduling, and so on. Automated recording procedures are a challenge requiring the construction of structured minutes that can be searched like data and that recognize, for example, off-the-cuff remarks. Speaker aids include an electronic blackboard and pointer, an electronic slide projector, provisions for joint document production and joint data retrieval, and other imitations of the technology of the common conference room.

Although work on this system is continuing at the Defense Advanced Research Projects Agency, I have been devoting my attention lately to a new effort because I have decided that if the construction of a "hi-fi" teleconferencing system was completely successful

and it was possible to perfectly mimic the effectiveness of
meetings held in a room, the success would be, in a sense, a failure. I believe meetings currently held in a room are not very
effective, either for communication or for group decision making.
A more ambitious goal would be a teleconferencing system that
supported teleconferences that were dramatically more effective
than conferences in a room, exploiting the geographic distribution
of the participants as a feature, rather than a bug in the system.

It is important to step back and understand the nature of the
problem of poor group effectiveness. I believe that society is increasingly faced with larger and larger problems. There are a
variety of ways of trying to get those problems solved. Clever
selection tests could be developed for identifying the best people
to solve large problems. Advanced training courses could be devised
for increasing the problem solving capabilities of individuals.
Computer aids for decision making could work symbiotically with
individuals. Artificial Intelligence computer programs could completely replace human problem solvers at some unknown time in the
future. And, finally, groups of people could be used to solve
problems that were too difficult for any of the individual members
of the group. None of these approaches seems to provide a realistic
solution to the problem.

The problem with groups is that effectiveness scales upward
very poorly with increasing membership. A group of ten people is
usually not ten times as effective, ten times as "smart", as an individual. In fact, some people think that any member of a problem
solving group might do better than the group as a whole! This is
not a universal truth, in that groups scale upward very effectively
for simple tasks, such as fact retrieval. For more complicated
problem solving and decision making, the upward scaling is considerably worse. To make matters even worse, problems facing groups
increasingly require interdisciplinary solutions. This means that
the groups are heterogeneous containing, for example, an agricultural scientist, a physicist, a meteorologist, a politician, and
so on. These people have different knowledge, use different languages (even if they think it is English) and communication style,
(most insidiously) have different methods of reasoning, and different personal goals and views of national goals. The problem is
worst when groups are required to make decisions under crisis circumstances (where, incidentally, the members of groups cannot typically be brought to one place quickly enough, thus requiring teleconferencing services). Under crisis circumstances individual
abilities are degraded by stress, the bandwidth of information assimilation required is usually dramatically increased and the members
of the group are frequently strangers needing some initial "station
identification" time to get calibrated.

What can we do to improve the upward scaling of group effectiveness, reduce the difficulties of interaction in heterogeneous
groups, particularly of strangers, circumvent the demand of high
information bandwidth and reduce the effect of stress? We can in-
vent "lo-fi" or active teleconferencing systems that add to, subtract
from, control and distort the communication stream among members of
a distributed group in order to assist the group decision making
and problem solving process.

I have tried to consider what functions could be performed by
an active teleconferencing system by introspecting about the kinds
of things that translators do at international meetings, that arbi-
trators do at bargaining and negotiation sessions, and so on. I
have encountered a number of ideas. Some are clearly practical,
while others are clearly impossible with current technology. Some
of the ideas might provide major improvements in meeting effectiveness while others might have less dramatic impact. I want to quickly
outline a selection of ideas, not because any specific idea is particularly noteworthy but in order to communicate a "feel" for what
I mean by a "lo-fi" active teleconferencing system.

1. When people compose text, they write in a personal style
    that may be as distinctive as a fingerprint, and they can
    probably read their own style of text better than anyone
    else's style. It should be possible to write a computer
    program that would transform text from style to style,
    using various rules of paraphrasing, synonyms, and so
    on.

2. In trying to reach a consensus, there are many ways of
    arguing a point, and the particular form of argument
    depends on the audience and the nature of the desired
    conclusion. A computer program might serve as an ad-
    visor, suggesting different forms of argumentation and
    forecasting potential impasses. The advisor might keep
    a record of the past argumentative behavior of the con-
    ference participants and adapt its advice during the
    proceedings.

3. Abbreviations and jargon are real aids to communication
    within a group, and significant impediments to communi-
    cation among groups. A computerized abbreviator/
    deabbreviator could serve to transform abbreviations
    among groups and construct abbreviations for frequently
    used words and phrases within a group.

4. An automated diplomat might be able to detect profanity,
    insults, and snide remarks in the communications stream
    and filter them out so as to improve the flow of the

conference. The computer would thus serve a function
similar to that of translators at international meetings.

5. When people deceive, they give off a number of clues to
their deception. Some of these clues can be measured
psycho-physiologically, while others can be revealed
through small body movements. Still others can be seen
in the style of the communications, such as the specific
words and phrases that are used. A computer program
would serve as a deception detector to help a conference
participant during sensitive bargaining and negotiation
sessions.

6. Some people speak quickly, while others speak slowly.
It is likely that fast speakers are fast listeners,
and slow speakers are slow listeners. A computerized
teleconferencing system could speed up and slow down
communication among participants to match idiosyncratic
communication rates.

7. The contents of most communications are thoroughly predictable. The important parts are those that are counterintuitive. It should be possible to construct a counterintuitive filter for each conference participant, revealing
to him only those parts of the conference that he will
find unusual, unpredictable, or interesting. Needless to
say, there would be some probability of missing an important statement, but that potential loss must be weighed
against tremendous reduction in cognitive load. A
special case of the counterintuitive filter would be a
counterintuitive briefing, containing only new or surprising facts.

8. There is a set of rules for effective brainstorming,
which includes eliminating criticism, removing a leader
from the proceedings, generating ideas by analogy, and
so on. An active teleconferencing system could suggest
or impose rules of brainstorming on groups that are
trying to suggest alternative solutions to a problem.

9. Each member of a decision-making group has his own
decision model. Each decision model could be communicated through the teleconferencing system to be shared
and discussed among the participants in some explicit
fashion. Sharing decision models could lead to more
rapid consensus by pinpointing dramatic differences in
utility or probability estimates.

10. People who are more beautiful or handsome and who are
    better groomed are more credible and believable. A
    video-based conferencing system could contain an image-
    processing program that might be called a beautifier,
    which would make it easier for a conference participant
    to make his point.

11. People have idiosyncratic approaches for storing informa-
    tion in their memories, and it is reasonable to believe
    that people can better remember information presented
    to them in a form that is congruent with their memory
    structures. A computer program in the teleconferencing
    system could transform the structures of messages to
    achieve that congruence.

12. The "mission impossible" effect is that communications
    are more believable if they are presented in a form
    that is apparently expensive to create. That is, a
    neat document is more expensive than a sloppy document,
    and so it is more believable. A movie is more expensive
    than a text story and is therefore more believable.
    Human intuitions about cost lag behind computer technology,
    for it is insignificantly more expensive to create a neat
    computerized document than a sloppy computerized document.
    Thus, a future teleconferencing system might contain a
    "neatener" to increase the credibility of communications.
    Computer programs to transform stories automatically from
    text to cartoon sequences are not out of the question.

13. A teleconferencing system should aid a speaker in
    analyzing the reaction of his audience. The raw facts
    entering the analysis could be explicit audience comments
    or, based on the concept of the instrumented audience,
    could use non-verbal behavior. As previously described,
    this could aid in choosing a sequence of presentations
    in a briefing, choosing a particular argument during a
    presentation, and so on.

We are trying to build a "lo-fi" active teleconferencing system
and to measure the improvement in group problem solving effectiveness that could result from its use. The Defense Advanced Research
Projects Agency pursues no in-house research, but works exclusively
through grants and contracts to companies, universities, and other
research institutions. I have received only one proposal in the
area of "lo-fi" teleconferencing (it was funded), and hence progress
on system development has been exceedingly slow. More high-quality
proposals would be welcome.

Many of the ideas in this paper were suggested by: A. Freedy,
R. Hayes, J. Licklider, W. Mann, J. Markowitz, and R. Nickerson

DISCUSSION OF PAPERS BY DORMOIS AND FIOUX, JULL, SHULMAN AND WISH

Discussant and moderator: Norman Gleiss

The discussion centered on the motivations for research in
the area of telecommunications applications in Business and Public
Administration; a dichotomy was seen between research based
on technical development and that based on communication needs.

Wish felt that more emphasis should be put upon communication
needs and that social science should play an active guiding role.
Questions of technology are meaningless unless basic concepts are
understood. Cowie thought that both aspects were important but
that there are many barriers to be broken down in an area of 'hybrid
research'. For example, the engineering and social science departments in universities often have little contact with each
other and this separation also applies in national forums which
plan research. Moss suggested that research directed toward
theoretical issues be distinguished from problem-focused studies
which are oriented toward policy making. Jull said that research
in Canada was entering a new phase of organizational objectives.
This sets new constraints on the existing pattern of communication,
e.g., relocation and decentralization had changed attitudes towards
face-to-face meetings. The new programme of research must address
these issues and at the same time produce a system which technically
fits the demands of users. Gleiss suggested that research was not
using the available information to define the market. There is
a wide gap between the apparent demand and actual use. Goldstein
relied that this 'gap' is an important marketing question;
laboratory experiments will not discover why people buy a system.
We must gain more understanding of how and why people communicate
in certain ways and study the channels of organizational communication. Wish agreed that much research is needed into the organizational context, needs, the value of time, decision makeup and
the processes of meetings, and practical studies of group communication. Shulman pointed out that the need to take in the organization's structure and goals is evident; without such a context we
would, in all probability, continue to be able to explain only
about five percent of the behaviour in question. The problem, however, is in defining what we mean by organizational context for the
individuals, for the groups and for the organizations. Standard
taxonomies of context need to be evolved. Some of the presenters
have initiated this task. However, there is a long way to go.
For all those who are getting excited about bringing in the organizational context, he noted that these organizational contexts have
yet to be worked out. You will not find a ready-made taxonomy.

Tyler illustrated a 'barrier of adaptive effort' which must be crossed by telecommunications users; research has not fully realized the significance of this. Gleiss then posed the question 'At whom should the research be aimed ' Goldstein answered that it is the managers who make the decision to buy telecommunications, so they are the level at which to aim. In addition, we should look at the needs at clerk/client level as these are frequently the actual users. Rockoff pointed out that in the health industry the decision to purchase is made by health system providers (for example, doctors) and managers while many of the benefits (such as reduced in-person referrals) accrue to patients and third party payers, not the providers. Others agreed that this is also the case in business applications. Brownstein felt that the aim of research should be to give managers new choices. Tyler suggested a two-level approach—the needs of the individual on the one hand and organizational interaction on the other. It was generally agreed that research should move towards a more field based, practical study of the market-place and of the communication patterns and needs of users.

DISCUSSION OF PAPERS BY CONRATH, IRVING AND JOHANSEN AND
STATEMENT BY FIELDS.

Discussant and moderator: Percy Tannenbaum

The session began with a discussion of the advantages and
disadvantages of teleconferencing. Tannenbaum felt that conference
calls can be very efficient and do not waste time, but only so long
as everyone has done his homework. More preparation is needed than
for a face-to-face meeting. However, he missed the 'gossip' and
personal interaction: to use teleconferencing we need to overcome
old habits and adopt a new set of expectations and practices. Researchers should use the systems themselves in order to appreciate
where they do people some good and where possible harm. Those
present were asked if they had actually used systems for business
and for personal interactions. (Result: about half for business
and a few only for personal use.) One participant said that the
public does not know of existing conference call facilities in the
United States. Ohlman felt that the telephone is seen as a dyadic
medium and that this is a barrier to conference use; another barrier is the lack of confidentiality and the extra time needed for
set up also discourages use. Lucas suggested that combined systems should be developed. Rockoff added that graphics facilities
should also be included. Conrath said that complex complementar-
ities were involved, both simultaneous and sequential. A face-to-
face meeting and mailing is necessary before a successful teleconference can take place. Tannenbaum said that there is a 'cost'
to the individual in using any teleconference system or systems--
an incentive is needed to overcome the initial extra effort. Wish
said that this incentive exists in such aspects as the increasing
costs of travel. Most groups using teleconferencing, in his
experience, were using it regularly. Elton felt that the most
important question is whether it is easier to do the business as
a whole by teleconference. It was generally felt that this could
best be achieved by combining different systems.

There followed a short discussion on CB radio. Ohlman asked
if any sociological studies were being made. Shulman felt that it
was being studied out of context of other communications systems.
Ohlman saw CB as developing under a new group psychology. In the
future, CB and mobile-telephone markets may merge, providing for
both mass and individual voice-communications needs within the
same system. Shulman suggested that it was not a new psychology;
CB users now communicate by other modes e.g., by face-to-face
meetings. Tannenbaum pointed out that no one would have predicted

the phenomenal growth of CB. Burns suggested that this growth is
due to the fact that CB fulfills a need and is very easy to operate.

    The discussion then moved to the contribution by Fields. This
was felt by Tannenbaum and others to be a very specialized case,
relating to a position of maximal need for such a system. Fields
said that, in this case, there is actual advantage in distance--
an active need to 'put something between' the actors involved. It
is also exceptional in that it is for crisis situations in a purely
military context. A further distinction is that in such cases cost is
an insignificant factor; only relative cost-effectiveness has to be
considered. The ensuing discussion questioned the effect of crisis
on use. One cannot simulate crises for purposes of design and
evaluation; the real test is whether or not the system is used.
Shulman said that repeated crises are no longer crises, but routine
matters. Organizational structures can be set up to make crises a
routine. Fields replied that the alogrithm he had described is
based on people's preferences and makes decisions based on these.
Irving remarked that the kind of information required to make a
decision can only be established if previous knowledge of the
crisis is available. This is unlikely as there are no patterns in
crises. Fields described the weighting system and explained that
the scheme used is adaptive and incorporates automatic filtration.
The criteria required for the linear model do not vary dramatically.
Irving argued that the issue is not the weights, but whether the
criteria underlying the weights are relevant. Fields countered
that the test was the high degree of agreement between automatic
and personal selection. The machine can do the same job as the
individuals given the same information--but is faster, cheaper, and
more reliable. Ohlman pointed out that the penalty for being wrong
in these circumstances was enormous. Tannenbaum felt that some risk
must be taken, however a decision is made. Fields pointed out that
the system would be used in cases of terrorism, in which there is
little time for decision-making. He added that organizational
acceptability of the system depends on: (1) the desire to keep and
control power; (2) the desire to solve the crisis. A balance must
be found between the optimal speed of decision-making and the
conventional hierarchial structure. Tannenbaum concluded that no
valid generalizations about teleconferencing systems can be drawn
from such a specialised system and that researchers should beware
of this fact.

Section Six

NEW SERVICES

The common element in this section's papers is the description of new telecommunications services. Each paper, however, approaches its subject from a very different perspective.

Björn Fjaestad and P.G. Holmlöv describe plans for the pilot test of a public switched broadband network in Sweden to be used for picture telephones, high speed facsimile transmission, video-conferencing and the interconnection of security television systems. They report on the first ("before") wave of an intended series of market research studies seeking to monitor needs, resources and attitudes toward telecommunications systems. Results are discussed. It seems that relatively few organizations are interested in participating in the trial; high speed facsimile is regarded as the most attractive of the services and electronics companies exhibit more interest in the trial than do other types of organization.

Bruno Drioli and his colleagues are concerned with a possible European system for teleconferencing via satellite. One part of their paper describes issues of technical design and the proposed system. The other part reports a study seeking to estimate the size and nature of the market for its services. A substitution perspective is adopted and, relatively speaking, very high estimates of demand are obtained. The study was conducted before much of the research described in the preceding section was published.

Viewdata is a new information service developed by the British Post Office for which market trials are scheduled for mid-1978. It makes use of the telephone, with a push-button facility, to call up information directly from a computer, and of the television set, with an adaptor, to display the retrieved information. The Post Office would act as the middleman between providers and consumers of information, whether in the office or in the home. The flexibility, capacity and promised economies of scale have aroused a great deal of interest in many countries. (See, for example, the comment about Viewdata in Thompson's paper in the next section.)

Samuel Fedida has led the design of the system. His paper,
atypically technical for this volume, explores various networking
designs which would remove residual needs for human involvement in
the operations of the system.

Murray Turoff and Roxanne Hiltz provide a comprehensive overview of computer conferencing and of possible services which may
come to be associated with it. They go on to discuss the problem
of assessing its impact and to raise a number of policy issues.
They conclude that such services "should be open to the widest
range of investigation and experimentation with the greatest possible
incentives to encourage individuals and organizations of all types
to be involved."

We start this section with a synopsis of the after-dinner
presentation by Edward Goldstein on the Bell System's experience
with visual communications systems, in particular with the use of
Picturephone® Service in the criminal justice system and with the
Picturephone® Meeting Service. The predecessors of these services
were not the instant successes originally envisioned; now, however,
"the future of visual communications services appears to be very
favorable." As Director of Product Marketing for AT&T,
Mr. Goldstein is uniquely placed to summarize the Bell System's
experience in this area and to comment on the future outlook.

BELL SYSTEM VISUAL COMMUNICATIONS SYSTEMS

E. Goldstein

American Telephone and Telegraph Company

Basking Ridge, New Jersey 07920

BACKGROUND

AT&T is presently engaged in a market exploration program to determine what future action the Bell System should take in regard to switched visual telecommunications services. This program consists of market research, market trials, technical developments and long range planning. These activities include a number of integrated projects involving both intracity and intercity visual services. We are investigating the visual communications needs of such markets as criminal justice, advertising, education and health care delivery. These projects have progressed to varying stages but the most advanced are the market trial for criminal justice applications and an intercity visual conferencing (PICTUREPHONE® Meeting Service) project which is studying the use of intercity visual conferencing. Picturephone Meeting Service involves video equipment substantially different from Picturephone Service. This paper summarizes our experiences to date with Picturephone® Service in the Criminal Justice System and Picturephone® Meeting Service. Included also are some comments about the potential outlook for visual communications for the next 5 to 10 years.

PICTUREPHONE® SERVICE

Introduction

Picturephone Service never developed into the overnight success originally envisioned. In the late 1960's it was estimated that by 1980 there would be one million Picturephone stations in service,

a one percent penetration of the 100,000,000 main telephones.
The service was first offered in Chicago, Pittsburgh and
Washington, D.C. By 1971 the market expectations were not
being realized and the future outlook was not good. Active
promotion was discontinued in 1972 and today there are
approximately 400 Picturephone stations installed in Chicago
primarily for intercom use. The evaluation of our experience
with Picturephone reflected several reasons for the lack of
success: the system design was inflexible, intercity service
was never provided, and customers perceived insufficient value
for the cost of the service as it was configured. These findings
resulted in our revamping our market approach.

In June, 1973 we began a market exploration program aimed
at answering these two questions: 1) What are the characteristics
of visual communications services that seem to meet market
needs? and 2) What are the size and economic characteristics
of the market?

The customer market trials have provided the greatest
insight to us in answering these two questions. One of the
first steps taken in implementing the market exploration trials
was to change our technical standards from 1MHz quality to
4MHz. This provided television quality transmission for both
face-to-face and graphics communications. Commercial television
hardware could now be added to provide flexible station configurations designed to meet customer needs. This flexibility was
critical to the implementation of our most extensive market
trial which was conducted in Phoenix, Arizona with the Maricopa
County Criminal Justice System.

Criminal Justice System Market Exploration Trial

The criminal justice system was studied for several reasons,
the first being its apparent high need for visual content in
day-to-day communication. In addition, the experience of
marketing Picturephone Service in Illinois revealed considerable
interest for visual communications among the courts and law
enforcement agencies. Lastly, the Law Enforcement Assistance
Administration was interested and willing to provide assistance
in placing and evaluating a visual communications system in an
actual criminal justice environment.

The market trial applications to be evaluated were selected
on the basis of the potential benefits and their transferability
to other jurisdictions. Special adjunct equipment was provided
with several of the Picturephone installations to meet unique
needs of the users including: large screen monitors, hard copy
machines, video tape recording equipment, tripod cameras and
various lenses. A total of seventeen Picturephone stations
were installed.

In general, the applications implemented are covered by
three broad categories: 1) inmate interviews, 2) administrative
functions and 3) court functions. The strategy was to implement
the less complex, non-adversarial applications first and then
the more complex. Some examples of the applications explored
include: interviews by public defenders and probation officers
with their clients in jail; remote access to police records by
police officers and other authorized personnel; intra and
interagency conferences; arraignments of defendants in custody;
motion hearings conducted by a 3-way video conference call
involving the judge, prosecuting and defense attorney; and
testimony in Superior Court and a Justice Court by witnesses
remote from the courtroom.

Toward the conclusion of our project, the criminal justice
Picturephone network was being used for approximately 1000
calls per month. At the end of our involvement, the criminal
justice community requested that eight of the more heavily used
stations be continued. They are being financed through a
combination of local and federal funds. Actual users of the
system commented that they benefitted by more timely intra and
interagency communications, saved travel and waiting time,
increased productivity and improved service to their clients.

PICTUREPHONE® MEETING SERVICE

The commercial name for the intercity visual conferencing
project is PICTUREPHONE® Meeting Service. In essence,
Picturephone Meeting Service is a switched audio-visual service
that permits a group of people in one location to conduct
meetings with a group of people in another location.

A four city network has been established for the market
trial. The service is now available in New York City, Chicago,
Washington, D.C. and San Francisco. The service is being
offered under an experimental Federal Communications Commission
(F.C.C.) tariff filing. Hourly rates for meetings range from
$150 to $390 depending on the mileage between the cities.

Picturephone Meeting Service calls are placed from public
conference centers or customer Picturephone Meeting Service
rooms specially designed for such meetings. Public Conference
Centers are conveniently located in each city and accommodate
up to six active conferees and a number of other participants
who can sit in the back of the conference facility. The centers
are designed for maximum comfort, privacy and efficiency. Each
conference center is equipped with several cameras which provide
for face-to-face communications and the ability to transmit
graphics. The face-to-face cameras are automatically switched
by voice activation. There are two large screen monitors in
each room which show the incoming and outgoing picture trans-

mission. The room design provides a range of graphic capabilities
that permit conferees to transmit easel or blackboard presentations, 35mm slides, transparencies, small objects, video
tapes and closeups of sketches, and handwritten or typed material.
Today's design represents the fifth generation of room configuration in our attempt to meet user needs.

To date we have experience from over 1600 visual conferences.
Approximately 55% of these conferences involved customers and
the remaining number represent Bell System Corporate usage.
Users mention the following benefits to them most frequently:
saves executives' time; speeds up the business decision process;
allows support personnel to attend (who would not have travelled);
enables urgent meetings to be held quickly; saves money associated
with travel expense.

Our market trial experience with Picturephone Meeting
Service has emphasized the importance of the service addressing
specific customer needs and indicates that the organizational
and behavioral aspects associated with the usage of this system
must be addressed.

FUTURE OUTLOOK

The future of visual communications services appears to be
very favorable. Customers are becoming increasingly aware of
the potential of such services and have begun to incorporate
such services in their future planning. Many businesses have
invested in closed circuit television systems (primarily in-
house systems to produce and show tapes for training and employee
information) which may naturally evolve to an integrated communications system and network.

There are several technical areas that need to be addressed
before visual services could be available on a widespread
basis. A major problem in the past has been the high cost of
long haul transmission because of bandwidth requirements.
Recent work indicates that digital bandwidth compression technology may be able to reduce the effective bandwidth requirement
by at least an order of magnitude which would result in lower
costs. A second area that needs further evaluation is the
potential of satellites.

With the availability of new and lower cost technology the
potential of visual communications services becoming more ubiq-
uitous in the next 5-10 years is enhanced. In that timeframe

visual communications services probably will continue to be primarily oriented toward business applications; however, in the more distant future we may see many forms of these services become increasingly important to residential users. Over the next 10 years, much of society will be exposed to various forms of these services and visual communications may become an accepted and expected means of communications as the telephone, mail and in-person visits are today.

D.C.

E.G.

THE SWEDISH MARKET FOR A PUBLIC SWITCHED MULTI-PURPOSE BROADBAND

NETWORK

Björn Fjaestad and P.G. Holmlöv

The Bank of Sweden Tercentenary Foundation and The
Economic Research Institute at the Stockholm School of
Economics. Stockholm, Sweden

BACKGROUND

In order to learn about, and prepare for, a possible nationwide, public, switched broadband system, the Swedish Telecommunications Administration (Televerket) plans to introduce a small-scale
test network in 1979, with approximately one hundred extensions in
Stockholm and in Norrköping, 200 km south of the capital. The
bandwidth will be 5.5 MHz, which will allow such technologies as
videotelephone, conference television, security television and
superfast facsimile.

Today there is no public, switched, broadband network in
Sweden. However, there exist many small, private systems, a few
of them switched. They are used mainly for security, personnel
control and internal education, and to some extent, for conferences.
The main types of organizations already using broadband technology
are transport companies, banks and manufacturing corporations.

In June, 1975, a working group at Televerket suggested that a
study should be made of the market for services offered via a
public network, to which most of the existing installations would
be connected. The group emphasized the need for a test project,
and the fact that the new network must be compatible both with
existing systems and with the regular telephone network.

During the spring of 1976 Televerket invited just under one
hundred organizations from Stockholm and Norrköping to demonstrations of the broadband technology. An attractive brochure was
distributed, describing the different services offered and the

costs involved. During the fall of 1976 and the following winter
Televerket's broadband project group conducted sales calls and
calculated investment costs. A decision to proceed with the
construction of the test network was reached in May, 1977.

CHARGES TO TELECOMMUNICATION USERS

The expected basic fee mentioned to the potential subscribers
for connection to the test network varies (in U.S. dollars) from
$1,500 (one-way security television and facsimile) to $3,500
(videotelephone and conference television). Basic fees for
installation of rented equipment also vary widely: a black and
white camera is $350, a videotelephone $1,400 - 2,500, a video exchange $2,500, and a superfast facsimile machine $3,500 - 7,000.
For connection and equipment there are also quarterly rental fees.
They are approximately 4 percent of the connection fee and 15
percent of the equipment fee. Finally, there are traffic charges.
One-way traffic is 25¢ per minute within Stockholm and $3.50 to
Norrköping. Two-way traffic costs twice as much. After the investigation was concluded Televerket cut all charges to 50 percent,
largely because of unpromising reactions.

By way of comparison, one may consider today's charges for
telephone, sending documents and using Confravision. A local telephone call costs 3¢ for an unlimited time; from Stockholm to
Norrkoping 13¢ per minute. The connection fee is $70 and there is
a quarterly fee of $9.

To mail a document inside Stockholm or to anywhere within the
Nordic countries costs 25¢ if it is a letter weighing less than 20
grams, 15¢ if it is printed matter. A letter is almost always
delivered the morning after the day it is mailed. Slow facsimile
is a service offered by the Post Office. It costs $2 per page
between Stockholm and Norrköping. If one has one's own slow facsimile equipment, a document is delivered in less than five minutes
anywhere in the world; the traffic fee for such a transfer from
Stockholm to Norrköping is 50¢. The price for a slow facsimile
machine (Rank Xerox Telecopier) is $2,100, or it can be rented for
$80 a month.

Confravision in Sweden is available today only between Stockholm and Malmö, approximately 400 km south of the capital. The
cost is $185 for the first hour and $45 for each subsequent half-
hour. The system is a commercial failure.

OUR INVESTIGATION

We became aware of Televerket's broadband project in April, 1976. It appeared to us as the first in a possible chain of developments that may come to have great influence on the communication patterns within and between organizations. In order to have baseline measures (i.e., to interview a number of potential users before they had very much knowledge of the broadband technology), we had to move very fast. The study, therefore, was conducted under severe time pressures and without thorough planning.

The most interesting focus for our investigation is not, we feel, the test network itself, but rather the attitudes of the organizations toward the possibilities of using and learning from the communications potential presented by a public, switched, test network. It is, to a large extent, these attitudes - based on continuous assessments of needs and resources - that will decide the composition of future institutional communication channels.

To achieve a compromise between depth and breadth in our study we decided (1) to interview a handpicked sample of organizations prior to their visits to Televerket's demonstrations and (2) to distribute a questionnaire to all organizations present at these demonstrations. These questionnaires were mailed back to us.

In our interviews we first asked about the organizations' current level of communications technology, if they had any problems in this respect and if they had actively sought alternate solutions. We then queried future communications needs and what role, if any, broadband technology might play. The decision process regarding participation in Televerket's test network was discussed, and the interviews ended with questions about expected positive and negative consequences of this new technology.

The mail survey questionnaire was highly structured. We asked about currently available communications facilities, about interest in different broadband services, and about the probabilities of participation in the test network and in a future national network.

Figure 1 shows how the present study and its planned continuation aim to monitor the long-range processes of planning, decision-making, implementation, learning, and change that lie ahead.

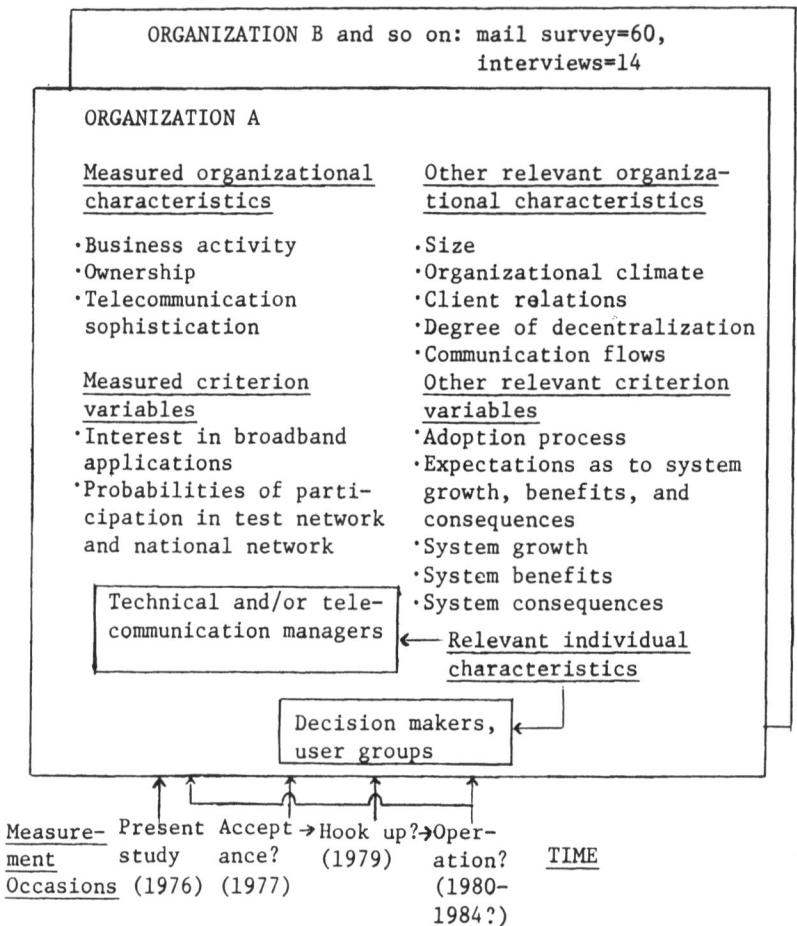

Figure 1:   RESEARCH PLAN

ORGANIZATIONS PARTICIPATING IN THE INVESTIGATIONS

In the mail survey the organizations studied were those invited
by Televerket to attend the demonstrations of the broadband applications. The sample can, therefore, be taken to reflect Televerket's
views as to what would characterize organizations willing to take
part in the test network. The one hundred organizations invited by
Televerket were reduced to a sample of 72, since many of them were
subsidiaries of parent companies also invited, or were inter-agency
project groups set up mainly to hold conferences and write papers.
Of these 72 organizations 60 - some 83 percent - replied to the
postal survey.

The collection of the organizations' written answers to the questionnaire took a fairly long time, almost two months, from the middle of May 1975 to early July. This amount of time - coupled with occasional postal and telephone reminders - was necessary in order to limit the number of non-respondents. The extensive duration had other advantages too. It was repeatedly pointed out to us that the reason for organizations delaying their answers was neither reluctance nor idleness but arose because several members of each organization had to get together, after returning from vacations, business travel, etc., in order to give true and representative statements.

Some characteristics of the organizations involved in the postal survey are set out in Figure 2. Because of lack of time we confined our description of the organizations sampled to the background variables shown. As can be seen, the modal organization is privately owned and the modal activity is the "media businesses." (The label "judicial and defense" does not refer to the armed forces themselves, but to agencies regulating or undertaking research for defense.) Only five of the 60 organizations are situated in Norrköping, the other 55 being located in Stockholm, thus making it unwise for us to single out the opinions of the Norrköping organizations.

Figure 2: THE BUSINESS ACTIVITY AND OWNERSHIP OF ORGANIZATIONS PARTICIPATING IN THE MAIL SURVEY

Business activity	Number	
Media businesses	12	
Education	5	
Banking and insurance	7	
Computer and telecommunications consulting	5	
Judicial and defense	7	
Security	3	
Electronics	5	
Other manufacturing	6	
Petroleum	4	
Transport	6	60

Ownership		
Private	30	
Cooperative	8	
Government corporations	8	
Government agencies	14	60

Of the 72 organizations surveyed 14 were selected and visited
before the distribution of the questionnaire - in all but one case
even before the demonstrations set up by Televerket. At these 14
organizations personal and informal interviews were held mostly with
managers responsible for telecommunications or data technology, but
sometimes with as many as 5 people from quite different departments.
The organizations in this subsample were selected to represent different kinds of activity and ownership as well as deeper experience
in their particular fields of action. The organizations thus chosen
are three news organizations, three companies handling bank and
insurance services, one agency with the prime purpose of buying
technical and other equipment for universities, one regulatory
agency, one electronics corporation, two other manufacturers (one
of highly complex products but with a fairly stable product line,
the other with a very broad product mix and production spread all
over Sweden) and, lastly, three quite different transport companies
mainly handling passenger transport.

The persons chosen for interview were those who had reported
an intention to attend the Televerket demonstration. In a few cases
others had to be interviewed instead; in many instances others were
present as well. At all interviews, both investigators were present,
taking turns in posing questions and taking notes, respectively.
No use was made of a tape recorder since we feared that - consciously
or otherwise - this might have inhibited the participants' free flow
of thoughts and words.

Some time after the interviews, typewritten and edited records
were sent to one person at each organization. This was done chiefly
in order to make sure nothing had been lost or misunderstood, but
also as a first means of feeding back results to participants and
to anchor the aims of the survey work with those people. This approach seemed worthwhile: many mistakes were corrected and only in
one instance, where his blasphemous language had been edited, did
an interviewee object.

Present Level of Telecommunications Sophistication

To obtain an idea of the telecommunication resources of the 60
surveyed organizations without forcing them to go into detail, we
presented a list of 72 different means of telecommunication and
asked them to check the ones in current use. Those means used by
more than half the organizations are listed in Figure 3.

The most advanced operations in this respect belong to the
media businesses. Banks and transport companies are not far behind.
Not counting quality, but merely number of devices, manufacturing
corporations take the lead. Of the 10 different industries,

Figure 3:   NUMBER OF USERS OF VARIOUS MEANS OF
TELECOMMUNICATIONS

Means of telecommunication	Number of users (out of 60)
Regular telephone	60
Intercom	50
Data transmission (between offices, branches)	49
Pocket dictaphone	49
Rented telegraph or telephone lines	47
Stationary dictaphone	46
Telex	46
Loudspeaking telephone	44
Internal television for education and training	42
Mobile radio	35
On-the-wall paging system	34
Portable paging system	33
Mobile telephone	32
Internal television for security and control	31

judicial and defense agencies (not the armed forces themselves) and, surprisingly, education are least advanced in regard to quality as well as quantity. We calculated an index based on the number of means in operation. The index is not shown here, but is used later in the paper as a background variable.

INTEREST IN BROADBAND SERVICES

A client-oriented organization or any other open system has different kinds of information needs. Questions concerning subjectively felt needs may lead to answers different from an outsider's "objective" assessment of system needs. There would probably also be incongruent results if one were to compare answers to questions concerning detailed and specific needs with those concerning what might be called overall needs.

We interpret the interest in broadband applications stated by each organization to be that organization's felt need for such applications (or, rather, the need of the departments to which our informant belonged). The stated probability of taking part in the future test network is different, and we understand this to represent what might be named the organization's (or one of its parts) compromised need: the felt need made visible as an actual demand (although, of course, what is initially measured here is only an indication of a future demand). This demand might or might not be satisfied by the network; it is certainly likely to change over time as also are objective needs and felt needs.

This section and the next are concerned only with mail survey results. In a later section, these results are compared with some scattered evidence from the personal interviews. (This might seem inconsistent, since the interviews were actually carried out before the postal inquiry. We regard the results from the latter, however, as more thorough and important, although they certainly have to be interpreted by using crucial clues from the interview data.) In this way we will discuss probable reasons for stated acceptance or rejection.

How interested are the organizations in the different kinds of services? This may be ascertained from Figure 4 which shows the mean number of organizations indicating different degrees of interest in each application, across four different combinations of communication partners. Figure 4 is a compact way of tabulating responses to the third question in the questionnaire, where respondents had to specify their organization's interest in each of the four broadband services described (videotelephone, superfast facsimile, conference television and security television) for communication in each of the following four situations: inside their own organization and in one city, inside the organization but between two or more cities, with other organizations in the same city, and with other organizations located in other cities. Four possible answers were offered, but because there were few answers in the categories "Very great interest" and "Fairly great interest" these two were collapsed into "Great interest".

Figure 4:   THE MEAN NUMBER OF ORGANIZATIONS
SHOWING DIFFERENT DEGREES OF INTEREST IN THE FOUR
BROADBAND APPLICATIONS, ACROSS ALL FOUR
PARTNER-PLACE COMBINATIONS*.

Degree of interest	Video-phone	Superfast facsimile	Conference television	Security television
Great interest	12	16.8	5.5	15
Little interest	13.5	15.8	13.8	22
No interest	33.2	27	39.2	23

* Sums not totalling 60 are due to internal nonresponse.

There is a clear difference in the participating organizations'
interest in, on the one hand, fast facsimile, security television
and videotelephone and, on the other hand, audiovisual teleconferencing. The dominating answer, however, is "no interest". The
reasons for this attitude can be inferred from the interview data
and will be discussed later in the paper. Those few - mostly private
manufacturers and legal authorities - indicating a keener interest
in videotelephone seem to be aware of the advantages of connecting
the videotelephone to microfilm files. Manufacturing companies and
legal authorities are also strong supporters of superfast facsimile,
though here they are surpassed in their interest by data and telecommunications consulting firms. The lack of interest in audio-
visual teleconferencing is almost unanimous. Security television,
finally, seems to be of interest mainly to those in the electronics
or security industries.

Figure 5 maps the attitudes of the organizations in the reverse
direction: the mean number of organizations marking their interest
in each combination of interaction partner and area, across the four
different services.

The main conclusion to be drawn from Figure 5 is that organizations have a relatively greater interest in broadband communications
with other cities.

Figure 6 is an attempt to summarize the results. Business
activities and ownership attributes are ranked according to the
median interest indicated in the four broadband applications and the
four partner-place combinations. Electronics companies are the most
interested organizations.

Figure 5: THE MEAN NUMBER OF ORGANIZATIONS SHOWING
DIFFERENT DEGREES OF INTEREST IN THE FOUR PARTNER-
PLACE COMBINATIONS, ACROSS ALL FOUR BROADBAND
APPLICATIONS

Degree of interest	Inside own org., in-side one city	Inside own org., be-tween two cities	With other org., in-side one city	With other org., be-tween two cities
Great interest	8	16.7	7.7	13.3
Little interest	13	17	12.3	15
No interest	39	25.7	38	30

Figure 6:   ORGANIZATIONAL TYPES RANKED ACCORDING
            TO THE INTEREST SHOWN

	Business activity	Degree of interest
1	Electronics	Fairly great interest
2	Other manufacturing	
3	Judicial and defense	Little interest
4	Data and teleconsulting	
5	Banking and insurance	
6	Security	
7	Media	Little or no interest
8	Petroleum	
9	Transport	No interest
10	Education	

Ownership

1	Private	Little interest
2	Government authorities	Little or no interest
3	Government corporations	
4	Cooperative	No interest

The index measuring the organizations' possession of technolog-ical equipment for communications purposes was mentioned above. Figure 7 demonstrates the relations between, on the one hand, this index and some of the devices and, on the other hand, indicates interest in the various broadband applications. A priori, it seems plausible that there would be strong relationships between the pos-session of communication resources and the interest in new communica-tions technology, since present technological sophistication indi-cates at least one of the following organizational characteristics: technological knowledge, innovativeness, high risk toleration, will-ingness to allocate money towards acquisition of communication devices of any kind. Furthermore, we expect relationships between present use of a certain resource and interest in a certain applica-tion when these two bear a functional likeness or could be used together synergetically. When we have judged this to be the case, we have underlined the corresponding cell in Figure 7; so the lines shown are there for theoretical reasons.

For the sake of clarity, only correlations (Pearson product-moment) larger than .20 are tabulated. Also, in order to make any patterns more visible, the presentation is restricted to median cor-relations, across all four partner-place combinations. This may sound unorthodox, but the procedure makes sense since the different correlations not shown are quite similar to their medians.

Figure 7: CORRELATIONS BETWEEN PRESENT (1976)
TELECOMMUNICATIONS CAPABILITY AND INTEREST IN
FOUR BROADBAND APPLICATIONS*

Means of telecommunication	Video-tele-phone	Superfast telefac-simile	Confer-ence te-levision	Security television
Quantity index	26	35	32	45
Rented telegraph or telephone lines				20
Direct access to extension tele-phone	22	33	25	35
Loud-speaking telephones		20	—	
Conference tele-phoning inside the organization			—	
Conference tele-phoning with other organizations		25	27	
Slow facsimile inside the organization	23	35	36	24
Slow facsimile with other organizations		—		
Telex	20	36	20	32
Data trans-mission		23	26	21
Internal tele-vision for sec-urity & control	36	26		38
Microfilm equipment	23		26	

* Median correlations, multiplied by 100; only
correlations of at least .20 are shown. Underlined cells
indicate relation (or lack of such) of theoretical relevance.

The correlations are all positive and many are fairly strong.
The conclusion to be drawn is, as was expected, that organizations
already better equipped show more interest in new means of communication. Four currently used devices seem to be relatively useful
predictors: slow facsimile inside own organization, direct access

to extension telephone exchange, telex, and internal television for
security and control. One possible reason for this state of affairs
may be that, at present, only the **wealthiest and/or most knowledgeable**
organizations can use these means of communication. (This suggestion
is supported by the fact that very few organizations have implemented
direct access to extension telephone exchanges, through which it is
possible to communicate with employees without contacting their organization's exchange, and therefore this device is not tabulated
in Figure 3.)

On the whole, our theoretical guesses are supported by the
data, though not to an impressive extent, and there seem to be some
even stronger relationships than those predicted. This is probably
due not to our concepts, but to our way of measuring them. The
questionnaire indicates only the use or non-use of various communications means, not whether the use has shown the communications technique in question to be useful and profitable. Thus the possession
of a certain communications facility may have resulted in negative
feelings which could be expected to lead to a lack of interest in
that facility's broadband counterpart. Certainly, however, this
does not explain why the users of slow facsimile for communication
with other organizations show no interest in obtaining the superfast
broadband version, since facsimile equipment could be used both
outside and inside one's own organization, but would not be used in
external communication if this were judged to be unsatisfactory.

PROBABILITY OF PARTICIPATION IN THE TEST NETWORK
AND THE NATIONAL NETWORK

While organizational interest in broadband applications provides an indication of possible benefits, questions about probability
of participation force the questionnaire respondents to consider
these benefits in relation to the financial resources they would
have to forego in order to achieve them. In Televerket's brochure
detailed cost information was given and, thus, each organization
responding had an opportunity to carry out rough cost/benefit calculations.

Each organization was asked to state: the probability that it
would participate in the test network at all, the probabilities of
using each of the four proposed applications and, finally, the probability of participation in a possible future, nation-wide broadband system. The answers were requested on an eleven-point scale
from 0 to 100 percent in 10-percent intervals. A summary of the
results is given in Figure 8.

Almost half the surveyed organizations - chosen by Televerket
on the premise that they would be interested in broadband communication - stated their probability of connecting to the test network

Figure 8: NUMBER OF ORGANIZATIONS (OUT OF 60)
STATING DIFFERENT PROBABILITIES OF PARTICIPATION IN
BROADBAND NETWORKS AND OF USING CERTAIN APPLICATIONS

Probability	Test net-work	Video tele-phone	Super-fast fac-simile	Conf-erence tele-vision	Security tele-vision	Possible nation-wide network
60-100 %	10	9	10	4	4	20
10-50 %	24	22	23	17	22	34
0 %	26	29	27	39	34	5
Median percentage	10%	10%	10%	0%	0%	40%

would be zero.   Of the 10 organizations leaning towards participation, i.e., giving a higher probability than 50 percent, three are
from the media businesses and three from the electronics industry.
Fairly high medians can also be reported for security companies
(40 percent) and judicial and defense agencies (30 percent). Private enterprises, which probably have faster and easier access to
money for risky investments, do not show higher probabilities than
other forms of organization.

For the four different applications, higher probabilities of
use were given for superfast facsimile and videotelephone than for
conference and security television.

Since the test network cannot offer all the advantages of a
full-scale national network we included a question about the latter.
We asked organizations to indicate the probability of their participation in a network reaching all the larger cities and with a total
of about 10,000 subscribers (for a reference point, Sweden has 8
million inhabitants). One-third give probabilities of more than
50 percent. The industries with high probabilities are (in order
of magnitude) electronics, security, manufacturing, banks, data and
telecommunications consulting, and judicial and defense agencies.
Relatively low probabilities are reported by cooperatives.

An indication of the degree to which present telecommunications
sophistication predicts the stated probability of connection to the
test network is given by the correlations between these two groups
of variables (Figure 9). Correlations (or lack thereof) of particular interest are underlined.

Figure 9: CORRELATIONS BETWEEN PRESENT (1976)
TELECOMMUNICATIONS CAPABILITY AND STATED PROBABILITY
OF PARTICIPATION IN THE TEST NETWORK AND
THE NATIONAL NETWORK*.

Probability of participation in

Means of tele-communication	Test net-work itself	Video-tele-phone	Super-fast fac-simile	Conf-erence tele-vision	Secur-ity tele-vision	Nat-ional net-work
Quantity index			28		23	30
Rented telegraph or telephone lines						
Direct access to extension telephone exchange			26	23	22	
Loudspeaking telephones			22	—		22
Conference telephoning inside the organization			26	—		
Conference telephoning with other organizations			26	—		
Slow facsimile inside the organization	28		<u>28</u>	24		23
Slow facsimile with other organizations			<u>—</u>			
Telex			<u>29</u>	20	23	32
Data transmission	21			23		
Internal television for security and control	34	25	26	27	<u>30</u>	40
Microfilm equipment		—				

* Only correlations of at least .20 shown, multiplied by 100. Underlined cells indicate relations (or lack thereof) of theoretical relevance.

The single best overall predictor is present use of security television. Other fairly good predictors are telex and use of slow facsimile within the organization. Of the six criterion variables in Figure 9 superfast facsimile has by far the largest number of correlations with the predictors. On the other hand it is very dif-ficult to predict connection to the test network or use of video-telephone from present telecommunications resources. Only a few of the expected – underlined – correlations are present. Present use of microfilm does not predict an interest in videotelephone, nor does present use of loudspeaking telephones or telephone conferencing predict probability of installing conference television.

DISCUSSION

So far we have shown that few organizations are really interested in being connected to the broadband test network, even fewer con-sider this hook-up probable. Figure 10 is one way of summarizing this and at the same time visualizing some of the differences be-tween felt needs (indicated interest) and stated demand (probabi-lities). It shows instances of potential but unsatisfied interest. One crucial question when planning communication systems must be the question raised by such an interest: What forces some prospec-tive users into rejection?

Strikingly, one-sixth of the organizations are interested in using some broadband technique and yet feel participation is en-tirely out of the question. From the discussions we had during the interviews, two reasons stood out. (1) The dimensions of the test network are far too small encompassing only two cities, thus making it too costly to operate both with it and with present communication channels. It would have been better if at least the three largest cities in Sweden had been wired. Viable alternatives are available for communication all over Sweden. This is especially true for superfast telefacsimile, the success of which is said to depend on the system's reach: its present competitors are telex and (for

Figure 10: INSTANCES OF SOME INTEREST IN, BUT NO
PROBABILITY OF USING, EACH BROADBAND APPLICATION

	Video-telephone	Superfast telefac-simile	Confer-ence tele-vision	Security television
Number with at least "some interest" but "probability: 0"	11	14	14	12

pictures and tables) slow telefacsimile. (2) The proposed costs
seem to be far too high, at least when it comes to the initial costs
in relation to liquidity or chances to raise new money. Maybe a
better pricing strategy would have been one of penetration, i.e.,
low initial costs and high unit costs. Televerket, however, has
traditionally advocated a policy of high entrance fees, which con-
flicted with user organizations' desires to try something new without risking too much money. Televerket's pricing is not particularly perspicuous. The two prime reasons are sometimes supplemented
by others. (3) Not only is the size of the network important, the
participating actors are as essential. Almost all organizations
have frequent contacts with (a) business partners and the like, and
(b) their own regional branches. The latter's participation must
be guaranteed; if only one regional branch were not connected, the
whole system would be unprofitable since it would have to be used
jointly with other communication systems. (4) For a few observers,
already familiar with the techniques considered, the picture quality
of Televerket's transmission is felt to be inadequate; most of the
equipment has never been put to a field test. Especially vulnerable
to bad quality appear to be videotelephone and facsimile. That
quality is inadequate is emphatically denied by Televerket. (5) Some
old systems are thought to be too good to be made obsolete. Most
organizations have implemented their own security systems, mostly
for operation only at one place, which would make connection to a
broadband network functionally unnecessary (but maybe profitable).
(6) There is also the attitude that existing means of communication
offer synergy. Conference travel is often combined with other
duties, helping to create and maintain valuable contacts. New communication systems could thus destroy existing personal relations
at and between different work sites. (7) Two or three informants
complain about the way in which Televerket has previously furnished
them with spare parts or maintenance services. However, these persons pointed to the fact that Televerket is most likely to coordinate
and intensify its efforts when it comes to creating a new network.

The advantages of a broadband network are that new services
would be offered, old services performed faster and costs lowered
in the long run thanks to the sharing of lines. What, then, are
the opinions that tend to accompany a wish to join the test network?
(1) The informant and his organization show some sense of adventure:
they would want to try it and see, and they estimate that risks are
low since a government utility is proposing the project. (2) The
organization wants to maintain its good relations with suppliers
and agencies concerned with telecommunications (and Televerket is
at the same time an agency and a supplier). (3) The organization
sees a potential new market which it would be advantageous to skim
or penetrate before other competitors enter it. This description
corresponds to attitudes shown by electronics corporations, data
and telecommunication consulting firms (when it comes to fast facsimile), and security companies (only in regard to security tele-

vision). By keeping pace with the developing technology, these organizations can intensify the sales of their goods and services and
even find new market segments. (4) The informant believes that one
need felt by him could be matched by at least one broadband application, at any rate according to Televerket's description of those
services. This felt need seems to be related chiefly to the management of contingencies: to speed up the delivery of spare parts or
the handling of complaints, to correct errors quickly, to have excellent communications system for crises, and so on. Some informants
describe the system as matching their needs for regular meetings and
education. (5) Another kind of need fulfillment is related to an
organization's present possession of resources. Some of these could
be combined with broadband applications and synergy produced. For
instance, videotelephones could be connected to a microfilm file.
Other suggestions are shown as underlined cells in Figures 7 and 9.

On the whole electronics companies seem to be the ones most
interested in the broadband services, although not even they feel
complete dedication. The reasons for their greater interest could
be of two different kinds. On the one hand, electronics firms are
technological gatekeepers and, as such, relatively well-informed
about the potential of the technology and the rate at which prospective users will find or develop new needs to be satisfied by the
applications. Seen in this light, the electronics firms' interest
may be well founded, while other organizations are uninformed lag-
gards. On the other hand, their knowledge stems from the fact that
they have vested interests in this market. Certainly, electronics
companies need not worry about the profitability with which the test
network equipment could be used inside their own organizations. The
primary purpose of their participation is to show the hardware and
software as a reference system for those of their customers who want
to try it and see its versatility.

It is surprising that rather little overall interest is indicated by transport companies with their scattered operations, by
educational bodies with huge flows of information, and by media
businesses with an even larger and much faster transmission of messages. The reasons appear to be that the transport and media businesses have already developed means of communication which they do
not regard as obsolete relative to the broadband technology (maybe
because the older means are less complex and the newer technology
has not yet been fully tried); and that universities and schools
lack money.

Two other ingredients in some organizations coming to appreciate benefits in the broadband system are time and information.
This is a conclusion to be drawn when we compare the attitudes
verbalized in our initial informal interviews with the written responses to the questionnaires which were sent back after the round
of demonstrations. During this time the organizations grew less

skeptical. We feel that our informants, in attending Televerket's
demonstrations, in seeing the physical equipment to be used, and in
discussing advantages with individuals from other organizations and
from inside their own organization, became aware of the network's
potential uses and gratifications.

By whom, then, has Televerket's information been received? We
do not know this for sure, but it appears that Televerket has not
tried to mobilize all prospective influencers inside each organization. Televerket initially invited the people they saw as probable
decision-makers, managing directors and the like. However, in the
main technical people were sent to the demonstrations, certainly
not individuals who would make the formal decision, rather technological advisers who would influence it. The end users and people
from buying departments do not seem to have been informed, at all
events certainly not by Televerket. Furthermore, the goals for
this campaign may seem unclear or dysfunctional. Televerket claims
their one aim is to set up a finite number of connections in 1979 -
nothing is said about which segment would be most profitable to have
as customers, what organizations would influence others to buy the
system, how the benefits of the operations are to trickle down to
non-users, how Televerket will create and, more importantly, maintain favourable attitudes and increase their beneficial effect.
What is important to Televerket is to make customers see their own
communications needs.

The application most likely to be accepted quickly appears to
be superfast facsimile. To be sure, it has its disadvantages:
available substitutes, uncertain quality, restricted reach inside
the test network, and high initial costs. But several factors point
to its possible success. Quite a number of organizations are interested in the fast facsimile. This interest is relatively easily
predicted from their present use of communications facilities:
those already communicating in writing and on paper want to increase
their transmission speed. Furthermore, the unit costs of superfast
facsimile are lower than those of mail or slower facsimile, a high
transmission volume thus neutralizing high investment costs.

Superfast telefacsimile is actually an old service, thus having
the advantages of being understandable and compatible with existing
practices. Instead of offering new services, facsimile equipment
performs old services faster and cheaper, which would be a clear
and distinctive advantage. That telefacsimile stands out as the
most wanted service could, however, be taken as evidence of organizations failing to realize the synergy benefits of the network
being a multi-purpose system.

RECENT DEVELOPMENTS AND PRESENT PLANS

The data presented in the paper were collected from April to July, 1976. Since then Televerket has conducted sales talks by telephone and in person, with most (though not all) the potential customers: in effect, with 32 organizations. Its board has formally decided to proceed - though a total of only 19 organizations (4 in Norrköping) would want to connect to the planned network; this would represent some 40 to 50 extensions (6 in Norrköping). Electronics companies are, again, the most interested. The board has further decided to cut installation charges and long distance traffic charges by half.

We shall monitor the negotiation process in order to understand the real reasons for wanting or not wanting to join. We also intend to conduct a series of interviews with Members of Parliament, civil servants in relevant ministries and agencies, makers of telecommunications equipment, and other parties with interest in the field in order to examine their respective attitudes, knowledge, intentions and actions. Finally, we plan to survey communication patterns and organizational climate in a few organizations before, as well as after, the actual introduction of the test network.

A POSSIBLE EUROPEAN SYSTEM FOR TELECONFERENCING VIA SATELLITE

B. DRIOLI AND L.A. CIAVOLI CORTELLI; J.L. JANKOVICH

TELESPAZIO S.p.A.; INTERNATIONAL BUSINESS INSTITUTE

ROME, ITALY; LEYSIN, SWITZERLAND

Video teleconferences are becoming a reality. In Europe the use of satellites could represent a means of providing such a service. This paper presents the main results of a preliminary systems study performed by Telespazio for the European Space Agency in 1975 with the collaboration of Dr. J.L. Jankovich in the market study and of G.T.E. (Societa Generale di Telefonia ed Elettronica) in the definition of the earth terminals.

INTRODUCTION

Various experimental and operational video teleconference (TC) systems already exist around the world. In general, these systems permit the connection of only two studios located in cities far from each other by bidirectional audiovideo links in a terrestrial network. Experiments with teleconferences via satellite have also been conducted using a satellite simply for point-to-point connections. The system considered in this paper is based on the possibility of connecting more than two studios, taking full advantage of the intrinsic point-to-multipoint capability of satellite systems.

If teleconference services come to be accepted by users our proposed European system would represent a flexible solution which could be implemented in the next decade.

EUROPEAN MARKET ESTIMATES

The marketing team's main concern was to analyze TC market potential in the 1975 to 1985 period, identify potential users and their needs, and then translate those needs into technical specifications for investigation of cost-effectiveness.

The chief interest of the study centered on arriving at order-of-magnitude demand figures, the relative size of various applications and user groups, and their timing.

Our study reviews the TC market for selected medical, educational and business applications as follows:.

    1 - medical:
      a - conferences of international health organizations
        HQ - sponsored by WHO World Headquarters
        EU - sponsored by WHO European Headquarters
        reg. - sponsored by other European intergovernmental health
               organizations
      b - conferences of national medical associations
      c - medical university education
      d - consultations on special cases

    2 - educational:

      a - regional and international conferences
      b - conferences of national educational authorities
      c - classroom CCTV and program exchange in universities

    3 - other CCTV:
      a - intergovernmental organizations
      b - international business organizations
      c - Congress Centers.

The numerical estimates are derived by selecting an illustrative sample within each group of applications; collecting and analyzing statistical data about the selected sample; estimating the size and trend of the pertinent user population; and extrapolating the sample findings to the entire group.

Thus, the results of the market study reflect subjective estimates, to be verified by more detailed studies and practical experience with sample user groups.

The number of conferences in the various categories is given in Table 1. as reported for 1973 and projected for 1975, 1980 and 1985.   (The basis for the projections is explained on the following page.)

Application Group	1973	1975	1980	1985
1a HQ	117	133	167	214
1a EU	44	47	66	88
1a reg.	117	126	154	187
1b	56	68	109	175
2a	221	248	332	445
2b	24 000	26 000	31 700	38 600
3a	12 890	14 436	19 335	25 908
3b	33 000	35 700	43 420	53 800
3c	3 580	3 870	4 715	5 740

Table 1

Number of Conferences Per Year

In order to select the part of the potential market most suitable for teleconferences, four criteria have been considered: duration, practicability, feasibility and user acceptance.

It seems unlikely that conferences which normally last several days can be carried out by telecommunications because of either service cost or people's difficulty in concentrating for long periods of time in front of the monitors. On the basis of reported data, conferences which last one day or less, that is, 1/4 of a day (2 hours), 1/2 of a day (4 hours) and 1 day (6 to 8 hours), are listed in Table 2 as percentages of all conferences in the various categories.

Application	Conference Duration (in days)		
Group	1/4	1/2	1
1a HQ	1	1	2
1a EU	2	4	5
1a reg.	12	22	34
1b	-	-	-
2a	-	-	1
2b	15	35	40
3a	1	3	3
3b	12	23	50
3c	-	5	12

Table 2

Short Conferences as Percentages of
All Conferences; 1973 to 1985

Teleconferencing is considered to be practical for only a certain fraction of all conferences: routine, periodic, technical discussions, or for the extension of the reach of professional conferences, etc. (see detailed discussion below).

The corresponding percentages are estimated to be as follows:

Application Group	Percentage Practical
1.a	0.5
b	0.5
2.a	0.5
b	0.35
3.a	0.35
b	0.5
c	0.1

Table 3

Percentage of All Conference Situations
where TC is Practical; 1975 to 1985

Under feasibility we consider the probability that the necessary TC terminal equipment (not the satellite earth station) will be procured by the conference's sponsoring organization and that it will be available in time and in proper condition, if wanted. Then, we estimate the probability that the participants will accept and use the teleconferencing facility, if and when practical and available, as a substitute for the traditional way of conferences.

Our numerical estimates are based on what behavioral scientists call the "adoption process"; it explains how new ideas are learned and accepted. For the market segments considered in our study, the process is characterized by an S-shaped curve which shows the percentage of adoptions as a function of time by various types of groups and organizations. For instance the first group to adopt, the innovators, make up 3 to 5 percent of the market; business firms in this leading class are usually large and rather specialized. (McCarthy, 1975)

Actual teleconference usage can now be expressed as the product of the appropriate elements in the matrices of practicality, feasibility and user acceptance. Table 4 provides estimates as percentages of the total number of conferences that will take place.

Application Group	1975	1980	1985
Group 1.a	0.003	0.075	0.36
b	0.003	0.075	0.36
Group 2.a	0.000875	0.014	0.126
b	0.000875	0.014	0.126
Group 3.a	0.000875	0.014	0.1575
b	0.005	0.113	0.428
c	0.00005	0.00125	0.016

Table 4

Potential TC Market as Percentage of All Conferences

The numerical estimates of short TC market demand in 1980 and 1985 (Tables 6 and 7 below) are obtained by multiplying the number of yearly conferences (Table 1) by the appropriate product of percentages (Tables 2 and 4).

SELECTION OF DATA BASE

Our investigations concentrated on regular conferences of major organizations based in Europe and omitted several feasible but more futuristic applications in order to provide a conservative estimate of traffic demand. In each application group a growth factor was chosen according to the slowest growing member of the group.

In medical applications, for instance, the actual yearly number of all major congresses is 2 to 3 times higher than the figure retained on the basis of periodic meetings alone. Thus, although the annual number of congresses of the British Medical Association has quadrupled over the last five years, a 10 percent p.a. growth factor has been retained for Europe as a whole.

In educational conferences, even regular meetings in the chambers of ministerial officials are disregarded because no detailed statistics about them are available, although one estimate puts their number at least equal to all the other meetings in larger conference rooms. The number of meetings with out-of-town attendance has increased a hundredfold over the past 5 years in the Dublin Ministry of Education and still continues to grow, as contrasted with the 4 percent p.a. growth retained for our European total.

In business and industry, the total time spent on all out-of-town conferences is ten times more than the estimates retained on the basis of strictly periodic meeting statistics alone. Also, the potential needs of three-quarters of the multinational companies working in Europe are not included in our estimates (i.e., companies with less that $1000 million annual sales as of 1971) because:
a) no detailed statistical meeting data are available about them;
b) they are thought to have less rigidly set regular meeting schedules; and, c) they are less likely than the larger companies to introduce teleconferencing early in our considered time period.
In certain Congress Centers, the annual number of congresses is growing at a rate of 10 percent per year, as compared to the more conservative factor of 4 percent which we have retained.

Yet, the applications actually retained reveal an important potential need for teleconferencing and also provide a broad enough statistical basis for a few useful observations.

The average conference characteristics, as defined for the various application groups, are summarized in Table 5.

Examination of the table shows that average conference characteristics are not homogeneous within major application fields, such as health or education; however, we can discern similarities between fields.

Medical consultations and university teleconferences have similar characteristics not only statistically but, also, in that a substantial portion of both can be handled with some time delay and outside the peak traffic hours. Conferences of a day or shorter duration, with twenty or so participants, appear as typical in both ministerial and business situations. Three-to-four-day long conferences are often found in medical, educational and intergovernmental organizations. And very large conferences, extending over several days with hundreds of participants, occur in major international organizations as well as Congress Centers.

Application Group	Partici-pants	Duration	Volume	Bases (Countries)
1a	25	4 days	100 man-days	12
1b	333	3 days	1000 man-days	1
1c	5	2.5 hours	12.5 man-hours	5
1d	5	2.5 hours	12.5 man-hours	5
2a	33	3 days	100 man-days	18
2b	20	6 hours	120 man-hours	1
2c	5	2.5 hours	12.5 man-hours	5
3a	33	3 days	100 man-days	18
3b	20	1 day	20 man-days	7
3c	400	3 days	1200 man-days	7

Table 5

Average Conference Characteristics - 1973

Our findings suggest that, on the basis of the statistical data, it is more logical to consider the above four categories as main components for design, rather than the disciplinary groupings of medical, educational and other CCTV applications.

It must be kept in mind that our final figures reflect the relative weighting assigned to various user groups in the transition matrix. The numbers actually used are subjective estimates and can be verified only after more detailed studies and practical experience with sample user groups.

POSSIBLE EUROPEAN TELECONFERENCES IN THE 1980's

Our survey concentrated on the more or less regular conference habits of major organizations based in Europe and includes multi-country applications, domestic long-distance and the European segment of intercontinental TC needs.

Application Group	No. of all Conferences	No. of possible TCs of duration of			
		¼ day	½ day	1 day	1 day or less
1a HQ	167	.12	.12	.25	.49
1a EU	66	.10	.20	.24	.54
1a reg.	154	1.40	2.50	3.90	7.80
1b	109	-	-	-	-
2a	332	-	-	.04	.04
2b	31700	66.57	157.33	177.52	399.42
3a	19335	2.71	8.12	8.12	18.95
3b	43420	588.77	1128.48	2453.23	4170.48
3c	4715	-	.29	.70	.99
Total:		659.67	1295.04	2644.00	4598.71

Table 6

Number of Possible Teleconferences in 1980

Application Group	No. of all Conferences	No. of possible TCs of duration of			
		¼ day	½ day	1 day	1 day or less
1a HQ	214	.77	.77	1.54	3.08
1a EU	88	.63	1.26	1.58	3.47
1a reg.	187	8.07	14.81	22.88	45.76
1b	175	-	-	-	-
2a	445	-	-	.56	.56
2b	38600	729.54	1702.26	1945.44	4377.24
3a	25908	40.80	122.41	122.41	285.62
3b	53800	2763.16	5296.07	11513.20	19572.43
3c	5740	-	4.59	11.02	15.61
Total:		3542.97	7142.17	13618.63	24303.77

Table 7

Number of Possible Teleconferences in 1985

The final results provided here disregard several feasible but more futuristic applications, as well as those which are likely to be only incidental to the situations which are considered (i.e., 1c, 1d, 2c.) This selection was made deliberately, in order to provide a conservative estimate of traffic demand. For the same reason in each application group the growth factor was chosen according to the slowest growing group member.

Yet, the applications which have been retained in the estimate reveal important potential needs for TC and also provide a broad enough basis for a few useful observations. For instance our analysis shows that the average conference characteristics are not homogeneous within major application fields, such as health or education. The greatest demand stems from conferences of not more than one day's duration with 5 to 25 participants attending from several locations; meetings of this type are typical in ministerial, intergovernmental and business situations.

The general area of this traffic covers Northern, Western and Southern Europe, and the Mediterranean Basin with North Africa and the Middle East. The heaviest traffic concentration appears in the zone defined by the Quadrilateral of London, Hamburg, Milan and Marseilles.

TRANSMISSION REQUIREMENTS AND CONSIDERATIONS

The geographical areas considered in the study are those shown in Figure 1. Following some considerations on Radio Regulations, the 12.5-12.75 GHz band for the down-link and the 14.3-14.4 GHz band for the up-link with the addition of part of the 14.0-14.3 GHz band shared with terrestrial radionavigation service, were assumed to be appropriate for teleconferencing via satellite.

The channels needed have been sub-divided into two categories:

a) video channels, with associated audio and visual aid channels; and,

b) auxiliary channels; i.e., audio channels not associated with video, as well as voice, data and signalling channels.

Video channels. Digital techniques may be particularly promising for teleconferences since they could allow a considerable reduction in the required band width and power. We refer, for example, to methods for digitizing video signals, by using a frame memory for reduction of the interframe and the intraframe redundancy.

However, besides the fact that these methods are still at the development stage, the present costs of the necessary equipment would exclude their application within the very near future, for a system with many small stations, such as the one considered here. Frequency modulation has, therefore, been assumed for the video channels.

The conclusions of the market study, regarding the characteristics required by the users for video information, and subjective tests carried out under both "standard" conditions and "collective viewing" conditions, show that a standard color TV system and an unweighted S/N = 33 dB for the satellite link, would generally satisfy user requirements for most types of teleconferences.

An optimization of RF bandwidth/satellite RF power was performed. It leads to the conclusion that:

a) in order to obtain a link quality equal to 33 dB for the unweighted S/N for 99% of the time, a 29-30 dB C/N in 1 MHz must be maintained as a minimum for 99% of the time; the RF bandwidth should be equal to about 25 MHz;

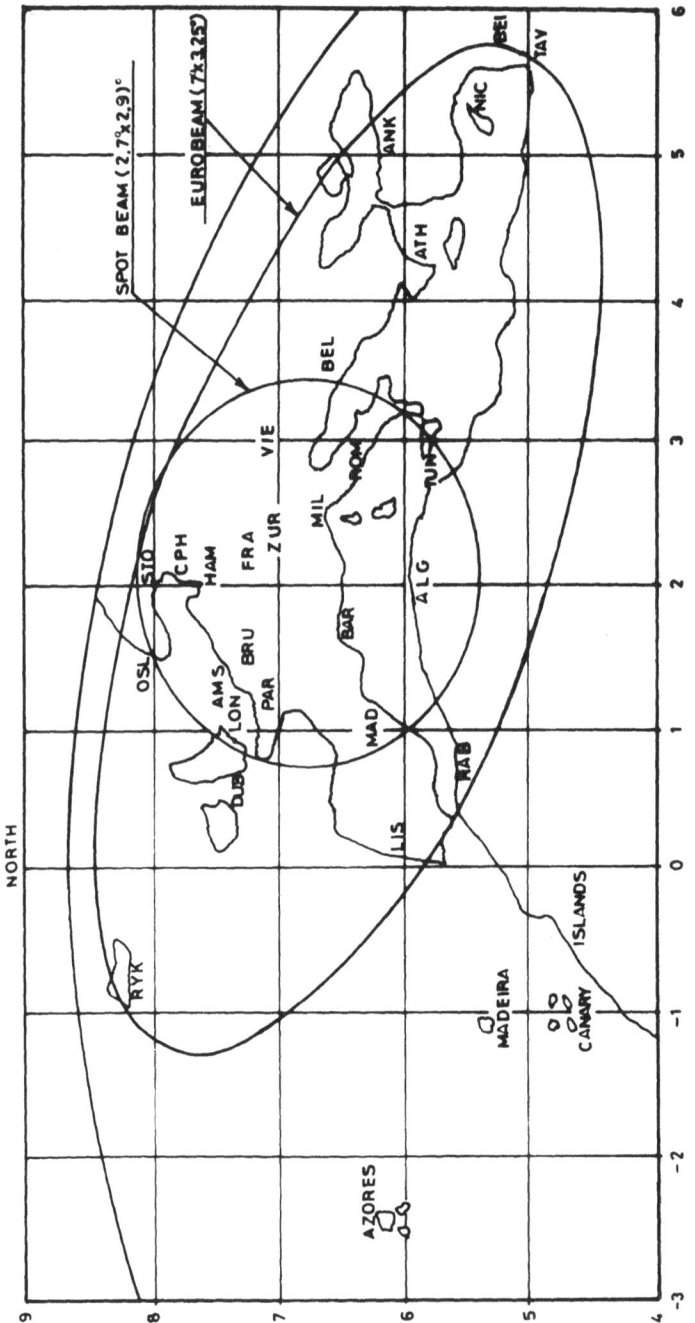

Fig. 1 Coverages (Pointing Error: 0.2; Sub. Satellite Point: 10°W)

Coverage	Ground Antenna	Paramp noise temp. (°K)	RF Bandwidth (MHz)	(C/N)1 MHz (dB)	G/T Ground St (dB/K)	Ground Antenna Diam (m)	Up-link Noise Contr. = 0.5 dB		Up-link Noise Contr. = 1 dB	
							Ground TX power (W)	Sat. TX power (W)	Ground TX power (W)	Sat. TX power (W)
Spot beam	Fixed	125	24	29.5	22.9	3	3300	70	1700	75
"	"	280	25	29	20.9	3	3000	95	1550	105
"	Tracking	125	24	29.5	24.7	3	2000	45	1050	50
"	"	280	25	29	22.7	3	1800	60	950	70
Eurobeam	Fixed	125	24	29.5	22.9	3	8900	180	4500	210
"	"	280	25	29	20.9	3	7900	250	4100	290
"	Tracking	125	24	29.5	24.7	3	5100	120	2800	140
"	"	280	25	29	22.7	3	4600	170	2500	190
Spot beam	"	125	24	29.5	29.6	6	700	15	350	16
"	"	280	25	29	27.6	6	600	21	300	23
Eurobeam	"	125	24	29.5	29.6	6	1800	41	950	45
"	"	280	25	29	27.6	6	1600	55	850	60

Table 8

Parametric Link Calculations

b) the above values would also ensure the required service
availability for 99.9% of the time.

The link calculation was made parametrically in order to allow the
overall system optimization to take into account the costs of the
earth and space segments.

Different alternatives were considered:

- satellite antenna coverage; Eurobeam and Central Europe beam
- parametric amplifier noise temperature : 280°K and 125°K
- ground antenna : fixed and with step tracking
- up link noise contribution : 0.5 and 1 dB

Some of the results of the calculations are shown in Table 8

Auxiliary channels. Digital systems were chosen for modulation.
The particular transmission techniques chosen were: high-quality
audio channels - PCM/20PSK/SCPC; telephone channels - adaptive
Δ/20 PSK/SCPS; CSC channel - PCM/20PSK/TMDA. For the auxiliary channels system comprising 25 carriers (8 for high-quality channels; 16
for telephone channels; 1 for CSC), calculations showed that 12 W
and 60 W of satellite TWT power respectively would be required for
3 m and 6 m ground antennas and about 400 W and 2 KW of power would
be required from ground transmitters.

BASELINE SYSTEM AND COST ESTIMATES

The estimated potential traffic, which is mainly made up of
business meetings involving people from different European countries,
was so high that the limitation to the capacity of the satellite
teleconference system which could be realized in the 1980's appeared
to be the satellite itself. The use of a satellite working at frequencies above 10 GHz and having a mass at the beginning of its
life of 450 Kg. was considered to be realistic.

Two geographical coverages were considered: (1) global coverage
including Northern, Western and Southern Europe and the Mediterranean
Basin with North Africa and the Middle East and (2) spot coverage of
Central Europe. The transmission study and the system cost optimization have shown that by using earth terminals having a 6 m. tracking
antenna, and by assuming FM modulation with a single-channel-per-
transponder, four high-quality color TV channels (with no eclipse
capability) can be obtained: two for global coverage (40 W TWT's)
and two for spot coverage (14 W TWT's).

The system provides for the use of at least one television channel per teleconference assigned to the studio where the speaker is located; it would be received by all the other participating studios. For each television channel the associated audio channel and a visual aid channel are transmitted by time division insertion techniques, (the latter can be used for the transmission of hard copies to all the participating studios). A second television channel could also be used; it could be assigned permanently to the studio from which the teleconference was chaired to help in improving the environment of the meeting, or it could be used part-time for showing diagrams, drawings etc. Assuming that the teleconference is being chaired by a single person it is sufficient that each participant request the floor by means of a special signalling channel. If many calls reach the chair one after another, they are sent to a memory and the floor is assigned according to the order of requests.

From an operational point of view there are several ways of controlling the progress of the teleconference. We have proposed that a control center should remotely control the starting and stopping of the transmitters using telecommands provided by computer. For a teleconference involving several studios the use of "auxiliary" channels is necessary. A distinction can be made between the "dedicated channels" (that is audio quality and telephone quality channels which transfer information among the participants) and the "operational channels" (that is, those voice, telegraph and data channels used for technical control, e.g., telemetry and telecommands).

With each of these channels adequate signalling is associated when necessary. The dedicated and operational channels are shown in Figures 2 and 3. It should be particularly noted that high quality audio channels are foreseen for audio-only interventions (while the video is still assigned to another studio). The auxiliary channels can be transmitted on a separate transponder (12 W TWT) with Eurobeam coverage, PSK modulation and single channel per carrier access.

The baseline system is composed of the following main elements:

a) a geostationary satellite with its telemetry and telecommand station (the satellite has 5 full, redundant transponders: 2 for the TV channels on spot coverage, 2 for the TV channels on global coverage and 1 for the auxiliary channel on global coverage);

b) 15 high-traffic receiving and transmitting earth terminals with 4 connected studios, located in the major European cities;

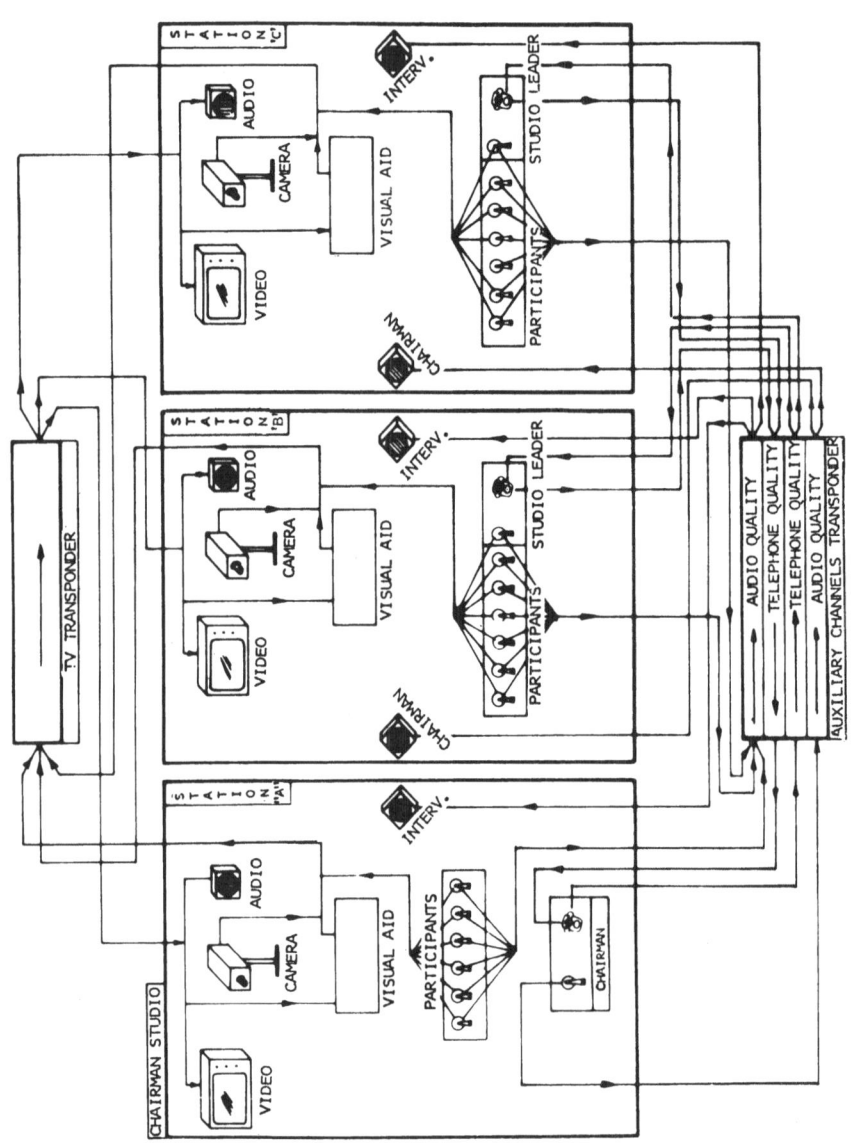

Fig. 2 Diagramm of Dedicated Channels

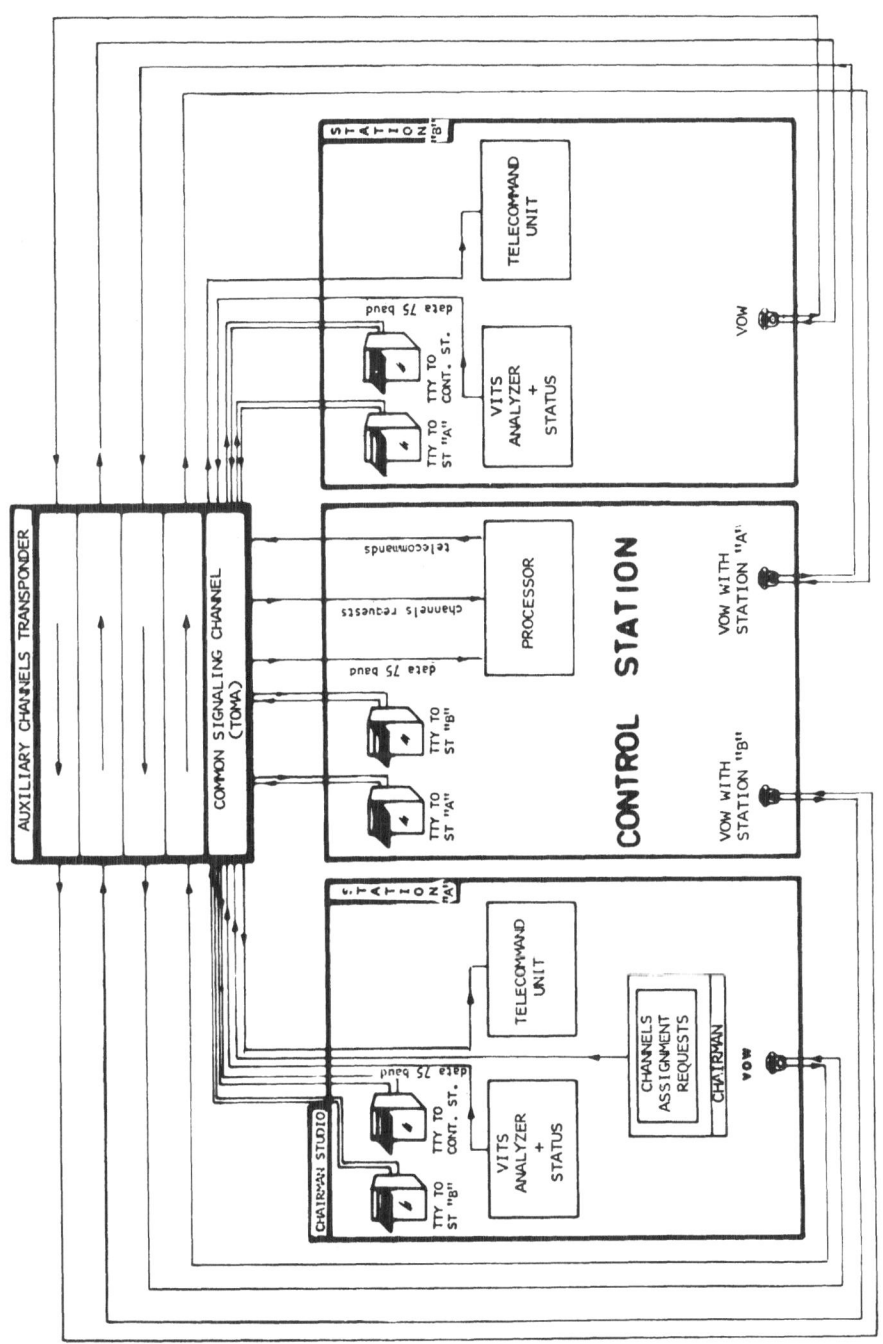

Fig. 3   Diagramm of Operational Channels

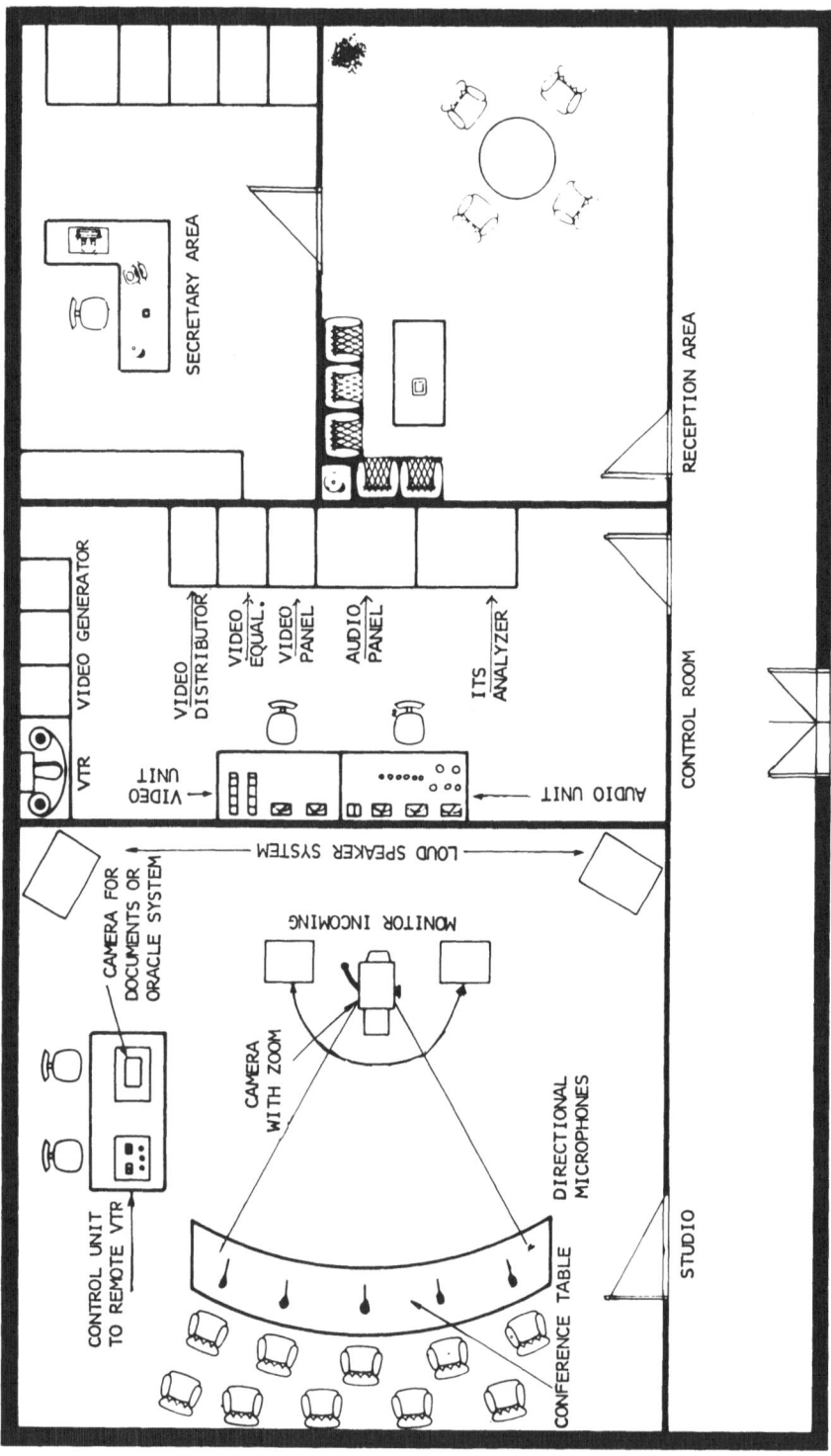

Fig. 4   Studio Type 1

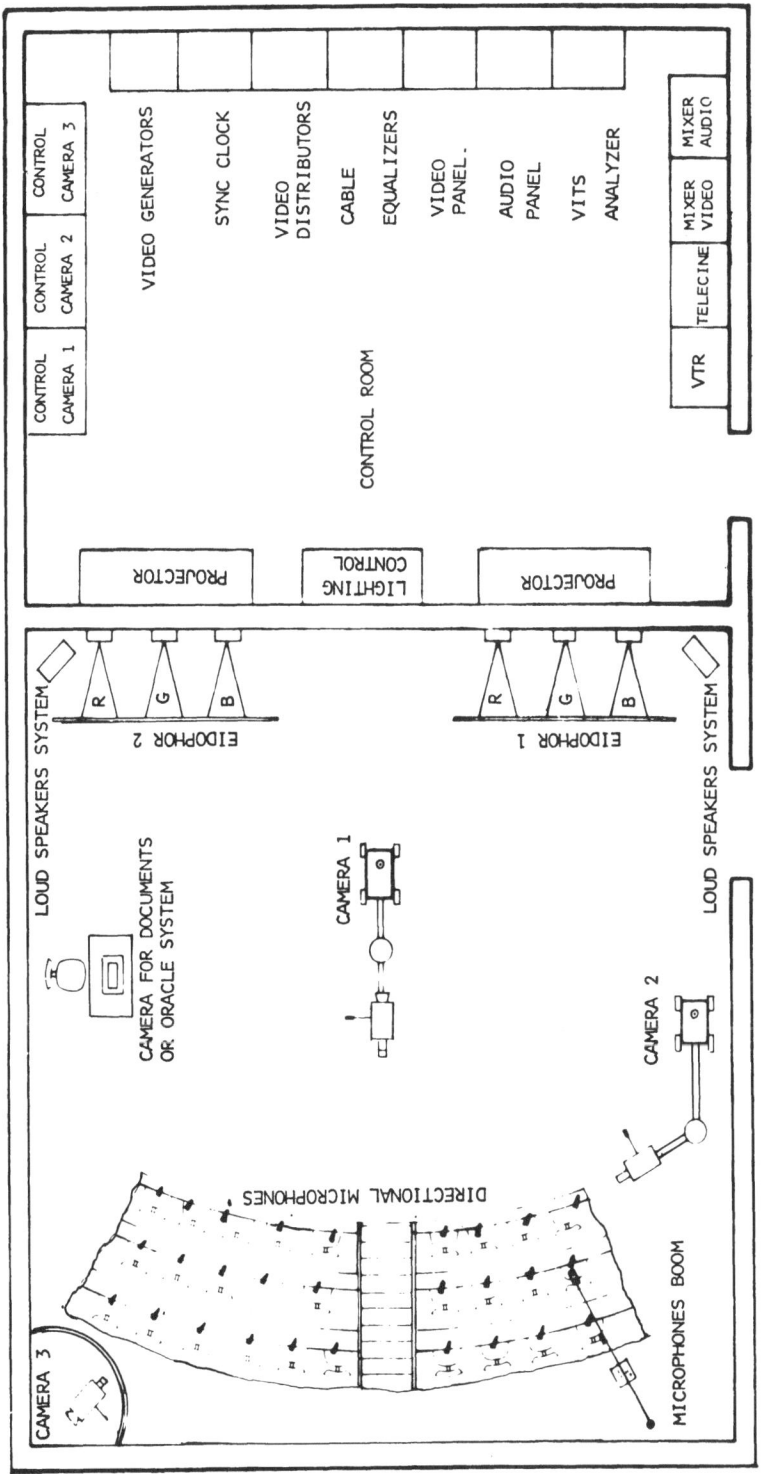

Fig. 5 Studio Type 2

c) 25 low-traffic receiving and transmitting earth terminals with 1 connected studio located in minor European cities, large European organizations, companies, congress centers etc.;

d) control of satellite access, control of the quality of the signals received by all studios, and control of the assignment of the channels to the studio where the speaker is located.

e) a European coordinating center for administrative matters. Two types of studios have been studied. The first type can accommodate up to 10 participants (fig. 4) and the second up to 50 participants (Fig. 5).

The estimated yearly investment costs for the overall system are as follows:

a) space segment - about $10 million

b) earth segment (25 high traffic earth terminals at $441,000 each, 15 heavy traffic earth terminals at $269,000 each, 85 studios at $77,000 each, 1 control center at $1.032 million) - about $7 million.

c) operating costs (185 persons for the studios and the earth terminals, 15 persons for the control center, and 20 persons for the European coordinating center, plus studio operation, plus materials and utility consumption, plus administrative expensives) - about $8 million.

CONCLUSIONS

Our preliminary market investigation has shown that in Europe and in the Mediterranean countries there will be a high traffic potential for audio-video teleconferences by the 1980's. It has also shown that this traffic potential is constituted mainly by business meetings.

The baseline system proposed in this paper makes use of high quality television channels and could represent the natural development of point-to-point teleconference systems currently being experimented with in Europe.

The main advantages of the system proposed are the possibility of achieving teleconferences between more than two studios, by taking advantage of the point-to-multipoint capability of the satellite, and the possibility of reaching any place within the satellite coverage area using small earth terminals. The proposed system foresees the use of a satellite which utilizes the technologies now being developed in Europe for the OTS and ECS satellites and the use of earth terminals with antennas not exceeding 6 m. in diameter.

High quality color television channels have been considered and all the costs have been charged to the teleconference system. The hourly cost we have estimated is, therefore, somewhat pessimistic.

Assuming 1,500 hours of service for each video channel per year the hourly cost per channel is about $4,100. This is an estimate of the cost and not of the price to the user. Nevertheless, in a first approximation, this figure can indicate the order of magnitude of the price.

By assuming an average cost for a person attending a meeting in Europe of $450 the teleconference hourly cost is comparable to the cost of a traditional meeting with the participation of about 15 persons if very short conferences are considered.

It must be recalled that teleconference costs have been estimated for a system which uses full-time high quality television channels and that all costs have been charged to the teleconference service.

We can assume that by using black and white television channels, image processing techniques and videophone images, and by sharing the cost of some parts of the system with other services such as audio-only teleconferences, data transmission, interconnection of small TV networks, the hourly rate could be considerably reduced. Further studies to examine these possibilities would be extremely useful.

References

S. Brofferio (Politecnico di Milano, Milano), U. Lauro Grotto (SIT Siemens, Milan), P. Grattoni, A. Raccu (Ist. Elettrotecnico Nazionale, Turin), E. Paolillo (Telespazio, Rome): Redundancy Reduction of the Video Signal: Implementation of a Prototype of an Inter-Intra-Frame Compressor, presented at Eurocon 77, Venice 3-7 May 1977.

Jankovich, J. Leslie: Business and Politics: The Challenge and Opportunities of Satellites and Communications, JBIS, Vol. 29, No. 1, p. 57, 1976.

McCarthy, E. Jerome: Basic Marketing: A Managerial Approach, R. Irwin, Illinois, Fifth Edition, 1975.

Telespazio S.p.A. per le Comunicazioni Spaziali: "European Teleconference System via Satellite", 1975.

VIEWDATA NETWORKS

Samuel Fedida

Post Office Research Centre
Martlesham Heath, Great Britain

SUMMARY

This paper examines the configurations of Viewdata centres and
their arrangements in Viewdata national and possible global networks
from the point of view of operation which would be completely unattended, yet capable of providing a high degree of service reliability.

Such systems must include facilities for overall monitoring and
control; the acceptance, distribution and maintenance of the distributed data base; the retrieval of information; the transmission of
messages; and many of the necessary internal functions such as the
gathering of statistics, billing and so on.

1 INTRODUCTION

Viewdata is a new computer-based visual information and communications medium using the domestic TV set as the display device.
It was developed by the Post Office with a view to providing a
nationwide service to members of the general public and to professional users.

The use of computers for information retrieval and for the
discharge of specific communications tasks is not new. However,
computer-based systems for information dissemination have so far
been designed mainly for use by the specialist and generally require
a high degree of training and expertise for effective use. They
are also costly both in terms of user's hardware (the terminal),
in terms of supporting computer equipment and hence in terms of
usage costs. Because of these factors they can only service small
numbers of specialist users.

The concept of Viewdata on the other hand is entirely new. By appropriate design of the component parts, a system has been developed which is both economical and capable of being used by the general public.

A Viewdata system consists essentially of the following components:

(a) A low cost, rugged and reliable terminal, based on the domestic TV receiver and hence suitable for use in the home environment or in the office.

(b) A set of interconnected computers, ideally situated within a local telephone call's reach of the majority of users.

(c) An information structure which is easy to understand and simple to use.

(d) A computer protocol which is naturally adapted for use by human beings and which requires no training whatever on the part of users.

(e) Computer software which minimises the cost of information retrieval, by minimising the amount of computation per transaction.

(f) The dial-up telephone network which enables any user to connect to a local Viewdata computer.

The system may also be used readily across national boundaries and successful tests have recently been carried out from Belgium, Denmark, France, Germany, Italy, Sweden and Switzerland.

As well as being used for information dissemination, Viewdata lends itself to a host of other applications such as person to person messages, education in the home and communication, at a distance, with the deaf.

In some respects Viewdata has similarities with the Teletext Information Service (Ceefax - Oracle) developed by the broadcasters. Both systems use compatible display standards on the TV set, in order to simplify the manufacture of sets and to reduce their costs, and both supply information to the general public.

Unlike Teletext however, Viewdata has a virtually unlimited information capacity. 70,000 pages of information for a small Viewdata centre, are accessible with a maximum waiting time of

2 seconds, against a maximum of 800 pages on Teletext with a maximum
waiting time of 24 seconds.

In addition Viewdata is interactive, i.e., the user may respond
to the information supplied by the computer or may enter his own
data. Hence the system lends itself to the handling of electronic
mail, education in the home, and a host of other applications such
as ordering goods and services from the home or the office, calculations, and so on.

The Viewdata concept, the range of applications and many of
the technical considerations pertaining to the system have been
published recently (Reference 1-9).

2 THE COMPUTER NETWORK

The computer network required to support Viewdata has to fulfill
a number of requirements, amongst which low cost of usage is of the
first importance. This is a function not only of the cost of equip-
ping, running and maintaining the computer installations and data
bases as such, but also of the cost of the telephone connection
giving access to the customer. Hence one of the first assumptions
made is that the Viewdata network would consist of a number of
computers, with their associated data banks, each sited within a
'local call distance' of the major user concentrations.

The second important requirement is to enable the 'Information
Providers' who are intended to provide and maintain the bulk of
the information comprising the Viewdatabase, to have ready access
to one or more points in the network for the purpose of inserting,
editing, updating and distributing the data for which they are
responsible. Thus the concept of a computer hierarchy, as shown
in fig. 1, was developed. In this arrangement local centres,
marked L, provide database access to the users.

Nearby local centres are joined to a Regional Centre (RC) via
a star network, and regional centres are joined to a National Data
Centre, in the same way, the network hierarchy being arranged in
such a way as to provide the means of accepting and distributing
national data (i.e., data and information of interest to all users)
from the national data centre to all the local centres via the
regional centres. Similarly a regional centre could accept and
distribute data and information of regional interest to the dependent cluster of local centres. There could also be provision for
entering data alone at the local centres.

In order to minimise costs local centres and regional centres
have been designed on the basis of completely unattended operation.

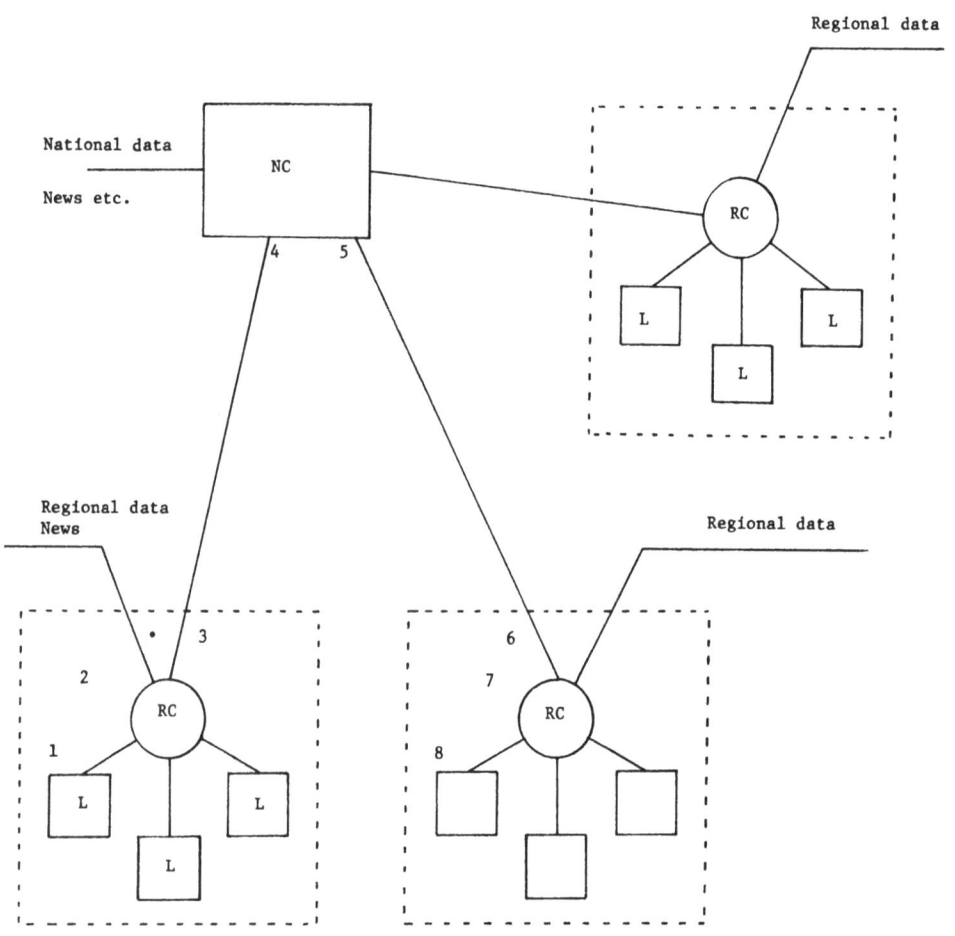

Fig 1 Network of Viewdata Centres

It was initially proposed that only the National Data Centre would
be manned, but this would entail a substantial amount of additional,
and possibly unnecessary, expenditure, particularly if manning were
to be carried out on a multi-shift basis to cover the greater part
of the 24-hour day.

This concept has been re-examined to determine whether, in the
light of the considerable experience which has now been obtained
in the running of the Viewdata Pilot Trial, which began in September
1975 and is still in progress, the proposed arrangement described
above may be improved upon, and costs reduced further by adopting
unmanned operation throughout.

2.1   The Hierarchy of Computers

The proposed hierarchy of Viewdata centres was designed to
meet six objectives:

1) The administrative control and the monitoring of the
operation of the whole system, from a single location,
(e.g., the National Data Centre).

2) The acceptance and distribution of data (including
updates) of interest to all or several local centres.

3) The archiving of all data bases.

4) The production of bills.

5) The gathering of statistics.

6) The maintenance and enhancement of software, and
applications programs for the whole Viewdata system.

Some of the functions listed above implied that the National
Data Centre had of necessity to be manned, e.g., for the operation
of magnetic tape decks, used for archiving purposes.

In view, however, of the very great cost of providing attendance
at the National Data Centre, it is important to re-examine the whole
concept to determine whether or not there could be an alternative
means of fulfilling the above requirements, while organising the
network on the basis of completely unattended centres.

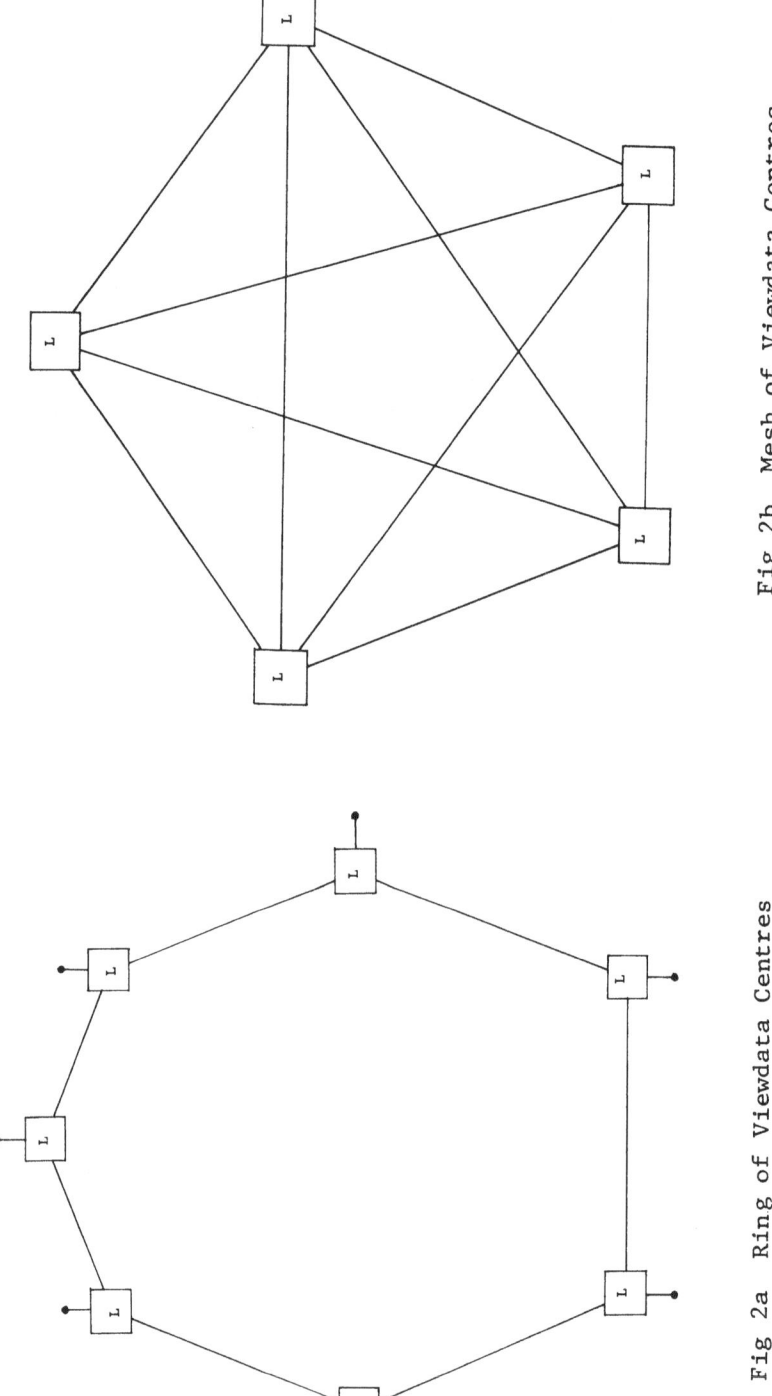

Fig 2b Mesh of Viewdata Centres

Fig 2a Ring of Viewdata Centres

2.3   The Ring and Mesh Networks

An alternative to the two-tier star based hierarchy is the
ring structure, shown in fig. 2a. This envisages a number of local
centres, L, all identical or nearly so, connected together in a
closed loop.

The mesh network, fig. 2b, is a cross between the star and the
ring, wherein cross connections are established between the individual stations of a ring, in addition to the loop connections.

The ring structure has a number of advantages over the star
network. It avoids placing an excessive reliance on the few star
points of the star network, through which much of the internal
traffic must flow, by providing alternative communication routes.
Thus failed stations may be by-passed, until such time as they are
brought back into service.

The mesh network provides in addition a number of direct interconnection links, to avoid the excessive traffic that might need
to be routed through all the stations in a ring.

These three alternatives are examined from the point of view
of completely unattended operation and overall reliability in order
to provide some guidelines to the establishment of a nationwide
Viewdata system, with the potential of global interconnections.

3   THE COMPLETELY UNATTENDED VIEWDATA SYSTEM

The completely unattended Viewdata system will now be examined
in the context of the functions listed in Section 2.1.

3.1   Administrative Control and Monitoring

This function includes overseeing the operation of the whole
system, receiving fault reports, and ensuring that maintenance is
carried out promptly and effectively. Some of these administrative
and control functions may well be delegated to the regions; many
advantages would derive from this, e.g., faster response and knowledge of local conditions. Fault conditions would be reported
automatically to a local centre, say to the Regional Managers'
office.

This function clearly does not need to be located in the National Data Centre or indeed in any one of the Viewdata computer centres.
An arrangement of ring or mesh networks is therefore entirely capable
of meeting this requirement.

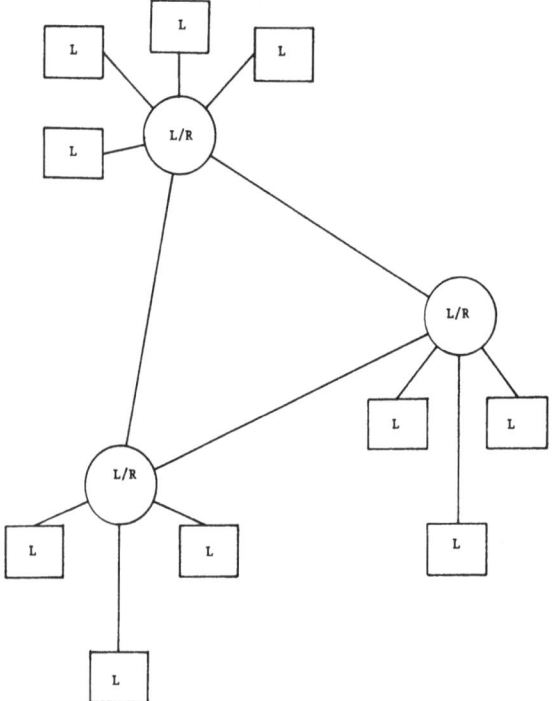

Fig 3   Viewdata Intracommunication Network

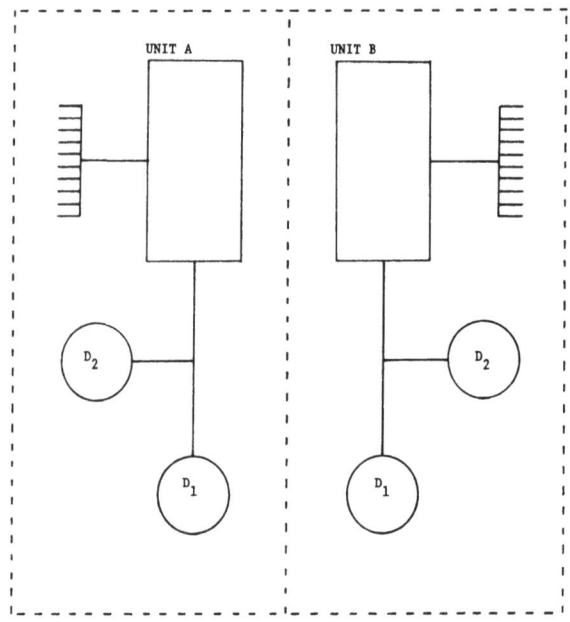

Fig 4   Configuration of Local Viewdata Centre

3.2 Acceptance and Distribution of Data

The acceptance of data, including national or local updates, may be carried out at any centre. Its transmission and distribution to the other centres, be it in the same region or on a nationwide basis, may be carried out quite simply using the existing message service suitably modified.

This service which has been developed during the Viewdata pilot trials, provides the facility for a user to enter, at a local centre, a message intended for another customer of the same local centre. When message entry is completed, the Viewdata centre calls the addresses automatically, using the telephone network. When the addressee responds the message is presented to him on the screen. Additional facilities exist for leaving an indication at the user's terminal if no response is obtained to the telephone call. The message may then be obtained when the addressee dials Viewdata.

A variant of this service consists in sending messages from one local centre to another. The initiating user might be an Information Provider inserting an update. This is then communicated from one local centre to another and used to update the data base.

In this arrangement any local centre may be the point of entry for an Information Provider.

Indeed this development will need to be implemented for the extension of the message service between any two users. A typical arrangement is shown in fig. 3. Three local Viewdata centres, shown as L/R, have been nominated regional data centres for the purpose of data distribution while supporting the Viewdata service for their own customers.

Data may be entered at any centre L or L/R and its distribution would be carried out, first to the local regional centre L/R, from the latter to the remote regional centres, and finally to the remote local centres. Communications between all centres are carried out by means of private wires at 4.8 or 9.6 Kbits per second.

3.3 Archiving

The purpose of archiving is to ensure that should data at one centre be corrupted it may be reconstructed with the minimum of delay. To some extent the configuration of a local centre, proposed from the outset on the ground of simplicity, low cost and reliability and shown in fig. 4 provides this facility.

A local centre consists of two identical systems A and B:
each with its own central processor unit (CPU); multiplexers for
line access; data base disk store D1 which may consist of one or
more 70 MB disks, and small disc D2 which contains the IPL software
(Initial Program Load); directory for direct jumps to enable users
to go straight to the page of information they require; and data
base updates for archiving purposes.

Both systems are available to the user through the dial-up
network on a single telephone number, with the outlets interleaved
so as to ensure, as far as possible, even loading of the two machines.

Each machine is designed to service about 200 ports, if the
whole data is one disk store only, but more ports may be serviced
if the data load is distributed over more than one disk. Current
system tests are being carried out to ascertain equipment capabilities in terms of simultaneous ports.

The archiving principle is as follows:

When a page is updated (or added, deleted, etc.),
only the changes are recorded in disc D2, in both
machines, and the data bases on discs D1 are modified.

A copy of the entire data base is also kept in a
different location to ensure a satisfactory level
of security.

If the information on one of the disks D1 should
become corrupted, it could be reconstituted from
the other disk.

If both copies of the data base on disk D1 should
become corrupted or destroyed at the same time,
then the record of updates on disk D2 could be used,
in conjunction with the separately archived copy
of the entire data base, to reconstitute the local
data bases.

Finally, if the entire set of disks should become
corrupted, a most unlikely occurrence, then the
separately archived copy of the entire data base
would be used instead, but the updates on disk D2
would be lost. The reconstituted data base would
be only partly up-to-date.

There are a number of ways in which a separately archived copy of the data base may be provided, namely:

1) Archive on magnetic tape at a separate location, central or local.

2) Archive on a disk cartridge at a separate location, central or local.

3) Archive available on-line at a separate location, central or local.

4) Archive available at the information providers' premises.

In order to keep capital costs and labour costs to the minimum it is preferable to avoid archiving on magnetic tape which invariably requires manual handling at the local centre. Thus in the case of (1) some means must be available at the archiving centre to transfer the archive from magnetic tape to disk, the reconstituted data base then being transferred physically in a new disk pack, or being transmitted electronically to the local centre when the latter has been restored to working condition. Similarly, means must also be available at the archiving site, to transfer the data base from disks, in the local centre, to magnetic tape, preferably by accepting the updates from disk D2, say once weekly, again either using physical means of transporting a disk pack from D2, or by electronic transmission.

This is the origin of the requirement for a national data centre, which would clearly need to be substantially manned to cater for the archiving needs of the whole system, but not necessarily on a 24-hour basis. The updates from disk D2 could be transmitted at night from the various local centres, to be stored on one or more large capacity disks, which would then be used, under operators' control, during working hours to update the archives on tape.

In the case of (2) a copy of the entire data base may be prepared, say once a week on a blank disk pack installed in one of units D1, by transferring information from the other unit D1. To do this during working hours, would require that the system be unavailable for service during this time, which is estimated to be about 15 minutes say about once weekly. This job could be done during a routine inspection visit by local staff, but would not require constant attendance. The archive copy of the disk pack would then be stored locally, but at a different location from the computer centre. Air-conditioning is not required in the local store.

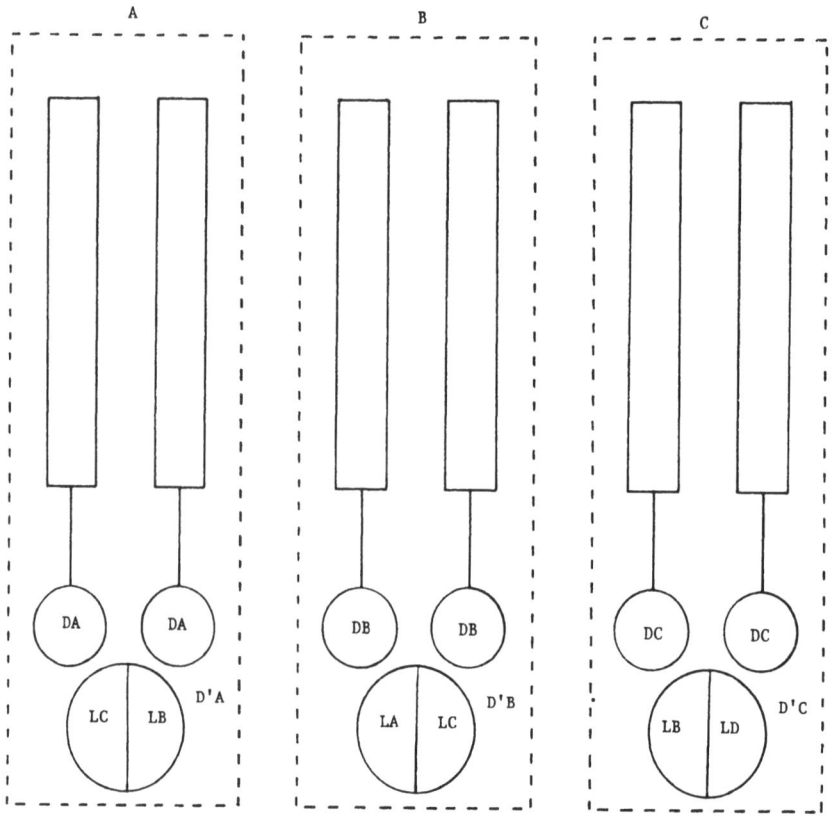

Fig 5   Archiving in Contiguous Centres

Alternatively it could be carried out on one machine only, using the updates of unit D2, the other machine being available for service, but not for editing purposes or messages.

It is, however, possible, as explained below, to do much better than this in reducing costs, to provide unattended operations and to improve reliability.

3.3.1 Archive on-line at contiguous centres with duplicated installations. The archiving of the data base on-line to a separate location raises a number of interesting and potentially promising possibilities.

The principle involved is illustrated in fig. 5. This examines the case of three contiguous local centres, each centre being made up of two identical computer systems operating autonomously, but providing each other with a substantial measure of standby protection. Each member of a pair of disk units DA, DB, DC contains a data base identical to that in the other member of the same pair. Note that a separate disk for storing updates and IPL has been dispensed with, all data and programs being stored in the above disk units.

The data base in Unit B, for example, will be partly made up of a national component which is the same as the national component of all other pairs DA and DC, and a local component which is different from the local components in DA and DC.

It may be useful to note here that the local data base component at a Viewdata centre consists not only of information of local interest to users, but also of details of all users allocated to that centre, such as identification and special requirements normally required for billing, message delivery, etc.

For the price of the third disk unit D1, which is used exclusively for archiving purposes in each centre, a duplicated remote archiving facility, available on-line could be provided at all the local centres. For example, in local centre B, disk unit D'B would contain the local components of A and C (LA, LC) and its own local component (LB) would be available at both local centres A and C, and of course on its own disks DB in two copies. The local components LB at centres A and C are updated on-line at the same time as the data base in centre B.

The cost of archiving disk unit D' is a small percentage of the cost of the centre.

A variant of the above scheme can achieve further useful cost reduction in addition to providing duplicated remote archiving. This variant is shown in fig. 6. It is based on the assumption that the

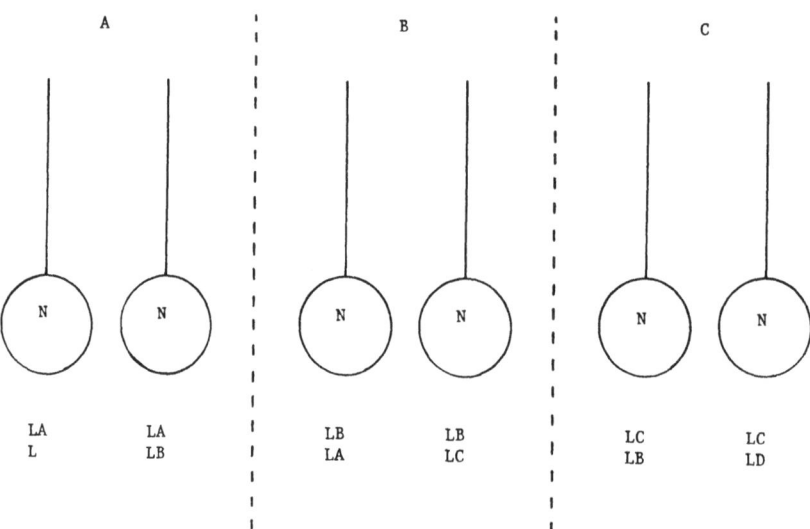

Fig 6   Archiving in Contiguous Centres, with On-line Service
        (duplicated centres)

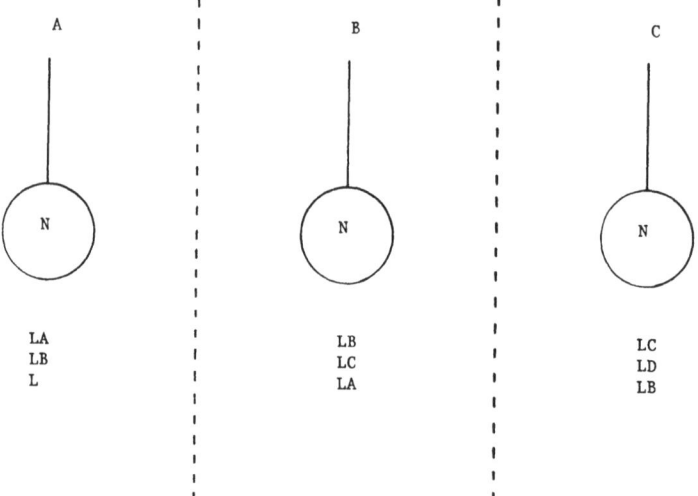

Fig 7   Archiving in Contiguous Centres, with On-line Service
        (unduplicated local centres)

total data base in a local centre is accommodated by design in a
single disk drive (i.e., it consists of rather less than 70,000
pages). In general the local data base would be a small proportion
of the national data base, say about 10%. It is then possible to
keep on one disk the national data, which is identical in all the
local centres, one copy of the local data base of the local centre
and a copy of the local data base of the adjoining centre. Thus in
centre B disk N contain the national data, its own local data base
(LB) and a copy of the local data base of centre A (LA). Indeed,
one of the disks could contain a copy of the local data base of A
and the other a copy of the local data base of C.

Thus in this arrangement duplicates of the national data base
exist at all centres while the local data base is duplicated within
a local centre, and available at two other contiguous centres.

What is more, in the event of a partial or total failure in
centre, B, centres A and C are able to support fully the activities
of centre B. The only requirement is that provision should be made
for users to connect to A or C in the event of a failure at B.

3.3.2 Archiving on-line at contiguous centres with unduplicated
installations. A further derivation of the scheme which can provide
additional cost-benefits is shown at fig. 7.

In this scheme, local centres are unduplicated, i.e., they consist of a single set of equipment as shown in fig. 7. In the event
of failure at centre B, centre A or centre C would take up the additional load, as if the two computers were co-located. Duplicated
national and local files now exist in at least three separate centres,
so that no further archiving needs to be done anywhere else, and
no manual intervention is needed except in the event of a failure in
a local centre, when the automatic warning system would come into
operation and initiate the necessary corrective action.

In a system, such as that of fig. 6 or 7, the routing of calls
to B, for example, should be capable of automatic diversion to A
or C. This may be done, for example, in the call set-up unit in
the terminal by providing for the automatic dialling of three numbers
instead of just one.

One considerable advantage of the system shown in fig. 7 lies
in the opportunity it offers to reduce the number of lines which
need to be connected to individual centres, thus reducing traffic
congestion and telephone planning difficulties. However, the facility of being able to reconstitute a duplicate data base at a centre,
to replace a corrupted unit at a contiguous centre, implies that
two disk drives must be available at all local centres (as shown in
fig. 8), to allow the copying to be done while maintaining normal
service. This additional disk drive can increase the traffic capa-

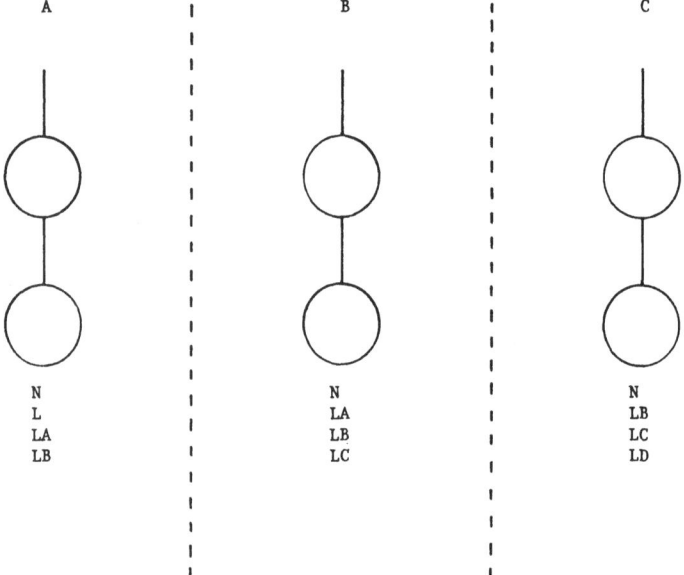

Fig 8   On-line Archiving at Contiguous Centres — Unduplicated
         Installations and Twin disc drives

bility of a local centre by approximately a factor of two, thus
bringing it much closer to the traffic capability of a duplicated
centre, but at a substantially lower cost.

In the initial build-up of a Viewdata system, the choice
probably lies between (i) a configuration such as that of fig. 6
providing a duplicated installation at a single site with a third
installation at a remote side, providing the archiving back-up*;
and (ii) a configuration such as that of fig. 8, providing two unduplicated installations at two different sites, with local facilities for copying and reconstructing data bases. Alternative (ii)
can provide almost the same level of service, reliability, data
base security and throughput capability as (i), provided users are
able to access all centres on a local call basis. If, however,
this is not so, as for example in the initial build-up phase when
Viewdata centres are likely to be rather sparse, then option (i)
is clearly the one to use.

3.4 Billing

The collection of customer billing information may be done
quite simply at the local computer centre, since all the transaction
details are known at this point. Indeed, details of transactions
(i.e., connection time, cost of pages accessed, etc.) are stored
in the user file, which consists of a frame stored permanently in
the local data base, and available to the user for examination at
any time.

During the course of a Viewdata call, the billing details are
held in the main computer memory and the user's file is updated
either when the user terminates his call, or when he accesses his
statistics page, which contains the billing details of his user's
file.

All billing information may, therefore, be collected from each
local centre, by a central or local accounting office, at pre-deter-
mined intervals, say once a quarter. The bills would then be produced and despatched to the customers.

It had been intended that this function be discharged by the
National Data Centre. It is, however, possible to discharge this
function more economically and speedily at the local centre in a
number of ways.

One possibility consists in fitting each customer terminal

* This third installation must have two disk drives to allow
for the copying of a data base.

with a simple low cost print-out unit, which would print out bills
and a request for payment at present intervals, the bills giving
details of usage to the extent judged necessary. This is clearly
an attractive long term solution. However, it may take some considerable time to implement, and is, therefore, not of interest
in the initial build-up phase of a Viewdata service.

An alternative would be a form of self-billing, which could
be implemented quickly and economically subject to meeting certain
legal requirements. It consists in sending a pre-formatted payment
request to each user from a local office (which could be the Telephone Manager's Office). The handling charges involved in this
operation are very small. On receipt of the pre-formatted bill the
user would be required to call a Viewdata page number to find billing
details in his or her user file, enter them in his or her pre-formatted bill form and return the duplicate bill and payment to the
central office in a prepaid envelope.

To make this procedure simple to use, it would be possible
to show separately in the statistics page (the user's billing information) the amount due to the end of the billing period, in
addition to the cumulative information on usage.

On receipt of the duplicate bills and cheques at the central
office, the accounts clerk would check that the details are correct,
delete the appropriate entry in the user's file, (insert optionally
that cheque has been received) and thus complete the transaction.
Note that any manual work involved here is that which would be
needed even if the billing system were implemented on a central
computer.

Additional mechanisation could clearly be introduced here, if
required, to check bills automatically against a user's file, by
making use of the computer power available at the local centre.
It would be possible, for example, to use an automatic cheque
handler (with manual entry of cheque amount), connected on-line
to the local computer centre.

As far as Information Providers (IPs) are concerned it is
necessary to accumulate two counts, the one of the total number
of pages which have been accessed and the other of the total amount
due to the IP's. The rental charged to the IP's is subtracted to
give the net amount owed or due. The preparation of bills or payments should be done manually in this case, since only a relatively
small number of IP's are likely to be involved.

We may assume for the purpose of this discussion that IP's
enter information at the nearest or any convenient local centre or
regional centre. The page usage corresponding to each IP is maintained at each one of the local centres purveying this information.

It is then transmitted periodically, say once a week or even once
a day (preferably at night when the system is lightly loaded), to
the centre at which the IP's account is maintained. This centre
could be point of data entry of the IP or any other local centre
convenient to the accounts staff.

3.5 Statistics

Statistics are required from time to time to investigate a
number of parameters. Amongst these are the following:

> System response time
> User response time
> Number of users accessing the system,
>   and their distribution by time of
>   day and time of year.
> Distribution of access by page number
>   (to analyse user requirements).

It is proposed that statistics be taken on-line on a sample
of users, chosen at random. For each user so selected (the selection is random by port number), the system response time is noted
for every transaction and a moving average maintained throughout
the session on this particular port. Maximum and minimum times
may also be noted, together with the total number of transactions.
Thus for every port selected 4 or 5 numbers only need to be stored
to give the information required.

As in the case of other distributed data, such as billing
information for Information Providers, the concentration and consolidation of the required statistics at a convenient local centre
may be done automatically in the quiet traffic hours.

3.6 Access to Rare Data

Some of the data stored in the Viewdata system may only rarely
be accessed from any one local centre, but the totality of accesses
may be large enough on a national basis to warrant its being made
available to Viewdata customers. This data is clearly 'national'
data in the sense that it is of interest to many users, scattered
over the whole network, but the frequency and total number of accesses is not high enough to justify its storage at every local
centre.

There is a number of ways in which this information may be
made available, but it must be recognised and accepted from the
outset that the cost of retrieving rarely needed information will
be rather higher than the cost of retrieving frequently needed
information.

We shall assume here that all transactions are carried out by
a user through his or her local centre, to ensure that identification
and billing are carried out easily and at minimum cost, and to
comply with requirements of the short range LTU (Line Transmission
Unit), incorporated in the Viewdata receiver, which has been designed
to provide satisfactory communication to a local Viewdata centre at
minimum cost. Thus 'rare items' will have a point of entry only
at local centres, but the bulk of the information will be located
at a single centre.

On accessing such a page, the user will be informed that the
information he or she seeks is to be obtained from a distant centre.
The local centre will then fetch the required data in one of two
ways:

(a) By using the public switched telephone network (PSTN)
direct from the local centre to the centre where the data
is available.

(b) By routing the request through the associated regional
centre, across to the regional centre associated with the
local centre in which the rare item is available and thence
to the local centre itself.

It is likely that the communications costs and waiting times
for (b) could be rather less than those for (a), but some load will
of necessity be imposed on the various computer centres through
which the traffic is passing. Since, however, this is occasional
traffic, there should be no difficulty in providing for it. Nevertheless, the total number of computer ports occupied in (a) is
three, as against a maximum of 7 for (b) assuming the configuration
of fig. 3. Data transmission costs and computer cost differ between
the two cases.

Computer connect costs are in the ratio of 7:3 when case (b)
is compared with case (a), since they are related to the number of
computer ports occupied. However, transmission costs may be much
less in case (b) than in case (a). Whereas in (a) the costs are
made up of a local call and a long distance call (depending on the
distance to the centre in which the desired information is stored),
in (b) the costs are of a local call and up to three computer to
computer calls, using the internal computer network. These calls
could well be cheaper marginally costed on a facility which must
exist anyway for other purposes.

Once the connections are established, by either means, there
are two possible options available for transmitting the information
for retrieval.

3.6.1 <u>Page by page system</u>. In the page by page system the
normal Viewdata protocol is observed, the user response being transmitted from centre to centre. If we assume that 4.8 KB links are
installed between centres, the transmission time per page per link
is about 2 seconds (1 second on the average). The transmission
times in the two cases are pretty much the same, as shown below:

In case (a) transmission time is:

      Local centre to remote centre
       (PSTN)   8 secs peak

      Disk access at remote centre
                                <u>2 secs peak</u>
                        10 secs Maximum waiting
                                     time to start of
                                     writing on screen

In case (b) transmission time is:

      Regional Centre to Remote   2 secs peak
        Centre (4.8KB)

      Regional Centre to Regional   2 secs peak
        Centre (4.8KB)

      Regional Centre to Local   2 secs peak
        Centre (4.8KB)

      Disk access at Remote Centre   <u>2 secs peak</u>

                              8 secs peak

                              Maximum waiting
                              time to start of
                              writing on screen

3.6.2 Anticipation system. In the anticipation system, the page asked for and the ten filial pages are transmitted at the same time. The transmission or response time is as follows:

Case (a)

    Transmission time

    1st page   8 sec maximum

    10 filial pages 80 secs

    Disk access time
    for first page   2

    Disk access time
    for filial page 2

Maximum response time is then 10 seconds, for the first page and anything between 2 seconds and 10 seconds for subsequent pages, according to whether the required 'filial' page has arrived or not.

Case (b)

    Transmission time

    1st page   8 sec maximum

    10 filial pages 20 sec maximum

    Disk access time
    for first page   2

    Disk access time
    for filial page 2

Maximum time is 8 seconds for first page and nearly 2 seconds for filial pages and all pages thereafter.

Thus the anticipation system for access to rare information should be quite satisfactory from the points of view of cost and response time.

3.7   Distributed Data Bases

The case of a distributed data base may be handled in a closely similar way. A distributed data base, for example, a national tele-phone directory, might consist of area directories which are stored at local centres. The totality of all area directories would con-stitute the national directory.

The majority of directory enquiries are to local numbers, say 80%, while some, about 20%, are for out-of-area numbers.

As in the case of rare data, the protocol for access to out-of-area numbers, would follow closely the lines of the anticipation system, which has the clear advantage of minimising response time, and possibly initial connect time. Costs vary according to configu-ration, but these also could be competitive with the direct con-nection via PSTN.

4   VIEWDATA NETWORKS

There are a number of alternative ways of inter-connecting a Viewdata system. They fall in three categories as follows:

1)   A hierarchy of star networks as in fig. 1

2)   A hierarchy of ring networks as in fig. 9

3)   A hierarchy of mesh networks as in fig. 10.

In both figs. 9 and 10 Viewdata centres in a region are inter-connected as an independent cluster of identical stations. In each region one of the local centres is given a nodal role in addition to its normal roles. All Viewdata nodes are inter-connected in the same way as the cluster. Within the three categories of inter-connection, there are large numbers of variants using a suitable mix of star, ring or mesh connections.

While figs. 9 and 10 only show two levels in the hierarchy, one would clearly have more levels, for example, for the inter-connection of national networks. Nevertheless, the inter-connection of such networks need not be carried out strictly as a hierarchy; for example, two large multi-cluster networks could be inter-connected via a local centre.

These networks may be examined on the basis of the following five parameters:

1) Security against link failure

2) Security against equipment failure

3) Port occupancy and number of communications
    links

4) Addition of new stations

5) Communications load.

4.1 Network Security

It will be assumed that the communication lines and the corresponding installation are completely duplicated. Equipment
duplication may be carried out either in the same site (co-location)
or at different sites as discussed in para. 7. For purpose of this
discussion, a pair of single installations at the same or contiguous
sites will be considered to be a single Viewdata Centre. For example, a cluster of ring-connected stations, as illustrated in
fig. 9, is in fact connected as shown in fig. 11.

Security against link failure at the cluster level is higher
in the case of a ring structure than for a star structure, since
there are two directions of transmission around the ring available,
both duplicated. In the mesh structure security is even higher.
Indeed, the level of security is related to the number of computer
ports available for inter-station communication. This number is
4, 6 and 10 per centre for the star, ring and mesh networks respectively. The number of ports for the mesh connection holds for the
configuration of fig. 10, and increases rapidly as more stations
are included in the mesh.

Security against link failure at the network level also is
higher for the ring and mesh structures than for the tree structure,
again because of the same port relationship.

If we assume that the probability of failure of a single computer to computer link is p, we can deduce the following failure probabilities in a number of typical cases. Let the failure probability
of a duplicate link between two stations = a, and between two contiguous stations

in a star network = b

in a ring network = c

in a mesh network = d

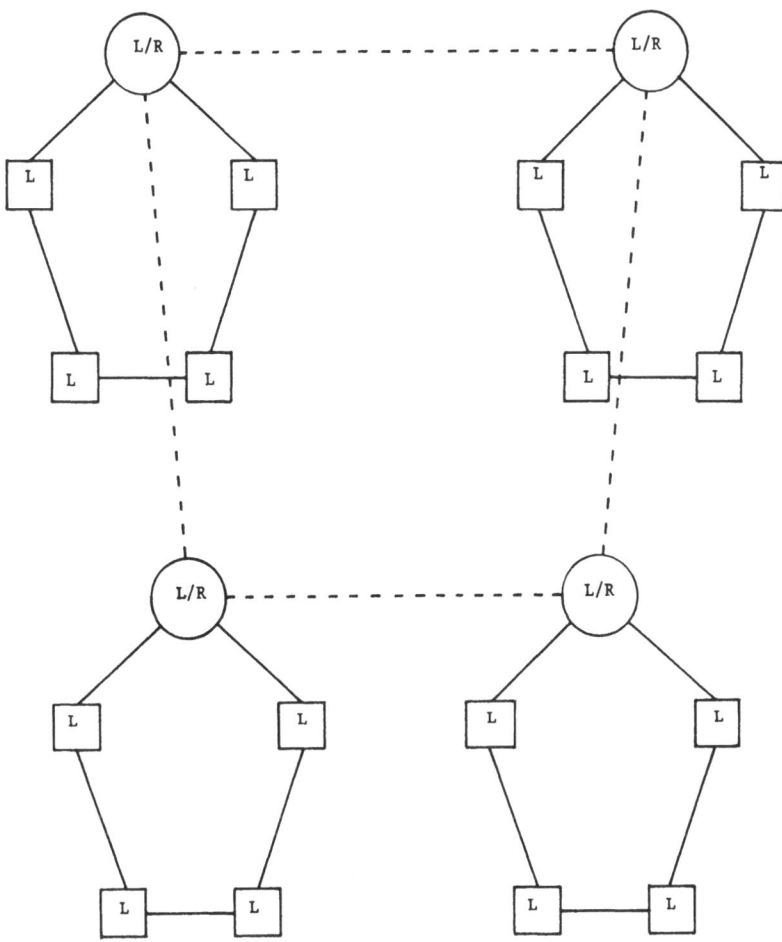

Fig 9   Hierarchy of Ring Connected Viewdata Centres

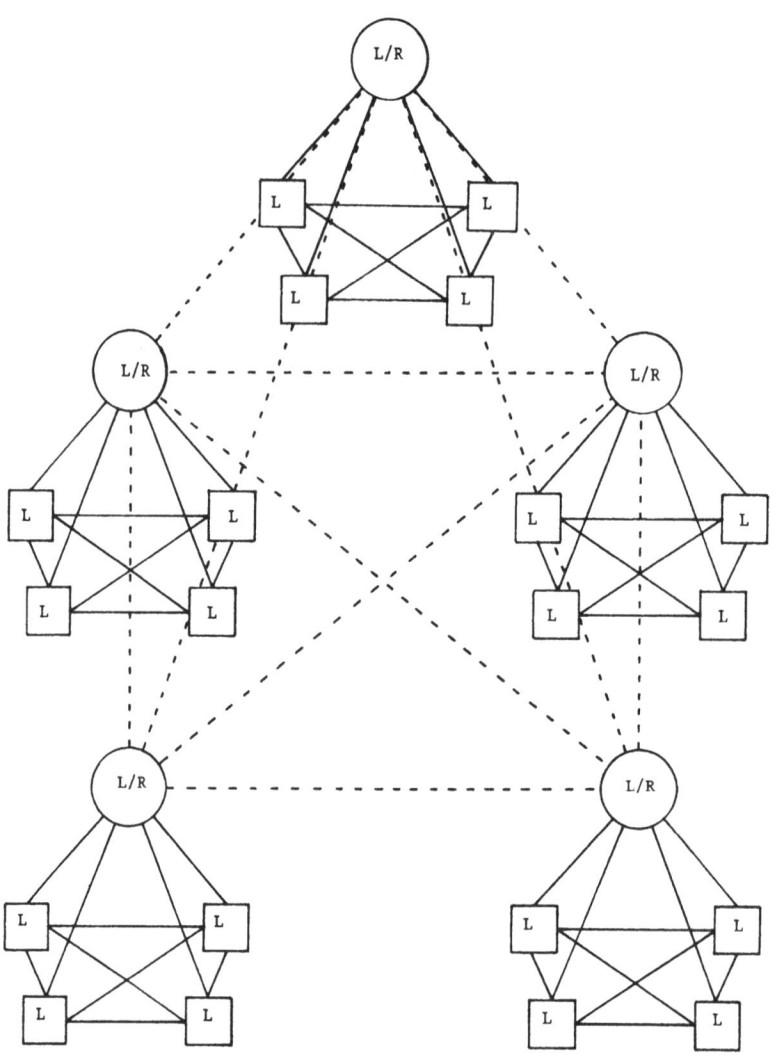

Fig 10 Hierarchy of Mesh Connected Viewdata Centres

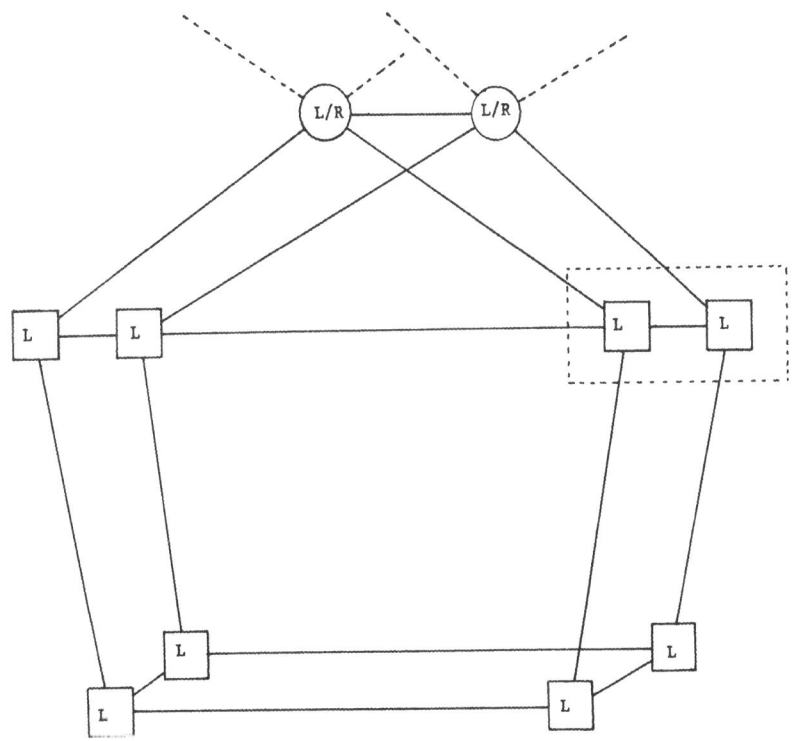

Fig 11   Ring Connected Cluster, Showing Interconnection of
         Duplicated Installations

Assuming the network configurations of fig. 1, 9 and 10, we obtain the following results:

	Duplicated	Unduplicated
Computer to computer link	$a = p^2$	$a = p$
Star Network	$b = 2p^2$	$b = 2p$
Ring Network	$c = 4p^4$	$c = 4p^2$
Mesh Network	$d = 145p^{12}$	$d = 145p^6$

For the configurations shown the unduplicated links in the mesh network give far greater reliability than the duplicated links in the ring network and we shall see later they may present other advantages, with no increase in the number of communications links.

To give orders of magnitude regarding this, if the reliability of a computer to computer connection is, say, 0.99 ($p = 10^{-2}$) then the failure probabilities are as follows:

	Duplicated		Unduplicated	
Computer to computer link	10^{-4}	(1 hr)	10^{-2}	(100 hr)
Star Network	2.10^{-4}	(2 hr)	2.10^{-2}	(200 hr)
Ring Network	4.10^{-8}	(1 sec)	4.10^{-4}	(4 hr)
Mesh Network	145.10^{-24}	($3x10^{-15}$ sec)	145.10^{-12}	($3x10^{-3}$ sec)

Times of interruption per annum are shown in brackets.

4.2 Security against Equipment Failure

The level of security against equipment failure is of the same order (about 0.99) in all the systems considered, at the cluster level and at the overall system level.

The enhancement of system reliability obtained by duplicating equipment and through the use of various network arrangements is similar to that obtained for the case of failure of communications lines, at least for the case of intracluster communications.

As far as inter-cluster communications are concerned, the
junction stations have a crucial role to play. The provision of
duplicate equipment at these stations should be adequate to provide
an acceptable level of system reliability. Any failure at this point
will affect primarily inter-cluster information transfers.

4.3 Port Occupancy and Number of Communication Links

The enhanced level of communications security shown in 4.1 is
obtained at the cost of additional expenditure on communication links
and computer ports. The port occupancy is two per communications
link and the latter, including duplicates, are as follows:

Star network: $2(N - 1)$ links per cluster

Ring network: $2(N)$ links per cluster

Mesh network: $N(N - 1)$ links per cluster

when N is the number of stations per cluster.

The increase in communication links and corresponding ports
is very steep (e.g., from 10 to 20 for a 5 station link) as we upgrade the system from a ring to a mesh network, but much less so if
we upgrade from a star to a ring network (8 to 10 in the case above).
At the same time we have seen in 4.1 that for about the same number
of communications links and computer ports, the mesh network gives
a higher degree of reliability than the ring network.

4.4 Addition of New Stations

The changes to the pattern of communication links induced by
the addition or deletion of a Viewdata Centre differ in the three
cases.

The case of the star network is the simplest: only the link
joining the station to the star point is affected. With the ring
network changes are required to the two adjoining stations in the
ring, while with the mesh network changes to all the stations in
the cluster are needed. But in the latter case only unduplicated
links are involved given equal levels of network reliability.

4.5 Communications Load

The three categories of network configurations show consider1ble differences as regard the handling of intercomputer traffic. Thi' traffic may be classified under the following headings:

Archiving

IP traffic

Access to out-of-cluster information

Messages.

4.5.1 <u>Archiving</u>. Archiving may well represent the heaviest communications load since the statistics of every transaction incorporated in the local data base must be communicated to the two contiguous stations for archiving purposes. The ring network lends itself well to this traffic (since every station is linked to the one on either side of it), as does the mesh network but at a higher cost.

If M is the number of archiving messages per hour originating at the local centres, then the number of messages per hour handled by the various stations is as follows:

Star network:
Each local station sends M messages per hour to its
star point and receives 2M messages per hour from
its star point.
Total 3M messages per hour.

The star point receives M(N-1) messages per hour from its
local stations and transmits 2M(N-1) messages per hour to
its local stations.
Total 3M(N-1) messages per hour.

Ring network:
Each station sends M messages per hour to each of its two
contiguous stations and receives M messages per hour from
each of its two contiguous stations.
Total 4M messages per hour.

Mesh network:
As for ring, mesh connection not used for this purpose.
Total 4M messages per hour.

It is clear that the star network reduces the communications load of local stations in the ratio of 4 to 3, but increases very considerably the load of the star point, since the latter must handle all the archiving traffic arising from its local stations. Obviously the larger the number of stations dependent on the star point, the larger is the communications traffic.

4.5.2 IP Traffic. Let M messages per hour be the IP traffic at an input station. The traffic is made up of updates, edits and new data. We shall assume that the total number of IP messages for the whole network is Q messages per hour (mph) and examine two scenarios in turn. In scenario I IP traffic is evenly distributed across all stations in the system, except the junction stations (i.e., star points in star network, and mesh and ring junctions in the other networks). In scenario II it is concentrated exclusively at the junction points.

We shall also assume that there are N stations in a cluster and a total of N clusters.

Thus in scenario I the number of IP messages per hour at an input station is $\dfrac{Q}{N(N-1)}$ while in scenario II it is $\dfrac{Q}{N}$.

Star Network - Scenario I:

Each input station receives a total of Q mph of which $\dfrac{N}{N(N-1)}$ are direct inputs.

For each direct input it transmits a message to the local star point.

Thus total traffic at input station is: $Q + \dfrac{Q}{N(N-1)} \cong Q$

Local Star Point:

Each one receives a total of Q mph of which

$\dfrac{Q}{N(N-1)}$ x (N-1) $=$ $\dfrac{Q}{N}$ are generated in the local

cluster. Each one of these causes a message to be transmitted to the main star point N and to (N-2) of the local stations. In addition each message received from the main star point (total $\left. \dfrac{Q(N-1)}{N} \right)$ is repeated to all the local stations in the cluster. Thus total traffic is:

$$Q + \frac{Q}{N} + \frac{Q(N-1)}{N} + \frac{Q(N-1)^2}{N} = \frac{Q(N^2-1)}{N} = QN - \frac{Q}{N} \cong QN$$

Star Network - Scenario II with IP inputs at junction stations only:

Input at star points only:

local stations: Total traffic Q mph
star points   : Total input Q
                 output to main star
                 point $\dfrac{Q}{N}$

output to local cluster $Q(N-1)$

Total traffic $QN + \dfrac{Q}{N} \cong QN$

Ring Network - Scenario I:

Each station in a ring receives Q mph of which $\dfrac{Q}{N(N-1)}$ mph are
received direct and $Q - \dfrac{Q}{N(N-1)}$ relayed from the earlier contiguous
station in the ring. Directly received traffic is retransmitted to
the next ring station. Relayed traffic is transmitted on in
$\dfrac{N-1}{N}$ cases.

Thus total traffic is $Q + \dfrac{Q}{N(N-1)} + \dfrac{QN(N-1) - Q}{N(N-1)} \times \dfrac{N-1}{N}$

$Q + \dfrac{Q}{N(N-1)} + \dfrac{Q(N)(N-1) - Q}{N^2}$

$Q + \dfrac{Q(N-1)}{N} + \dfrac{Q}{N(N-1)} - \dfrac{Q}{N^2} = 2Q\dfrac{(N-\frac{1}{2})}{N}$

$\cong 2Q$

Each junction station in a ring receives Q messages per hour
$\dfrac{Q(N-1)}{N}$ represent externally generated messages and

these are relayed totally in the local ring, and a proportion
$\dfrac{N-1}{N}$ to the main ring. $\dfrac{Q}{N}$ which represent internally (to the ring)

generated traffic is relayed in its entirety to the main ring and
all but $\dfrac{N-1}{N}$ in the internal ring. Thus the total traffic at a

junction station is:

Input $\qquad\qquad$ Q

Output to local ring $\dfrac{Q(N-1)}{N} + \dfrac{Q}{N}\dfrac{N-1}{N}$

Output to main ring $\quad Q\dfrac{N-1}{N}\dfrac{N-1}{N} + \dfrac{Q}{N}\dfrac{N-1}{N}$

Total $\quad Q + Q\left[\dfrac{N-1}{N}\right] + Q\left[\dfrac{N-1}{N}\right]^2 + \dfrac{2Q}{N}\dfrac{N-1}{N} = 3Q\left[1 - \dfrac{N+1}{3N^2}\right]$

$$\cong 3Q$$

Ring Network – Scenario II:

Each local station receives Q messages per hour and transmits Q mph to the next ring except for the last station, which receives only.

Thus total traffic $\qquad\qquad$ 2Q

$\qquad$ last station $\qquad\qquad$ Q

Each junction also receives Q mph, $\underline{Q}$ direct and $\underline{Q(N-1)}$ from the
$\qquad\qquad\qquad\qquad\qquad\qquad\quad$ N $\qquad\qquad\qquad$ N

other junctions in the ring.

All IP messages received directly at a junction are relayed to both local ring and main ring. IP messages received from the main ring are relayed to the local ring, and only $\underline{N-1}$ are relayed to the main ring. $\qquad\qquad\qquad\qquad\qquad\qquad\qquad\qquad\qquad\qquad$ N

Thus total traffic is

$\qquad$ Message received $\qquad\qquad\qquad$ Q

$\qquad$ Relayed to local ring $\qquad\qquad$ Q

$\qquad$ Relayed to main ring $\qquad\qquad Q\dfrac{(N-1)}{N}$

$\qquad$ Total traffic $\qquad\qquad\qquad\quad Q\left[2 + \dfrac{N-1}{N}\right] \cong 3Q$

Mesh Network – Scenario I:

Each input station in a mesh receives Q messages per hour. Of these $\dfrac{Q}{N(N-1)}$ are received direct and relayed to the junction station.

Thus total traffic at local stations $Q + \dfrac{Q}{N(N-1)} \cong Q$

Each junction station receives Q messages per hour of these $\dfrac{Q}{N}$ are
received internally and retransmitted to N-2 stations in the local mesh and to N-1 station in the main mesh; thus making a total of $\dfrac{Q}{N}$ (2N-3) mph. $\dfrac{Q(N-1)}{N}$ are received from the main mesh and retransmitted to N-1 stations in the local mesh, thus making a total of $\dfrac{Q(N-1)}{N}$ (N-1) mph.

Total traffic of junction stations

$$Q + Q\left[\frac{2N-3}{N}\right] + \frac{Q(N-1)}{N}\ (N-1) = Q\ \frac{N^2 + N - 2}{N}$$

$$\cong QN$$

Mesh Network – Scenario II:

Each local station receives Q messages per hour. Each junction station receives $\dfrac{Q}{N}$ messages direct and $\dfrac{Q(N-1)}{N}$ messages through the main mesh. All messages received at a junction station are transmitted to the local stations – and the direct messages are transmitted to the main mesh stations.

Thus total traffic received is   Q mph

   Transmitted to local stations   Q(N-1)

   Transmitted to main mesh $\dfrac{Q(N-1)}{N}$

$$\text{Total } Q + Q(N-1) + \frac{Q}{N}\ (N-1) = \frac{Q}{N}\left(N^2 + N - 1\right)$$

$$\cong QN$$

4.5.3 Access to out of cluster information. In the case of the star network access to out of cluster information (assuming a two-tier hierarchy) involves a totality of 8 messages for one-way communication between a local centre accepting an out of cluster request, and the remote local centre in which the information is stored. With the mesh network this is reduced to 6, while with the ring network the number of transactions is a minimum of 6 and a

maximum of 6 + 2(n-2) transaction pairs, where N is the number of
stations excluding junctions in the rings involved in the transfer.
Hence obtaining out of cluster information could be substantially
more expensive in the ring network than in either the star or mesh
network, clearly because many more stations are involved in the
transmission of messages than in the first two cases.

4.5.4 Messages. Messages involve similar considerations to the
archiving of out of cluster information, and the amount of traffic
generated is identical on a per transaction basis. In the case of
in-cluster messages, we also see the same difference between star
and mesh networks, on the one hand, and ring networks, on the other.
The situation is illustrated in table 1.

	Message traffic caused by IP inputs			
	Scenario I IP input distributed evenly over network		Scenario II IP input at junctions only	
	Local station	Junction	Local station	Junction
Star Network	Q	QN	Q	QN
Ring Network	2Q	3Q	2Q	3Q
Mesh Network	Q or 2Q	QN or 3Q	Q or 2Q	QN or 3Q

Table 1

Traffic for retrieving out of cluster information (per micro
transaction):

    star network: 8

    ring network: 6 + 2(N - 2)

    mesh network: 6

Traffic for messages (per transaction):

    Out of cluster messages: as for out of cluster information.

In-cluster messages (not within the local area)

      star network: 6

      ring network: $4 + 2$ (N-2) N in number
                                      of stations involved in
                                      transaction

      mesh network: 4

In-cluster messages (within local area): 2.

4.6 Network Comparisons

The network configurations examined in this section differ in their degree of reliability and the amount of traffic the stations have to convey.

As anticipated, the ring and mesh network do not exhibit the bottleneck of the star network for archiving traffic, where the star point must carry the archiving traffic of all the stations in the cluster.

Information Provider traffic is handled slightly better in the mesh than in the ring network and much better than in the star network, where once again the bottleneck at the star point is evident.

Messages and out of cluster information are best handled in the mesh network.

With regard to network reliability the mesh network scores well against both the star and the ring network, even where communications links in the mesh network are unduplicated. In the latter case, provided the number of stations in the mesh is not too large (about 2 or 6), the number of communications links and ports is about the same as in the ring network.

Thus, on balance, the mesh network is the preferred configuration provided the number of stations in a cluster is not too high.

5 CONCLUSIONS

The many functions to be provided by a Viewdata system can be implemented economically with sets of unattended clusters of individual centres, joined together in a network of meshes, each centre being connected to all the other centres in the same mesh.

The essential functions ensuring the reliability of such a system, such as archiving and reconstituting corrupted data, may be provided readily without the need for continuous manning.

ACKNOWLEDGEMENTS

The author wishes to thank the Director of Research, Post Office Research Centre, Martlesham, for permission to publish this paper. He also wishes to thank his many colleagues in the Viewdata Team who have contributed with ideas and discussions.

REFERENCES

1 Fedida, S., (1975), Proc. European Computing Conference on Communications Networks, 261-282.

2 Fedida, S., (1976), IEE Colloquium on Broadcast and Wired Teletext Systems - Ceefax, Oracle, Viewdata.

3 Fedida, S., (1976), Screened Information at the touch of a button, PO Telecommunications Journal, Winter 1975/76, Vol. 27, p. 4.

4 Fedida, S., (1976), Viewdata - Development of Computer-based Information Media for the General Public, 2nd International Symposium on Subscriber Loops and Services, 3-7 May 1976, The Institution of Electrical Engineers.

5 Fedida, S., (1976), Viewdata - An Interactive Information Medium for the General Public Using the Telephone Network, 6th International Broadcasting Convention, 20-24 September 1976, The Institution of Electrical Engineers.

6 Fedida, S., (1977), Viewdata - The Post Office's Visual Information and Communication System.

Part 1 Background and Introduction, Wireless World, February 1977, Vol. 83, p. 32.
Part 2 Applications of the system, Wireless World, March 1977, Vol. 83, p. 51.
Part 3 Terminals and Codes, Wireless World, April 1977, Vol. 83, p. 65.
Part 4 The Viewdata terminal, Wireless World, May 1977, Vol. 83, p. 55.

7 Fedida, S., (1977), Viewdata Display Characteristics and Future Enhancements, Eurocon 77, Venice.

8 Fedida, S., (1977), Viewdata System Optimisation, International Conference on Communications – Chicago, 1977. Conference record Vol. 2, p. 41.

9 Fedida, S., (1977), VIEWDATA – A Post Office Interactive Information Medium for the General Public. Electronics & Power, June 1977, p. 467.

COMPUTERIZED CONFERENCING: A REVIEW AND STATEMENT OF ISSUES*

Murray Turoff and Starr Roxanne Hiltz**

New Jersey Institute of Technology and Upsala College

* Many of the points made in this paper grow out of the early months of a computerized conference devoted to a research review of computerized conferencing, sponsored by the Division of Computer Research of the U.S. National Science Foundation. In particular, we would like to acknowledge the valuable contributions of the following workshop organizers: Raymond Panko (Stanford Research Institute), Jacques Vallee (Institute for the Future), Dick Wilcox (U.S. government), and John McKendree (U. S. government). Among the participants who contributed ideas and comments reflected in this paper are Harold Bamford, David Penniman, Elaine Kerr, Ronald Uhlig, Ronald Wigington, Robin Crickman, Haig Kafafian, Barry Wellman, Robert Bezilla, Robert Johansen, Andy Hardy, and Julian Scher.

Development of the EIES System was supported by grants from the Division of Science Information Services, National Science Foundation.

The opinions stated in this paper are the personal views of the

** Murray Turoff is Associate Professor of Computer Science and Director of the Computerized Conferencing and Communications Center, New Jersey Institute of Technology. Starr Roxanne Hiltz is Associate Professor and Chairperson, Dept. of Sociology and Anthropology, Upsala College; and associate, Center for Technology Assessment, New Jersey Institute of Technology. Prof. Hiltz's work on this paper was made possible by a faculty fellowship from the National Science Foundation.

authors, and do not necessarily represent the views of the
National Science Foundation or of other participants in the projects described.

I.   INTRODUCTION

A computer mediated communication system for dispersed human
groups was first designed and implemented in 1970, at the Office
of Emergency Preparedness at the Executive Office of the President
of the United States. Since that time, a variety of computerized
conferencing and related systems have been designed and implemented, and we have begun to understand the opportunities, limitations, and issues which are raised when one uses the computer to
facilitate and structure complex human communication processes.
In this paper, we will review the nature of computerized conferencing systems, past, present and future, and then focus on a few
key policy, research and evaluation issues which are raised by
this new form of telecommunication. In treating these issues, we
will give our own, often controversial views. Because of space
limitations, we will not attempt to present opposing points of
view.

Computerized conferencing can be broadly defined as the integral use of computers to structure and facilitate communication
among a group of people. Existing computerized conferencing systems are based on written entries made through a computer terminal
connected to a host computer through a telephone network system
such as those operated by TELENET, Inc., or Tymshare, Inc. At its
simplest level, the computer assigns a number and date to the
entry and then stores it for delivery to recipients the next time
each one accesses the system. Thus, the sending and receiving of
material may be separated in time and space.

II.   COMPUTERIZED CONFERENCING: PAST AND PRESENT

The nature and range of potentials for computerized conferencing becomes somewhat clearer if we briefly review the capabilities
of some of the current systems and utilize that base as a take off
point for future developments.

There are a very large number of what might be termed pure
"electronic mail" or message systems such as Scientific Time-
sharing's MAILBOX, which attempts to replicate for the user a post
office system. These systems simply store an entry and deliver it
after which it disappears. Thus electronic mail is essentially a
one-to-one form of communication, like the letter or the telephone
call.

Conferencing systems, by contrast, are built to facilitate a group discussion. They entail a common space in which all of the discussion or data on a topic is stored and available for subsequent review. The original "Party-Line" and "Discussion" systems developed at the Office of Emergency Preparedness, the current PLANET system, the BELL CANADA system, the Canadian MEMO from Turner Systems and the CONFER system recently developed at the University of Michigan represent simple computerized conferencing systems, utilizing essentially a free form discussion structure. All these systems allow messaging as an adjunct to the conference process and not as the primary component. (See Johansen, et. al., 1977, for a more complete review of existing systems.)

Many of these simple conference systems offer various features or options made possible by including the computer in the communications loop such as voting, association methods among items, special items such as agendas, roles for a moderator or chairman of a discussion and various retrieval and summary features.*

At least two conference systems, the Delphi Conference developed at OEP in 1970 and the IFF FORUM, represented the option of integrating a high degree of structure into the allowed communication process. Both these systems were developed to emulate Delphi like structures in an electronic environment. Perhaps these were both ahead of their time in terms of both the software technology and the degree of user readiness to depart from conventional communication methods. They do represent, however, a demonstration that a highly structured communication process can be implemented on a computer. They also highlight the point that we should not limit our view of what computerized conferencing is by the images of the free form discussion systems now coming into wider use.**

The Emergency Management Information System and Reference

---

* In practice, the distinction between computer mail systems and conferencing systems has become blurred, as electronic mail systems have added many of the capabilities inherent in a conferencing system.

** Until we have a good computer language for making the design and implementation of a human communication structure an effort of only a few weeks duration, there will not be widespread use of electronic forms of highly structured communication processes. The requirements for and design of such a language is one of the research efforts of the Computerized Conferencing and Communications Center.

Index (EMISARI) developed at OEP in 1971 represented a Management
Information System designed as a pure communication process.
Messaging and simple conferencing represented a subset of that
system. The primary innovations in that system were the ability
to set up alternative communication forms such as collections of
numeric estimates, tables of numbers, situation report forms,
etc., and to have these assigned as a permanent responsibility to
some member of the communication group who would supply the information (with his name attached as the author) on a regular
basis. There was also considerable ability to reassociate items
so that text could be used as footnotes to data and sequences of
items of various types could be built up into reports. All **this**
was under the control of a human monitor who could tailor the
communication structure and the reporting responsibilities as a
funçtion of the problem at any time. The result was a single
communication environment in which the people responsible for
reporting, analyzing, interpreting, or setting policy and taking
actions could work together in a cooperative manner.

Until recently, EMISARI has represented the most sophisticated example of any conferencing system in existence, and is
more illustrative of the range of potentials one infers from the
ability to structure human communications than those examples
typified by the simpler conference systems. This is not to say,
however, that simplicity does not have its own benefits in the
right situations.

The Electronic Information Exchange System (EIES) put into
operation at NJIT in 1976 and supported by the Division of
Science Information of the National Science Foundation represents, we believe, an example of a new generation of conferencing systems. On the surface EIES appears to the user as a single
communication environment offering four fundamental types of
communication services in an integrated manner. These are:

Messaging	A private communication space
Conferencing	A common (group) communication space
Notebooks	A personal communication space that may be shared with co-authors, for the developing and editing of reports or papers.
Bulletins	A public communication space (an on-line newsletter or mini-journal).

Superimposed on this is a membership directory, a group structure to affiliate members into sets of common interest and concerns, and elementary and advanced word processing. At this level EIES appears to be no more than collecting the desirable elements of previous systems into a common man - machine interface design and developing it as a dedicated mini-computer operation, as opposed to the time sharing environment of the earlier systems.

We believe current computerized conferencing systems have merely begun to take advantage of the possibilities offered by the computer for structuring and supporting human communication. In order to offer a set of concrete examples of the ways in which systems of the future may open up a whole range of possibilities that are fundamentally different than simply a computerized message switching system, we are going to summarize some of the features which are being designed for incorporation into EIES.

The content of the communications on EIES can be either actual text or it can contain any mixture of "procedures." These procedures provide a programming language power to the text in EIES so that the text delivered to an individual can be utilized to completely change what the user perceives as the system. At the simplest level, procedures can be used to record any sequence of operations that the user frequently employs and establish that sequence as a personal command of the user's own choosing. The more interesting aspects are:

1) Computer Assisted Instruction (CAI)

The PILOT CAI language and extensions to it have been incorporated as a subset of the procedures available in EIES. Obviously this allows an EIES user to develop lessons for use in an educational or training application and incorporate them as a subset of the communication process. What is also possible is that general EIES users can take advantage of this to write "adaptive text." In other words, a writer can incorporate questions to the reader such as "Do you wish to have more detail on this point?" and then let the reader answer such questions to determine how much more or less of the text will be printed out. Options of this sort imply that basic changes in user writing styles could occur as a result of long term use of such systems.

2) Form Generation and Collection

Procedures have been incorporated that allow individual users to generate (describe) forms, to have those forms sent to other individuals to be filled out, and to have automatic sending of the completed or gathered data to the appropriate recipients.

Basically this is no more than extending the text editing capabilities and capitalizing on the communication aspects of the
design. Being able to systematically collect data from individuals in specified formats has many and varied potential applications.

3) Interface Masking

Procedures are available that allow the complete masking of
the current user interface and provide the ability to make the
system appear very simple in terms of features and capabilities.
This can be employed to facilitate controlled communication and
gaming experiments.

4) Basic Language

It is planned to incorporate the BASIC programming language
as an additional segment of the procedure ability. This will
further the ability for selected groups to incorporate analyses and
modeling routines directly into their communication process as
the need occurs.

5) A Computerized Member to Talk to Other Computers

Another capability being incorporated into EIES begins to
give us an insight into the role a computerized conferencing
system can play in the area of resource sharing. A fairly
sophisticated microprocessor (a Zeilog Development Unit) with its
own automatic dialer has been programmed to engage in the conference system as a full fledged member, with the same powers of
interaction as any human member (Hal Zilog as it/she/he is referred to). This entity may perform any of the following tasks:

(a) It may enter EIES and receive or send messages or
     retrieve and enter items into the other components
     of the system.

(b) It may exercise certain analysis routines or display
     graphics on data provided it by other EIES users, and
     return to them the results.

(c) It may phone other computers and select out data from
     existing data bases or obtain the results of a model
     to send back to any designated group of EIES users.

(d) It may drop off and pick up communication items from
     other conference and message systems.

As a result of the microprocessor's capabilities, a conference system can become a central node allowing a human group to bring together and utilize a host of differing computer resources. The particular person knowledgeable about a given information resource now becomes, with the aid of the EIES system and Hal, the transponder for the group as a whole. The result is a mechanism for a group to produce a collective wisdom or knowledge base.

The concept of computerized conferencing embodied in EIES is the incorporation of the user as a integral part of a computerized information system. By integrating normal computer, information, and analysis services into a communication structure we allow the individual and the group to contribute to the objectives of an organization and tailor the system to fit these objectives. This is a considerable advance over the design philosophy in current information systems. Such a view also dashes any hope of maintaining a distinction between communication and information systems without curtailing the advantages to be gained from their integration.

It is widely recognized that all of the existing systems are very rudimentary in terms of what might be possible in the future. It is also generally accepted that current costs for the computer and the communications network (as low as $8 per hour for on-line time from anywhere in the United States on EIES, for instance) will drop substanially in the next few years. As a result, it is obvious that cost factors alone will stimulate the widespread proliferation of such systems over the next decade and their attempted application in many kinds of organizations and for many kinds of communication purposes. (See Turoff, 1976, for a complete cost analysis).

III. FUTURE POSSIBLITIES

Once one perceives that the computer can be used to tailor a human communication structure and consequently create new human systems, the range of possibilities is endless. A few concrete examples, most of which have not been attempted as yet, will serve to illustrate the point. This section of the paper is admittedly speculative or visionary in character, in order to indicate the range of possibilities.

Translation

A communication structure between human translators and conference participants with multiple proceedings kept in different languages would be a structure useful for various types of international meetings or discussions in bilingual communities. The computer can offer design options to insure that misinterpretation does not occur. As suggested above other analysis and data base capabilities can be incorporated to facilitate understanding across differing cultural backgrounds. For a meeting that is ongoing over a long period, a history of accepted concepts and their relative

translations can be built up for the benefit of both translators
and conferees.

Modeling and Simulation*

Computerized conferencing can potentially play three key roles
in the area of improving the utility of modeling and simulations.
The first is in the formative process where a wider range of
expertise can be brought to bear and the opportunity to provide
better interfaces between modelers and decision makers presents
itself. The requirement here is for data processing tools to make
the specification of model assumptions and structures easier and
to analyze, for feedback purposes, inconsistencies and differences
of judgments among the discussants. For example, one such tool
currently being incorporated into EIES through Hal is Warfield's
Interpretive Structure Model (ISM), for producing graphical relationships among interacting concepts.

The second area is in the actual executive of a simulation.
The objective here would be to emulate the real world communication and decision processes associated with the system being
modeled. The third area is the interpretation and validation of
the results of models, where a group of people comprising a
variety of backgrounds is usually required.

Project Management and Coordination

We see computerized conferencing as the only communication
mechanism that can allow a geographically dispersed group to
coordinate and cooperate on a complex project. By complex we mean
one that is not subject to the usual approach of reductionism and
division of labor, but in which a change in one of the elements has
to be reflected in the others. Construction projects as well as
large computer software development projects are excellent cases
in point. Obviously communication structures for this purpose
need scheduling and action tracking facilities incorporated in
conceptually the same manner as was done in the EMISARI system.

Educational Systems

In its simplest terms this is the concept of embedding CAI
systems within communication structures. For the situation of
geographically dispersed students and teachers the communication
aspects become dominant. A current example of this is homebound
handicapped youngsters who lack the peer group reinforcement of
a classroom environment. However, we also feel there is con-

---

* See Scher, 1977, for a further discussion of future possibilities in this area.

siderable possibility for bringing more individualized instruction back into the "classroom" environment by careful use of CAI
and computerized conferencing in conjunction. In the future we
can begin to conceive of educational, training, rehabilitation,
consulting and similar educationally oriented services being generally available over these conferencing systems. Certainly for
people currently holding full time jobs or living in rural areas
there are obvious benefits to this concept.

The inability of penal systems to obtain educators to come
to prisons or reformatories is an additional area of potential
application. Obviously these applications require the merging of
communication structures with the data bases and programmed instructional aids that are associated with the educational process.

Considering the continual rise in the cost of a college
education, there exists a growing educational gap in the context
of which one suspects the concept of the college without walls
and buildings will emerge. One can easily imagine a complete
college education being made available via this communication
method. With the cost of college educations rising and the cost
of computer-communications systems falling, it is very likely that
interactive lessons designed by the best teachers in a field
could soon be provided to students much more cheaply than traditional classroom instruction.

People Finders

The unique characteristic of communication via a computer is
that the content of the message can be utilized to determine the
address. Since people can choose to enter a conference on the
basis of the topic under discussion, without necessarily knowing
any of the other participants, we have a highly efficient
mechanism to enable people to form groups representing areas of
common interest and concerns. This one application alone is
likely to have far reaching consequences upon current societal
structures and social relationships. One of the primary bottle-
necks in the political process is the time and costs involved in
forming lobby groups. Publicly available systems would make that
a trivial and inexpensive operation for any group in the society
seeking to gather political support.

The above represent only a few of the possibilities for future
applications and impacts of computerized conferencing systems.
Each application presents very different design requirements and
raises different regulatory and evaluation issues. The point we
have tried to make is that the free discussion systems are but one

small example of what can be expected over the next decade. They represent a far too narrow perspective upon which to base policy or decisions governing the future of this area.

Let us turn now to a brief discussion of some of the potential societal impacts of present and future applications of these systems, and to a consideration of the difficulties involved in their evaluation. We will then be prepared to consider the key policy issues that are raised by this form of communication on the national and international levels.

IV.   IMPACTS AND THEIR EVALUATION

To properly evaluate the impact of computerized conferencing systems we must take a more "holistic" or "systems" view than is usually the practice. Communication is the fundamental process by which interactions among the elements of a human system takes place; it is the exchange process by which goals are formulated, decisions made, and "work" accomplished. It is inevitable that changing the form of communication used by the members of a group will affect the goals, interaction, cohesion, productivity, etc. of that group and their relationship with the rest of society. Despite the fact that we intuitively know that the use of computerized conferencing might have very fundamental impacts upon groups which use it, there has been very little research into the nature and processes by which such impacts occur. For instance, unlike the excellent program of experiments with audio and visual teleconferencing carried out by the Communications Studies Group (see Short, et. al., 1976) there has not been a single controlled experiment to date which can help us to understand the ways in which computerized conferencing as a form of communication may change the amount, kinds, and outcomes of communications among the members of a group, compared to other communications forms. Evaluation has frequently been limited to field trials incorporating collection of some rudimentary statistics on amount and type of use of a conferencing system, and short questionnaires administered to participants after-the-fact. What is needed at this point is a commitment by the research community engaged in the development and implementation of computerized conferencing systems to undertake all such efforts only in conjunction with a thorough, multi-method program of research and assessment of the effects of various features or characteristics of computer conferencing systems and alternative ways of providing training for users. Ideally, such programs would include a loosely integrated set of controlled experiments, field experiments, observation and automatic monitoring of patterns of

use, questionnaires, unstructured interviews, and perhaps
simulations.*

There are at least four different levels of the impacts of
this new communication medium.

1. Personality level: For example, does working on these
systems make people feel relaxed or nervous; "in touch" with
others or isolated; "in command" of the system as an able
servant, or prisoner of a system s/he feels unable to fully
understand and control?

2. Small Group level: Impacts have to do with the outcome
of reliance on this medium as a primary means of communication,
upon such factors as the productivity, cohesion and morale of
the individual work (or play) group. For instance, does the
system tend to support or to undermine the dominance of a single
leader? To increase or decrease equality of participation in
discussion and decision-making processes?

3. Organizational level: This is focused upon such factors
as the interrelations among the various components of a larger
organization. For instance, how does such a system affect centralization vs. decentralization of decision-making? Does it
encourage greater or less lateral communication among branches or
staff groups on the same level, or more vertical communications
up and down the hierarchy? (See Hiltz and Turoff, 1976, for a
fuller discussion of the potential organizational impacts of
computerized conferencing system.)

4. Societal level: The potential impacts include such things
as the substitution of telecommunication for transportation; and
effects upon the relative power and equality of well being among
various groups within the society.

If one wishes to distinguish a world level from the societal
level, then questions such as effects upon the balance of power
among nations, or the flow of scientific and technical information
among nations, are raised.

---

* The Institute for the Future has tried to take such a multi-
method approach; see, for instance, Vallee, et. al., Vols. II
and III; and Johansen, 1976. However, Johansen argues against
the use of controlled experiments.

The particular impacts which one is going to find depend upon
a complex interaction among at least four sets of factors:

1. What one is looking for, and how, and for how long.
That is, by choosing a level of impact and factors within it to
focus upon, one is probably a priori ruling out the possibility
of finding other types of impacts. What one finds in a study is
partially dependent upon how long it goes on; certainly, the
behavior of users and the impacts of such use will be much
different after five years than after a short two hour experiment.
(Though this is merely speculation; no one has collected user
profiles for longer than a two year period.) Finally, and most
importantly for this set of factors, one's findings are going
to be partially an artifact of the evaluation methodology chosen
(the controlled experiment, the field experiment, the field
trial, questionnaires and interviews with users; participant
observation in and/or content analysis of the proceedings of conferences; or simulations).

2. Features and characteristics of the system itself, and
its implementation. This includes the complexity, flexibility
and style of user interface of the system; and the print speed
of the terminal used.

3. Application areas: that is, the kinds of groups which
are using the system; for what purposes or services; and in what
type of environment. (For example, work at home, remote meetings,
scientific communication, social or educational services, etc.)

4. Characteristics of the user and the immediate environment.
(User attitudes and motivation towards using a computerized communication system; user skills - reading and typing speeds, relative skill and preference for spoken vs. written communication;
type of role played by conference moderators, or other human facilitators on the system; and the total communications and work load
of the user.)

To treat all of these factors as simultaneously interacting
(which they do) generates a matrix which only a computer can
handle! It is perhaps too much to ask for any study in the near
future to treat such a large number of variables; but those who
undertake research in this area should at least be aware of
the variables and methods which they choose to exclude, as a conscious choice in the design of a study.

Each application area has potential impacts at each level.
Usually, evaluation studies select data gathering methods and
variables focused upon the intended consequences at one specific

level, whereas the most important impacts may be unintended ones
at a different level.

V. THE EFFICIENCY QUESTION: AN ILLUSTRATION OF THE NEED FOR EXPERIMENTATION

The appalling but true situation is that we have no empirical
foundation for stating with certainty some of the most basic
impacts of computerized conferencing upon its users. For example,
it would seem very straightforward to determine whether or not
this medium is more or less efficient for group communication than
face-to-face conversations or audio-conferencing, based on the
number of words that can be exchanged among participants in a
given amount of time. Yet, all we have are two conflicting sets
of opinions, both based on inadequate evidence.

Turoff has approached the question in terms of a mathematical formulation. The assumptions are that the average participant in a synchronous (simultaneous on-line participation) computerized conference can type at a rate of half a word a second;
and that the print speed/reading speed is 6 words per second.
Since reading speed is much faster than listening to the spoken
word, and since participants in a computerized conference can all
be writing or reading simultaneously, it is possible for a sizable
group to exchange much more information in a given amount of time
than a face-to-face group, at any group size greater than four.
At about 12-13 participants, the mathematically derived theoretical word exchange rate is about twice as large for computerized
conferencing than for face-to-face discussion. (See Turoff, 1972,
for a complete derivation.)

However, Johansen et. al., (1977) reach a completely opposite
conclusion, on the basis of projections and generalizations from
experiments conducted by Chapanis and his colleagues (1972, 1973,
1974). Chapanis used two persons "talking" to each other via
"slaved typewriters;" that is, whatever one participant typed
simultaneously printed out on the second typewriter; and only one
person could type at a time; and, of course, reading speed was
limited to typing speed. Under these conditions, it was found
that typing was much slower than speaking. Johansen et. al., use
these and similar experiments to conclude, "Both audio and face-
to-face allow many more messages to be exchanged in a given time
period than does typing." (1977, p. 31).

Now it should be obvious that, first of all, "real" people
may not behave in a way that matches the assumptions in Turoff's
formulae. Among other things, they will undoubtedly spend a lot
of time sitting around neither reading nor writing, but thinking

or daydreaming or trying to figure out the mechanics of how to
send their messages. On the other hand, a communication system
of two or three people attached to slaved typewriters bears practically no relationship to one consisting of eight or ten people
working independently through terminals on a conferencing system;
to generalize from one condition to another is unwarranted.
What is obviously needed is some direct experimentation which
measures the word exchange rate of the same groups of people
(with various sized groups), communicating about similar subjects
or problems for the same amount of time over two or three different media.

At the point at which we have some direct empirical evidence
of this type on which to base assertions about relative quantity
of word exchange, of course, it will then be time to try to go on
and answer the next question: in terms of "quality" of the words
exchanged, which is more efficient? That is, it is both time to
solution and quality of solution that have to be taken into
account in measuring "efficiency." It may be, for instance, that
computerized conferencing, with all of the rich non-verbal cues
missing, produces so many misunderstandings that many words must
be exchanged in order to accomplish the same goal. On the other
hand, it may be that the ability of people in a computer conference to "think before they write," and their natural tendency to
be brief when having to express themselves in the more laborious
typewritten manner, will mean that they need fewer words to accomplish the same or better solution or outcome. We simply do
not know.

VI. POLICY ISSUES: THE THREAT OF PREMATURE REGULATION

The first policy problem may, on the surface, appear trivial
to those of us who are technocrats. It is the problem of the name
of this beast we have been discussing. In the literature we find
such alternatives as:

Computerized Conferencing (CC)
Computer Mediated Interaction (CMI)
Computer Based Conferencing (CBC)
Electronic Information exchange (EIE)
Computer Assisted Teleconferencing (CAT)
Computer Mediated Teleconferencing (CMT)
Teleprocessing (TP)
Electronic Mail (EM)
Computer Communications (CC)

In addition to the general names we also have a host of
special names applied to specific systems:

EMISARI, PARTYLINE, DISCUSSION, CONFERENCE, FORUM

PLANET, CMI, EIES, CONFAB, MAILBOX, CONFER, etc.

The result is confusion on what is being talked about.
This is par for the course in terms of a professional area that
is newly developing. However, because the area is one that has
a large commercial potential in the near future some real problems are produced by the use of names.

This is no small issue in the minds of those who may not
understand some of the fine design and technical distinctions we
can employ to attempt to justify one or more of these alternatives.
Consider, for example, the image each name might conjure up for a
legislator or a lawyer working in a regulatory body or some particular company in either the computer or the communications industry. The use of the prefix "tele" immediately presumes in many
minds that this is to be an animal treated exactly like we treat
the phone system. A name like electronic mail automatically
limits the perception of what the technology is really capable of
doing. The word "computer" provokes antagonistic reactions on
the part of a surprisingly large number of people these days.

In many discussions on this form of communications individuals get wrapped up in the terminology problem. This phenomenon is symptomatic of a desire to define such systems as either a
computer system or as a communication system. If this can be
done, then we apparently believe we will know how to handle it and
all problems with it will miraculously disappear. It is the legal
fallacy that the question is one of definition. While this is the
common approach, it highlights why we have problems with effective
regulation. People are able by this preoccupation with definition
to sidestep the real issue: that is, that the objective and purpose of regulating is promoting the "public good." What debates
in this area should really be about is the degree to which the
public interest is served by either regulation or non-regulation;
and if regulation is needed, are the current regulatory mechanisms
and laws really appropriate to this situation?

To fully grasp the policy implications of computerized conferencing one must take care to neither view this area too narrowly
nor to make the mistake of viewing it on a strictly relative basis
compared to other communication alternatives. Traditionally there
are two basic approaches to justifying a computer development:

1. To show that what you have been doing can be done more
    cheaply or in greater volume or more accurately.

2. To do something you have never been able to do before.

The first approach is the common way that one justifies the
development of a computer system. Usually the second method is
too difficult to utilize as a justification process because a
great many people have a problem in perceiving things that have
not been done before and must conceptualize them in terms of what
is being done now. However, most of the major impacts of computer systems, and particularly the unanticipated ones, derive
from computers being able to do things we could not do before.

The specific issues that arise in the U.S. are very different
from many other countries that have historically decided that
communications are to be rigorously regulated or under government
ownership. By such policies, it is our view that many of these
countries may place themselves in disadvantageous positions in
terms of the benefits offered by this technology. Some of the
issues now evident in the U.S. are:

1. Is the regulatory model of these services the single
    phone or mail system, and is this model still an effective
    one even for phones or mail?

2. Are new regulatory mechanisms needed to deal with com-
    puterized services in general?

3. If computerized conferencing is interpreted as a com-
    munication service under the laws governing the FCC,
    does this mean the right of censorship provided under
    the legislation will also apply?

4. Should we view this medium as one for business, or take
    a view that it must be made available in the future to
    the average citizen?

5. What about communications of this type integrated into
    EFT (Electronic Funds Transfer) for bid and barter,
    stock and commodity exchanges? Who are the regulators
    now?

Because of the huge potential for "public good" that these
systems provide, a review of regulatory policies in many areas is
needed. In a sense other computer applications have already
stimulated this controversy in terms of actions now taking place
in the regulatory, executive and legislative bodies in the U.S.

Computer Mediated Communications

Whenever a useful new technology is developed, one policy

question which should be vigorously pursued is how to make it
available to those who cannot afford to buy it themselves.

We believe a key area that is not yet incorporated in the
current debates is the possibility of public use and the resulting
benefits. Besides such things as work at home options and the
resulting savings in energy from the use of information services
integrated into C.C. systems, one obvious area is the benefits
for the disadvantaged in our society.

At first glance, the advent of computer-mediated communications systems would seem to be just one more technological factor
which will be used only by the well educated and well-to-do in the
overdeveloped nations, who are ready with the skills and the money
to take advantage of the opportunities they offer. However, like
Sesame Street on U.S. educational television (another high-cost,
high technology medium), an enlightened and purposeful public
program might be designed to make the new communications medium
serve the disadvantaged rather than compound their disadvantages.

For example, a computerized conferencing system, accessed
through portable terminals that can be used in the home or even
in a prison cell, and including a CAI component, can be used to
bring education, counseling, peer support, and job opportunities
to the mobility-limited aged, the physically or mentally handicapped, or prisoners. (See Hiltz and Turoff, 1976, for a full discussion of this issue.) Taking the physically handicapped child
in the U.S. as an example of possible applications for computerized conferencing systems, there are many special terminal interface devices which enable almost any type of physically handicapped person to write and read from a computer terminal. However, unless funds and encouragement are allocated to such applications by governments, such non-profit applications of computerized conferencing systems are not likely to develop.

We believe that the right of access of citizens to such systems will be necessary for the public good, and the manner in
which a citizen is able to function in a free society will be
dependent on that access. In effect the citizen will need the
same ability to access information systems as he now has to access
the telephone system. Currently, for example, the pricing structures on computer systems and digital networks often cater only
to organizational users by imposing volume minimums. The rather
weak enforcement of interface standards further inhibits the possibilities for citizen users.

The key question of regulation comes down to whether or not
your national post office will make available one conference sys-

tem with only one design, and prohibit other conference systems, including the use of those based in other nations. If the automobile industry had been regulated in this manner, it might still be true that consumers could have any kind of car they wanted, as long as it was black, started with a crank, did not travel over 30 miles an hour and was not used for any international travel.

SUMMARY

At the moment this is an area that should be open to the widest range of investigation and experimentation with the greatest possible incentives to encourage individuals and organizations of all types to be involved. These systems should evolve in such a manner that users have the greatest opportunity to pick and choose from alternative services. There is not the knowledge in existence today, nor will there be for a long time to come, to design systems that could be considered to be the "best." Nor is there any need to, since the implementation of a computerized conferencing system involves only a software system sitting in a single computer. It can be completely divorced from the issues associated with establishing a common communication network providing service for the common good. Such a network can have hundreds of alternative computerized conferencing systems in many different computers satisfying the requirements of many different human groups with differing needs and problems. In terms of the public good there is no justification for artificially limiting the diversity of computerized conferencing and communication systems until we have a great deal more knowledge about the possible applications and impacts of such systems.

We believe that existing computerized conferencing systems resemble the future possibilities in this area as much as the Wright Brothers' little biplane resembles the jumbo jet. However, if premature regulation limits the development and assessment of a wide range of alternative designs and applications, the technology may never get off of the ground.

The specific points that we would like to emphasize are the following:

- The most important impacts will result from new types of applications, rather than from mere substitution for existing communications.

- There is an almost unlimited range of capabilities and applications for computer-structured human communication systems.

- Current names may be inadvertently producing incorrect images of these systems for policy and decision makers. There is considerable danger of mistakenly initiating policy and decision making based upon the characteristics of existing, simple systems.

- What is needed now is a period of innovation, proliferation, experimentation and assessment, during which maximum incentives operate to create a diversity of available systems from a variety of sources.

REFERENCES

Day, Larry. Computer Conferencing: An Overview; in Views from ICCC 1974, edited by N. Macon, International Council for Computer Communication, P. O. Box 9745, Washington, D. C. 70016, 1974.

Hiltz, Starr Roxanne. "Computer Conferencing: Assessing the Social Impact of a New Communications Medium." Paper presented at the American Sociological Assn. meetings, New York, September 1976. Forthcoming in Technological Forecasting and Social Change.

Hiltz, Starr Roxanne and Turoff, Murray. "Computer-Mediated Communications and the Disadvantaged," unpublished paper available from the Foundation at NJIT, 1976.

Hiltz, Starr Roxanne and Turoff, Murray. "Potential Impacts of Computer Conferencing Upon Managerial and Organizational Styles," 1977. NJIT, unpublished.

Hiltz, Starr Roxanne and Turoff, Murray. The Network Nation: Human Communication Via Computer. Reading, Mass.: Addison Wesley, advanced book program, forthcoming.

Johansen, Robert. "Pitfalls in the Social Evaluation of Teleconferencing Media," in Lorne A. Parker and Betsey Riccomini, eds., The Status of the Telephone in Education, Madison: University of Wisconsin Extension Press, 1976.

Johansen, Robert, Vallee, Jacques, Spangler, Kathleen and Shirts, R. Garry. The Camelia Report: A Study of Technical Alternatives and Social Choices in Teleconferencing. Menlo Park, Cal.: Institute for the Future, 1977.

Kupperman, R., Wilcox, R. and Smith, H. "Crisis Management: Some Opportunities," Science, (Feb. 7) 1975: 404-410.

Linstone, Harold A. and Turoff Murray. The Delphi Method:
    Techniques and Applications. Reading, Mass.: Addison
    Wesley Publishing Co., Inc., 1975.

Nilles, Jack M., Frederick R. Carlson, Paul Gray, and Gerard
    Hanneman. "Telecommuting - An Alternative to Urban
    Transportation Congestion." IEEE Transactions on Systems,
    Man and Cybernetics, Vol. SMC-6, No. 2 (Feb) 1976: 77-84.

Panko, Raymond H. The Outlook for Computer Message Services:
    a Preliminary Assessment. Menlo Park, Cal.: Stanford
    Research Institute, 1976.

Scher, Julian. Communication Processes in the Design and
    Implementation of Models, Simulation and Simulation-Games:
    A Selective Review from the Vantage Point of Computerized
    Conferencing. Newark, N.J.: Computerized Conferencing
    and Communications Center, NJIT, Research Report #4, 1977.

Short, John, Williams, Ederyn and Christie, Bruce. The Social
    Psychology of Telecommunications, London: John Wiley &
    Sons, 1976.

Turoff, Murray. "Party-Line" and "Discussion" Computerized
    Conferencing Systems. Proceedings, ICCC 1972.

Turoff, Murray. "The Cost and Revenues of Computerized
    Conferencing." Paper presented at the 1976 meetings of
    ICCC, Toronto, Canada.

Vallee, Jacques, Robert Johansen, Robert Randolph and Arthur
    Hastings. Group Communication Through Computers. Vol.
    2: A Study of Social Effects. Menlo Park, Cal.:
    Institute for the Future, Report R-33, 1974.

Vallee, Jacques, et. al. Group Communication Through Computers.
    Vol. 3: Pragmatics and Dynamics. Menlo Park, Cal.:
    Institute for the Future, 1975.

DISCUSSION OF PAPERS BY DRIOLI, FEDIDA, HILTZ AND HOLMLOV

Discussant and moderator: Stuart Yerrell

Yerrell began the discussion with a table depicting the characteristics of the systems presented in the papers:

System	B/W	Basic Use	Main User	Typical Contact Time	Typical 'Trans^n' Time
U. K. Viewdata	Narrow	Information	General Public	secs/ mins	Indefinite
Computer Conf.	Narrow	Information & Communication	Profes- sionals	hrs/ days	secs/min
European Tele- Conferencing	Broad	Conferencing	Profes- sionals	days	Instan- taneous for most services
Public Switched Broad Band System	Broad	Communication	Profes- sionals & Public	minutes	

Next Yerrell outlined what he felt were the major issues contained in the papers to help structure the discussion:

VIEWDATA:
- Develops general network theory
- Methodology of evaluation used

- Pilot trial
- Next stage
- Theoretical base

SWITCHED PUBLIC BROADBAND SYSTEM:
- Actual results of prospective study
- Methodology (sample size, control groups, etc.)
- Contribution to overall knowledge

COMPUTER CONFERENCING

   - Poor empirical foundation for analysis of impacts especially
     on users--the need for research
   - Danger of premature regulation
   - Nomenclature and taxonomy

EUROPEAN TELE-CONFERENCE (Satellite)

   - Actual results of market study
   - Methodology used: weaknesses, strengths

     Yerrell remarked that at first sight the only similarity among
the systems was that they all ran by electricity; he suggested that
the discussion focus on issues of their evaluation and the methodology
for this, rather than policy. With respect to the paper on Viewdata
networking, he suggested that the discussion focus on the methodology
used in evaluating Viewdata thus far, e.g., what are the criteria for
evaluation? He saw it as the largest new system in the near future
which would affect ordinary citizens as well as professional users.
The paper on computer conferencing also pointed to a problem in evaluation -- the poor empirical foundation for making decisions. Although
data are available on the macro-level effects of this technology, we
don't know why these effects occur. There was nothing equivalent to
the pioneering CSG work on face-to-face encounters. Another question,
raised by the switched broadband system paper was whether the results
were artifacts of the methods used.

     Cowie felt that the Viewdata paper addressed the possible tech-
nical development of the system and did not consider evaluation.
The market trial stage will provide more information about practical
aspects of the system, e.g., how people feel about using the system;
attitudes to costs; how sophisticated and unsophisticated users
react to the same system and the overall acceptance of the system.

     Conrath pointed to the underlying issue brought out by the
papers which is the need for cooperation and awareness among researchers of different systems. For example, Hiltz did not make
references to the work conducted by Englebart. How does this previous work fit in with Hiltz's findings? Other studies have also
been done relevant to the European teleconferencing system but
appear to have had no input into the paper. Or in the case of
Viewdata, can the instruments be used to evaluate that as well as
other systems? All this, in Conrath's view, points to the lack of
communication between communicators about research results and
evaluation.

Yerrell agreed that there was a problem of communication in the field and mentioned that this constitutes a problem on the national as well as the international level.

Another speaker then mentioned that the Viewdata technology, under another name, had been well accepted in Germany and that a market trial for payment will be run. This speaker suggested that we don't need new research. The French are contemplating a Viewdata system.

Ohlman said that the distinction between narrowband and broadband communications is more of interest to systems engineers than to users. Unless they are large users, who must deal with time-cost tradeoffs, most users are unaware of the type of channel over which they are communicating. Much more important to them is the distinction between "mass" and "class" communications systems. For example, Life magazine may have vanished because its market was based upon the voracious appetite for pictures by the general public, which became better satisfied by television. However, specialist magazines giving emphasis to hobbies, sports, consumerism, entertainment, etc., have flourished. In the future, the "class" extensions of television (cable, interactive games, teletext, etc.) will attack and likely displace many of these lucrative information markets.

It is a truism that new systems cannot flourish without good market research and adequate promotion. It is doubtful if such potentially dramatic and far-reaching innovations as telephone conferencing have been adequately researched and promoted, tacked on as they are to communication networks never designed for their switching and fidelity requirements.

Bernemyr said that data show very little demand for video but he doesn't believe that data because we don't know anything about the real demand until we have practiced the product. He pointed to television as an example. We didn't expect color to add much to that system but no one would suggest return to black and white. Adding picture facilities to audio systems may well be analogous to color television. Do, Bernemyr asked, researchers have methods to forecast the future impact of technology once it penetrated the society? We tend to fix on the costs of today's technology--but should focus instead on tomorrow's new and low cost technology.

Baudazzi mentioned that EDP, telecommunications and electronic mail appear to be converging, and that may make it necessary to reorganize how PTTs approach these services. Several nations have seen the decrease in the volume of some classes of mail along with a rise in data communications. It will be important in the 1980s to reconsider postal services in more general terms in light of

these technological developments taking place in the field of
electronic communications.

Cowie said that new technology may enable the better provision
of existing services or give opportunities for new services. In
the latter case, a commitment to a new capability may generate
demand levels which justify the decision but this is not guaranteed
and it is expensive to get it wrong. Hence the providers try to
reduce the risks with as much experimentation as possible around
an evolving programme of new capabilities.

Wells asked Cowie for an explanation of the purpose behind
Viewdata, and said he understood it was to decrease the under-
utilization of domestic lines. If the minimum charge was one call
unit, this might be too much if the information is easily available
elsewhere. On the other hand, the perceived costs of the service
to the consumer may be more important than the actual costs.

Cowie said that a simple charging formula would apply to
Viewdata customers but detailed costing information was not
available. He thought that the original concept of Viewdata was
that it would primarily be an information retrieval service to the
residential user but its potential in the business community and
for other applications was now being recognized.

Hiltz stated that Viewdata can be seen as a limited subset
of computerized conferencing. Specialized systems like Viewdata
offer only a few services; she suggested that users will soon want
more interactive capabilities than the Viewdata system currently
provides. In responding to a comment on secretaries, she stated
that this is a problem; that males who do not know how to type and
who interact only through secretaries tend to see the system as
more cold, formal, and limited in usefulness than those who at
least occasionally get on line themselves. A researchable question is how to resocialize those who are not accustomed to computer
terminals and who resist using them. To some extent, experimental
evidence suggests that this is an age problem as well as a sex-
roles problem.

Irving suggested that people also tend to fear computers and
that we must accustom people to the technology. We don't even
know, he said, what the impacts of the technology are, much less how
to measure them. While many researchers talk about impacts, research on impacts is almost non-existent. What is needed is impact
research over time; in other words, "before and after" studies.

Hiltz agreed, stating that all field trials should include an
attempt to monitor the full range of possible impacts, and that
multi-method approaches are necessary to do this.

Mason commented on the utility of forecasting, saying there
are techniques and models that seem to work. These are particularly

useful when one is replacing technologies which perform the
same function. Unfortunately, in many cases, it is a problem of
replacing old technology with one which performs more functions.
The additional functions may be as important in determining
effects or impacts as the original functions. Mason suggested that
if technology is feasible, if it doesn't harm anyone *a priori*; and
if there is a demand such that it will be purchased, then the
technology almost certainly will be developed. In such cases, we
may know what the secondary impacts are only after they have occurred.

Fields said that future technology costs are important and not
impossible to estimate. And he added that systems like Viewdata
may be rejected because the data is poor.

Holmlöv talked about attitude testing for the Swedish system
and gauging the interest of potential users. He remarked that
the probability of demand for service referred to a test network,
while the measured interest referred to a larger, full-scale
network. Holmlöv agreed with Conrath that instruments should be
shared.

With respect to Fields' comments about not needing research,
Hiltz stated that if the system works, Fields may be right. But
what if the system doesn't work, and research hasn't been done?
Then we have no way of knowing why it failed; and whether some
relatively minor change in design, training of users, or implementation might produce a "success."

Clark said we can construct more technology than we can use--
what's lacking is the ability to identify market needs and use
existing technology to satisfy them wherever possible. New tools
for estimating potential demand are urgently required.

Drioli suggested that the problems with long-range, futures
oriented studies is that you don't know the needs or how they may
change over time.

Yerrell reviewed the discussion, finding two recurrent themes.
First, it was evident that communication among peer groups is poor.
Second, work was needed to find ways to explain the impacts and use
made of communications systems which did not rely on specific technologies.

Bernemyr concluded the discussion by saying that if we look to
the theory of human communications, we find that human communication is more durable than the particular technology used to
communicate. He suggested that the software in communications should
also be considered. For example, with respect to Viewdata, the
quality of the information must be considered--whether it is reliable, current, and whether it is the information the consumer wants.

S.Y.

R.B.

Section Seven

THE INFORMATION SOCIETY

The papers in this section are gathered from three different
sessions of the Syposium. They are linked by a concern with the
changing role of information in society. In his paper, delivered
from Canada using a somewhat rudimentary audioconferencing arrangement, Gordon Thompson distinguished between two types of innovation in terms of the impacts they produce. For the intensive class,
first order impacts are the most significant; notions of efficiency,
productivity and labor release predominate. For the extensive class,
higher order impacts are the more important; here notions of transformation, wealth creation and labor absorption are predominant.
He suggests that in the interaction between information technology
and society the extensive class is, to our detriment, being inhibited,
and discusses three constraints which may be responsible. The paper
concludes with a discussion of strategies which may redress the
imbalance between the intensive and extensive classes.

The paper by Cohen, Conrath, Dumas and du Roure, presented by
Gabriel du Roure, focusses upon the growing overlap between communication systems and information systems. The authors discuss the
desirability of wholescale integration between the two and the
reasons why it has not come about more quickly.

In a highly technical paper Jean-Guy de Chalvron and Nicolas
Curien contribute to research on the information economy, a subject
which has aroused a great deal of interest in the USA over the last
two years, largely as a result of the work of Marc Porat. They
develop Leontief's model of a national economy, so that a distinction can be maintained between informational ("organization") and
other ("realization") work when using the model to examine flows
between different sectors of the economy. ("Organization" is
associated with channeling information and "realization" with
channeling energy.) Illustrative results are presented and discussed.

Money is a form of information which can be radically affected by developments in the electronic processing and transfer of information. Mick Williamson was invited to present a paper on the subject. He provides a broad review of current and prospective developments in the field of electronic funds transfer (EFT), pointing out that physical processing and transmission of notes, coin, checks and so on will be progressively replaced by the automated processing and transmission of information through data networks. He notes, however, that public discussion of the subject is frequently ill-informed and misleading, wrongly suggesting that the cashless and checkless society is imminent. (About 98% of all payments by individuals in Britain and in the USA are still in cash.)

After reviewing changes in banks' payment and associated systems, he turns to change outside the banks, but within the total payment system, remarking that banks' costs are probably less than half the costs of the total system. He next points out that payment systems are only parts of other systems. After discussing the pace and determinants of change in funds transfer systems, he draws some conclusions about the implications of EFT for developments in telecommunications. He suggests that the implications of the social, technological and economic changes giving rise to EFT will be more significant than EFT itself.

There is no discussion summary at the end of this section. Most of the papers were grouped with those of the next section for discussion purposes.

INFORMATION TECHNOLOGY AND SOCIETY

Gordon B. Thompson

Bell Northern Research

Ottawa, Canada

In the world of contemporary physics, the researcher must content himself with knowing only the effects of particles, for the particles themselves can never be observed directly. The investigator must content himself with a track of condensed droplets in a cloud chamber or the ionization of a counter. The physicist can only deal with the effects produced by the particles rather than the particles themselves.

Before the hunter can deal directly with the game he is stalking, he must deal with the effects or tracks produced by his prey. We, in the telecommunications industry, have been negligent in not assessing adaquately the effects of our actions, or innovations.

Previous communications innovations, that have been recognized as being revolutionary, such as the phonetic alphabet, produced extensive impacts that were far reaching and wide ranging througout the host societies. Havelock, 1963, reports on the profound effects of the adoption of the phonetic alphabet in Athens. These effects were very extensive, in that they penetrated far beyond merely increasing the efficiency with which records could be kept.

Activities that produce impacts can be placed in two classes. In the first class, termed the intensive class, the impacts of significance are the primary or first order impacts. In the second, or extensive class, the higher order impacts exceed

the importance of the primary impacts. In the intensive class,
the significant impacts cluster about the activity while in the
extensive class, the important impacts diffuse, propagate, or
extend, throughout other parts of the socio-economic system.

For the proposes of this analysis, innovations will be
divided between the two classes, the intensive class and the
extensive class. The intensive class of innovations involve those
innovations where the principal effect relates to intensifying a
particular ongoing activity. In this class, the notions of
efficiency, productivity, and labor release are predominant. In
the extensive class of innovations, the higher order impacts are
seen to be ones of significance. Here, the notions of
transformation, wealth creation, and labor absorption are the
significant ones.

The interactions between a whole technology and a
socio-economic system can be classified as either intensive or
extensive. The industrial revolution of two hundred years ago in
Great Britain had both sets of interactions present. The
industrial technology impacted agriculture in a largely intensive
way, and resulted in massive increases in agricultural
productivity and attendant reductions in the agricultural labor
force size. That same technology also produced whole new kinds of
businesses and ways of creating wealth, and it is these extensive
interactions that give the industrial revolution its significance.

The performance of the British economy during the latter
part of the 18th century would indicate that there was a
reasonable balance between the intensive and extensive sets of
interactions between the industrial technology and their
socio-economic system. In spite of a population increase of some
40%, the period from 1755 through 1805 saw no significant
unemployment and an average inflation rate of only 1.4%. By a
mere decade later, over one-third of the labor force was deployed
in the industrial sector, and that sector had been recognized as a
major economic wealth creating activity. Gilboy, 1972, reports
that in Lancashire increasing real wages occurred simultaneously
with advancing prices. She goes on to suggest that this can occur
when extraordinary industrial and technical changes are underway.

The performance of our present socio-economic system
under the influence of information technology has been
considerably less satisfying. In spite of our belief that this
technology is significant, it has not produced the same grand
economic benefits as were achieved two hundred years ago by
industrial technology.

If one hypothesizes that the intensive class of

interactions between information technology and society is
proceeding quite well, and that the extensive class of
interactions is being inhibited, a possible explanation for the
lack of satisfactory performance of information technology can be
formulated. Such an hypothesis would suggest that
telecommunications and other information technologies are used by
and large for cost reduction, productivity increases and similar
efficiency oriented uses. Examples of extensive interactions
between information technology and our socio-economic system,
either in the recent past or being planned for the future, are so
rare as to be non-existent. With the exception of the telephone
and broadcasting, both of which are now historic, it is difficult
to find innovations that belong to this extensive class. Such
innovations would be recognized by their characteristics of labor
absorbtion and new business creation.

It is much easier to cite example after example of the
application of telecommunications and information technology in
areas where the principal impact is deemed to be one of increasing
efficiency, increasing productivity, or reducing costs. The usual
"shopping list" of new services put forth by futurists in the
telecommunications area usually consists of new ways of doing old
things, ways which hopefully will be more cost effective.
Intensification!

By re-aggregating the U.S. National Income and Produce
Accounts so as to recognize formally defined Information Sectors,
Porat, 1976, constructed an Input/Output model and examined the
impact on that data of doubling the investment in the Information
Sector, with an accompanying prorata decrease of capital
investments in other sectors. The results showed that the
information sector is a rather shallow one, and such a policy
would have produced more unemployment and less total output. If,
as hypothesized above, we have a propensity to use information
technology principally in intensive ways, ways that release labor,
increase efficiency etc., it is not difficult to see how such
results would come about. The ripple effect from action in the
information sector is not as significant as it is in other sectors
of the economy. The impacts do not appear to be extensive.
Rather, they appear to be intensive.

If there are constraints inhibiting the significant
development of the extensive interactions between information
technology and our socio-economic system, what might be the nature
of the constraints? Three possible candidates for such a role have
emerged.

The first inhibiting factor may well be the product of
information technology itself. The availability of large

computers to operate exquisitely fine tuned accounting systems may
have had a significant impact on our ability to use this new
technology creatively. Precision management, resulting from the
use of these tools, may well have led us to concern ourselves more
with the pine needles than the trees, while missing the forest
entirely.

The second retarding factor is more intractable.
Parker, 1976, points out that information is a different kind of
good and may require a different kind of economy in order to grow
successfully. In shifting from its mercantile economy to an
industrial economy, Great Britain, two hundred years ago, faced no
such economic inconsistency. The gentle beginnings of the
industrial revoltion fitted precisely into the mercantile economy
of the day. No great policy or other changes had to be effected
in order to make the tender beginnings grow.

Western industrial societies have an economic
underpinning based upon ownership of property and scarcity. Goods
are exchanged for money, the money compensating the original owner
for loss of goods. In a very real sense, this process relies upon
the scarcity of goods. To have item A, one must relinquish item
B. The discipline of this exchange becomes less meaningfull as
scarcity is removed. On the other hand, information, when
exchanged, multiplies, and so is not scarce in the limit. The
original owner still has the information even after the new owner
has acquired it. Information is self-replicating. There need be
no shortage. Such a fundamental inconsistency makes prices
difficult to establish. In the past, the physical manifestation
of the information, e.g. the book, was what tagged the
transaction. Telecommunications has now ephemeralized the
physical carrier of information, and such tagging is now
inappropriate.

What seems to be needed is a new set of rules, rules
that would encourage the widespread creation and exchanging of
information for profit. Such an activity presents us with the
opportunity of building a society where the direct conversion of
labor into capital becomes an important characteristic. Galt
McDermott worked for two weeks to produce the score for the
musical 'Hair'. The royalty payments he has received are far in
excess of a million dollars, it is said. They can be thought of
as interest payments from a block of mythical capital created
while he was writing the music. Industrial technology never
offered this kind of capital creation directly from labor. The
only matching circumstance occurred in the various land acts used
to colonize frontier land in North America. In this case, title
to the land was granted after the settler had made certain
improvements and occupied the land for a period of time. Labor

was converted into capital. The ultimate social significance of
the technology that makes this conversion of labor into capital
widely available could be very great indeed. We have configured
neither the rules nor the technology to foster this kind of
activity.

A third negating aspect may be far more fundamental and
therefore more limiting. This factor is linguistic in nature, and
stems from the inability of our information technology toys to
understand language fully. Bar-Hillel, 1964, suggests that if we
cannot teach the dolphin to understand, what chance have we to
teach a machine? He states the impossibility of being able to
create reliable and effective management systems, language
translation systems or general purpose information retrieval
systems using computers. But, he goes on to suggest that we
cannot afford to stop trying. He presents us, in this way, with a
conundrum. He foresaw, many years ago, the limitations of using
machines for the intelligent processing of general or commonplace
language.

Consider the applications of information technology in
the office. Today most of these applications center around
automating the typing task, and are directly aimed at reducing the
cost of producing information. The likely result is a greater
output of lessening quality.

With all this technological help for the author, what
about the intended reader? Firstly, his in-basket will be
overloaded as never before. Secondly, little can be done to aid
him in the reading task beyond giving him a good chair, a lamp and
suitable glasses!

Because each reader of a given document brings a unique
background, interest, and need, the meaning of the document is
unique for each reader. It is this idiosyncratic uniqueness that
is the womb of creativity. To rely on the caprices of a
mechanized information sort performed by an information engine
means the potential loss of most creativity. People differ from
machines in that with machines, garbage in always leads to garbage
out. People, on the other hand, can occasionally produce a gem
from a garbage input.

These problems are illustrative of the difficulties and
dangers we face when attempting to use machines to deal with
meaning. Our natural languages are just not amenable to complete
and reliable machine understanding at this time, or in the
forseeable future. I suspect there is more hope for ESP than
there is for Space Odyssey's HAL 9000 computer!

The three inhibiting factors boil down to two
constraints, one economic and one linguistic. The economic one is
related to how we both perceive and measure utility, while the
linguistic one constrains utility itself. The economic one may be
amenable to policy type fixes while the linguistic one may offer
considerably more resistance. What kind of strategies might be
employed to circumvent these constraints?

The new service bing developed by the British Post
Office, VIEWDATA, may be an example of one such strategy. By
providing a means whereby entrepeneurs can assemble information
services and retail these services through the Post Office
VIEWDATA system, the system could create a flourishing information
service marketplace. Such a development would see a shift from
the idea of "author" to "information entrepreneur". The investment
required of such an entrepreneur, to launch an information
business, is largely his own time and effort in creating and
assembling an information rich service that has utility for large
numbers of users. An income tax expert could assemble a program
on filling out income tax forms. A lawyer could put together a
program to assist people in purchasing and selling real estate. A
history teacher could create a program that simulates a letter
interchange with your great great grandfather! To the extent that
these services get used, royalties would be paid to the creators
and so the opportunities for a repetition of the Galt McDermott
syndrome would increase. To the extent that this kind of activity
in fact builds, VIEWDATA may well be one of the rare examples of
an extensive interaction between information technology and our
socio-economic system. Hopefully the purveyors of VIEWDATA will
encourage such activities.

Our telecommunications networks today require that you
know the explicit address of the recipient for your messages.
Surely with the advent of cheap processing, it should be possible
to build communication networks with far more intelligence than
just the ability to connect well defined ports. Perhaps by
incorporating a library of Computer Aided Learning material, the
network could ascertain the emerging interest profiles of users,
and so bring together learners with shared interests. Such a
profile would be far more dynamic and alive than one that was
determined from a list of categories. During periods of
heightened interest and learning, the vocabulary one uses is
somewhat fuzzy, and a profile determined from behavior in a
Computer Aided Learning environment would be far more useful than
one composed from a standard list of subjects. An intelligent
network could easily accomplish this task, given access to a data
base created by peoples' wanderings through its stored riches.
The provisioning of our communication networks with sufficient
intelligence to develop profiles based on our use of the resources

stored in that system evades the linguistic constraint, for the algorithms used in this process are really quite straight forward. There remains a set of strategies that meet the linguistic thing head on.

The probability of being able to develop machines that understand language may be very low and consequently little hope should be placed in the immediate prospect of some large research organization really being able to produce Space Odyssey's HAL computer. Another more attractive strategy seems open to us at this time. Computer graphics could capitalize upon Western society's propensity for graphical communication. Already, in Ontario schools, Bliss symbols, (Bliss,1965), are being used to allow non-verbal children the benefits of communication. A computer system in Ottawa "talks" in Bliss symbols. However, these symbols do not make significant use of the powers of our modern information technology.

A totally new iconic language could be developed, based on the ability of the computer to produce "time varying icons". By using key-frame animation techniques, a range of a few hundred basic symbols could be used much as we use phonemes in spoken language. Thompson, 1975, has described such a language. Hopefully if such a scheme were pursued, we would consider the potential shifting from our spoken syntax to some alternative in order to free ourselves from the constraints that that syntax carries. The more facile interchange between space and time that characterizes Chinese syntax provides an interesting example, particularly as modern physics becomes more and more concerned with such interchanges.

Admittedly, such developments have little utility for you and me, for the impacts here would likely be reserved for future generations. The acute cash discounting we use for future benefits may well inhibit the careful development of such a strategy, and so the first inhibiting factor mentioned above, may prevent this kind of development from occurring. However one cannot help but wonder at the increase in creativity that could be accomplished by providing future citizens with such a powerful communications tool. Had the early Greeks used an accounting system with the sophisticated precision of those that we use today, and also used the future benefit discounting techniques such as we use today, it is unlikely that they would have ever considered adopting the phonetic alphabet. The cost of such a conversion would have just been too exorbitant in terms of the perceived benefits. Thank God the Greeks didn't have computers working for their accountants.

Yes, information technology certainly seems confined to

interacting with our socio-economic system in ways that merely
intensify what we are already doing. Even the bright shiny new
ideas generally seem devoid of the potential of creating impacts
that extend widely throughout the socio-economic system, rippling
through the whole system.

Unfortunately, except for one or two small exceptions
that just might turn out to be very important, VIEWDATA for
example, most strategies that would redress this imbalance between
the intensive set of innovations and the extensive set involve
such long delays before pay out that our present view of future
discounting gives them little chance for active and concerted
development. Such a prospect is disturbing in the face of
presently accepted popular wisdom. Such misalignments should
surely signal the need for good research.

References

Havelock, Eric A. "Preface to Plato" Belknap Press of Harvard
    University Press, Cambridge, Mass. 1963. PP 208-209.

Gilboy, Elizabeth W. "The Cost of Living and Real Wages in
    Eighteenth Century England" Appears in "Europe and the
    Industrial Revolution" edited by Sima Lieberman,
    Schenkman Publishing Company, Cambridge, Mass., 1972,
    PP 193-204.

Porat, Marc V. "The Information Economy" Volume I, Institute for
    Communication Research, Stanford University, Stanford
    California, 1976. PP 152-155

Parker, Edwin B, "Social Implications of Computers/Telecom
    Systems", Telecommunications Policy, December 1976.
    PP 3-20

Bar Hillel, Yehoshua, "Language and Information" Addison-Wesley
    Publishing Co. Reading Mass. 1964

Bliss, Charles C., Sementography (Blissymbolics)", Second
    enlarged edition. Sementography
    (Blissymbolics) Publications, (non-profit),
    P.O. Box 222, Coogee 2034, Sydney, AUSTRALIA.

Thompson, Gordon B., "Instant Communications in the Future
    Wired City", National Telecommunications
    Conference, New Orleans, December, 1975.
    I.E.E.E. Cat. No. 75 CH 1015-7 CSCB.

INFORMATION AND COMMUNICATION: IS THERE A SYSTEM?

Jean-Claude Cohen, David W. Conrath, Philippe Dumas &
Gabriel du Roure

Institut d'Administration des Entreprises

Université d'Aix-Marseille, Aix-en-Provence, France

INTRODUCTION

The technology and utilization of information systems and communication systems have come to a point where the two are beginning to overlap. For example the digital computer, a core ingredient for large scale information systems, is becoming an equally important factor in the switched telecommunications network. From the other perspective, we note that networks of computers and the construction of large time-shared information systems are dependant upon the availability of communication networks. Thus, it is not unreasonable to assume that one day the two systems may become a fully-integrated, comprehensive information-communication system which would be reached through a single multi-purpose apparatus available to everyone.

The realisation of such a prediction depends upon several things, not the least of which is the desirability of an integrated system. Thus we commence our paper with a discussion of its desirability, both on theoretical and practical grounds. But then, one might ask, if such an integration appears to be desirable, why has it not evolved more quickly? After looking at some of the reasons to explain the present existing differentiation, we consider, from economic, social and technical points of view, the work that should be done before the development of an information-communication system (ICS) takes place. The alternative is the development of a system through happenstance.

ARGUMENTS FOR INTEGRATION - THEORETICAL ASPECTS

One argument for integration stems from the two definitions of information systems (IS) and communication systems (CS) and their links with general systems theory (GST). Another results from theoretical considerations of the value of information.

IS is defined as the set of tools, techniques and methods used to collect, store, retrieve, manipulate and communicate data in an organization. CS is intended as the set of tools, techniques and methods used to transmit, over time and/or space, signals, signs and/or data. Both systems essentially deal with the same matter, namely data, although some nuances may be introduced between data, signals and information, especially with respect to the implications of "meaning" when one speaks of information. We note, however, that Shannon's theory (1948), which was a basis for the development of both modern communication and information systems theory, clearly states that it deals with data, not with meaning.

One might argue that it would be useful in practice to distinguish between the IS and CS subsystems of an ICS. Certainly there are times when one would wish to consider the data processing aspects of an ICS as distinct from those of the telecommunications network used to link users with a computer. But similarly, it is often useful to disaggregate the hardware of an IS from the software, and to disaggregate the software into its various functional components. Nevertheless, this does not negate the value of conceiving of an IS as an integrated whole. Likewise, the fact that there are reasons for considering subsystems of an ICS as distinct one from the other is not a sufficient reason for ignoring them as integrated components of a higher-level system. Finally, it is relevant to look at the work of Marshack (1968) on the economics of information. When calculating both costs and benefits, communication of data is treated in the same respect as the storage, retrieval and processing processes. More than just avoiding differentiation, Marshack treats communication and information as part of an interdependent and integrated whole.

Another common theoretical perspective is related to the study of organizational behaviour from the standpoint of General Systems Theory. Whatever the real impact of the systems approach has been on management theory and practice, and no doubt the impact often has been over-emphasized, GST does provide a framework for the formal definition of an information system. Interestingly, this formal definition either implicitly or explicitly includes communication. Thus, within existing theory one can find the concept of an integrated information-communication system. Let us see how this has emerged from the application of the systems approach to organizations and their utilization of information.

Following Forrester (1961), Blumenthal (1965) established that the information system of a firm is the system which coordinates, by means of information, the major flows processed. In the case of an industrial firm these would be people, raw material, goods, equipment, financial flows, etc. The coordination of such flows clearly implies the requirement of communication. Somewhat later Emery (1969) was more explicit. He presented his planning and control system as a CS, for he saw the plans, and the subsequent controls, as the basis for communication within an organization.

Ackoff (1974) and Dumas (1976) conceive of IS as the interface between the processing subsystem and the control subsystem of a general system, the latter being understood as a set of two or more interrelated elements assembled dynamically on behalf of a goal. In this sense an "information system" transcends the traditional perception of computer systems and communication systems. The IS becomes an information-communication system, the objective of which is not only to relate people to each other, but more generally to provide them with a means for controlling a goal seeking higher-level system. That higher-level system may be a workshop, a corporation, a government, or society as a whole.

THE PUSH TOWARDS INTEGRATION IN PRACTICE

We noted in the introduction that practical examples of the value of integrating information systems with communication systems already exist. If we look at things somewhat more closely, we find examples of the trend towards integration at three levels: the sharing of common technologies, the reliance of one system upon the other, and the present and proposed uses of ISs and Css.

The most obvious sharing of technologies is the use of the computer for both ISs and CSs. The actual concept of an IS, per se, would probably not exist today if the computer were not available to store and process large quantities of data. The use of the computer for CSs is more recent, but the trend is just as dramatic. The long range prospect is one of a completely digital transmission and computer switched telecommunications network that will be far more reliable and more economical than the present electro-mechanical network.

The use of the computer appears to be a movement from IS to CS, but the technology required for the computer was given some substantial support by technology originally developed for telecommunications. The need for very small electronic components, which led to the development of the transistor, was based on the need for very small and highly reliable repeaters for transmission

of telecommunication signals over long distances, especially from
continent to continent. These early results of miniaturization
made it feasible to build computers with a processing and storage
capacity that vastly exceeded that possible with vacuum tube
technology. The drive for miniaturization is now coming from
the opposite direction, and while the computer industry is reaping
the immediate benefits, the communication industry will benefit
over the long run. The existence of an all-digital, computer controlled network would not be feasible without the most recent
developments in electronics.

The increasing dependence of one system upon the other at the
network and control level is obvious to virtually everyone working
in either IS or CS today. One could argue that the concept of
traffic control of a communications network directly derives from
the IS concept of control. This in turn is perhaps best performed
through a computer-switched network where some processing is
available for optimizing flows - for example, the control subsystems of the French Transpac network. Conversely, the computer
controlled switching center can do a great many things with the
signals it receives, that are no longer mere transmission. For
instance, it can remember a telephone number which was dialled but
which led to a "busy" signal, and ring back the caller when the
required number becomes free. The switcher is also capable of
providing information independent of the signals that one might
expect to receive from the other party, such as the numbers of
who called whom, when, for how long, etc.

One other pressure for an all digital network is that more and
more of the signals being transmitted via the telecommunications
systems originate in a digital form. One computer is being put into
communication with another. In one instance we have the existence
of time-shared systems, with many users wishing to access a single
large computer. Such systems are quite common in large organizations today where many of the data bases are centralized, but users
of which are geographically dispersed. Another instance of the
dependence of IS on CS are the large computer networks that are
being developed, where users of one computer also have access to
many others if they so desire. The ARPA network is probably the
most dramatic example, but others approaching it in size, such as
Cyclades - the European Informatic Network, are being built in
several locations around the world. Clearly the existence of such
IS networks is totally dependent upon a supporting communication
network.

The existence of IS/CS networks is leading to another phenomenon. Computers are now being used increasingly as communication
devices. Users of these complex systems found that they had a
need to communicate with one another. The telephone, while often

adequate, did not always serve the purpose. This was particularly
true when two parties who wished to communicate were not available
to each other at the same time. The result was the development
of message systems which used the same computer network that was
being used for data processing and information transfer activities.
Only now, this information transfer reasonably reflects something
like a real-time mail system. The IS is becoming, in reality, a
CS as well.

     Another push towards integration is being given by the users
of both IS and CS. Because of the existing differentiation,
users have to deal with a variety of problems. First, dealing with
separate IS and CS doubles the efforts on various matters, such as
the choice of devices, agreement on buying or leasing, maintenance,
and so on. Secondly, the user has to worry about interface problems
with respect to the compatibility of communication and information
technology. This typically requires a special "interface" device,
and often also the onerous help of specialists in that field. For
the user at least, the splitting of these two technologies is perceived to be an unnecessary bother. This, somewhat amplified
through users' associations, has brought about a real effort in
defining international standards for Data Communication devices
through organizations such as ISO, CCITT, and so on. But this
is only a first step.

     The question then is why such integration has not yet taken
place, or at least why it is moving at such a slow pace. For
example, why have ATT and IBM been so close in technology and yet
so far apart from the user's perspective? Let us look at some
reasons which could explain the present differentiation.

                    REASONS FOR DIFFERENTIATION

     The reasons explaining the present separation between IS and
CS are complex and heavily dependent upon the historical development of the two fields. Let us examine how we have arrived at the
present state of affairs. We shall do this by commenting first
on the purposes that CS and IS were initially designed to serve
and on the markets that developed as a result. The different
markets in turn led to different responses by governments, which
further maintained differentiation, especially from the standpoint
of the suppliers. Throughout we shall indicate some conceptual
differences that aided and abetted the differentiation.

     Interpersonal communication systems have been in existence
for many centuries, and they have been recognized for what they are.
While one often thinks of the telephone network when one speaks of
a communication system, it is in fact a rather late arrival on

the scene. Systems of messengers have existed at least since the
time of ancient Greece, and reasonable effective postal systems
have been in existence since the Middle Ages. One characteristic
common to all of these systems is that they typically existed for
the purpose of permitting any individual to be put into contact
with any other individual. The concept of a complete system was
all-pervasive.

The fact that communication systems were presumed to be relatively global had several significant implications. Perhaps the
most fundamental was that the system should be simple enough to use
so that it could be available to almost anyone with the need and
the funds required to use it. Sophistication of the user could
not be required, and the design of the system had to take this into
account. In fact the value of the system was generally reflected
by the number of people who could be contacted by it, and thus value
and simplicity walked hand-in-hand.

The concept of a global communication system meant that it
existed externally to any given organization. An internal network
could be part of the overall system, but it was treated as only a
subsystem. The use and value of the total system existed because
it was external to any given enterprise. Therefore, it is not
surprising that governments have had an early interest in communication systems. Today virtually all communication systems of any
significance - those which cross organizational boundaries - are
either government-owned or government-controlled.

Information systems have had a very different history. The
concept of IS is recent as it grew out of the development of the
computer industry. The desire was to use the computer's processing
capability to do something more than just store and retrieve data.
There was the recognition that information per se had value, albeit
value usually restricted to a particular possessor of that information. Therefore, ISs were intra- rather than extra-organizational
in orientation. It was irrelevant that the user of the system might
require many years of training to use it well, if the value of the
system's output exceeded the costs involved in its use. Thus, IS
tended to become more and more specific to the user, in contrast
to the essentially common elements which existed for CS interfaces.

Because of specificity of an IS, governments have been far less
concerned with their regulation and with the market structure that
provides them. However, today the situation is changing, and for
reasons related to communication. Perhaps the most important one
is that ISs now exist which are almost as pervasive as most communication systems. Systems which contain data on one's credit rating
are one example. Perhaps of greater importance are systems linking

the financial structure within a country and those associated with the military. Thus the information industry is rapidly becoming a major concern for most governments. As yet, however, there is little evidence that they link this concern with that which they show for communication systems. Thus, the fact that the government has a vested interest in the development and control of each industry is no assurance that the two will be seen as parts of a higher level system, the ICS.

Perhaps one reason why governments, and users as well, have not viewed the two industries as merging into one is the way the two have been perceived. The goal in communications has been to reproduce the signal transmitted as faithfully as possible. Since the signal itself is relatively measurable, standards have been set with respect to the communications network. Subject to these standards, the emphasis has been on cost reduction. Innovation has been focused on doing the same things more efficiently. In fact, this has been the major argument leading to the acceptance of computer controlled switchers. The one significant exception, the visual telephone, has been notable for its failure. The presumption was that adding a visual image would provide a more faithful reproduction of the original signal than just the voice channel. However, it is now obvious that this increase in fidelity is marginal compared with the increase in cost. The increment in value was not substantial.

Information systems, however, have been designed and sold on the basis that they are value enhancing, something which is very difficult to measure. In the early stages of the computer industry, the emphasis was on the automation of manual processes. Since then the stress has been on developing hardware and software in order to perform tasks which were not previously being done, but which would have substantial value if done. Thus, CSs have evolved with the idea of duplicating reality, whereas information systems have been designed with the idea of increasing our comprehension of it.

But times are changing. The communications industry, especially telecommunications, is gradually beginning to realize that the ISs they are building to control the use of a network can also be used to enhance its value. The computer behind the switcher can do things that could not have been done before without it. It is only a matter of historically constrained insight that has prevented the development of a great many value-enhancing activities that could have been performed by these computers. But if the value is there, the developments will come - if not today, then tomorrow.

The computer industry is very aware of another fact. Not only is the computer a necessary backbone for most ISs, but, as the telecommunication people know, it is a very efficient switcher. In

fact, that is the technical basis of its operation. Thus a computer which is used to process data can also be used to switch signals
from one terminal to another, whether that terminal is a mini-
computer or a telephone. Hence, several computer manufacturers
now make digital switchers which can be used by the telecommunications industry. The differentiation which has existed in the past
is rapidly disappearing. Parts of the technology are merging,
markets are beginning to merge, and soon governments will be worrying. Perhaps then the remaining set of questions concerns what they
should be worrying about.

IGNORANCE MAY NOT LEAD TO BLISS

Most of the people involved in IS or CS fields would probably
recognize the key issue as being: "Who is going to operate and
control what?" Should a merger of the existing IS and CS industries
take place? Would it be more appropriate if one of them essentially
took over the other?

Such broad questions are undoubtedly going to be answered on
political as well as economic and technological grounds. However,
even political issues can benefit from the knowledge of the various
alternatives and the implications of each. Thus, let us examine
some of the things that should be done before such global questions
can be handled meaningfully.

A first step is the development of the theoretical grounds on
which one should examine questions relevant to the domain of an
ICS. More specifically, a conceptual basis is required before we
can answer such questions as: "What is an Information Communication System?" "What are its boundaries and basic components?"
Stress should also be laid on the control function of an ICS. This
would probably mean that functions like "Data Base Administrator"
and "Network and Switching Control Unit" should be included in these
considerations.

Such research themes require multidisciplinary teams, comprising communication or information specialists and social scientists,
as well as experts in technology. Perhaps their most important
function would be to establish measures and the means of measurement, common to both IS and CS, of system usage, of factors affecting usage and of performance. Measures of usage would include those
of capacity and content, and are needed to specify systems in
technical terms. Factors affecting usage would include tasks and
personal idiosyncracies. These are required to build models which
will predict and explain usage. Measures of performance would
have to be capable of determining the relative value of one system
to another, hopefully on a cardinal scale so that value can be related directly with costs.

Unfortunately, while these measures would appear to be needed now as well as for the future, no sets of measures even come close to meeting the requirements. More particularly, if one is to dis - cuss the integration of two very large industries, unless the discussion can be conducted in terms of a commonly understood basis - an agreed upon basis for determining needs and for evaluating how well they have been met - emotion rather than reason will prevail. To date, that seems to be what is happening.

The next steps have their roots in present practice. One consists of evaluating the social impact of an ICS through a quantitative analysis (using the tools defined in the previous steps) of what is currently happening in IS and CS. Such an investigation would also bring to light more evidence of the possible drawbacks which a technical integration of IS and CS would bring about. The other concerns the testing and evaluation of the new technologies, both in the laboratory under controlled conditions and in the field under operational conditions. Neither is sufficient, and an appropriate evaluation scheme has one feeding into the other. Particular problems found in the field should be sent back to the laboratory for more highly controlled analyses. Results from the lab should be further tested in the field to see to what extent laboratory results can be extrapolated to operational conditions. This is frequently done in terms of engineering design. Why is it not done in terms of the behavioural input into design? If it is because of a shortage of measures for determining usage and evaluating performance, let us ensure that these are developed so that this can no longer be used as an excuse.

After all the above is done, one should then be in a position to suggest how the system (an ICS, or separate ISs and CSs) might be configured, and the basis for its operation and control. But now another problem arises. If this decision must await the results of a lengthy research process, who would be motivated to undertake these tasks? The answer would appear to be reasonably obvious. In essence, it is the end users who require these answers, and they, in most instances, are effectively represented by the government. If the work were military in nature, the government would either do the research or pay to have it done, quite apart from the production of the final goods. This has been the case for governments of all political persuasions. Communication and information are about as essential to the well-being of a nation as the military. Furthermore, they are constructive rather than destructive. It is now time to bring about some coordinated research efforts to ensure that the construction is based on a firm foundation, rather than on poorly advised political decisions based on the whims of technology. Can we convince the various governments that it is in their best interests to do so?

REFERENCES

Ackoff, R.L. (1974): Redesigning the Future: A Systems Approach
   to Societal Problems. New York: Wiley.

Blumenthal, S.C. (1965): Management Information Systems: A Framework for Planning and Development. Englewood Cliffs, N.J.:
   Prentice Hall.

Dumas, P. (1976): General System Theory: A Behavioral Approach.
   Aix-en-Provence: Institut d'Administration des Entreprises.

Emery, J.C. (1969): Organizational Planning and Control Systems.
   New York: Macmillan.

Forrester, J.W. (1961): Industrial Dynamics. Cambridge, Mass.:
   M.I.T. Press.

Hayakawa, S.I. (1949): Language in Thought and Action. New York:
   Harcourt, Brace and World.

Marshack, J. (1968): "Economics of Inquiring, Communicating, Deciding," American Economic Review, 58, 1-18.

Shannon, C.E. (1948): "The Mathematical Theory of Communication,"
   Bell System Technical Journal, 27, 379-423 & 623-658.

INFORMATION, ENERGY AND LABOUR FORCE

J. G. de Chalvron and N. Curien

Ecole Nationale Superieure des Télécommunication

46, rue Barrault, 75013 Paris, France

INTRODUCTION

This study is part of an extensive research program whose objective is to discern the role in economic mechanisms of all the phenomena connected, on one hand, with information transmission (1) - (11) and, on the other hand, with energy transmission (12) - (19).

Here we are interested in the production process. We will try to describe this process as a succession of matter transformations by information and energy flows channeled by the labour force. The totality of these transformations enables economic goods and services to be elaborated.

The concepts of energy and information are ambiguous and can be defined in a multitude of ways according to the nature of the underlying economic realities being considered. For this reason, in Section 1, we give a precise meaning to the words "information" and "energy" for use in our description of the production system. We explain how organization of the production effort (which channels information) is complementary to the realization (carrying out) of the production effort (which channels energy).

With a view to carrying out a statistical study, we set out in Section 2 a theoretical model of the production system, explicitly emphasizing its labour flows. This model is taken from the Léontief model of an economy in static equilibrium (20) - (21) and is akin to the method known as "labour equivalent" (22/3). The model enables organization and realization work to be taken into account, and it emphasizes their joint contribution to economic products.

In Section 3 we set out the Input-Output table of the Fre .ch National Accounting System that enables us to implement the mode l introduced in Section 2. We then indicate the modification and pro cessing required for our study.

Finally, in Section 4, equipped with a theoretical model and adapted statistical data, we endeavour to evaluate numerically the respective roles of each of the two classes of activities (organization and realization) in the distribution of employment within each economic sector, on the one hand, and in the composition of each product in different forms of incorporated work, on the other hand.

We close by giving an outline of different types of products determined by their "labour-structure" and endeavour to discern the particular role of the economic sectors responsible for transmission of information (telecommunications and printing), energy (energy distribution) and matter (transport).

1. ORGANIZATION AND REALIZATION ACTIVITIES

Let us consider our description of the production system: the production process is a succession of matter transformations made possible by information and energy flows channeled by labour. Thus, first of all, we must define information channeling labour and energy channeling labour, then indicate how the total labour force can be broken down into two corresponding classes. In order to accomplish this, we first consider the most elementary level of the production system, i.e. that of the individual at work, then adopt a hierarchical approach by successively immersing the individual in the firm and then the firm in the National Economy.

1.1. Labour and the Individual

Every task calls for thought and organization, complemented by execution and realization. The phase of conceptual analysis supplies the necessary data for the phase of physically executing the job. It is in this sense that we say that the first phase consists of channeling information and the second one of channeling energy.

More precisely, the individual at work extracts the information (or messages) that he or she uses from a collection of prior attainments (educational knowledge or professional training) and

from all existing material data (field of physical possibilities),
verbal data (orders, instructions), or written data (work specifications), which he receives from his environment. Thus, the individual functions as an open system which processes information
(organization phase) before directing the energy (realization phase).

However, at this level of an individual's activity in relation
to other individuals, it is very difficult to divide the work-time
into time spent organizing and time spent realizing. The two are
often effected simultaneously, and cannot, therefore, be dissociated.

It is conceivable, for example, that a very close ergonomic
study could, for a well-defined work station, distinguish periods
of creative activity and exchange of information from periods of
direct work execution. However the discriminatory analysis of a
job which has a more complex structure, that of an engineer for
example, would be more difficult. For this reason, in order to
facilitate the evaluation of the organization/realization activities, we have chosen to observe the productive system through a
magnifying glass instead of through a microscope and to take as
the elementary agent a department of the firm (in relation to other
departments) rather than a sole individual (in relation to other
individuals). However, we will retain the fundamental concept that
information and energy are dual notions refering to two types of
labour - organization and realization - of one agent in relation
to other agents. This concept is independent of the level of
observation of the production system and is transposed without
difficulty from the "individual" level to the "firm department"
level.

1.2. Labour and the Firm

As a national intermediary between the individual at work
and the National Economy, the firm represents an appropriate field
for study. It is complex enough to constitute a real production
unit and limited enough to enable each of its constituent agents
to be clearly identified. Of course, on this scale the elementary
agents are not individuals, who are too minute to be distinguished,
but departments associated with the fundamental functions, i.e.
manufacturing, commercial, administrative, research and development
....functions. Let us, very schematically, isolate the manufacturing function, the task of which is the physical creation of the
product. The objective of all the other functions is to create,
to transform or to transmit all useful messages to the manufacturing organization; thus we can say that these functions weave
the firm's information web, an infrastructure necessary in the
execution of the energetic process of manufacturing.

Thus, we separate the different functions of the firm and re-
group them into two main aggregate functions, respectively linked
to information and energy: the organization of production and
the realization of production.

This conceptual model has already been the subject of numerous scientific studies. For example, information science applied
to organizations examines the inter-relations between the realization
function (manufacturing) and the organization function, in liaison
with the graph of hierarchical relationships within the firm (24).
By extending such a sociological analysis (25) it should be possible
to interpret the evolution of information and communication systems
at a national level. In a fully determined social framework, is
the development of these systems accompanied by a rise in overall
productivity of the National Economy or, on the contrary, does it
give rise to waste and message redundancy, an amplification of
"noise"? (10)

While conscious of the interest of such an approach essential
to the understanding of the mechanisms that drive the economy, we
will limit ourselves here to viewing the economy from an accounts
perspective. This approach permits evaluation of the respective
importance of information and energy in the production process.

1.3. Labour and the Productive System

The production system, as a whole, can be considered as a
collection of economic sectors. Each sector associated with the
production of a class of goods or services, is an aggregation of
firms. This particular description, which comes from National
Accountancy, makes the production system appear as a huge firm producing all economic goods and services. In order to carry out
production, this firm requires the two main functions that we have
just set out.

At the level of the individual firm, the distribution of the
work force among the different departments is indicative of the
information-energy balance in the production structure. Analytical
accountancy makes it possible for the price of the product to be
broken down into realization cost (manufacture) and organization
cost (management, marketing,...) and thus we can observe how the
two functions contribute to the value of the product.

We now propose to carry out these same studies at the level
of the National Production System, which, as we have seen, is simply
a multi-sector firm. We shall therefore endeavour

- first, to examine the labour structure (the distribution of organization and realization labour) in each sector.

- secondly, to study how the inter-industrial exchange process (described by Léontief's model), which creates inter-sector labour-flows (labour fixed in intermediate consumption), leads to the incorporation of labour of each of the two types into economic goods.

2. LABOUR AND THE PRODUCTIVE SYSTEM - A THEORETICAL MODEL

A model is needed to describe the way in which inter-industrial exchanges lead to the incorporation of labour into economic products. The idea which first comes to mind is to use Léontief's classical model of the economy in static equilibrium, the basis for National Accounts, and to adapt it to the requirements of our study, i.e. to take into account the labour flows.

2.1. The Léontief Model - Equilibrium in the Space of Goods

Let us consider, therefore, a Léontief model of production in static equilibrium, i.e. an economic model with the following characteristics:

There are n goods (or products). Each is produced by one and only one technique (or sector), which produces nothing but this good. Each technique requires the same application time, which we choose as a unit period. Finally, to produce its good, each technique employs factors selected from the n goods as well as labour; we suppose that there is no substituability among the factors. Thus, to produce one physical unit of the good j (j=1, ...n), one must consume in the associated production process, the quantities $a_{1j},...a_{nj}$, called the "technical coefficients" of the goods $1,...n$, and the quantity t_j of labour (measured, for example, in manpower or in working hours). Let x_j by the quantity of the good j produced by the technique j during one period; x_j also measures the level of activity of technique j. Presuming that the returns are constant, the consumption per period by technique j of good i is $a_{ij} x_j$; its consumption in labour is $t_j x_j$.

During each period the quantity y_j of the production of good j of the preceding period, which is not used in the form of intermediate consumption, is destined for final consumption. There is no stock accumulation.

Let :

$$A = \begin{pmatrix} a_{11} & \cdots & a_{1n} \\ \vdots & & \vdots \\ a_{n1} & \cdots & a_{nn} \end{pmatrix}$$

$$X = \begin{bmatrix} x_1 \\ \vdots \\ x_n \end{bmatrix} \qquad Y = \begin{bmatrix} y_1 \\ \vdots \\ y_n \end{bmatrix}$$

$$T = \begin{pmatrix} t_1 & & \\ & \ddots & \\ & & t_n \end{pmatrix}$$

For equilibrium of the physical quantities, during the successive periods production activities and final consumption must be repeated in exactly the same way as in Diagram 1. This implies that for each product the total production for each period will be equal to the amount of intermediate consumption plus that of the final consumption in this same period. Thus:

$$x_i = \sum_{j=1}^{n} a_{ij} x_j + y_i \qquad\qquad i = 1,\dots n$$

or, in terms of matrices :

(1) $X = AX + Y$

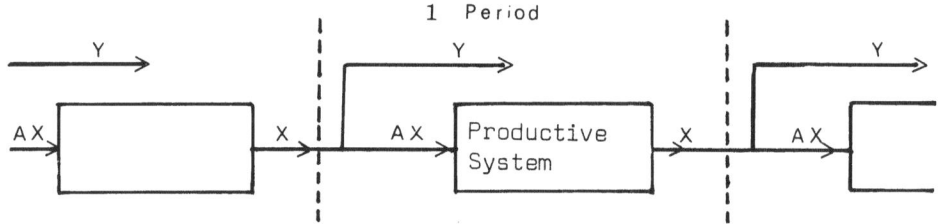

Diagram 1 : Static Equilibrium

When the matrix A is productive*, an hypothesis which is in practice not very limiting, we can show that matrix $(I - A)^{-1}$ exists and that all its elements are either positive or zero. Thus we can

---

* Matrix A with positive or zero elements is said to be productive when there is a X_0 vector with positive or zero elements in such a way that vector $X_0 - AX_0$ has strictly positive elements.

write (1) in the following form:

(2) $X = (I-A)^{-1} Y$

Any vector of final consumption Y (with positive or zero elements) can thus be satisfied by an exclusive production vector X (with positive or zero elements).

Equation (2) can also be interpreted in an inter-temporaral way. When A is productive, all its eigenvalues are of modulus less than 1, and the series $I + A + A^2 + \ldots + A^n + \ldots$ is convergent and its sum is $(I-A)^{-1}$. Thus we can write:

(3) $X = Y + AY + A^2Y + \ldots + A^nY + \ldots$

The above equation (3) simply shows that to bring the final consumption out of the production system during each period, both present and in the future, it is necessary at the start of the present period 0 to set the quantities of goods:

Y to satisfy final consumption in period 0

+ AY to produce in one period, Y destined for final consumption during period 1

+ ....

+ A^nY to produce in n periods, Y destined for final consumption during period n.

+ ....

2.2. Equilibrium in the Space of Labour

The labour consumed by a given sector during a period is incorporated into the product of this sector in the same way as intermediate goods. In fact, it is this very consumed labour which creates the added value supplied by the sector. Since the product of each sector is used by the other sectors as intermediate goods, we can see that through the inter-industrial exchange process each product incorporates labour quantities emanating from all sectors.

Let z_{ij} be the quantity of labour issuing from sector i, incorporated in one unit of product j and let:

$$Z = \begin{pmatrix} z_{11} & \cdots & z_{1n} \\ \vdots & & \vdots \\ z_{N1} & \cdots & z_{nn} \end{pmatrix}$$

We try to determine the Z matrix which is taken to be constant during the successive periods in our static model. For this let us write the balance for sector j (j= 1, ...n) in labour issued from sector i (i= 1, ...n); this balance is illustrated in diagram 2.

Total labour issuing from sector i incorporated in product j = labour issuing from sector i "fixed" in intermediate consumption of the sector j + labour brought directly from the sector j, if, and only if, j = i.

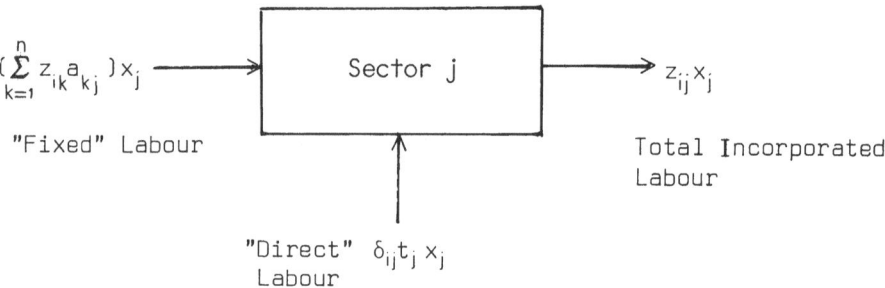

Diagram 2 : Labour-Balance of Sector j in Product i

Thus we have*:

$$z_{ij} x_j = (\sum_{k=1}^{n} z_{ik} a_{kj}) x_j + \delta_{ij} t_j x_j$$

in matrix terms:

$$= ZA + \hat{T}$$

then supposing A is productive :

$$(4) \quad Z = \hat{T}(I - A)^{-1}$$

---

* δ_{ij} designates the Kronecker symbol : $\delta_{ij} = 0$ is $i \neq j$
                                                        $\delta_{ij} = 1$ is $i = j$

Equation (4) determines the Z matrix calculated from the known matrices $\hat{T}$ and A. This equation can also be written as follows:

(5) $Z = \hat{T} + \hat{T}A + \ldots + \hat{T}A^n + \ldots$

from which we obtain an inter-temporaral interpretation: the total labour incorporated in products during the present period 0, represented by the Z matrix, can be identified with the sum:

$\hat{T}$ labour brought directly during period 0

$+\hat{T}A$ labour brought directly during period -1 and transmitted by intermediate consumptions

$+$ ....

$+\hat{T}A^n$ labour brought directly during the period -n and transmitted by intermediate consumptions

$+$ ....

Before demonstrating how this model enables us to take into account organization and realization labour, we will show how the link can be made with the classical Léontief equations in the space of values, on one hand, and with the method of labour equivalent, on the other.

2.3. Comments

2.3.1. Link with Léontief equilibrium in the space of values.
Let us suppose that in the sector j, (j = 1,...n) labour is remunerated at the rate w_j and let:

$W = (w_1 \ldots w_n)$

Let p_j be the price of the good j, and $P = (p_1 \ldots p_n)$. The value of each product results from the accumulation of labour value incorporated in this product. Thus:

$$p_j = \sum_{i=1}^{n} w_i\, z_{ij} \qquad j = 1,\ldots n$$

or,

$P = WZ$

By pre-multiplying the equation (4) by W we obtain:

(6) $P = W\hat{T}(I - A)^{-1}$

The equation (6) is simply the classical equilibrium equatio'
in the space of values.

In fact, (6) can successively be written:

$P = PA + W\hat{T}$

$P_j = \sum_{i=1}^{n} P_i a_{ij} + w_j t_j \qquad j = 1,...n$

This system of equations indicates that for each sector, the production price of one period is exactly equal to the sum of the direct
labour value and the intermediate consumption values.

   2.3.2. Link to the method of labour equivalent. The above model
is closely related to the method of labour equivalent presented in
(22/3). The essential difference lies in the fact that we are adopting an intertemporaral point of view: the intermediate consumption
of one period is elaborated during the preceding period. The labour
equivalent method, on the contrary, takes into account a single
period, during which, through successive labour injections, the
production system elaborates the intermediate consumptions before
using them in the production phase itself. Our model gives only a
very imperfect representation of reality. We presume, in particular,
that the economy is stationary (a simple reproduction process).
Nevertheless, we believe it has the advantage of adapting easily to
the periodic statistical data supplied by the National Accounts.

   Finally, let us note that certain American authors have used a
model very similar to ours in the study of energy flows within the
production system. (13), (18)

2.4. Organization and Realization Labour

   The discriminatory analysis set out in Section 1 makes it possible to divide the labour carried out in each sector into organization and realization labour. The results of 2.2. can thus be applied
without alteration to organization labour or to realization labour
alone. For this, each of the matrices $\hat{T}$ and Z is separated into two
matrices related respectively to the two types of labour:

$\hat{T} = \hat{T}_0 + \hat{T}_R$ division determined through discriminatory analysis

$Z = Z_0 + Z_R$ division to be determined to separate the products of organization and realization labour.

Thus, according to (4):

(7) $Z_0 = \hat{T}_0 (I - A)^{-1}$
(8) $Z_R = \hat{T}_R (I - A)^{-1}$

The analysis of matrices Z_0 and Z_R will enable us in Section 4 to outline a morphology of the products according to their labour structure.

However, before proceeding to our results, we first set out the statistical data which have enabled us to implement our model.

3.   THE STATISTICAL DATA AND THEIR PROCESSING

3.1.   The Input-Output Table

The French National Accounts divide the productive system into a given number of branches, each one defined as the set of activities producing a type of goods or services, with input divided into the 38 classes. (26)

Once the goods and services are produced, some may be used for intermediate consumptions, the rest are destined to meet final demand. This final demand corresponds either to consumption itself, or to the formation of capital (investments and stock variation) or to exports.

All these flows are expressed as values (quantity x current price) in the input-output table, the structure of which is derived directly from the Léontief model (Section 2).   This table is set out in a simplified way in Diagram 3.   (27)

In row i, different uses of product i are stated: as an intermediate consumption of a branch 1 ($1 = 1,...n + 1$), as capital formation, as final consumption by households, or as exports.

In column i, the origins of the resources used for product i are stated: as intermediate consumptions of product 1 ($1 = 1,...n$), as added value by the branch i, as commercial margins on sales, or as imports.

DIAGRAM 3 - INITIAL INPUT/OUTPUT TABLE

$[C+Y]$ Branches

$(C+Y)+(F+\varphi)+(Y+\eta)+ex = e$

	1	i	l	n	n+1					
Products i			$C_{il}+Y_{il}$		C_{in+1}	C_i+Y_i	$F_i+\varphi_i$	$Y_i+\eta_i$	ex_i	e_i
k		$C_{ki}+Y_{ki}$								
n										

$\bar{C}+\bar{Y}$	$\bar{C}_i+\bar{Y}_i$		$\bar{C}_{n+1}$	$CI+\Gamma$	$FBCF+\phi$	$CF+H$	EX	E
$+V$	V_i		$M-\bar{C}_{n+1}$	PIB				
$= X$	X_i		M					
Y	Y_i		$-\Gamma$					
$+\varphi$	φ_i		$-\phi$					
$+\eta$	η_i		$-H$					
$=\mu$	μ_i		$-M$					
i	i_i		0	I				
$X+\mu+i = r$	r_i		0	R				

$[F+\varphi]$

	1	i	l	n	n+1		
i			$F_{il}+\varphi_{il}$		F_{in+1}	$F_i+\varphi_i$	
k		$F_{ki}+\varphi_{kl}$					
n							

$\bar{F}+\bar{\varphi}$	$\bar{F}_i+\bar{\varphi}_i$	$\bar{F}_{n+1}$	$FBCF+\phi$

It is to be noted that the table includes an (n + 1)th column "Commerce" to which there is no corresponding row. In fact, the production of the branch "Commerce" is "sold" under the form of margins when the products are marketed. Thus, it is as though the ficticious product "Commerce" were transferred directly into the value of other products. The "Margins" rows of the I/0 table enable this transfer to be effected.

Finally, we also use a table supplied by INSEE giving the separate value, by branches, of investment or capital goods. This table is joined to the I/0 table of Diagram 3.

The notation used in the diagrams 3, 4 and 5 is as follows:

i) Aggregated values at a national level

CI : Intermediate consumption
FBCF : Investments in capital goods
Φ : Margins on FBCF
CF : Final Consumption by households and administrations
H : Margins on CF
EX : Exports
E : Total uses, including margins
E' : Total uses, excluding margins
Γ : Margins on intermediate consumptions
M $=$ $\Gamma + \Phi + $ H : Total margins
I : Imports
R $=$ E : Total resources including margins
R' $=$ E' : Total resources excluding margins

ii) Non-aggregated values

C_{ij} : Intermediate consumption in product i by branch j excluding margins, constituting matrix (C).
γ_{ij} : Margins on C_{ij} constituting matrix (γ)

$$\bar{C}_i = \sum_{k=1}^{n} C_{ki} \qquad i = 1,\dots n+1$$

$$\bar{\gamma}_i = \sum_{k=1}^{n} \gamma_{ki} \qquad i = 1,\dots n , \quad \gamma_{n+1}=0$$

$$C_i = \sum_{l=1}^{n} C_{il} \qquad i = 1,\dots n$$

$$\gamma_i = \sum_{l=1}^{n} \gamma_{il} \qquad i = 1,\dots n , \quad \gamma_{n+1}=-\Gamma$$

V_i : Added value in branch i, i = 1, ...n+1

$X_i = \bar{C}_i + \bar{\gamma}_i + V_i \quad i = 1,\ldots n \ , \ X_{n+1} = M$

$\eta_i \quad : \quad$ Margin on final consumption in product i   i=1,...n

$\eta_{n+1} = \ -H$

$\mu_i = \gamma_i + \phi_i + \eta_i \quad i = 1,\ldots n, \ \mu_{n+1} = -M$

$i_i \quad : \quad$ Imports in product i

$r_i \quad : \quad$ Total resources in product i, including margins

$r'_i \quad : \quad$ Total resources in product i, excluding margins

$F_{ij} \quad : \quad$ FBCF in product i constituted by branch j (Matrix (F))

$\phi_{ij} \quad : \quad$ Margin on F   constituting matrix ( )

$$\phi_i = \sum_{l=1}^{n+1} \phi_{il} \qquad i = 1,\ldots n \ , \ \phi_{n+1} = -\phi$$

$$\bar{\phi}_i = \sum_{k=1}^{n+1} \phi_{ki} \qquad i = 1,\ldots n \ , \ \phi_{n+1} = 0$$

$Y_i \quad : \quad$ Final consumption in product i

$ex_i : \quad$ Exports in product i

$e_i = r_i : \quad$ Total uses in product i including margins

$e'_i = r'_i \quad : \quad$ Total uses in product i excluding margins

iii) Vector notation

        Vectors are designated by the same letter (without index) as their generic elements.

3.2. Modifications Applied to the I/0 Table

    In order to obtain from the I/0 table a Léontief matrix A, adapted to the requirements of our study, we have had to apply two basic modifications.

    3.2.1. Creation of a "Commerce" product. In order to create the intermediate consumption matrix, we first had to create a "Commerce" product. The intermediate consumption of each branch in this product is defined as the sum of margins added to the intermediate consumptions of the branch. Thus γ_i (i = 1,..n+1) (cf. Diagram 3). To maintain a balanced table, we then exclude from each flow the commercial margins which were included, and remove the margin transfer rows as indicated in Diagram 4. On a practical basis, in order to proceed with these corrections, we have supposed that the margins are distributed prorata according to the value flows with which they are associated. In this way we have calculated from the including margins data appearing in the initial I/0 table, the excluding margin quantities (cf. Diagram 4):

$$C_i = (C_i + \gamma_i) - \gamma_i$$
$$F_i = (F_i + \phi_i) - \phi_i$$
$$Y_i = (Y_i + \eta_i) - \eta_i$$
$$C_{i,j} = C_i (C_{ij} + \gamma_{ij})/(C_i + \gamma_i)$$
$$F_{i,j} = F_i (F_{ij} + \phi_{ij})/(F_i + \phi_i)$$

3.2.2. Inclusion of the FBCF into intermediate consumption.
In evaluating the labour supplied by each sector, as is noted by
the Centre of Employment Studies (23), firms' investments, which are
closely linked to the production process and therefore to the labour
incorporation process, appear more as intermediate consumptions
than as final consumptions. Thus, we reintegrated the investments
into the intermediate consumptions so that final demand originates
only from household and administration consumption and from exports.
In order to obtain a balanced table (cf. Diagram 5), we have had
to cut off the counterpart to this FBCF from the added value:

$$V'_i = V_i - \overline{F}_i - \overline{\Phi}_i \quad i = 1,..n$$
$$V'_{n+1} = M - \overline{C}_{n+1} - \overline{F}_{n+1}$$

3.2.3. Aggregation. In order to reduce the number of
production branches appearing in the National Economy, thus facilitating the analysis, we have aggregated the 38 branches into 12
sectors in accordance with Diagram 6.

3.3. Processing

From the I/O table modified as just shown, and from a table
distributing employment by branches, we must now calculate the
matrices Z_0 and Z_R of the model introduced in Section 2.

3.3.1. To determine the Léontief matrix A. From the modified
table obtained in subsection 3.2.3, we can immediately calculate:

$$a_{ij} = (C_{ij} + F_{ij})/X_j \qquad i,j = 1,...12$$

a_{ij} is the intermediate consumption in product i, necessary to
manufacture one unit of product j.

3.3.2. To determine the direct labour matrices $\hat{T}_0$ and $\hat{T}_R$.
We initially have a table Θ, giving for each of the 38 I/O table
branches (set out in columns) the distribution of the wage-earners
according to 62 categories of employment (set out in rows). (28)
Then, we carry out the following operations on Θ_0:

i) Aggregate the columns of Θ_0 (62 x 38) to obtain table ,
(62 x 12). This conforms to our sector aggregation (cf. 3.2.3.).

ii) Aggregate the rows of Θ, (62 x 12) to obtain table $_2$,
(2 x 12). This conforms to our distinction between organization
and realization (cf. 1.2.). The first/second line of Θ_2, give
the number of organization/realization wage-earners for each
sector.

DIAGRAM 4 - I/O TABLE AFTER CREATION OF A COMMERCE PRODUCT

$[C]$ Branches

	i	n	n+1	C	+	F	+	Y	+	ex	$=$	e
i	Cil		$Cin+1$	Ci		Fi		Yi		exi		e_i
Products k	Cki											
n												
n+1	$\overline{\gamma}_i$		O	Γ		ϕ		H		O		M

$\overline{C}+\overline{\gamma}$	$\overline{Ci+\gamma_i}$	$\overline{Cn+1}$	$Cl+\Gamma$	FBCF $+\phi$	$CF+H$	EX	E
$+V$	Vi	$M-C_{n+1}$	PIB				
$=X$	Xi	M					
I	Ii	O	I				
$X+I = r'$	$r'i$	M	R				

$[F]$

	i	n	n+1		Fi
i	Fil		$Fin+1$		Fi
k	Fki				
n					
n+1	$\overline{\varphi}_i$		O		ϕ

$\overline{F}+\overline{\varphi}$	$\overline{Fi+\varphi_i}$	$\overline{Fn+1}$	FBCF $+\phi$

DIAGRAM 5 - I/O TABLE AFTER CREATION OF A COMMERCE PRODUCT
AND INCLUSION OF THE FBCF INTO INTERMEDIATE CONSUMPTIONS

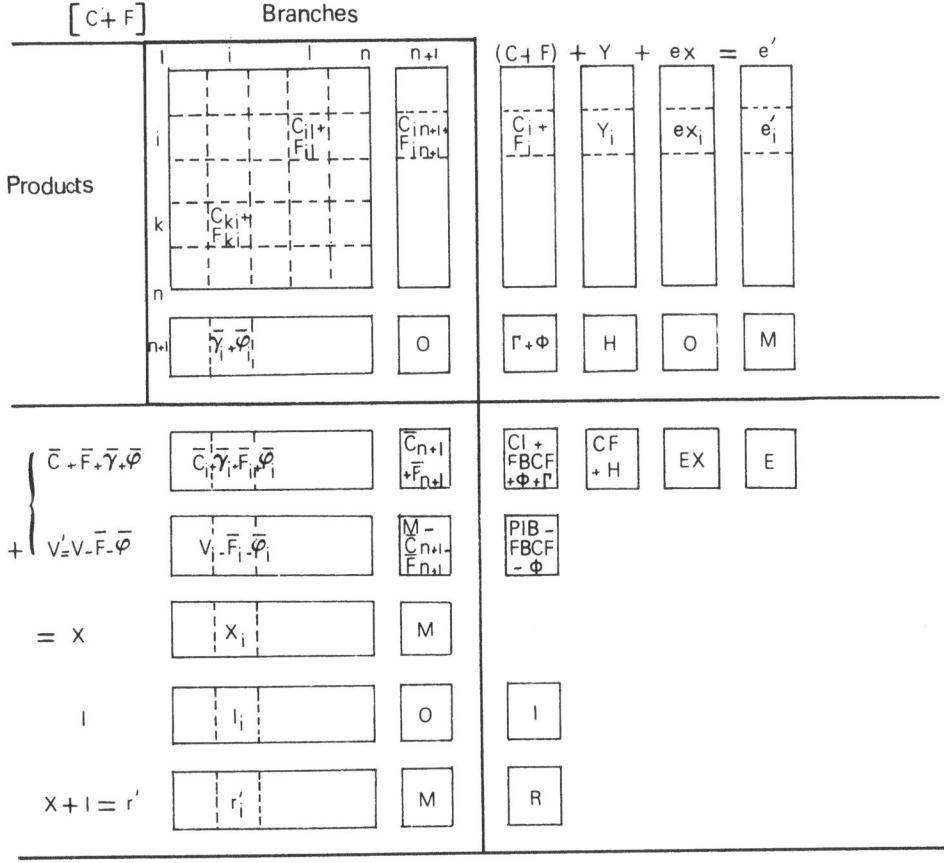

DIAGRAM 6 - THE AGGREGATION

No.	Aggregation (12 sectors) Sector	No.	National Accountancy (38 branches) Branch
1	Agriculture and food industries (IAA)	01 02 03	- Agriculture, forestry, fishing - Meat and dairy products - Other agricultural & food products
2	Energy (ENG)	04 05 06	- Combustibles, minerals, coke - Oil industries, natural gas - Electricity, gas and water
3	Transformation Industries (INT)	07 08 09 10 11 23	- Ferrous metal ores - Non-ferrous metal ores - Building materials - Glass - Chemical industries,synthetic fibres - Rubber, plastics
4	Consumer goods (BDC)	12 18 19 20 21	- Pharmacutical industries - Textiles, clothing - Leather and footwear - Wood and furniture industries - Paper and cardboard
5	Equipment industries (EQU)	13 14 15A 15B 16 17	- Foundries and metalworks - Mechanics - Electrical material - Household appliances - Car industry, land transport - Shipbuilding, aeronautics
6	Printing (IMP)	22	- Printing, press, publishing
7	Building (BTP)	24	- Building, public works
8	Commerce (COM)	25/8 29 30	- Commerce - Car sales and mechanics - Hotels, cafés, restaurants
9	Transports (TRA)	31	- Transports
10	P.T.T. (PTT)	32	- Post and telecommunications
11	Private Services (SMD)	33/8	- Renting, insurances, other services
12	Administration (SAD)	37/8	- Financial and public services

Labour	Employment
Organization Labour	Personnel specializing in organization, administration management, data processing, human sciences ...
Realization Labour	Engineers, executive technicians, technicians, skilled and unskilled staff directly concerned with the production of goods or services.

Diagram 7: Organization and Realization Labour

Diagram 7 sums up the distribution of employment between the two types of labour.

iii) Transform the wage-earning workforce into the two classes of labour. Knowing the total workforce for each of the 12 sectors, we make the (restrictive) hypothesis that the organization/realization structure of the wage-earners is the same as that of all workers. Thus we can deduce from θ_2, through linear extrapolation, an analogous table θ_3 (2 x 12) relating to the national workforce.

iv) $\hat{T}_0$ / $\hat{T}_R$ is the diagonal matrix (12 x 12), the spectrum of which identifies with the first/second row of θ_3.

3.3.3. To Calculate Z_0 and Z_R. These are calculated from the matrices A, $\hat{T}_0$, $\hat{T}_R$, by applying equations (7) and (8).

4. PRESENTATION OF RESULTS

We first examine the sectors for an analysis of the distribution of organization and realization labour. Thus we outline the morphology of labour flows which feed the production system. We then look at the products and see how they incorporate the organization/realization labour issuing from the sectors. In this way we divide the products into three groups, each group being characterized by its distribution of labour: material goods, services (non-material) and communications.

Finally we seek a global understanding of the production system by studying inter-sectoral labour exchanges. We take a close look at the specific role of each group of products and we emphasize the key position of the communication network, whose function is to direct matter, energy and information toward the production units.

4.1. Labour Structure of the Production Sectors

Table 8 (table θ_3 of section 3.3) shows the distribution between organization and realization labour for each of the 12 sectors (cf. 3.2.3.).

When we pass from the industrial and agricultural sectors to the services sectors, we can see from Table 8 that the percentage distribution of organization/realization labour differs very significantly. The former, producing physical goods, employ most of their work force in realization tasks. The latter, producing non-material goods, employ only a minority of their work force in actual realization tasks (mail, reports, interviews ...), and a majority in organization tasks (enquiries, research ...).

However, within each of the two groups (Industries/Services), the distribution is modulated according to the nature of each sector. The size of the production units of the transformation industries (INT) and the complexity of the production process of the equipment industry (EQU) explains the level of organization labour in these sectors. In post and telecommunications (PTT) and in the transportation (TRA) sectors the mobilisation of an important part of the work force to carry out the services explains the relative

%O	32	40	23	15	16	11	35	64	59	59	84	85
Labour												
%R	68	60	77	85	84	89	65	36	41	41	16	15
Sector	IAA	ENG	INT	BDC	EQU	BTP	IMP	COM	TRA	PTT	SMD	SAD

Table 8 - Labour Structure of the Production Sectors

importance of the realization labour.

N.B. The values relating to the two sectors are subject to problems of bias:

In the agriculture sector (IAA) the labour structure cannot be defined effectively by the wage-earning work force (cf. the hypothesis in 3.3.). The wage-earners are in fact a small minority attached mainly to large units, the structure of which is more industrial than agricultural. Hereafter we omit this sector from our comments.

In the PTT sector the distribution concerns only telecommunications. Consequently organization labour proportion is most probably over-estimated.

4.2. Labour Structure of Products

The upper part of Table 9 represents the vector $\bar{Z}_0$, the sum of the row-vectors of matrix Z_0. The lower part represents the vector $\bar{Z}_R$, the sum of the row-vectors of matrix Z_R (cf. 2.4.). Thus the table gives the distribution between organization and realization labour incorporated into products.

It can be seen that the inversion that appeared in the study of the labour structure of the sectors becomes less significant. The inter-sectoral exchange process gives rise to a certain regularization of the organization/realization labour distribution.

% O	23	25	26	32	38	39	50	52	57	67	67
Labour											
% R	75	75	74	68	62	61	50	48	43	33	33
Product	BTP	BDC	EQU	INT	IMP	ENG	TRA	PTT	COM	SMD	SAD

Table 9 - Labour Distribution in the Products.

Moreover it can be noticed that:

The transformation industries' products (INT), consumer goods (BDC), equipment goods (EQU) and, to a lesser degree, printing, press, and publishing (IMP), extract from their intermediate consumption an important amount of organization labour. However, the realization labour incorporated into these products is still dominant (about 70%).

Conversely, the products supplied by administration (SAD), private services (SMD) and, to a lesser degree, commerce (COM), extract from their intermediate consumption significant realization labour. Still, the organization labour incorporated into these products is dominant (about 66%).

The labour structure of energy (ENG), transportation (TRA), post and telecommunications (PTT) products differs only slightly from that of their associated sectors. However, the transfers which take place in the constitution of these products tend to reduce the difference in level between organization and realization labour, giving rise to a balanced structure (about 50/50).

Thus, the labour structure emphasizes three groups of products which are relatively homogeneous:

(i) Products (INT, BDC, EQU, BTP, IMP) which are relatively heavy users of realization labour. They constitute the material goods manufactured by the production system.

(ii) Products (SAD, SMD, COM) which are relatively heavy users of organization labour. They constitute the services supplied by the production system.

(iii) Products (ENG, TRA, PTT) which use organization and realization labour in roughly the same proportion. They constitute the production system's communications network (which includes transportation).

N.B. We have chosen to incorporate Printing (IMP) into the first group and Commerce (COM) into the second group. The difficulty in classifying these products underlines the ambiguous character of their role. In fact, seen from inside the production system, they belong to the communication network (IMP - information transmission, COM - matter transmission). On the other hand, seen from the outside, i.e. from final demand, their characteristics are akin to those of the other outputs (goods or services).

We will now take a closer look at the origins of the labour incorporated into each of the three groups of products we have just formed by analysing the inter-sectoral exchanges of labour.

Pr. Sec.	ENG	INT	EQU	BDC	IMP	BTP	COM	TRA	PTT	SMD	SAD
ENG	4.2	0.5	0.2	0.2	0.1	0.2	0.2	0.4	0.1	0.1	0.2
INT	0.4	9.8	1.5	1.1	0.5	1.7	0.5	0.6	0.6	0.3	0.7
EQU	1.4	2.1	12.7	1.3	1.5	1.9	1.6	2.7	3.8	1.0	2.2
BDC	0.2	0.4	0.4	13.2	2.5	1.0	0.4	0.4	0.4	0.6	0.8
IMP	0.1	0.1	0.0	0.1	13.7	0.1	0.1	0.0	0.1	0.4	0.2
BTP	1.3	0.9	0.7	0.7	0.7	15.4	I.1	1.5	1.6	1.7	4.0
COM	0.7	1.1	1.0	1.1	1.1	1.0	22.7	1.3	1.4	0.9	0.7
TRA	0.6	0.8	0.5	0.4	2.2	0.6	1.0	14.6	0.6	0.3	0.5
PTT	0.1	0.2	0.2	0.2	0.5	0.2	0.4	0.2	25.0	0.4	0.6
SMD	0.6	1.1	0.9	1.1	0.9	1.5	0.8	0.6	1.0	12.0	1.6
SAD	0.3	0.5	0.5	0.5	0.5	1.3	0.5	0.5	0.3	0.7	24.6

Table 10 : Matrix Z.

Table 10 represents the matrix Z (cf. Section 2.2.) which gives the total labour quantities (Organization and Realization labour) issuing from each sector (rows) and incorporated into each product (columns). These labour quantities are expressed per million francs produced.

Since we are interested in the exchanges which take place, we focus our attention on the non-diagonal elements. Three significant exchange currents appear.

The first current is issued from the material goods sector (mainly from BTP) and is incorporated into the services (COM, SAD, SMD). It simply indicates the importance of infrastructures essential to the supply of services and does not shed much light on the real nature of the service production process.

The following two currents, on the contrary, make it possible to outline the respective roles held by material goods and services in the production system. The material goods (INT, BDC, EQU, BTP, IMP) use an important contribution from the energy (ENG) and transport (TRA) sectors. The production process of these goods thus appears as transformation of matter (conveyed by transport) through an energy flow (channeled by the energy sector). The services (COM, SAD, SMD) require a significant contribution from the PTT sector (and to a lesser degree a contribution from transportation located in Commerce). Thus, one can see the specific character of

services, consisting in the creation, transformation and distribution
of messages through information flows.

Therefore, we can see that the communication network (ENG,
TRA, PTT) plays a major role in elaborating economic goods and
services.

4.3.   Communication Network and Productive System

We can see from Table 10 that the constituent products of the
communication network incorporate sectoral labour contributions in
a highly composite way. This is underlined by a balanced realization/
organization structure. In fact, the management of all transfers
of goods and services requires an important organization labour
component. The establishment of the infrastructure necessary to the
running of the network demands an equally important realization
labour contribution.

Diagram 11 shows the main inter-sectoral exchange currents and
brings out the key position of the communication network in the internal regulation of the production system.

The control exercised by the communication network on the inputs into matter, energy and information indicates the importance
of the post and telecommunications, energy and transportation sectors
in a production system which has been severely unbalanced by the
economic crisis.

The model we have put forward enables a detailed exploration
of inter-sectoral relations and should make it possible to tackle
the essential problem of substituting information for energy. We
would like to open up a discussion of the PTT's particular function
in this redirection of the production system.

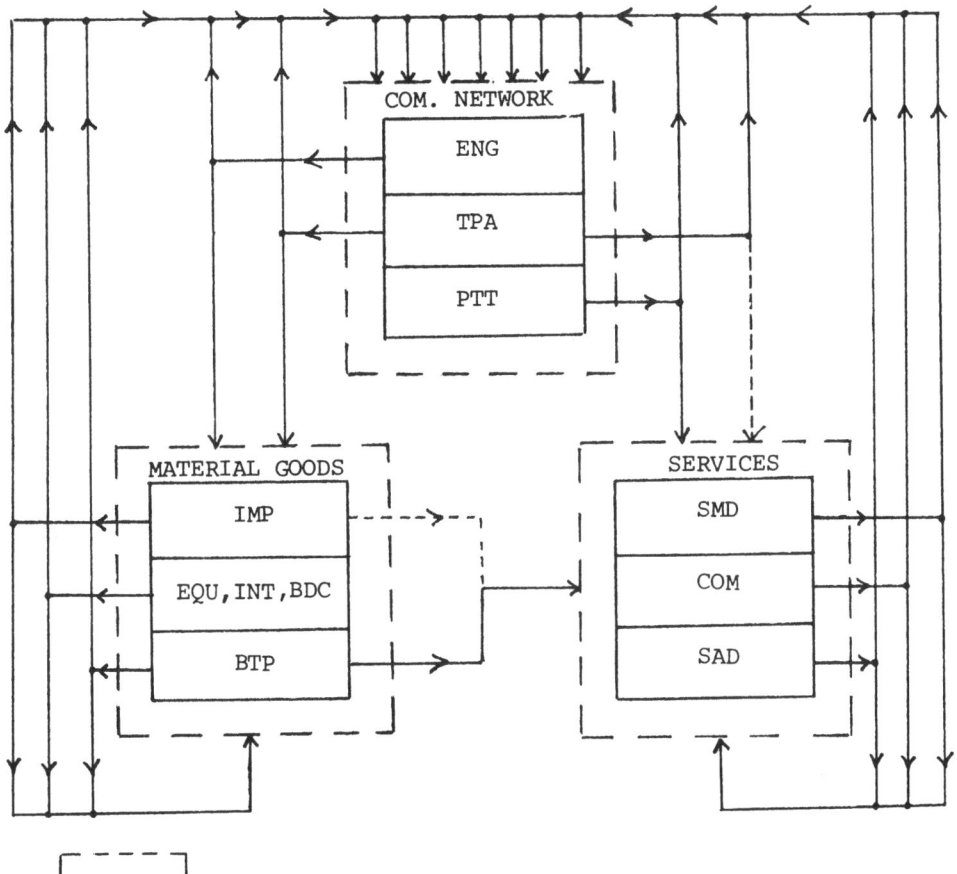

⌐ ⌐	: Group of products
▭	: Production sector
⟶	: Main labour contribution
⟶	: Secondary labour contribution

Diagram 11 :   Labour Flows in the Productive System

REFERENCES

(1)   :"The production and distribution of knowledge in the United
        States", F. MACHLUP. Princeton University Press, Princeton.
        1962.

(2)   :"Workers who produce knowledge: a steady increase 1920 to
        1970", F. MACHLUP and T. KRONWINCKLER. Review of World
        Economics, No. 4, 1975.

(3)   :"The coming of post-industrial society", D. BELL. Basic
        Books, New York. 1973.

(4)   :"Social implications of computer and telecommunications
        systems," E. B. PARKER. Conference on computer/telecommunica-
        tions policy, O.E.C.D., Paris, 4th-8th February 1975.

(5)   :"Where are we in the theory of information", J. HIRSHLEIFER.
        American Economic Review, Vol. 63 No. 2, May 1973.

(6)   :"Information and economic analysis", T. M. HAVRILESKY.
        Annals of American Academy, Vol. 412, March 1974.

(7)   :De l'éducation, de la recherche, de l'information et des
        communications dans la croissance économique", J. VOGE.
        Publication ENST, Paris 1973.

(8)   :"La parole et l'outil", J. ATTALI, PUF, Paris 1975.

(9)   :"Reflexions a l'occasion du 7e plan", L. VIROL. Revue
        Française des Télécommunications, No. 16, Juillet 1975.

(10)   :"Sur un modèle thermodynamique de la croissance et de la
        maturité économique", J. VOGE. Publication ENST, Paris 1976.

(11)   :"The information economy", M. PORAT. Institute for communica-
        tion research, Stanford University, report No. 27, August 1976

(12)   :"Energy and Information", M. TRIBUS and E. C. MAC IRVINE.
        Scientific American, Vol. 225, No. 3, pp. 179-185, Sept. 1971.

(13)   :"Mass-Energy based economics models", R. L. TUMMALA and
        L. J. CONNOR. IEEE, trans. on syst. man and cyb., Vol. 3,
        No. 6, pp. 548-555, Nov. 1973.

(14)   :"Energy, manpower and the highway trust fund", R. BEZDEK and
        B. HANNON. Science, Vol. 185, No. 4152, pp. 669-675.

(15)   :"Energy per Dollar Value of consumers goods and services",
        J. H. KRENZ. IEEE, trans. on syst. man and cyb., Vol. 4,
        No. 4, pp. 386-388, July 1974.

(16)   :"Affluence and energy demand", R. A. HERENDEN. Mechanical
        Engineering, Vol. 96, No. 10, pp. 18-22, Oct. 1974.

(17)   :"Energy and Economic Myths", N. GEORGESCU-ROEGEN. Southern
        Economic Journal, Vol. 41, No. 3, pp. 347-381, Jan. 1975.

(18)   :"Energy impact of consumer decision", C. W. BULLARD and
        R. A. HERENDEN, Proceedings of the IEEE, Vol. 63, No. 3,
        pp. 484-493, March 1975.

(19)   :"The flow of energy in an industrial society", E. COOK,
        Energy and Man, technical and social aspects of energy,
        IEEE press, New York, 1975.

(20)   :"Studies in the structure of American economy", V. LEONTIEF,
        Oxford University Press, New York, 1963.

(21)   :"A Dynamic Léontief Model for a Productive System",
        W. G. VOGT, M. H. MICKLE, H. ALDRERMESHIAN. Proceedings of
        the IEEE, Vol. 63, No. 3, March 1975.

(22)   :"Equivalent travail d'une production, nouvelle méthode de
        calcul et de prévision", J. MAGAUD, Revue Population,
        22e Année, No. 2, Paris, March-April 1967.

(23)   :"Essai d'application de la methode de l'équivalent-travail
        d'une production", N. DUBRULLE, J. DUMARD, M. PILOY,
        P. RANCHON. Revue Française des Affaires Sociales, 28e Annee,
        Paris, 1974.

(24)   :"Un modèle de mesure de l'information dans les organisations",
        G. du ROURE, I.A.E. d'AIX Marseille, note de recherche S.I.
        No. 30, May 1975.

(25)   :"L'approche développementielle de l'Etude d'un système
        d'information", J. C. COHEN, I.A.E. d'AIX Marseille, série
        "Recherche", Juillet 1976.

(26)   :"Présentation de la comptabilité Nationale Française",
        B. BRUNHES, Collections de l'INSEE, série C, No. 51,
        December 1976.

(27)   :"Rapport sur les comptes de la Nation", Collections de
        l'INSEE, Série C, No. 49, Paris, Sept. 1976.

(28)   :"Structure des emplois en 1971", Collection de l'INSEE,
        Serie D, No. 15, Paris, Sept. 1972.

M.W.

N.C.

ELECTRONIC FUNDS TRANSFER IN PERSPECTIVE

J MICHAEL WILLIAMSON

DEPUTY DIRECTOR

INTER-BANK RESEARCH ORGANISATION, MOOR HOUSE,
LONDON WALL, LONDON EC2Y 5ET

SUMMARY

Future developments in electronic funds transfer (the transfer
of computer based information instead of money) and in automation
in banking are generally thought to have significant implications
for the development of telecommunications. However, discussion is
often confused regarding the supposed move to a cashless and
chequeless society. The paper summarises the developments taking
place in the banks' money transfer systems and in related systems.
The physical movement of cash, cheques, etc. is being slowly but
progressively replaced by the movement of information between
computers. The implications for telecommunications of electronic
funds transfer itself may not be as significant as generally
supposed; the implications of the social, technological and
economic changes giving rise to electronic funds transfer may be
more significant.

INTRODUCTION

Electronic funds transfer systems are being developed by banks
in most countries throughout the world. At the Fifth International
Conference on Payment Systems in Mexico City in June 1977 about 150
senior bank staff from 24 countries met to consider together developments in the banks' money transfer systems and in electronic
funds transfer. It is significant that all the delegates had a
common interest in assessing the implications for their banks and
their customers of particular aspects of electronic funds transfer
although the countries represented at the Conference varied considerably in size and even more in the extent of their economic development.

The development of electronic funds transfer is much discussed
in newspapers, on radio and on television. It is of general interest to the public as well as being of interest to those concerned
with the future of telecommunications because it is affecting the
lives of ordinary people and of benefit to them. For example, in
the past people would receive their wages and salaries in bank
notes and coin from employers and would visit, say their electricity
supplier to pay their electricity bill in cash. Now people can
(and many do) receive their income direct into their bank accounts,
and their electricity bills can be paid by transferring information
between bank computers (by pre-authorised automated credit transfers between bank accounts) - no cash or cheques need be used.

However, public discussion of the subject is frequently ill-
informed and often very misleading. Comment over the past five to
ten years, especially in the USA but also in the British press and
journals, has suggested that the 'cashless and chequeless society'
is imminent. Even recent television programmes give the impression
to general audiences (and also to some bankers and those concerned
with telecommunications developments) that the 'plastic card
society' will be able to do without that dirty stuff called cash.
Paradoxically, bank staff close to branch activity, cash handling
and cheque handling operations, observe more and more notes and
coin, cheques and other paper to be handled. One of the main
purposes of this paper is to present electronic funds transfer in
a better perspective.

It is inevitable that with the variety in the relative size
of countries, the extent of their economic development, their
institutional structure and their regulatory frameworks, that the
banking systems of individual countries and their payment systems
will be at least as varied. This is illustrated in the simple
table below.

Country	No. of Banks	Cheque Volumes per year	Payment Volumes per year
Britain	10	1750 m	2600 m
USA	14000	27000 m	33300 m
Japan	158	400 m	3000 m

Comparative Statistics on Banks and Payment Volumes

Several problems arise in preparing such a table. First the problem of defining banks. In some countries the number of banking institutions providing payment services is clearly defined. In others there is the complication of savings banks, co-operative banks, mutual associations, giro systems and Post Office involvement.

Second, the problem of defining the accounts on which payments are made. In the USA the "checking account" is clearly established; there is also the usage of credit card accounts. In Japan the majority of transactions are on "ordinary" accounts and are recorded in pass books.

Third, the problem of establishing the total volumes of payments. The payment services most commonly used in some countries are cheques (USA and Britain) whilst in Japan they are transfers on pass book accounts, and in other countries are transfer orders processed through banks and giro systems.

Nevertheless, in spite of all the differences between countries and between banks and banking systems, there is the same concern with the development of systems which enable payments to be made without the use of cash (notes and coin) or cheques. It is becoming clear that overall trends in payments systems are likely to be very similar as a result of long term technological and economic trends.

THE LONG TERM TREND

The rate of change in computer and communications technology is without precedent. The capacity is being developed to process and transfer information in amounts and at speeds which were unthinkable only a few years ago, and at much lower costs than hitherto. This, together with the increasing volume of information processing that banks are being called upon to do and the difficulty and rising costs of processing such volumes of information by existing systems, clearly indicates the long-tem trend.

The physical processing and transmission of money in the form of notes, coin, cheques, etc. and of information on paper, will be progressively replaced to a significant extent by the automated processing and transmission of information using computer networks. The only questions are : at what rate will this occur? To what extent will it take place? The answer to the first of these questions will depend on the country and on the current state of development of bank systems. The answer to the latter is likely to be very similar in all countries.

The emphasis in the statement should be on the word
"progressively". From analysis of banks' and customers' interests
it is clear that the process of change will be evolutionary and
not revolutionary. In many countries customers appear to be content with existing money transmission services. In general, banks
do not impose change on customers to force them to adopt automated
payment methods. To bring about change requires banks to provide
even better and more attractive services (and to reflect the
relative cost-effectiveness of those services in bank charges).

PATTERNS OF CHANGE IN BANK PAYMENT SYSTEMS

The banking system of each country has its own unique starting
point from which it has to consider developments in its payment
systems. Some are overburdened with cheques and other paper and
seek to avoid paper handling. Some countries seem to have largely
avoided the paper handling problem but face other issues

However, in each country there is a pattern to the change
that is taking place and this can be illustrated by reference to
developments in Britain.

From the time when there was little more sophisticated technology in the branch than an adding machine, we have come a long way.
It is now expected to have terminals linked to computers. Computer
systems are accessed by pressing keys on a piece of equipment,
data is transmitted over wires; transactions on accounts are
initiated, enquiries are made. The position has already been
reached where bank staff could not cope with the amount of payments
and other banking business without the aid of the bank's computer
systems; indeed, without those systems the growth of the bank's
business would have been limited. And in addition to the development of the banks' own systems there have been a number of inter-
bank systems developments.

As soon as it was technically possible, attention turned to
the development of an automated clearing house because of the
large volumes of payments cleared between banks. We have a notable
success story in Bankers' Automated Clearing Services Limited
(BACS). BACS' role initially was that of a bureau enabling the
clearing banks to process centrally information about standing
order payments on magnetic tape. Now it facilitates the processing,
distribution and collection of large volumes of payments for more
than 2,300 corporate customers as well as the banks. The volume
of payments processed automatically by this means has grown from
none in 1967 to over 260 million in 1976 - about 13 per cent of
all non-cash transactions involving banks. Such payments include,
for example, salaries and pensions by automated credit transfer,
and insurance premiums and mortgage repayments by standing order

and by direct debit. We can expect substantial further developments as more customers find it convenient and cost effective to
use the services based on this automated clearing house.

Another inter-bank systems development of significance for
international activity, which is now coming into operation is SWIFT
(Society for Worldwide Inter-Bank Financial Telecommunications).
Initially this will link 500 banks in more than 15 countries by
means of a computer-based communications network.

For some time the possibility has been studied of an automated
system to handle payment items of high value currently cleared
through a limited system in the City of London called the Town
Clearing, on a same-day basis. These items account for about 90
per cent of the value of the clearings, although they are less
than 0.5 per cent by number. A system is now being developed
called CHAPS (Clearing House Automated Payments System) in which
clearing banks, the Bank of England and other participant banks
are connected directly by transmission lines to a central computer
system. By means of messages introduced into such a system
through the terminals of a participating organization, payments
will be made by the participant efficiently, with better information
about the flow of payments and improved control, by means of
accounts with the clearing banks, with automated same-day settlement at the Bank of England.

The CHAPS development will not be operational for some time
but already it is conceivable that a CHAPS system could eventually
exist in which participants would not have to reside in the City
of London. The development of computing and communications is
likely to be such that, technically, payments could be made through
a CHAPS system on a same-day basis from any part of Britain.

Whatever other inter-bank systems developments take place,
the general cheque clearing and the credit voucher clearing will
remain for some time to come, together with the problems at branch
level of handling the flow of paper. In 1976 more than 1,700
million cheques were drawn on the London clearing banks, which also
handled more than 350 million credit transfer vouchers. The
majority of these passed over branch counters and through the
clearings. Whilst every effort is being made by banks individually
and collectively to encourage the use of alternative automated
services (such as those based on BACS), the need remains to find
better ways of dealing with the growing flood of paper. In terms
of technology the obvious solution is to 'capture' the data on the
cheques and other vouchers at the branch where they are paid in by
the customer and thereafter process this information only, using
the banks' computer networks. But account has to be taken of the
branch staff's and customers' interests in the vouchers, as well

as legal and other requirements. The way ahead is certainly to
automate the information handling, but this can only be step by
step, dealing with particular aspects of the cheque and credit
processing which lend themselves to this treatment. Some of thes
developments will be matters solely for individual banks to
decide; others will require the banks and perhaps other organizations to agree the standards and procedures to be adopted.

Perhaps the most important type of future systems development concerns customer-operated banking terminals. These devices,
generally linked to a bank's computer network, are usually
operated by a customer by inserting a plastic card and keying a
secret identity number. We already have many of these in
operation in the form of cash dispensers. Some versions facilitate
customer enquiries. Technically, such terminals can be developed
to include several of the functions of a branch and give rise to
the concept of the automatic branch where cash and cheques can be
deposited (for bank staff to deal with later), cash can be withdrawn, transfers made automatically, cheque books or statements
ordered or enquiries about balances dealt with. Technical possibility and economic viability are still a long way apart and the
early forms of these devices provide more limited but, nonetheless,
effective services.

The nature of a customer-operated banking terminal will depend
to some extent on its location. They can be found within branches
- and in the outside walls. They could also be where a need exists
and where most convenient for customers e.g. at the workplace of
customers (as cash dispensers), in large stores (as point-of-sale
terminals), in public places (as auto-tellers) or our homes (as
part of our telephones). There can be little doubt that developments along these lines will take place.

Clearly the pattern of change and the pace of change will be
different in different countries. Those who depend extensively on
cheques and paper-based systems will probably be experiencing a
similar pattern of change to that in Britain described above. Those
who instead depend extensively on pass book accounts, giro systems
and so forth are experiencing a different pattern of change. In
all cases, I believe, the customer operated terminal is likely to
be a significant feature of future developments together with the
automated clearing house.

RELATED BANK SYSTEMS DEVELOPMENTS

The development of electronic funds transfer in Britain and
elsewhere seems likely to be accompanied by other systems changes
relevant to bank staff. In addition to the steadily improving

reliability of systems, and the improved services available to
customers, bank staff will have available to them better information about the business with which they are concerned, presented
in a way that makes their work easier. This development of
extensive management information systems and of what some describe
as 'management decision support systems' will also enable banks to
move further into a new area of business - the provision of
financial information and advisory services. Already there have
been significant developments in Japan in this area.

CHANGE OUTSIDE BANKS

Change in banking and in the technology of bank payment systems
is accompanied by change of other kinds. In the first place banks
are not alone in making use of computer and communications technology in their information handling and processing; all corporate
customers are doing the same to some extent. Secondly, the
business climate is changing and corporate customers are becoming
more aware of the costs they incur in money transfer activities and
aware of related costs such as float, customer convenience, etc.
Third, and perhaps most important of all in some countries, there
is substantial social change : the values and expectations of
people whether as bank customers or bank employees, are changing.

In considering the development of electronic funds transfer
we must consider not just the banks' payment activities, but also
those of customers : we must look at the Total Payment System.

TOTAL PAYMENT SYSTEM

The "total payment system" is the system of all participants
in a payment transaction - payer, payee, payer's bank, payee's
bank and other intermediaries, where relevant. Let us consider
some examples :

(i) Purchase of a book by an individual by cash

    Participants: buyer, seller only
                        (bank involvement is only indirect, in
                        providing cash for buyer and accepting
                        deposit of cash from seller).

(ii) Purchase of a book by cheque

    Participants: buyer, seller, buyer's bank, seller's
                        bank - simple non cash transaction.

(iii)   Payment for service

    Participants: consumer, provider of service, postal
                           system, consumer's bank, supplier's
                           bank.

(iv)   Payment of wages by cash

    Participants: employer, employee, bank provides cash.

(v)   Payment of salary through bank

    Participants: employee, his bank; employer, his bank.
                           Automated Clearing House.

(vi)   Payment of Pension through bank

    Participants: pensioner, his bank; pension scheme and
                           its bank; automated clearing house,
                         postal system (for advice of payments).

In looking to the development of electronic funds transfer we have to consider trends in the costs and the benefits as perceived by all participants in the total payment system. This includes, in addition to banks and bank institutions such as Automated Clearing Houses, both payers and payees, individuals and organisations who make and receive payments. We cannot look at electronic funds transfer from a bank point of view only. We must necessarily concern ourselves with the way costs and benefits are distributed in a total payments system.

Banks' costs are probably no more than half of total payment systems costs and possibly much less. As well as considering trends in bank costs, we must consider the costs as perceived by corporate customers - their own operations costs plus bank charges plus float. Customers, particularly the larger corporate customer with a large payment activity (whether issuing or collecting payments) are as active as the banks in their own attempts to deal with economic trends in their business.

The question of benefits from payment systems, and more precisely the distribution of benefits, is more complex. Benefits can be pecuniary - measurable directly in £s, or $s; they can also be non-pecuniary - timesaving, convenience, better control of financial affairs, better service to a corporate customer's customers, etc.

One thing is clear. Unless the benefits to each participant in a total payments system are more than, or at least commensurate with his costs, the operation of that system in its present form will be short lived. What is more unless the net benefits to each

participant in a proposed new payments system are more than, or
at least commensurate with the net benefits of existing systems -
the proposed new system will stay on the drawing board or, if it
is implemented, it will not operate for long in its proposed form.

Let us consider an example : Point-of-Sale (POS) developments
in retailing. A service to whom? : Who will bear the costs? Who
will enjoy the benefits. The average cost of a retail cash transaction is small and cash payment is a simple system used by all.
By contrast, the cost of POS includes capital cost of system
(including development), cost of operating the system and cost of
running more than one system. However, POS can be seen as a
natural extension of other retail systems developments. This,
together with social change suggests it might well be viable.

We can draw two conclusions from this example alone:

(i)   the analysis of the costs and benefits of alternative payment
      methods is far from easy;

(ii) payment systems are only parts of other systems.

PAYMENT SYSTEMS ARE ONLY PARTS OF OTHER SYSTEMS

In considering the way in which costs and benefits fall on
parts of the total payment system, some other points were hinted
at. They concern related systems.

It is rare that a payment is made without any other flow of
recorded information. A purchase of, say, a book from a friend in
the street by cash might be such a case. In all other cases,
there are information flows; sometimes these information flows
and the associated systems prove as important as the payment
itself. If the purchase of the book was at a shop, there would
probably be a ticket produced at the cash till with details of the
purchase(s) and the total payment (including state, city and other
taxes where relevant); and in many stores now the ownership of
such a ticket is proof of purchase (against accusation of theft
from the shop).

Similar points arise in more complex systems. When a bill is
received by post for electricity supplied during the previous
quarter it is not only a bill but also information about the
quantity of electricity consumed, the rate of charge, other charges
(e.g. purchase/rental of equipment) and the total for the period.
Similarly, when a cheque is sent in payment for the electricity
with the appropriate part of the bill, this enables the electricity
company to identify the payer, the amount due and the payment, and to

record the receipt of payment (receipt of the cheque is generally taken as indication of receipt of payment subject only to advice from the banks on an exception basis that the cheque is invalid).

It can be seen that the payment itself is only part of an information system. Without adequate information flows in a convenient form either the payer or payee or both will have less information about the transaction than they need and this could well militate against acceptance of the method of payment. (And it is not only advice to the payer that matters. Increasingly, in automated systems such as POS, and customer operated banking terminals in general, it is necessary for several reasons - but principally to combat fraud - to build in enquiry facilities ; at the initiation of a transaction it is necessary to establish that the payer is who he says he is and that he is in a position to pay the bill; an enquiry must be made of an information file, his balance, or of credit worthiness before the transaction is authorised.)

THE PACE OF CHANGE IN PAYMENT SYSTEMS

In Britain the future appears to be one in which there will be further steady growth in the number of bank customers, of accounts and of business to be handled. This will mean more cash handling transactions at bank branches, more cheques and other payment items. This pattern of development will certainly not be economically viable without new systems. Costs will be unacceptable. There will have to be further development of automated systems and of the use of those systems by customers. To accommodate the greater volume of business, banks will increasingly depend on the use by customers of customer operated banking terminals, for routine transactions such as cash withdrawal, payments and enquiries, and on the use of Automated Clearing House for processing the very large volume of regular payments. Even so, the remaining volume of cash, cheque and other paper processing at branches will call for substantial improvements in cash handling and cheque handling systems within each bank over the next five to ten years.

However, the pace of change as a whole will be slow because even if it is technically possible to develop new systems, it does not follow that society as a whole will change its behaviour. Four points are relevant :

(i)   cash (notes and coin) is a convenient and cheap medium of payment for many people for many purposes (about 98 per cent of all payments by individuals in Britain are in cash; the position in the USA is similar);

(ii)   the banking habit is spreading and in total more cheques
         are being drawn in spite of the increased use of automated
         methods of payment;

(iii)   although automated payment methods are potentially much
         more efficient, many businesses prefer to use cheques
         because of the greater control and flexibility they allow
         at present;

(iv)   the rate of increase of use of more recent methods of
         payment such as credit cards is misleading; the use
         of credit cards accounts for less than 4 per cent of
         all non-cash transactions by individuals.

        Taking into account these facts and the increasing use of
automated payment services, it is clear that in Britain cash will
continue to circulate in very large amounts throughout the foreseeable future.

        Furthermore, the numbers of non-cash payments are increasing
steadily. IBRO's view of likely future developments in the use of
existing money transfer services, on the basis of current trends,
is illustrated overleaf. In particular it is expected that the
number of cheque payments will grow for some time (to, in 1985,
perhaps, 70 per cent more than at present).

        In addition to the growth of use of existing non-cash payment
media, we expect to see further development of the use of cash
dispensing machines. More generally, the use of customer-operated
banking terminals in Britain at the point-of-sale (in retailing)
or point-of-payment (in public utilities), is expected to begin to
take effect in the early 1980s; by 1985 perhaps 2-3 per cent of
non-cash transactions could be dealt with by means of these terminals with the proportion increasing substantially thereafter.

        The position is similar in other countries. For example,
cheque use is expected to increase in the USA. "In 1970 the annual
volume of checks was 21.5 billion, not including another billion
checks written by the government. This year the volume will be 27
billion with the forecast of up to 45 billion by 1980". (Nilson
Report : May 1977).

        Why the continuing increase? Two reasons : more accounts and
more account activity. More personal accounts means that more
organisations can make/receive payments from individuals. (In the
UK the increase in account holding and account use is related to
trends in the payment of wages and salaries. In 1969 about 75% of
the working population (income earners) were paid in cash; in
1976, less than 60%.)

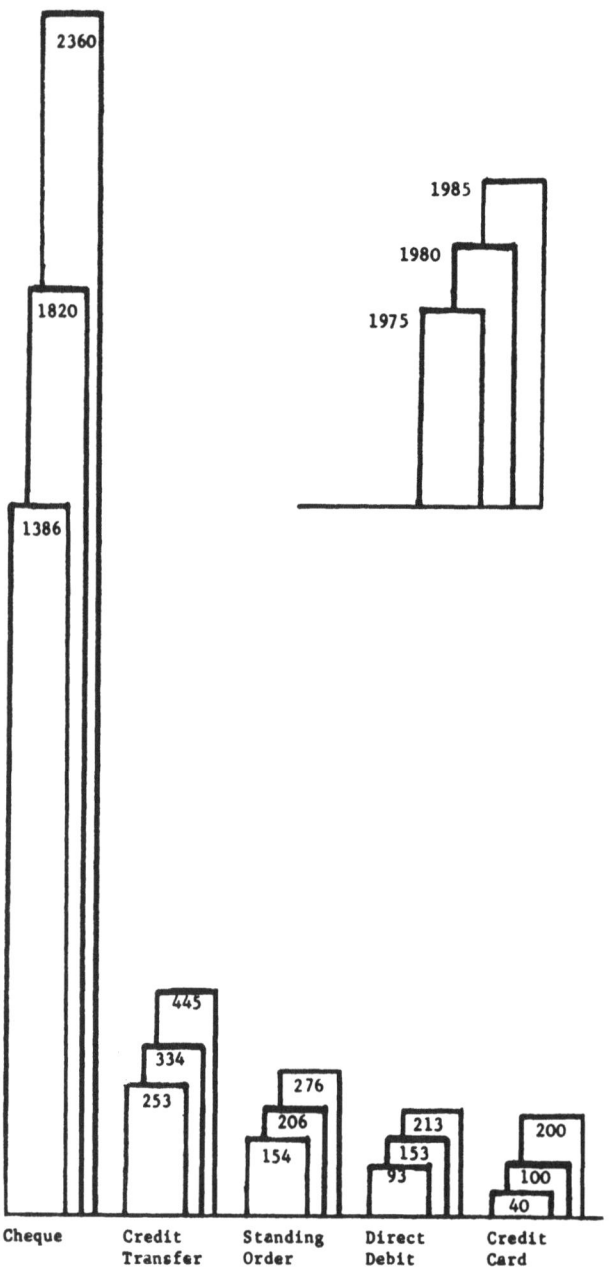

Expected Future Growth of Non-Cash Payments in Britain

EXPANSION OF THE BANKING HABIT

One of the most fundamental questions regarding economic trends in the total payment system concerns the extent of bank account ownership.

Do the banks want a large increase in accounts over the next few years from those who have not got accounts at present, who are likely to be less well paid than others? Under what conditions? Should this business be allowed, even encouraged to go to savings institutions and better arrangements developed for transferring payments between institutions of all kinds? Should savings and similar institutions be regarded as competitors or complementary operators?

The cost of further expansion of current account ownership falls on banks and near-bank institutions to provide the accounts, the payment and related services. It may well be that under present arrangements the cost to the banks will exceed any benefit they could hope to enjoy. However, the spread of current account ownership is making it easier for employers to pay salaries/ pensions by bank media, easier for governments to pay social security benefits by bank media, easier for organisations, large and small, to make and to receive payments from individuals. Thus the corporate sector may well enjoy substantial benefits from such a growth in bank account holding. The expansion of the banking habit may therefore be accompanied by a redistribution of the benefits of electronic funds transfer by pricing policies.

DETERMINANTS OF CHANGE IN FUNDS TRANSFER SYSTEMS

The long term trend in payment systems has been identified. Change is inevitably towards electronic funds transfer but the pace of change is much slower as a whole than is commonly believed. What are the determinants of change?

The main factors determining the pace and pattern of developments in funds transfer systems can be classified as :

(i)   pressures;

(ii)   constraints; and

(iii) opportunities.

The pressures for change arise from such things as :

- problems with existing systems
- rising costs of staff and other resources
- non-availability of staff of the right kind prepared to do the work

- competition from other institutions
- deteriorating postal services
- customer demands for improved or new services
- actions by the authorities

The constraints on change, which may prevent or delay developments, include :

- the limited availability of resources, especially of qualified
  staff and of capital, to plan and implement the changes

- the need for change to be consistent with other developments
  (of existing systems, of other business activities)

- the price of or non-availability of adequate telecommunications
  facilities

- the need for inter-organisational agreement and, perhaps,
  concerted action

- customers' inertia

- restraints by the authorities (e.g. regarding prices for
  services; data communications)

The opportunities for change include:

- change in existing systems in order to benefit from technical
  developments in computing and telecommunications

- the development of new services to customers (made possible
  because of developments in technology)

Any bank or similar institution involved to a significant
extent in the provision of payment services has a large commitment
of staff, systems and equipment, and usually branches, to money
transfer systems. This has a significant effect on the structure
and pattern of growth of the costs of the bank. Payment services
are provided by banks to attract deposits and the value to banks
of these deposits enables them to subsidise payment services to a
large extent. The proportion of money transfer costs recovered
directly by charges is often small.

The recent experience of banks in several countries, is that
the total cost to banks of providing payment services is rising
and will continue to rise faster than inflation as a result of the
growth in account holding and in transaction volumes even with
continuing improvements in productivity. Furthermore, deposits
directly associated with payment services are not keeping pace with
inflation.

The conclusion must be that at constant interest rates, a
steadily increasing proportion of the banks' payment systems costs

must be recovered directly by charges or there must be further
cross subsidisation from other banking activities. Of course if
interest rates fall the need for recovery of payment systems costs
is increased - perhaps dramatically.

This is simply saying in part what economists have argued for
several years, that the time will come when a significant proportion
of the costs of payment services will be recovered directly by
charges and, perhaps, the banks will be obliged to pay interest on
demand deposits.

In these circumstances, action will be essential to ensure
that the total costs of handling payments - especially in high cost
areas such as cash handling and cheque handling - are kept to a
minimum consistent with the quality of service desired. This will
undoubtedly call for improvements in all bank systems - some of
which will place demands on telecommunications facilities.

Pricing policies will need to reflect the true costs to the
banks of handling payment transactions (and associated activity) by
different methods. Without adequate price differentiation there
is no way in which the customer can judge what is the most cost
effective payments service for his needs in terms of bank resources.
However, government policy, competition or other factors might
preclude this for some time to come.

In the light of the economic pressures there are certain to be
substantial changes in bank payment services and in prices for
services. How will customers react?

In general, corporate customers tend not to know their payment
costs. They know the bank charges they pay; they know the value
of payments made and their cash flows; they know their balances
with the banks/extent to which they are indebted to banks, but not
their payment costs. Corporate customers are not in the money
transfer business; they exist to provide other goods or services
at a profit or, if government departments, they have other objectives.
As a result the management time of corporate customers is devoted
to their business activities and not to payment systems.

Only two kinds of corporate customer will tend to be more
knowledgeable: those who have a sufficient volume of payments or
receipts to make the cost of payments activity an identifiable cost
centre to be considered in its own right; and those who, having
developed business systems appropriate to their needs, find that
they can extend those systems to include some payments activity
also.

The overall trends in the payments and related activities of corporate customers are the result of changes in the business of those organisations, their relationship with their own customers, technological developments in their own and in bank systems, legislation, labour costs, bank charges and other costs. It is not really surprising that only those with extensive payments activity are well informed about economic trends. However, all will be progressively influenced but in varying degree, by economic pressures, and the opportunities presented by new bank payment services.

But it seems unlikely that corporate customers as a whole will rapidly respond to economic trends in payments systems as such. For example, consideration by the government department in the UK concerned with the payment of social security benefits (1000m payments per year) is only now about to result in decisions to provide for the possibility of some payments to bank accounts in the early 1980s; other priorities and considerations precludes a quicker response to changed circumstances.

In the case of individual customers the determinants of change are quite different. For them economic trends in payment systems are reflected in the rising cost of postal services, less collectors/ collection offices, and bank charges. There are also discounts offered by suppliers, in some cases, for paying by one means rather than another - or free credit associated with credit cards. Increasingly, individual customers are becoming concerned with the value of funds and interest on deposits. They also put a higher value on convenience. Increasingly those who have not had a bank account before are opening accounts; those who have accounts already are using them more. All are affected by the payments systems developments by organisations regarding wages and salaries, pensions, social security payments, regular bills, etc.

CONCLUSIONS

Technological change will strongly influence the development of funds transfer systems; it will have a significant effect on bank systems and services and on banking business. Banks and their corporate customers will depend to an increasing degree on electronic funds transfer as a means of handling efficiently the ever increasing volume of transactions on accounts.

Technological change accompanies and is accompanied by social and economic change. The increasing spread of the banking habit and the changing values and expectations of people will provide a radically different opportunity for electronic funds transfer in future both for banks and their corporate customers.

However, the pace of change seems likely to be constrained by customers' inertia. Corporate customers' perception of trends in payment systems is, in general, poor and ill-informed. Yet their costs may well exceed those of banks in the total payment system. Customers are becoming more aware of their costs and this may well lead to greater concern about payments systems.

The efficient use of national resources in payment systems can only be achieved by processes of joint planning between institutions and between sectors : e.g. government and bank discussions of social security payment systems. The aim should be to co-operate in establishing a framework within which competition can take place more effectively. (Similarly, joint consideration is needed by banks and PTTs of future developments in telecommunications in the long term.)

The pace of change as a whole will be slow. Cash will continue to be used for the majority of transactions for many years to come. Similarly, the use of cheques will continue to grow in several countries for perhaps a decade or more. Change will be evolutionary; electronic funds transfer will progressively be developed; but the overall pattern of payments activity will change only slowly.

What may prove of more interest to those concerned with telecommunications developments are the related systems developments; first, narrative processing associated with payments and related business systems; second, management information systems in both banks and their corporate customers; third, information and advisory services by banks to customers.

Payments systems are only parts of other systems. In most organisations business systems developments precede the use of new payment systems. In the operation of new business and payment systems it will be seen as important in future to ensure that the participants in an automated payment transaction have all the information required by them for their respective purposes. This calls for much more information processing than simply passing a debit or credit item through the banking system.

Up until now automation in banking has been largely directed at the automation of routine payments and accounting activity. It seems possible that more emphasis in future will be placed on the development of management decision support systems. A similar conclusion can probably be drawn for many other organisations.

In future, banks will be in a good position to offer information and advisory services to customers. There has already been such a development in Japan. With easy access to a bank's

own data base and to external data bases this sort of development
could prove significant.

It is concluded, therefore, that the implications for developments in telecommunications of electronic funds transfer itself
may not be as great as sometimes supposed; the implications of
the social, technological and economic changes giving rise to
electronic funds transfer may be more significant.

Section Eight

DESIGN AND PLANNING

Seymour Mandelbaum provides an essay dealing with the design
of the community which designs telecommunications services. There
is good reason to question the functioning of this community. He
is concerned not so much with its coordination, as with the questions whether there are missing elements, whether the flow of information is adequate and whether the conditions of work are
appropriate. These issues are considered in relation to three
dilemmas which arise in the design of communication networks: (i)
the conflict between aggregation of communications policies and
penetration into the depth of on-going systems; (ii) where should
be the boundaries of participation in the planning process; and
(iii) the competing pressures for rigorous and for robust design
of particular projects.

Al Shinn was invited to present a paper on a subject of his
choosing. He chose to consider a methodological problem which has
loomed large over the last few years, the use of social experimentation in policy research. Social experiments, intended to provide
reliable cause-and-effect information, are distinguished from less
rigorously designed field trials and demonstration projects. Their
major disadvantages are identified and the conclusion is drawn that
telecommunications policy research is not ready to use them in many
situations. Support for this conclusion is provided in the discussion, which follows, of the problems associated with the disadvantages listed earlier.

Shinn then describes a well executed social experiment, Roger
Mark's project in which nursing homes in the Boston area were
connected to the Boston City Hospital. His final point is the need
to see experiments in their proper place as one of a variety of
research approaches. The paper ends with three major conclusions:
they concern the relationship between experimentation and theory,
the political purposes served by social experiments, and their need
for meticulous planning.

Two presentations at the symposium were concerned with the incorporation of telecommunications into central government's strategies for social and economic regional development. Bertil Thorngren summarized pioneering research in Sweden and extracted some challenging hypotheses from it. Unfortunately the paper is not available for inclusion in these proceedings.

The second paper was by Daniel Chauche. He describes recently initiated explorations now under way in France. He starts with relevant goals of the 7th Plan and their translation into questions concerning telecommunications. He then turns to three field trials, which are at different stages of development: "audiographic" teleconferencing, telephone-television information services such as Viewdata (see the paper by Samuel Fedida in Section 6), and TV screenings in public places for special interest groups. Finally he offers some thoughts on a coordinated strategy for the regional introduction of new communication services.

The final paper was presented by Chris Stockbridge using the rudimentary audioconferencing link with North America. While the preceding papers were concerned with global issues, he addresses the particular problem of deciding which locations to include in a multinodal field trial of a new telecommunications service. He describes a practical approach which was developed for the design of the trial of Picturephone[R] within the criminal justice system in Phoenix, Arizona. Subjective probabilities are used in a heuristic optimizing algorithm which balances cost against a measure of the expected yield of a particular trial configuration. Some statistics are included on usage levels through time in this and another Picturephone[R] trial.

THE DESIGN OF THE DESIGNING COMMUNITY

Seymour J. Mandelbaum

Department of City and Regional Planning
University of Pennsylvania

Some of us, like weekend gardeners, attend to the flowers and inevitable weeds of telecommunications evaluation and planning only in time squeezed from other activities. Others, make their living at these tasks. Weekend or weekday -- all of our work is channeled by the processes of a community as palpable as those we seek to enhance through the application of telecommunication technologies. The purpose of this paper is to assess this "designing community" of which we are all members. I might be on sounder empirical grounds if I described the firms, agencies, professional associations and individuals engaged in telecommunications design as forming multiple communities. I've settled on the singular form in order to emphasize the political hope that they may be able to communicate more fully in the future.

There is a certain intellectual justice in examining the designing community through the same lenses we hold to the world of business firms, schools, hospitals and the like: What is the design of this community? How does it work? How, most critically, may it be evaluated and improved?

It is more than a sense of fairness -- doing to ourselves what we do to others -- which encourages self-examination. At the risk of projecting my own feelings, it seems to me that a portion of the designing community suffers from a serious (though not, I hope, fatal) loss of confidence. Within this portion, many individuals have hoped that new technologies might be used to democratize (if not overnight, at least with discernable speed) public access to knowledge and to transform the relations between groups and the organizational and spatial arrangements of major social institutions.[1]

These hopes have infused a great many exciting beginnings:
clear analyses, interesting experiments, prestigious commission
reports. It is not clear to anyone now, however, where to go from
here. (One evidence of this uncertainty may well be the rising
interest in evaluation.) The major national communications policy
agencies -- and here I speak only with substantial confidence of
the United States -- seem incapable of sustained intelligence:
they do not appear to know what they have done, how the world
works or how to act to improve their performance. The major social
institutions in health, welfare, education and a variety of other
fields use new technologies but rarely in the comprehensive fashion
of which the researchers and reformers dream. There are a great
many innovations still to be crafted, exciting experiments to be
evaluated, but in all the turmoil, there is no sharply defined
developmental path, no Jacob's ladder from here to blue sky.

It would be a perverse form of self-deprecation to attribute
this loss of confidence to the design of the designing community.
The hopes may have been excessive and hence properly vulnerable to
deflation. Even if the hopes are sound in substance (if not in
timing) the world is certainly intractable and the problems
"wicked".[2] Clumsiness and frustration are both understandable
and even necessary. In some ultimate accounting, our shortcomings
will undoubtedly be forgiven. Nor, to continue with the theologi-
cal image, are we entirely free. It is only an act of analytic
convenience which separates designers from the systems they seek
to control. The stability of the major patterns of social communication largely depends upon the constraints they place upon those
who would alter them. It is difficult to read Comstock and Lindsey's study of television research or Kay's assessment of the early
work on social services and cable television without sensing how
much the research communities are dominated in both idea and form
by the major established characteristics of the objects of their
attention.[3] The design of empirical research catches the analyst
in a political agenda which characteristically assumes that the
major features of the system under inspection will persist. The
result may, as Comstock and Lindsey ruefully conclude, innocently
be a form of "unplanned obsolescence," in which researchers looking
backwards are constantly overcome by events.[4] Such obsolescence
also serves, however, to strengthen established institutions.
Even critical research ironically becomes a form of legitimation.
Wise evaluators have certainly come to understand how much their
external and objective posture cloaks their role in internal
systemic politics.[5]

Despite all of these difficulties, there is no viable
emotional alternative to insisting on the designer's form of exis-
tential freedom. The designing community must be treated as if it
were plastic, amenable to change through the forms of both

intellectual inquiry and organizational development. At least
the first approach to the loss of confidence should be through
introspection and self-reform.

There is an advantage in using the designing community rather
than a particular agency or firm as the context for introspection.
When we seek to enhance the planning intelligence of a bounded
organization (such as the Federal Communications Commission) we
tend quite understandably to look for clarity of purpose and coordination and concentration of effort. In our current quandary, we
are better off asking of the whole community: "Are the right people
talking to one another? about the right topics? under promising
conditions?" The community is not constrained by the limitations
of a single organization. Its members can simultaneously march
off in several conflicting directions without violating the good
sense of a rationalizing administrator.[6] It can protect enclaves
against premature tests of significance, relying upon time and
competition to settle matters which no judge can decide in
advance. The questions to apply to a community are not whether
it is "coordinated" nor even whether it is always "cooperative" but
rather whether there are missing elements, whether the flow of
information is adequate and whether the conditions of work are
appropriate. If (as I believe is the case) there are major flaws
in the design, then they will not be remedied by recapitulating
the early stages of policy development. In those early stages,
bold images of the future command attention and energize conversation. Once joined, however, it is the design of the conversation
and not its unique and prefigured outcome which is the source of
legitimacy and confidence. Organizational scenarios forecasting
the interaction within the designing community should replace
technological projection as the mode we use to find a way out of
the quandary, moving beyond the inevitably frustrating search for a
powerful developmental path.[7]

The straightforward task of describing the institutional map
of the designing community which operates in the United States
would push me beyond my alloted space. It might be possible to
untangle the overlapping responsibilities of Washington agencies[8]
but that would leave, hardly touched, the interaction between those
agencies, the major technology and service providers, research
institutions and organized users. If I tried to sketch a full map
and then matched it with similar ones drawn for six or a dozen
other nations (following the pattern of what is now an old-fashioned
form of comparative political science) I would occupy a very considerable monograph. Nor is it clear that such a series of maps
would serve a useful purpose. Insofar as the designing community
spans national units -- and it does so very powerfully indeed --
then national variations do not represent independent structures
but adaptations within the niches of a single system. Even

allowing for considerable independence of national forms, "config-
urative description" is not an adequate basis for evaluation and
change.[9] If we know that the U.S. Federal Communications Commission and the British Post Office are differently organized, does it
follow   that the FCC could or would take on the estimable qualities of the PO if it were reconstructed in the British mold?

In order to surmount both the constraints of space and the
limitations of conventional institutional maps, I have chosen to
present three dilemmas which severely test the design of the
community's networks: 1) the tension between aggregating communication policies and penetrating into the details of on-going systems,
2) the similar tension surrounding the boundaries of design concern
and participation and, finally, 3) the competing pressures to
design rigorously and to design robustly. In each case, I argue
that the response of the design community is flawed. Its structures
have not encouraged the appropriate **conversations** and conditions;
dilemmas, grasped at all, have been treated as if they were merely
"problems" to be resolved by a **sound** analytic answer and a coherent
program.

AGGREGATION AND PENETRATION

For those deeply involved in policy formulation and planning,
the idea of a communications network often takes on a concrete
form. The network is composed of channels, receivers and transmitters with physical dimensions, tested attributes (described in
an accompanying pamphlet or an attached metal plaque) and -- beyond all else -- a price. Occasionally, in the course of policy
deliberations, the concreteness fades. Regulators of computer-
communication links, for example, find the boundary between channel
and computer constantly shifting before their eyes.[10] For the most
part, however, such troubling problems are put aside. Policy deals
with such hard matters as spectrum allocation, channel investments
and "lowering the barriers to telecommunications growth."[11] The
values and range of concerns brought into policy deliberations
vary but there is little question about the nature of the immediate
objects of inquiry.

All of us, in our quiet moments, know better. The idea of a
communications network is a sophisticated analytic construct, tear-
ing apart and then resynthesizing the elements of human interaction
to separate the transmission, storage and manipulation of information from the tangle of other activities. Communications (compared, for example, to transportation or land use) planning is so
difficult precisely because it entails a rending of ordinary
configurations of activity, touches the guidance of complex systems
and is patently constructed around our own convenience. What, for

example, are the boundaries of the communication network which
serves the delivery of "health services." It's difficult even to
give a system a name without suggesting implicit boundaries but
sensitive planning requires that the ordinary terms (and hence
conventional organizational domains) be set-off in quotation marks
and subjected to inspection. At what level of aggregation should
the elements of a network be considered? Is a physician a channel,
a receiver/transmitter or an entire network?

An idealized designer chooses a system description with boundaries and defined levels of aggregation from a potential array
so as to serve a particular set of goals, adapting to resource
constraints and exploiting a feasible space for effective action.
Actual designers are not so free. They are torn by two tasks and
two largely discrete communities each pulling in an opposite
direction. On the one side, practictioners either pull them deep
into the tangle of the operating system or reject them as unfit to
penetrate. You can't hope to design a communication system for the
delivery of health services or vocational rehabilitation unless
you are a trusted member of the group, demonstrating through your
skill, your loyalty and your sensitivity to constraints that you
truly understand.

On the other side, system descriptions must be simplified if
diverse demands are to be aggregated and global policies defined.
National decision makers cannot be concerned with the requirements
of each individual clinic or school; only with the total demand for
channel space or particular devices.

The test of a designing community is how well it integrates
the often conflicting requirements of penetration and aggregation.
The simplest and most common form of integration asks each unit to
express its preferences directly in the ordinary terms of the
external decisionmakers: How many channels will you purchase?
How many telephones do you want? How much computer capacity do you
require?

Even private sellers of communication services (such as the
Bell system in the United States) recognize the limitations of this
form of integration. Many units will not readily articulate their
requirements in these terms without considerable assistance. The
readily articulated demands, moreover, will tend to serve established configurations. The administrator of a large hospital will
join a biomedical information system which reduces the time and
cost of current search activities but will not -- out of the same
impulse -- launch into an inquiry into the communication attributes
of the entire organization.

The purveyors of hardware and services, balancing costs
against anticipated sales, characteristically engage in some organizational development activities intended to penetrate beneath the
surface of readily articulated requirements. Their own obvious
self-interests, however, always render them somewhat suspect. In
virtually every one of the domains touched by telecommunications,
communities have developed which are largely independent of the
purveyors and which attempt to enhance the functions of penetration
and aggregation and their integration. Specialists in business,
biomedicine,   police and educational communications ply their craft
along the often difficult road which links practicitioners, purveyors and political decisionmakers. State and regional consortia
and specialized professional associations have also acted as
brokers and mediators.[12]

These communities differ in style and in capabilities. Across
the globe, by virtually every measure that I can imagine, the military community is characterized by the fullest penetration into
the world of practitioners and the most direct and insistent
claims upon the attention and resources of the aggregators. At the
other extreme (at least in the United States) the social welfare
community is poorly developed. It neither reaches deeply into the
operating agencies nor does it speak powerfully in national councils.

Agencies of national governments allied with these communities
have often invested in their growth. The biomedical communication
effort in the United States, for example, is largely a creation of
initiatives that Washington elicited and supported. The Federal
Communications Commission and the Office of Telecommunications
Policy have also helped to organize a policy research community.
The annual spring conference at Airlie House in Virginia (now something of an international event) illustrates, however, some of the
difficulties in meshing communities so as to enrich the policy discourse. Originally heavily dominated by economists working at the
most aggregate level, it has broadened its scope in each successive year to include more "penetrators". They often feel, however,
as if they were interlopers in a conversation whose terms were
already set.

For the most part, the high level policy agencies have
approached community development diffidently and without a self-
conscious strategy. The National Science Foundation acts (ill-
advisedly, I believe) as if the purpose of its "experiments" in the
use of two-way cable television were to test particular technological applications rather than to create a new community. A
variety of agencies persist in the fantasy that some great demonstration project will crack through the morass of organizational

resistance and open the way to a telecommunications take-off to
a self-sustained growth.

I do not mean to suggest that failure to consider the design
of communities which can bridge between penetration and aggregation is merely a product of inadvertence or misguided intellectual
development. There are both positive incentives for keeping the
demands of aggregation and penetration separate and grave problems
in linking them.

There are very strong international influences in the designing community which work to enhance three major planning themes:

> Long-distance transmission rather than local switching is
> the most powerful engine for communications change.
>
> The economics of mass-production and the opportunities in
> international trade encourage uniformity in products and
> the processes they serve.
>
> The concerns of major hardware and service providers
> (such as RCA, Xerox and IBM) frequently cross boundaries
> but many regulatory agencies deal separately with each
> code and transmission channel.

Communications (not merely telecommunications) planning which
penetrated deeply into the life of major social institutions would
strengthen a series of already important attacks on these themes.
In almost every actual setting, switching rather than transmission
is the most important obstacle to institutional change. Many
practitioners reject the substitution of extended range for local
flexibility. Similarly, sensitive planning tends to demand custom-
crafting and a willingness to engage in substitutions across media
and code boundaries. Deeper penetration would challenge not merely
the intellectual simplicity of aggregate telecommunications planning
but its economic, political and social intentions.

The intellectual and emotional demands of penetration are
also -- on the negative side -- enough to discourage even the
strongest and most committed will. In areas such as biomedicine
or vocational rehabilitation in which there is a strong federal
role in the U.S. in financing and guiding regional planning, the
communications advocates have proceeded warily, opening opportunities but never insisting lest they be rejected by their practic-
tioner hosts. I sometimes also wonder about the nightmares of
FCC commissioners. They must dread what would happen if someone
held them accountable for the impact of telecommunications on such
problems as the form of cities.[13] Even if they were absolved of
past sins, how would they start in the light of morning to grapple

with so many uncertain issues? The British Post Office, in a
setting which has long-supported centralized land use controls
and government intervention in the internal organization of firms,
has undertaken to think imaginatively about some of these matters
but always with considerable political caution.[14] In the United
States, caution is an even more common and understandable mantle
for decisionmakers.

THE BOUNDARIES OF DESIGN CONCERN AND PARTICIPATION

The penetration-aggregation conflict poses a genuine dilemma
for individuals and agencies. If they move a little in one direction, they lose (or appear to lose) their conceptual and political
bases in the other. This first dilemma is linked to a second which
revolves around the boundaries of design concern and participation.
If medical directors or city planners could speak directly and
fully with national telecommunication officials, there would still
be two insistent questions unresolved: What is the domain of
communications planning and where does the planner overstep the
boundary? Who should participate in the planning process?

Unfortunately, grappling with these questions and the dilemma
which underlies them - rather than confidently assuming a fixed
answer - only complicates the politics of the first dilemma and the
demands on the design community. Broaden the domain and the range
of participation, and penetration becomes more difficult; narrow
them, and you are trapped by unwelcome constraints.

The manner in which designers deal with this second dilemma
is linked to the character of their own introspection. There are
levels of self-doubt in every field. At the simplest level, the
eager advocate of innovation realizes that things are not going
well and turns in upon his or her own activities. If only the
message were clearer, the activity perfected, the experimental
controls tighter, the evaluation more rigorous, the report more
lucid, then everyone would listen. At a second level, the advocates adopt a more thoroughly critical perspective. Even the most
effective discrete project may fail if it is not adapted to the
implementation requirements of other organizational settings. Concern for the process of diffusion becomes part of the standards of
project design. Research and development contracts call for an
explicit strategy for diffusion and for subsequent institutionalization. You are required to specify how you will guide organizations from where they are to where you would like them to be.

The telecommunications design community appears to me to be involved in this second level of self-doubt. And it is the wiser for it. Despite an occasional waving of "implementation" as if it were a verbal talisman, the ideas of transition states and the hierarchy and costs of bureaucratic decisions are powerful additions to our ordinary design lexicon.[15] There is, however, a third level of self-doubt which requires a shift in the conceptual focus and the boundaries of the design community. As long as we are arranging "implementation strategies," control appears to be lodged firmly in our own hands. There are obstacles in the world but, if we are clever, they can be overcome. In level three our cleverness recedes before two complementary questions. First, are we straining unnecessarily, blindly oblivious to a structural obstacle? If that obstacle were removed, would agencies previously resistant to our efforts fall all over us, eager for the new telecommunications wisdom? That eagerness, as is the case in business communications, would generate its own implementation experts and would more than compensate for any lack of ingenuity.

The second question is more radical. Have we confused our own condition by generalizing our problems as those of resistance to innovation rather than rejection of the accurately recognized threat of specific communication technologies and organizations? If the answer is positive, then managerial issues of diffusion and implementation strategies must recede before plans for political conflict, bargaining and accommodation.

The first question about structural obstacles you will recognize as the characteristic query of political economists. In both socialist and non-socialist countries, many of the areas which might be powerfully influenced by telecommunications are dominated by service providers (such as schools and hospitals) that have three important characteristics: they have only vague criteria for success; they are relatively free from severe competitive pressure; and they cannot, by and large, recover from their clients the cost of most of the private benefits they offer them. They are also usually labor-intensive so that, faced with external political pressure to reduce expenditures or increase efficiency, they have a reservoir of organizational slack. This reservoir can be tapped by vigorous managers in a mode which preempts more far-reaching institutional change.

There is a good deal to be said for the Draconian solution to the market structure of the health, education and welfare industries: the introduction of competition and the creation of differentiated pricing mechanisms. Many communications planners in the United States are certainly awaiting with some eagerness the spread of pay-television systems which will allow social service providers to charge either viewers or third parties for programming and benefits rendered. Planners focused on public services have tended to confuse wide-spread private benefits with public goods. There is an enormous volume of information produced by profit-oriented firms for which consumers appear to be willing to pay. Pay systems will encourage competition by allowing new institutions with low fixed costs to intrude into the fields of local monopolies. The resistance to some of the open universities which depend heavily upon well-designed print (even more than video) material is a measure of the promise of this competition.

There are also limits to ordinary market mechanisms. The public character of a great deal of information requires collective processes to achieve broadly desired service levels.[16] Nor, even for private goods, does the "exit" option provided by competition (as Albert O. Hirschman has emphasized)[17] insure the responsiveness of firms or agencies. Exit without effective voice may not allow customers or clients to signal clearly the bases of their discontent.

The creation of voice opportunities, the chance for users to understand and influence directly the organization of service providers, is the point of linkage between the political economist's query and the second, political, question. The ordinary language of formal research proposals expressing a hope that telecommunications may increase the "efficiency and effectiveness" of a particular service is a protective shield. Goals (and means which are almost indistinguishable from goals) are altered in the course of a communications planning effort. There is no unique set of goals served or demanded by the technology of telecommunications. There are, however, a broad set of possibilities whose very appearance is threatening: the realignment of client-professional relationships, the shift in the balance of authority between center and periphery, the redefinition of "public" and "private" interchanges, the redrawing of organizational boundaries and the reconceptualization of eligibility.

When the aggregation-penetration dilemma is attacked by binding high communications policy tightly to the practitioner organizations, the designing community is impoverished. Buzzing around each of the institutional domains which conern us, there are communities of critics, reformers, politicians and researchers. Some of the participants in these communities are inside the direct service settings; others external. If the demands of penetration are met by talking only with the established insiders, the variety of possibilities is severely reduced and the bias of communications policy decided almost by inadvertence. The outside linkage, unfortunately, is professionally costly. It projects telecommunications into the center of often acrimonious debate and lowers the protective shield of a merely "facilitating" service.

It may seem nostalgic to emphasize the dangers of losing the relationship between broad social policy and telecommunications design. For a long time, the field has been dominated by a broad line of theorizing talk which went far beyond any immediate experimental or demonstration activity. There are now important advantages in professional relationships and in intellectual clarity to going beyond normative theory and getting to the tasks at hand. It's precisely at the point where there is an incentive to break the link, however, that it is important to recognize their centrality. If the major social issues surrounding health, education, welfare, criminal justice, work and industrial organization are not integrated into the telecommunications designing community they will not disappear. Conflict will out and it is only self-deception to ignore it.

RIGOROUS AND ROBUST DESIGN[18]

To this point, my reflections have dealt with general features of the telecommunications design community. The recommendations which might emerge from this analysis, once detailed and tested in particular national contexts, would extend to macro-issues of research funding, the organization of information networks, the strengthening of linking roles and the creation of community-building settings. The final dilemma in my triptych operates at a micro-scale and touches the intellectual focus and the social organization of particular projects. I describe this dilemma as the conflict between rigorous and robust design.

Communications planning is not an easy business. The ordinary tangle of action and symbol, energy and information, in a complex system is torn apart in order to reveal ways of enhancing performance through improved communication. Since the movement and manipulation of information supports the major guidance functions of systems, laying communication patterns bare always appears to risk a dangerous intervention into established practice.

There are three responses to this risk. The first is characteristic of much of traditional communications planning in its many guises. The apprehensive designer reduces the danger by weakening the intervention. Most transportation and land use planning (often guided by communication goal statements) is determinedly weak: you may link two neighborhoods with a road, school or playground but you really can't force the residents to interact. Similarly, in a more explicitly communications vein, you may provide a new capability as an addition to an established array but you don't try to reorganize the great mass of behavior. Most of the history of educational communications, for example, has taken the incremental approach and has, as a result, managed only to hang-on barely at the margins of schooling.

The second approach, characteristic of the ambitious research community which does not want to replicate a history of futile in-crementalism, takes up the challenge of strong interventions di-rectly. If changes in communication patterns are so influential, then there is a professional responsibility to provide a basis of knowledge which will sustain rigorous design. This responsibility underlies all of the detailed studies of the effect of one medium in contrast to another, of the technical requirements of diverse tasks and of the response of users to alternate configurations. The risks of rejection are high so one's efforts are bent at get-ting the new communications system right the first time. There are many examples of inept design in one place ruining opportuni-ties in others. The burden of research is to avoid a new wave of disappointments (as, for example, in the delivery of health ser-vices) which would replicate the experience of "fouling the nest" in education or in libraries.

The final response moves in the opposite direction. The attempt to get designs right is always doomed to a substantial measure of frustration. Possibly because communications is so critical to systematic functions, it encounters a compensating increase in sophisticated rejection or cooptation to counter its own improvements. A clumsy design is easily brushed aside. (The lone television set in the otherwise unchanged classroom is shut-off or falls prey to vandalism and is never repaired.) Well-designed communications interventions generate more powerful (though not always hostile) counter-responses which always demand compensating adaptations in the second round of change.

Faced with this perversity in the field, the designer must be sensitive to the continuing processes of the designing community within the organization. If the users are ignorant of the ration-ale and dynamics of the new system, if they are dependent upon out-siders for continuing support and adjustment, then they can only allow a thin surface for success. If, in contrast, they under-stand and control the system, then they can afford and rectify initial clumsiness. A telecommunications design is robust not be-cause the response terminals are tamper-proof or there is a back-up computer in case of malfunction but because it has strong social bases for correction and adaptation.

Phrased in this way, there is no designer who would ignore the continuing organizational community. We are all searching for robust designs which can withstand and hopefully control the pro-cesses of change. There is, however, a genuine dilemma. A great many things happen in the course of local community-building which do not fit easily into common forms of client-professional relationships and, ironically, into our hopes of developing a rigorous knowledge base for telecommunications planning.

Think, for example, of the experiments in the use of two-way television for the delivery of social services in the United States.[19]   It's possible to ask of each: "When does the experiment begin?" The answer seems to me to be clear: <u>When the experimenters go away</u>! All the formal experimental activity is a form of expensive, hand-crafted community building. The measurement of each system's success should follow the assumption of local control. At this point, however, it will be difficult on a continuing basis to describe the nature of the systems and their impact. Strict experimental protocols and robustness may not go hand in hand.[20]

This problem at the end of a process is paralleled by one at the beginning. The patient social worker's approach to community development does not commend itself as the way to introduce radically new possibilities and sophisticated technology into complex and highly professionalized systems. The designer's sensibilities suggest preserving control as long as possible, getting the system right and then training the users to employ it correctly. (Or, indeed, making it proof against their abuse.) Robust design processes, in contrast, seem (and are) sloppy, difficult to control, uncertain in their costs and in their outcomes. In hierarchical systems, portions of the client group will share these professional perceptions.

There is a genuine dilemma. Move towards rigorously designed and controlled projects and you get a high yield of formal knowledge but low effectiveness. Move towards the other extreme and you may get things to work but somehow you don't know much about them. The pressures of professionalization and the demand for accountable and "high quality" research within the design community all emphasize the value of rigor. At the risk of appearing to defend "low quality research" let me finish with a plea for preserving the tension within the dilemma. If national decisionmakers think of themselves as community builders, then evaluation will take on a concern for the enhancement of dialogue. The powerful rules of public scientific discourse will not be used to stifle conversations. At the micro-scale, community formation with its invariable false starts and less than perfect beginnings will be risked even though it requires excess channel capacity, underutilized personnel and a tolerance for mistakes.

IN CONCLUSION: THREE DILEMMAS

The body of this paper maps the general relationship between three pervasive dilemmas in telecommunications planning and the

organization of the designing community. In a field now caught in
a quandary of purpose and direction, resolutions will not appear in
grand visions of the future but in images of the processes of
inquiry and of societal deliberation. Any attempt to impose the
models of consolidated and intelligent organizations upon these
processes is bound to fail. The relatively independent research
community, which often legitimates initiatives more than it generates them, may play an important role in preserving and enhancing
a diversified and complex designing community or it may act to
weaken the community as the paradoxical cost of its own attempts
to enhance its special role within it. The choice is a real one.
It will not be apparent, however, in any great moment of policy
clarification but in countless small decisions.

NOTES

1. My own hopes are described in Community and Communications (New York: W.W. Norton, 1972). For an even more ambitious statement see Harold Sackman, Mass Information Utilities and Social Excellence (Princeton, N.J.: Auerbach, 1971).

2. H.W.J. Rittel and M.M. Webber, "Dilemmas in a General Theory of Planning," Policy Sciences, IV (1973), 155-169.

3. George Comstock and George Lindsey, Television and Human Behavior: The Research Horizon, Future and Present (Santa Monica, California: Rand, R-1758-CF, June, 1975); Cable Television Information Center, Social Services and Cable TV (Washington, D.C.: National Science Foundation, NSF/RA/760161, July, 1976).

4. Comstock and Lindsey, 44-5.

5. Carol H. Weiss, "Evaluation Research in the Political Context," in Elmer L. Struening and Maria Guttentag, eds., Handbook of Evaluation Research (Beverly Hills, California: Sage Publications, 1974) I, 13-26.

6. The same point was made forcefully by Professor William Capron, at the March 29, 1976 Washington, staff seminar of the Aspen Institute Program on Communications and Society: "We are still so far from the clear identification of policy options in this area that we need the protection of what is sometimes called redundacy and duplication, lest a single, powerful organizational voice - which will inevitably have its own blind spots and biases - lead us down a garden path which turns out to be full of brambles and checkholes." Refocusing Government Communications Policy (Washington, D.C.: Aspen Institute for Humanistic Studies, 1976), 26.

7. Arthur D. Little, Inc., Telecommunications and Society, 1976-1991: Report to Office of Telecommunications Policy, Executive Office of the President (Springfield, Virginia: National Technical Information Service, PB-256829, 1976) is a fascinating compilation of such scenarios.

8. There is a brief summary of one such map of the Washington-based policy community in Frank Lloyd and Ellie Koch, "The Hidden Network: Communications Policy Is Made In the Strangest Places," Access 48 (February, 1977), 4-5.

9. James A. Bill and Robert L. Hardgrove, Jr., Comparative Politics:
    The Quest for Theory (Colombus, Ohio: Charles F. Merritt, 1973).

10. I'm indebted for this illustration to Janet Taplin Thompson
    whose dissertation-in-progress at the University of Pennsylvania
    deals with the history of computing and communications.

11. See the volume of that name, edited by Douglass D. Crombie for
    the Office of Telecommunications, Department of Commerce,
    November, 1976.

12. See the image of this mediating role described by John Wither-
    spoon, President of the Public Service Satellite Consortium,
    June 16, 1976 in Refocusing, 49-52.

13. In the 1950's many analysts believed that the dense centers of
    such cities as New York would be sustained by their unique
    communication advantages. The doubts are beginning to accumu-
    late. See George Sternlieb and James W. Hughes, "Is the New
    York Region the Prototype?" in the volume they edited, Post-
    Industrial America: Metropolitan Decline and Inter-Regional
    Job Shifts (New Brunswick, N.J.: Center for Urban Policy
    Research, Rutgers, 1975), 101-137.

14. I'm alluding to the work both of the Communications Studies
    Group and the Post Office Telecommunications Long Range Studies
    Divisions.

15. The best introduction I know to these ideas is Jeffrey L. Press-
    man and Aaron B. Wildavsky, Implementation (Berkeley: University
    of California Press, 1973). See also Paul Berman and M.W.
    Mc Laughlin, Federal Programs Supporting Education Change, Vol.
    IV: The Findings in Review (Santa Monica, California: Rand,
    R-1589/4-HEW, 1975).

16. I don't mean to deny the value of new payment schemes. See
    Bruce M. Owen, Jack H. Beebe and Willard G. Manning, Jr.,
    Television Economics (Lexington, Massachusetts: Lexington
    Books, D.C. Heath, 1974), 78-81.

17. Exit, Voice and Loyalty: Responses to Decline in Firms, Organi-
    zations and States (Cambridge: Harvard University Press, 1970).

18. I've already described some of the ideas in this section in
    "Robust Approaches to Regional Communications Planning," a
    paper presented at the Fourth Annual Telecommunications Policy
    Research Conference, April 21-24, 1976, Airlie, Virginia, ERIC
    Number 122767.

19. CTIC, _Social Services and Cable TV_, describes the three experiments.

20. See _ibid._, III, 71-74, and my essay, "On Not Doing One's Best: The Uses and Problems of Experimentation in Planning," _Journal of the American Institute Planners_, XLI (1975), 184-190.

THE UTILITY OF SOCIAL EXPERIMENTATION IN POLICY RESEARCH

Dr. Allen M. Shinn, Jr.

National Science Foundation

Washington, D.C.

INTRODUCTION

Social experiments are a relatively new phenomenon. None of the examples which I have considered predate the mid-sixties, when several major experiments were undertaken as a part of the expansion of social welfare programs of the Johnson administration. Social experimentation was then a largely untried and undeveloped methodology, and much was expected of it. As with other innovations, it turned out that it could not deliver as much as it promised, and we are now beginning to understand its limitations. My purpose in this paper is to review the advantages and disadvantages of social experimentation in policy research, with emphasis on its use in communications policy work. I take a rather skeptical view of it, because it seems to me that its problems may easily be underestimated. These problems are serious. They need to be anticipated and solved, or they will ruin a lot of expensive research,

But there is a role for social experimentation in many areas of policy research, including telecommunications. Properly conducted, with sufficient groundwork and planning, they can provide information which is not available in any other way. I will conclude with a statement of what I take to be the proper conditions for their use.

Despite its newness, there is a substantial literature available which deals with social experimentation as a method. Much of this results from a series of studies done under the

general auspices of the Brookings Institution's Panel on Social
Experimentation. Four significant volumes have resulted from
this effort so far.   These include case studies of several of
the important early experiments,[1]/[2]/[3]/   with both descriptive
and thoughtfully critical material, and a volume dealing with
ethical and legal problems associated with experimentation.[4]/
There is also an important unrelated study of the New Jersey
Income Maintenance Experiment (NJIME).[5]/   These works are by
far the best source of information and criticism concerning
social experimentation which I am aware of. Finally, there are
a number of articles and miscellaneous reports which deal with
the subject in some way, and there are the reports which stem
from the projects themselves. This literature, plus my own
experience as Director of the Telecommunications Policy Research
Program of the U.S. National Science Foundation (NSF) for
several years, is the basis from which I have drawn the comments
and conclusions which follow.

WHAT IS A SOCIAL EXPERIMENT?

Social experiments are attempts to take the controlled methods
of scientific experimentation into field settings in order to
assess the effects of some innovation. Usually "effects" means
"costs and benefits," and the innovation is a piece of technology,
a new way of organizing, or some other new way of doing something.
In telecommunications policy research, the innovation is usually
advanced communication equipment of one sort or another. But it
is also the organization that supports and uses the equipment,
and confusion on this point can cause serious problems. Defining
just what the innovation is is not always easy.

Social experiments are related to demonstrations and field
tests, and to evaluations, but the differences are important.
Experimentation implies both theory and control: To do it right
one must specify hypotheses, define and measure theoretically
meaningful variables, and account for the effects of outside
forces. Hypotheses can't be specified without theory, and the
results can be trusted only if control is effective. Demonstrations and field tests involve little theory and attempt little
control; instead the approach is to "try it out and see what
happens." While the goal--determining the effects of the
innovation--is the same, the results are very different. Demonstrations and field tests can teach much about how to make an
innovation work, and can identify problems and provide impressions concerning effects, both expected and unexpected. But they
cannot provide cause-and-effect information, because they don't
test well-defined hypotheses, and they lack a mechanism for
making careful comparisons between situations with and without

the innovation. Evaluations generally involve proper controls, but tend to be weak on theory. Rather than ask "why" questions, and try to describe specific cause-and-effect relations, they tend to ask "whether" an innovation is associated with a desired outcome. The focus is on overall effects, rather than underlying mechanisms, with the result that understanding may still be lacking even if an effect is shown to be present. An evaluation which is theoretically well enough developed to specify hypotheses concerning underlying mechanisms, however, is indistinguishable for present purposes from an experiment.

Experimentation involves both enough theory so that underlying mechanisms are illuminated, and enough control so that the results are trustworthy. It is the dominant method of scientific investigation in fields where the requisite theory and control are available.

Taken into the field, the classical method of scientific experimentation offers policy makers a tool of potentially great value. Policy makers who have wanted to know whether higher welfare payments would reduce the incentive to work, or which of several teaching strategies was best, or whether advanced telecommunications systems would be worth their cost, have turned naturally to social experimentation. Unfortunately, they have not often found very useful answers.

SOCIAL EXPERIMENTS HAVE MAJOR DISADVANTAGES

While social experiments have some important advantages, there are so many serious problems associated with them that I think their use is indicated only in rather limited circumstances, which tend to occur quite late in the research and development process. In most cases, I think other methods will produce more knowledge, faster, and at less cost. The disadvantages are:

(1) Cost: Social experiments are costly, generally much more so than is initially expected. Their cost limits their numbers, so that they can only be justified when the central research question is important. Replication is generally not feasible. Their high costs lead to a tendency to underestimate costs and to promise more than can be delivered, in order to justify them to potential funders.

(2) Politics: Social experiments are deeply political in nature. Their cost makes them politically visible. They often arise from political needs, and their results are politically significant. Thus crucial

decisions in their conduct may be based on political criteria as much as or more than on scientific criteria, with resultant difficulties in interpreting the results.

(3) Time: Social experiments take more time to plan and execute than is ever expected. But costs rise with time, and results are generally needed quickly. This creates pressure to cut back on research design, and the validity of the results may be compromised.

(4) Noise: Once in the field, social experiments are subject to all sorts of perils. Schedules slip, equipment doesn't work, local officials object, funding fails, objectives change, designs have unanticipated flaws, and any number of other things go wrong. There isn't time to fix the problems, and it costs too much to start over. The result is more doubt about the validity of the results.

(5) Ethics: There are significant ethical problems associated with social experimentation. "Informed consent" is difficult to achieve, and may endanger results. Randomization is not always acceptable. People may be made worse off by an experiment, especially if they come to be dependent on an innovation which is withdrawn at the end of the experiment.

(6) Inconclusive results: Most social experiments yield inconclusive results. Most innovations don't work very well anyway, but even when the innovation works the research design may be too limited, because of time, cost, ethical, political, or other constraints, to detect the effects.

(7) Policy impact: For one reason or another, the results of social experiments often don't seem to have much effect on policy. The results may be flawed or the questions may not have been properly formulated to begin with. Events may have passed them by. Or the results may simply not seem very compelling when viewed in the context of other, more "political" factors.

(8) Bad theory: Most of all, social experiments suffer from inadequate theory. When attempted in areas which are theoretically poorly developed, it is difficult to define the innovation clearly and to design the experiment efficiently. More than any other factor, poor theory accounts for the disappointing results often obtained in social experiments. Hypotheses which

should have been refined in laboratory studies are
tested first in the field in vague forms that make
meaningful results difficult.

Against these disadvantages, social experiments offer the
cardinal advantage of providing a test of an hypothesis under
realistic conditions. The people involved are real, living
their own lives rather than playing roles. In principle, their
reactions to an innovation should be the most accurate available
measure of its success. There is also a secondary advantage in
that social experiments do sometimes yield unexpected insights
into side issues, something which generally does not happen in
laboratory work.

Under the right circumstances, social experiments can
surmount all their problems and deliver information obtainable
in no other way. But I think their role should be quite limited,
and I doubt that telecommunications policy research is ready to
use social experimentation in very many situations.

SPECIFIC PROBLEMS WITH SOCIAL EXPERIMENTS

I. Social Experiments Are Expensive

That social experiments are expensive, very expensive
relative to other forms of research, is indisputable. The New
Jersey Income Maintenance Experiment (NJIME) was expected to
cost $4 million, but actually cost $8 million. The budget for
research costs rose from an estimated $1.1 million to an actual
$5.5 million. Head Start Planned Variation cost $7 million.
The Educational Performance Contracting Experiment (EPCE) cost
$6 million for just one year. The telecommunications policy
research experiments supported by NSF, which were much less
ambitious efforts, averaged about a million dollars each. At
that level, they might be termed "mini-experiments."

The costliness of these projects has a number of undesirable
effects. It makes them politically quite visible, since they
often have to be explicitly justified to political bodies, and
this results in political factors inevitably entering into
decision making. Spending money for these projects means little
is left for other projects, which can easily result in unbalanced
research programs. Pressure to hold down costs can easily lead
to design weaknesses.

II. Social Experiments Are Political in Nature

Because of their size and visibility, and because of their
concern with politically sensitive questions, social experiments
are inevitably affected by political considerations. Because
they are expensive, they must be justified to political decision
makers. The justification may fail if the experiment does not
serve a political as well as scientific purpose. NJIME is
typical: "... [it was] clear that the experiment was conceived
by the OEO [The Office of Economic Opportunity] as part of a
broad strategy to obtain administrative and Congressional approval
of a negative income tax."[6] The educational experiment called
Follow Through Planned Variation became an experiment only when
Congress refused to fund it as an operating program. All the
large welfare and educational experiments I am aware of had
political goals as well as scientific ones, and often the
former took precedence. There is nothing wrong with this if the
political and scientific goals are compatible and kept clearly
in mind, but confusion on this point can severely affect the
value of the experiment. This is a major advantage of small
scale, and it may be that social experiments will be most useful
when they can be applied to policy issues that can be dealt with
on this scale, and that are not intensely political in nature.
The large issues of educational and social policy are not of
this sort, however, and ways must be found to deal with the
political problems if we wish to do social experiments in these
areas.

In general, it appears that the conjunction of political
forces that make it possible to mount a large scale social
experiment will also make the experiment politically sensitive.
The linkage between politics and science that one sees in large
scale social experiments thus seems inevitable.

Smaller experiments cost less and are less visible, and
therefore suffer political complications to a lesser degree.
The five National Science Foundation (NSF) funded telecommunications experiments seem to have been fairly free of political
problems, except on a local scale.

All social experiments and demonstrations are political
acts, in the sense that they all seek to influence action with
knowledge. They are undertaken in the belief that the innovation
to be tested is likely to be useful, which leads naturally to a
concern with the political effects of the research. This concern
is fine, provided that the scientific goals are clearly stated
and clearly kept in mind. In many cases the chief aim of a
social experiment probably should be political: to convince
decision makers and the public that the innovation works in a

real field setting. The social experiment has some advantages
in this, since it appears to be more credible than laboratory
experiments or analyses.7/ Viewed this way, social experiments
should be justified as much on political as scientific grounds.
Their contribution to important political goals may be more
important than their contributions to science. But it should
also not be surprising that when social experiments develop
cost and time overruns, get the local population upset, or
seem likely to endanger vested interests, political forces are
likely to take a hard, critical, and very unscientific approach
to them.

III. Social Experiments Take A Long Time

Social experiments always take more time than planned;
delays are the norm. The NJIME was planned for less than four
years, but took seven. The Educational Performance Contracting
Experiment (EPCE) was planned for only one year, when five
might have been more reasonable. NSF's experiments will have
consumed four years from conception to completion; the first
two encountered delays of a year or more beyond the original
schedule. Weikart and Banet argue that in education one
should plan on 10-12 years and about $6 million to carry one
significant innovation through from development, small scale
laboratory testing, revision and retesting, small scale field
testing, and finally large scale field experimentation and
dissemination. The schedule can vary from one field to another,
but I think these estimates are quite reasonable.8/

The consequences of these time requirements are substantial.
Cost increases with time, and so does vulnerability to political
pressures. Even worse, a large scale social experiment commits
one to testing one major idea, with one set of research questions,
for several years. If the questions turn out to be poorly
formulated, or theoretically unimportant, a major effort is
wasted, and the opportunity cost is high.

Even if the questions posed are theoretically important
and politically relevant, events can easily pass them by. By
the time NJIME was over, President Nixon's welfare reform
proposals were no longer under consideration. By the time
NSF's Miami telemedicine experiment had shown that two-way
video was not cost effective in the particular medical application
tested, laboratory research had shown that video was not likely
to be significantly better than audio in a variety of tasks,
and thus that the theoretical basis for the Miami experiment
had really not been present. The time required to plan and
execute a social experiment is a severe handicap, and argues
against them in many circumstances.

IV. Social Experiments Are "Noisy"

Even the best design cannot anticipate everything, and the circumstances of social experimentation allow much to go wrong. The ability to control events that exists in the laboratory is just not there in the field. In NJIME the State of New Jersey changed its welfare laws in the middle of the experiment, making it much more difficult to interpret results. In the Miami telemedicine experiment local contracting procedures caused nearly a year's delay, and important equipment just didn't work. These problems could often be worked out with more money and time, but money and time are usually in short supply, and the pressure is on to produce results. If the design can't be changed, and the experiment can't be suspended while a defect is remedied, the consequence may be severely compromised results.

V. Social Experiments Present Significant Ethical and Legal Problems

I don't want to dwell on this area, which is less likely to cause trouble in telecommunications policy than in social policy research. The best treatment by far that I am aware of is by Rivlin and Timpane.[4/] It is enough to note that any experiment which involves human beings is necessarily subject to a range of ethical and legal constraints which often constrict the design and complicate the experimenter's life.

VI. Social Experiments Are Often Inconclusive

In general, the most likely result of a social experiment is no difference between experimental and control groups. NJIME found that income maintenance did not have much effect on work response. Head Start and Follow Through Planned Variation found no differences between teaching methods. Performance Contracting did not appear to improve children's learning. The Miami tele-medicine project showed that video communications did not affect the quality of health care delivered (although it did increase costs). Similar findings are common.

There may be several reasons for this. In some cases, cost, time and political constraints on designs make them simply too weak to detect any differences. In other cases the experimental innovation may be too poorly defined, implemented, or measured for any effects to be demonstrated. This is essentially a problem of underdeveloped theory, which is treated below.

Many innovations simply don't work, of course. Mosteller studied thirty innovations tested in various settings, and concluded that only four resulted in "substantial improvement."[9/]

Results like this can't be blamed solely on social experimentation, but they may often be due to testing an innovation in the field before it is theoretically well enough developed. This has been a common problem in educational experiments.

In some cases, "no difference" findings are valuable, and may indicate success. In the Miami telemedicine experiment, the finding that video was not cost effective is worth something, if it saves the installation of a lot of unnecessary equipment. In NJIME the "no effect" finding was a positive outcome, since it was thought that this would reduce the fears of political conservatives that income maintenance would destroy work incentives.

The trouble with no difference findings, however, is that they are especially open to challenge, and the general messiness of social experiments contributes to this problem. Failing to demonstrate an effect is different from demonstrating that the effect is not present. Although the Performance Contracting experiment demonstrated no differences between teaching approaches, there probably are differences, and the experimental results are thus not very convincing. It can always be argued that video would improve health care if it were done "the right way," and so on. The point is simply that no difference findings can be due to any number of causes, and they are often not very convincing. This is especially true when it is remembered that the political impact of an experiment is often more important than the scientific impact.

VII. Social Experiments Don't Affect Policy Very Much

This seems like the unkindest cut of all, for the whole basis of social experimentation in policy research is to make better policy choices possible. Yet the record does not show many instances where a social experiment led in any direct way to a policy decision. Perhaps this is too much to expect since most non-experimental policy research also has no discernable impact on decisions. But it is easy to see why this is so often the case with social experiments.

One problem is timing. It often seems to turn out that the issues the experiment was set up to help decide can't wait. The political decision making process just refuses to wait for the experiment to finish up. This is sometimes referred to as "Timpane's Law,"[10] which, loosely stated, holds that whenever there is enough political backing to get an experiment started, the decisions are too pressing to await the results. This is an important result of the connectivity between policy research and policy making.

Another problem is that of no difference results. As noted already, they often are not convincing because of possible flaws in the design. They may also leave the impression that an experiment has "failed," and thus can be disregarded by policy makers.

An even more serious problem can be that the results of an experiment are simply unwelcome politically. Political conservatives are unlikely to be moved by NJIME's findings that income maintenance doesn't reduce the will to work, and those committed to performance contracting in education remain sure it would work if done properly. Against preconceptions of this sort, the results of a social experiment, with all its likely flaws, are not a strong weapon.

Experimental results can also be disregarded because they can't be generalized sufficiently. For all their messiness and their setting in the real world, experiments have a hard time simulating reactions to large scale implementation of an innovation. People won't use video telephones if no one else has them, for example, and no experiment could simulate well a situation in which most present telephone subscribers had video equipment. Thus even if an experiment meets all other tests, its results may be disregarded on grounds that they still don't apply to the real world for which policy makers make policy.

VIII. Social Experiments Suffer from Poor Theory

I think this is really the biggest problem that confronts social experimentation. It is linked with all the other problems, and exacerbates each of them. If good theory were available before planning began for a social experiment, many of the other problems would be much reduced in scope, and their solutions would be a lot more apparent.

Bad theory introduces confusion into experiments, which increases both monetary and time costs. It makes planning more difficult, since experimenters don't know what to expect. It exacerbates political problems, since poor theory means that the purpose of an experiment cannot be explicitly stated in scientific terms, and thus increases the likelihood that political and scientific purposes will be confused. Poor theory is often responsible for inconclusive results, for experiments can't easily be designed to measure effects which can't be predicted. Finally, bad theory often makes ethical problems more difficult to deal with, because the results of particular actions are harder to predict. The search for good theory is thus extremely important, and the development of good theory before planning begins is the most important condition for a successful experiment which I can identify.

Good theory means principally two things: That the innovation
to be tested be well defined, and that its effects be reasonably
predictable in advance. Both are serious problems, and both
require substantial prior work.

Defining the innovation can seem easy, but is often quite
difficult. Failure to do it effectively caused major problems
in all the education experiments. In the performance contracting
experiment, for example, the innovation seemed simple--firms
would be put under contract to produce specified educational
gains in their pupils, and paid only if these gains were achieved.
The experiment would simply determine whether or not the gains
were achieved. But it really was not anywhere near that simple,
for this concept of the experiment was entirely devoid of any
ideas concerning the ways in which educational gains, if any,
might be achieved. Without further specification, such a conception implies simply that teachers will try harder if they are
paid on a piecework basis. This may be true, but a useful test
of this approach would also have had to investigate just how the
teachers achieved gains, if in fact they did so. Did they
simply "teach to the test"? Did they motivate their students
better? If so, how? With better materials? Better teachers?
Fear of punishment? Obviously a very long list of such questions
could be drawn up, and just as obviously, no single experiment
could hope to answer more than a few of them.

As it was actually set up, the EPCE confused profit incentives
with tests of existing methods and the capabilities of particular
private firms. The notion of providing profit incentives wasn't
really much of an hypothesis, and no test of it could be conducted
in a short time. Conceivably, it might have been possible over
a long period of time to monitor the effects of shifting the
provision of public education from the public to the private
sector, with incentives for good performance. Over time, private
operators might, with a strong profit motive, have developed
better and more effective ways to teach. But there was no way
to test for such effects in the short run, so the experiment
became one of testing various approaches suggested by different
firms, and the "innovation" became a mix of the profit motive,
various new methods, and private organizations with varying
technical and managerial capacity. In such a situation, it is
most difficult to determine the meaning of an effect, even if
one can be reliably determined to exist.

Similar problems have plagued other social experiments.
The "planned variation" experiments with Head Start and Follow
Through had serious problems in defining each of the educational
approaches that were tested. Often these approaches were defined
in quite general terms, more in terms of a general philosophy of

education than a specific set of procedures. In practice, the philosophies turned out to be quite flexible enough to allow wide variation in procedures, so that just what was being tested was never very clear.

In telecommunications, this problem of defining the innovation may not be as severe as in other fields. Often the innovation is advanced technology of one sort or another, and we generally try to test, for example, "video" rather than a specific piece of video equipment. There is rather general agreement that it is important to look for differences between the effect of systems using video, audio, or written communications, or combinations of these.

This simplifies the problem, but it doesn't solve it. Technology is always embedded in organizations, and a given organization may or may not be optimally set up to exploit the technology. If technology and organization interact, then one is always testing some mix of the two, and it is difficult to justify a choice of a particular mix, or even to specify very clearly what the salient aspects of the organization under test really are.

The problem becomes still more complicated when one realizes that we are only just learning how to use this new technology effectively. Any test of the technology increases our knowledge of how to use it better, and it is important not to freeze either the technology or the way we use it too early. Yet no test can really take place unless we hold the technology and related matters of organization and practice constant. If we do so, and find the technology not especially advantageous (as is often the case), then how do we interpret the results? Do we conclude that the technology is not much use, when we know perfectly well that refinements would improve it? Such a conclusion would be foolish, yet we would also be foolish to assume that an "improved innovation" would be effective. The problem is that we always want to know what the effect of the latest and best developed form of the innovation would be, while the results of any social experiment necessarily can tell us only about the effects of the innovation as it existed when the experiment began. Because of the necessary time lag involved in experimentation, we are always learning about a form of the innovation in which we are no longer very interested. Lucas' suggestion of "serial experiments," elsewhere in this volume, may offer a partial resolution of this problem.

The EPCE illustrates another confusion about theory and its testing that may not be general, but is worth commenting on. This experiment was intended as a test of the hypothesis that

economic incentives might be effective in producing better
teachers. The results were essentially negative. Yet no one
would conclude from this that economic incentives are not
effective. We know perfectly well that such incentives are
effective, and if the test didn't show it then it's easy to
conclude that the test was inadequate. Even in education,
economic incentives are effective, as is evidenced by the obvious
fact that private schools, which get paid only if they satisfy
their clients, continue to do well even though they have to
compete with essentially free public schools. So the experiment
turns out to have been an investigation into the ways to make
use of economic incentives, rather than a test of the theory of
incentives itself. Confusion on such points does nothing to
improve the outcomes of social experiments.

The other principal component of good theory is some knowledge
of what the effects of the innovation should be. My own position
is that we should not get involved in field experiments until we
are quite sure that the innovation works well in the laboratory
and on a small scale. If we know its effects on a small scale,
we should be able to predict them reasonably well on a larger
scale. This knowledge should allow us to design efficient
social experiments--ones whose purpose is to confirm predictions
which are based on well-established laboratory results. Knowing
where to look is more than half the battle. Not knowing condemns
one to try to cover all contingencies, with attendant cost,
dilution of effort, and often failure to provide for an unanticipated effect.

Often, knowing the effects on a small scale could lead to a
decision not to do a social experiment at all. The Miami Prison
Telemedicine Experiment was based on the plausible (in 1972)
notion that video communications might so improve the provision
of health care in a prison setting that it would more than pay
for itself. The idea, essentially, was that video would allow
enough extra information to be passed along so that nurses could
often handle patients themselves, without the immediate physical
presence of physicians. As it turned out, the nurses did quite
well, but audio communications were generally sufficient. The
video added little but extra cost.

In 1972, we had no real theory that could tell us much
about the expected effects of video. By 1975, however, still
two years before the experimental results were available, we had
substantial evidence from Chapanis' laboratories at Johns Hopkins
that video probably would not be very effective.[14] Had we
insisted that the Miami experiment be better grounded in theory,
we might very well have concluded that it was not necessary at
all, and a million dollars would thereby have been made available
for more promising telecommunications research.

Knowing ahead of time what to look for can save a great
deal of time and money in the design of social experiments. In
NJIME, it was much more expensive to create cases involving
large amounts of subsidy than cases requiring only small amounts.
Lacking theory to suggest where the effects of subsidy were to
be found, the cheaper cases were emphasized, primarily for cost
reasons. As it turned out, having some more of the high cost
cases would have helped substantially in interpreting the results.

All the education experiments were hampered by a lack of
theory that could be used to develop appropriate measures of
effects. It wasn't clear whether the innovative approaches were
supposed to improve reading and writing skills, basic reasoning
and conceptual ability, better social adjustment and self appraisal,
or something else. Lacking this knowledge, it wasn't possible
to choose appropriate testing procedures to measure results, and
considerable argument developed when hoped-for improvements
didn't materialize. Lacking theory, the arguments were fruitless.
It wasn't possible to determine whether the problem lay in the
ineffectiveness of the teaching methods or the inappropriateness
of the tests. Marshall Smith concludes a discussion of the
problem of designing social experiments by saying, "The development of adequate theories and empirical data about the expected
effects of proposed intervention, though an extremely difficult
task, seems necessary. It involves a retreat to laboratory and
small scale field studies and finally a reexamination of the
underlying assumptions of treatments.[11]/

The relationship between well-developed theory and the
moral and ethical problems of social experimentation is perhaps
not widely appreciated, but the need to satisfy those who criticize
such experiments on ethical and moral grounds may serve as an
important practical stimulus to the development of better theory.
Peter Brown notes two continua on which social experiments can
be placed, and which define the essential ethical problems which
they face:[12]/

Full information ⟵——————————⟶ Poor information

No risk ⟵——————————————⟶ High risk

Experiments which fall towards the left end of both continua are
morally and ethically acceptable: They provide full information
to subjects, and subject them to little or no risk of harm.
Experiments falling towards the right-hand end of both continua
are clearly unacceptable. A high risk experiment might be
acceptable if full information were provided to subjects, or
information might be withheld in an experiment involving no
risk, and a case might still be made for the moral and ethical

acceptability of the project. <u>But only when theory is well</u>
<u>enough developed so that effects can be reasonably well predicted</u>
<u>will it be possible to assess risks accurately or provide reliable</u>
<u>information to subjects.</u> Thus theoretical development is clearly
prerequisite to the satisfaction of the moral and ethical criticisms which are lodged against social experimentation. This
fact is likely to become more significant as social experimentation becomes more widespread, and the need to ensure their
ethical acceptability becomes more generally understood and
accepted.

A WELL EXECUTED SOCIAL EXPERIMENT

A brief description of a well planned and executed social
experiment may serve to illustrate in a different way many of
the points I have made so far. In 1972, NSF funded a small
experiment[13]/ to study the effects of using nurses in conjunction
with some improved communications technology to deliver health
care to patients in nursing homes in the Boston area. The
innovation was primarily in the form of organization, which was
carefully specified. Nurses with special training in care of
the elderly were to be assigned to a group of nursing homes, but
were to be supervised by hospital physicians. The hospital was
thus to have responsibility for delivering both in-patient and
out-patient care. This was a sharp departure from previous
practice, which relied on neighborhood physicians for care in
the homes, and hospitalization for most tests and all significant
illnesses.

The benefits of the innovation were expected to be improved
care for the patients and reduced costs, the latter primarily
due to reduced hospitalization. As it turned out, both of these
effects were clearly demonstrated. The experiment solved or
avoided all of the problems which I have discussed. It cost
less than $300,000, so that funding was not a serious problem,
nor did the cost make it politically vulnerable.

The experiment had a political goal as well as a scientific
goal, but the two were not confused. The political goal was to
demonstrate to the Massachusetts health authorities the desirability of this approach to health care. The scientific goal
was to determine whether the combination of specially trained
nurses, an innovative form of organization, and some improved
communication equipment would have simultaneously beneficial
effects on both the quality of health care and its costs. These
goals were quite consistent, since both quality and cost effects
had to be demonstrated before the state would consider large
scale adoption.

The experiment took almost three years to design and execute, but this was acceptable under the circumstances. Health officials were willing to wait for the results, since there was no political crisis over the issue. Thus there was an excellent chance that the results of the experiment would have a significant effect on policy, and cost and time factors were under control.

Theoretically, the experiment was well grounded. The innovation was clearly specified, and it did not change during the experiment. Enough prior work had been done so that it was generally believed that the organization and technology were pretty well optimized, so there were no significant pressures to improve or change either of them during the experimental period. Success was clearly defined in terms of improved quality of care and reduced cost, and satisfactory measures of each were available.

The program avoided the problem of inconclusive results, because the results were quite predictable, and the research could be designed to measure them. The study groups were large enough to detect results reliably, but not so large as to drive up costs unnecessarily.

The experiment avoided any serious ethical problems, again primarily because theory made reliable predictions possible. The patients were not placed at risk because it was known in advance that the nurses could perform the necessary medical procedures properly, and were provided accurate information concerning how the system would operate. The major uncertainty concerning costs was not of direct interest to the patients themselves, since these costs were all borne either by the experiment or the state.

As it turned out, everything happened about the way it was planned.

The experimenters learned a good bit about the rates at which things happened, which was the major uncertainty beforehand. They were thus able to pin down costs for the innovative system much more closely, and they were able to document positive effects on quality. The results were convincing to health officials, who agreed to continue the program indefinitely on a regular operating basis.

Thus this experiment managed to avoid all of the perils which I have identified as having plagued other efforts. It was clearly a successful exercise in policy research. It is worth considering what we might do to insure more successes of this sort, and fewer of the very expensive failures which seem to have been so common.

SOCIAL EXPERIMENTATION IN CONTEXT

It is important that we view social experimentation in the context of other research approaches. It is only one of a number of research tools which we have available, and we would do well not to forget the others.

The most important alternative to social experimentation is clearly small scale experimentation in a laboratory. There has been a lot of good laboratory work in telecommunications policy research, of which Chapanis' work at Johns Hopkins is an excellent example.[14] John Short and others have reviewed much of this work in a recent book.[15]

Laboratory experimentation has virtues to match most of the major vices that trouble social experimentation. It is comparatively cheap and quick. Replication is easy. It is politically invisible. Its ethical problems are more manageable. If it, too, often produces inconclusive or negative results, it can better afford them; the cost of pursuing a blind lead can be several orders of magnitude less in the lab than in the field. Against these advantages are two major disadvantages: Laboratory experimentation is likely to have even less impact on policy than social experimentation, and it suffers from a lack of realism. The greater realism of social experimentation is its chief advantage, and accounts for its greater ability to influence policy makers. Clearly there are situations in which nothing else will do, and a social experiment is justified. But in some cases we could probably be more ingenious in designing laboratory experiments, and thereby increase their realism. One approach which has not been tried to my knowledge would involve combining research involving the simulation of complex negotiating and bargaining situations with telecommunications research. There are a number of well-developed games, for instance, which simulate such things as bargaining in international trade negotiation, quite realistically enough to be used in training diplomats and other professionals.[16] It should be relatively simple to combine plays of games such as these with different communications arrangements, and thereby considerably extend the range of the work which has already been accomplished, but which is limited to much simpler tasks.

We could also probably profit from greater use of computer based modeling and simulation of systems involving telecommunications. This is probably best done in conjunction with laboratory and social experimentation, since experimentation is usually necessary to provide empirical data to inform a model. What a computer simulation can do is allow an investigator to extend the analysis, and to force the experimental data to yield additional insights.

The telemedicine experiment I have described provides a good example of this. The experiment aimed at demonstrating a few very important facts, principally that positive effects on quality of care and cost were possible. But in the process, a great deal of quantitative data was collected, which made it possible to develop a useful computer model of a telemedicine system. This model was then used to learn a good bit more about the way such a system could be optimized. In this way the results of the experiment were extended well beyond the original goal.

CONCLUSIONS

By this point my conclusions should be apparent, but let me summarize them briefly for convenience.

1. The first, and most important, conclusion is that social experimentation should not be used until the subject is theoretically ready for it. Theory should be <u>developed</u> in other ways; social experiments are just too costly and slow. Before a social experiment is mounted, theory should be able to explain pretty well what is going to happen. Surprises can be afforded in the lab, but not in the field. The primary purpose of large scale social experimentation should be to confirm in a field setting what, because of prior lab work, is generally believed already to be true.

2. Secondly, it should be recognized that social experiments serve political purposes as much or more than scientific purposes. The questions addressed should be of central importance to policy makers, but not of such critical importance that decisions can't wait for the results. I am not sure that there are very many questions which can meet these conditions.

3. Thirdly, social experimentation requires meticulous planning to be successful. Six months to a year of detailed planning, once the broad outlines of an experiment are clear, will not be too much. Such planning may well identify unanticipated questions which should be answered by further laboratory experimentation or other research before the social experiment should go forward.

Social experimentation in many ways is like manned space flight. It's big, showy, expensive, time consuming, exciting,

politically volatile, satisfying, and fun. It's a grand thing
to do, a tremendous challenge. But it isn't necessarily very
cost effective in scientific terms. As with unmanned space
probes, it may turn out in telecommunications also that lower
cost, simpler, less complex research approaches pay greater
scientific dividends.

FOOTNOTES

1. Joseph A. Peckman and P. Michael Timpane, eds., Work Incentives
    and Income Guarantees: The New Jersey Negative Income Tax
    Experiment. Washington, Brookings Institution, 1975.

2. Alice M. Rivlin and P. Michael Timpane, eds., Planned Variation
    in Education: Should We Give Up or Try Harder? Washington,
    The Brookings Institution, 1975.

3. Edward M. Gramlich and Patricia P. Koshel, Educational
    Performance Contracting: An Evaluation of An Experiment.
    Washington, The Brookings Institution, 1975.

4. Alice M. Rivlin and P. Michael Timpane, eds., Ethical and Legal
    Issues in Social Experimentation. Washington, The Brookings
    Institution, 1975.

5. Peter H. Rossi and Katherine C. Lyall, Reforming Public
    Welfare: A Critique of the Negative Income Tax Experiment.
    New York, Russell Sage Foundation, 1976.

6. Pechman and Timpane, op. cit., p. 3.

7. See Leo Bogart, "Warning: The Surgeon General Has Determined
    That TV Violence Is Moderately Dangerous to Your Child's Mental
    Health," Public Opinion Quarterly 36(4), Winter 1972-73,
    pp. 491-521, for a discussion of the relative political credi-
    bility of large and small studies concerned with the effect of
    violence on television.

8. David P. Weikart and Bernard A. Banet, "Model Design Problems
    in Follow Through," in Rivlin and Timpane, Planned Variation,
    73-77.

9. Frederick Mosteller, "Comment," in Rivlin and Timpane,
    Planned Variation, 169-172.

10. After Michael Timpane, who noted the problem in an early
    article on social experimentation. "Educational
    Experimentation in National Social Policy," Harvard
    Educational Review, 1970 (40) 547-566.

11. Marshall S. Smith, "Design Strategies of Experimental
    Studies," in Rivlin and Timpane, Planned Variation, 144.

12. Peter G. Brown, "Informed Consent in Social Experimentation:
    Some Cautionary Notes," in Rivlin and Timpane, Ethical and
    Legal Issues, 85.

13. Roger Mark, et al, Final Report of the Nursing Home
    Telemedicine Project. Boston City Hospital, Mime, 1976.

14. Alphonse Chapanis, "Interactive Human Communication,"
    Scientific American, 232(3), March 1975, 36-42. This is a
    summary of a number of research reports by Chapanis and his
    associates.

15. John Short, Ederyn Williams, and Bruce Christie, "The Social
    Psychology of Telecommunications. London, John Wiley and
    Sons, 1976.

16. See, for example, G. R. Winham, "Negotiation as a
    Management Process," (1976); "Complexity in International
    Negotiations," (1976); and Winham and Glyn R. Berry, "Trade
    Negotiation Simulation" (1977). All are papers of The
    Center for Foreign Policy Studies, Dalhousie University,
    Halifax, Nova Scotia, Canada.

NEW TELECOMMUNICATIONS SERVICES AND REGIONAL DEVELOPMENT:

APPROACHES TO EXPERIMENTATION AND PLANNING

Daniel Chauche

Centre d'Etudes des Communications - C.A.F.

Levallois, France

1. INTRODUCTION

Concern for regionally balanced economic development has led to the creation of an overall plan to co-ordinate, at national and regional levels of government, the planning of economic and social-service development in the different regions of France.

The very low penetration of telephone service, per head of population in France, has recently created a strong incentive to invest heavily in developing a modern and comprehensive telecommunications network. Such a major investment effort is now under way; it includes provision for the introduction of new services and their extension once the basic network infrastructure has reached an adequate level of development, comparable to that found in other European countries.

This paper reviews planning studies that have been undertaken in preparation for the application of new services - mainly teleconferencing and teletext - and discusses three key questions for regional development:

- Is there a systematic operational way to decide priorities and plan the introduction of new telecommunication services in accordance with regional development aims?

- Examining the few experiments carried out to date, do we have data for evaluating their contribution to regional development?

- What are the main issues in introducing new
services advantageous to regional development, what
difficulties must be overcome and what are the implications for strategic planning?

2. AIMS: TELECOMMUNICATIONS PLANNING
IN RELATION TO REGIONAL DEVELOPMENT
GOALS AND NEEDS

Telecommunications systems are now considered in France as a
means to achieve more balanced development between the different
regions. Economic development, and the development of cultural and
social services, appear to depend on the availability of information
and communication services in urban areas and in rural areas. The
traditional centralisation around Paris is counterbalanced by a plan
for decentralisation which involves the dispersal of public and private management functions. Emphasis is being placed on the movement
of tertiary industry, government departments, banks and service organisations to the less developed regions. Regional development
programmes are selective and sequential, with a variety of incentives
and subsidies for decentralisation.

Improvements in telecommunication services are a new priority
and must be integrated with transportation facilities, retraining of
labour, and investment in social services and cultural activities.
The introduction of telecommunication services can be defined in
two ways:

- an overall plan for organising priorities for
the whole country

- research into the specific information and
communication needs of particular locations.

Information and communication needs arise from different activities (business, education, health, leisure, etc.). We intend to
translate the regional economic, social service and cultural priorities into telecommunication priorities (regional needs for telecommunication) and then to examine the existing telecommunication plant
and the possibility of developing new facilities which could be used
to achieve the aim of decentralisation and to meet specific regional
needs.

2.1 First Stage: Translating Decentralisation and the Overall Plan for the Regions into Telecommunication Facilities

The 7th Plan (1977-1982) results from the work of several Commissions and working groups during 1975. The Assemblies in each of the 21 regions have been consulted on preliminary orientation, as have regional social organisations, regional councils, economic and social services committees. The main aims of the overall plan can be summarised as follows. The policy of regional development must:

- contribute to national economic development by encouraging the full integration of the country into the European socio-economic system;

- improve every French person's standard of living;

- be part of the struggle to reduce inequalities in development by a constant effort to correct imbalances between regions.

The principles of the regional development strategy set up in response to these aims are to ensure the harmonious development of towns (four-fifths of the population), notably of medium and small towns, to stop the population drift from rural areas, and to ensure a more balanced division of industrial and tertiary activity between regions. Priority goals include:

· Development of industry in the West and in the Massif Central.

· Decentralisation of the service sector, government offices, and financial organisations.

· Improvement of the quality of town life, control of the regional development of urban agglomerations and of medium and small towns, and co-operation between local authorities.

· Creation of social services and jobs for the young people in rural areas and of the necessary infrastructure for education, health, and communications.

· Improvement of the communication network, notably to connect the West and the Massif Central to the other regions.

The overall aims have to be defined more precisely if they are to be used as a basis for decisions in telecommunication planning, and for research more specifically geared to telecommunication services.

2.2 Second Stage: Taking Advantage of the
New National Effort in Telecommunications
Through a Strategy of Testing New Services

Two simple priorities apply to the French telecommunications systems: the need to bring backward infrastructure up to date and the need to reduce inequalities between regions, accelerating the investment programme in the west and central regions.[1]

These two considerations were the basis for the implementation of the data network Transpac (operational in 1978). It is a packet-switched network which allows two-way computer communication applications (50 b/s to 48,000 b/s). To further decentralisation, this public network is accessible in any region (15-point switched network in 1978 and 25 in 1980), and its tariff will not be related to distance.

New services geared to regional development would choose to use first the existing networks, bearing in mind the prospect of complete conversion to digital working in due course [2], and to optimise the utilisation of data networks such as Transpac.[3]

To help those involved at regional level in identifying priorities, I envisage that the search for useful services could be carried out in a series of stages which can be conveniently represented in matrix form (Figure 1). One would evaluate the general trends in different areas and make clear the particular situation of each region, each with its own special character.

The next stage of the planning process involves the estimation of the application priorities by service categories. If we categorise communication services, not according to the technical means which is used, but according to the type of communication expected by each user (resident, employee, family member), we can group new telecommunications services in new categories. Different telecommunication systems within the classification scheme may be used either as substitutes for each other or in a complementary fashion. (Figure 2). In order to obtain a meaningful classification from the user's viewpoint, the scheme emphasises the degree of selectivity of information and the degree of interactiveness.

The main problem in defining priorities is to get the people involved at local and regional level to state their most urgent needs in each area of service provision. This requires a great deal of information and examples of practical services (with corresponding systems). We have used the scheme shown in Figure 3 for organising such data.

In order to define overall planning priorities, we need indices for defining service priorities combining the overall regional development aims of the country and the specific needs of each region. For new services the process for defining overall priorities is open to more discussion that it is for each specific region. The contribution of the new services may be more effective for intra-regional or for inter-regional communication improvement.

Besides the existing processes for national planning consultations, a participatory planning process must be set in motion, with an increase in the number of people involved and with the provision of more details concerning the types of service available.

This systematic approach is not yet a reality. The various examples of preliminary experiments and of operational introductions of new communication services reviewed in the following pages are expected to contribute to regional development, but a comprehensive approach to evaluation, which meets the criteria suggested here, has yet to emerge.

Region X / Applications	Urban area			Rural area		
	Capital	Medium town	Small town	Isolated	Urban attraction	Industrialised
Business activity Education Employment Socio-cultural activities Health						

Figure 1: Identification of main needs for each type of application, and most important areas of activity for regional development.

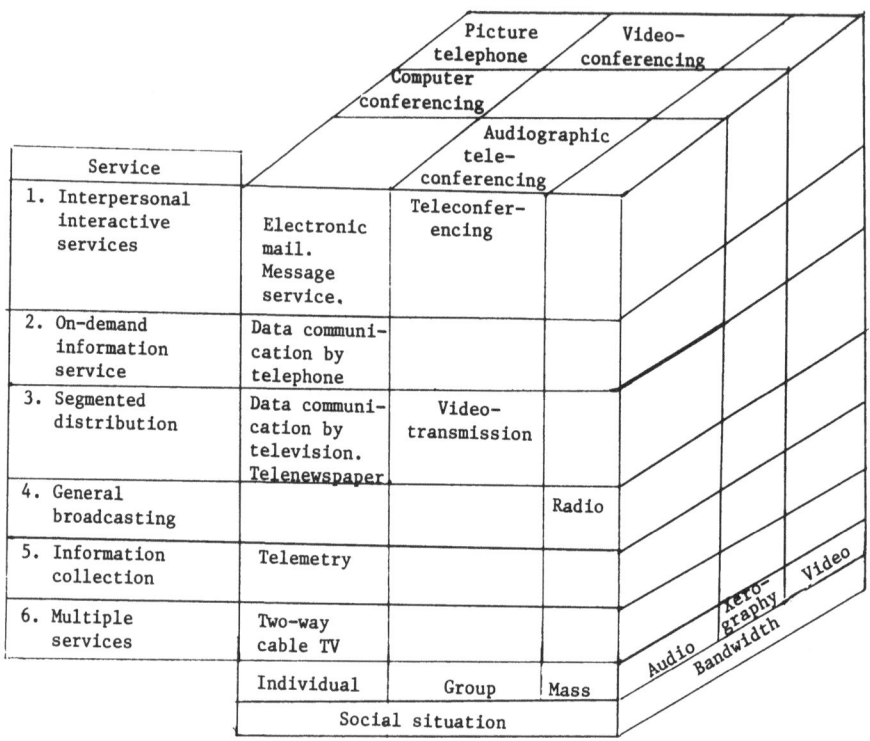

Figure 2:   User-oriented classification of services.

Applications Types of Service	Business activity	Education	Employment	Socio-cultural activities	Information
1. Interpersonal inter-active services					
2. On-demand infor-mation service					
3. Segmented distribution					
4. General broadcasting					

Figure 3:   Scheme for estimation of priorities by type of service.

3. FIELD TESTS: EXISTING EXPERIMENTS WITH NEW
TELECOMMUNICATIONS SERVICES AND THE REGIONAL CRITERIA

Some telecommunication services which seem particularly suitable for effecting decentralisation or creating a more attractive regional environment have received the support of the regional planning agency DATAR (Délégation à l'Aménagement du Territoire et à l'Action Régionale). As these applications are in their initial stages we cannot yet evaluate them; instead, we sketch their development and their possible impact on regional activities.

The three telecommunication systems receiving support from DATAR represent three categories of service:

Type of Service	Social Situation	Type of Information
1. Interpersonal communication service: the audiographic teleconference network-telecentre. (Started in April 1976.)	Group	Audiographic
2. Information services: data communication service by telephone and data communication service by television – teletext.	Individual	Alphanumerics and graphics
3. Segmented programming service: videotransmission system. (Tried in April 1977.)	Group or mass	Color video, moving picture

The few observations we have been able to make about these 'experiments' are based on a survey which aimed to explore the introduction of these new services in different sectors. They have been tested and organised first in regions which have a high priority for development (the West and the Massif Central).

3.1 Teleconferencing Service

A national network including both private conference studios and public telecentres, which contain public studios, has been in existence since 1976.

The components of this system are an audioconference meeting
table for six people in each location, with automatic switching
and speaker identification.   There are additional compatible terminals:

       · for telewriting   (tested)

       · for telefacsimile   (tested)

       · for microfilm teleprojector   (operational).

At present, the intercity links are provided by the circuit-
switched network, CADUCEE.   This offers capability for transmission
between subscribers in both analogue and digital form.   It allows
great flexibility for both slow and high data rates, and is capable
of adapting to a variety of terminals as technology evolves.   Access
is provided through special leased lines and from the public switched
telephone network.

The teleconferencing network allows any location to connect with
any other.   It includes a computer-controlled system for making
reservations between public telecentres.   The system can connect
public access telecentres and private studios, set up 'multiconfer-
ences' with four studios (public telecentres or private studios),
and 'patch in' links with several private telephone sets.

Next year there will be a limited complementary experiment with
a 'visiophone' conference system in three telecentres situated in
the West.

The teleconferencing service is intended to aid inter-regional
relations in the context of energy shortage and rising transportation
costs.   The objective is to have 50 cities linked together and 150
to 200 private studios operational in 1980.

A qualitative survey, to define the specific conditions for
applying the teleconferencing system in large companies, was conducted
in three firms with locations in several regions (an oil company, a
manufacturing corporation and an engineering company).   The analysis
has been mainly oriented towards questions of decentralisation with
an emphasis on relations between the head office (in Paris) and
local branches:

(i)   Identification of current communications and available means
      of telecommunications:

    · networks and terminals

    · communication flows between head office and regional
      locations (quantitative measurement).

(ii)   Analysis of communication flows:

    · Communication patterns

    · Frequency of meetings and use of different means of
      communication (travel substitution) distinguishing between:

      - regular ongoing activities and specific activities
        (e.g., connected with a new project)

      - hierarchical and non-hierarchical interactions

      - programmed communications and non-programmed communications.

(iii) Assessment of meetings where teleconferencing seems worthwhile:

    · replacement of existing practices; cost benefit assessment
      (value added)

    · possibility of generating new practices

    · identification of different forms of teleconferencing

    · cost analysis of particularly promising situations.

The preliminary study suggested that it is realistic to envisage the complementarity between public access telecentres and private studios, but that the setting up of private studios and public telecentres must be co-ordinated. Private conference rooms must be set up in head offices and in main departments, while other offices, in particular regional head offices (frequently eight) and various other offices, would do best to use the local public telecentre. Although the system points to decentralisation, its use in the regions depends on the head office, still usually in Paris, being equipped with private studios.

The most suitable situations for the use of teleconferencing involve technical activities and projects (marketing, new products, research and development, engineering workshops). These applications tend to favour the extension of international links. It seems important to connect to dispersed experts and to foreign operations.

Use of the service is being taken up slowly, because of insufficient user experience, because graphics terminals are becoming available to users only in stages, and because those at executive level (particularly regional heads) have not been sufficiently informed and prepared.

3.2 Data Communication Services
by Telephone and by Television

We regard these two concepts as complementary, allowing at
the same time:

· the passage from general broadcast information to
  personal services

· the passage from videogames (with a push-button pad) to
  computer message systems (with a full alphanumeric keyboard).

A preliminary analysis of how to introduce the services was
conducted in 1976. One part of the study was concerned with listing
national organisations which could participate as suppliers of information through members in different regions (e.g., the stock
exchange, the national office of scientific and technical information,
national employment agency).

Teletext is now being investigated by the stock exchange, and
a pre-trial study is also being carried out by the PTT. We have
conducted another survey which is specifically aimed at identifying
a site for a trial in a development-priority region. The first part
of the study identified the most representative regional organisation
and the best conditions for analysis and a possible trial. The
Chambre de Commerce et d'Industrie, which has 150 offices in France
was chosen because:

· it is in touch with all the sub-professional categories in
  the industrial, commercial and service (tertiary) sectors,

· potential users need information in their professional work,
  as much as they do as private individuals,

· the members of the Chamber of Commerce are natural links to
  an informal information network.

A study of its regional and local institutions was undertaken in
the following stages (Table 1):

· information about organisational features and types of
  information

· selection by reference to relevant criteria.

Table 1: Summary of the Criteria for Selecting Regional
Organisations for the Data
Communication Service Experiment

		Parameters	Indicators
11		Regional activity:	
	111	Overall planning priorities	No. of active population
		Regional priorities	No. in industry, commerce, services.
		Population involved	
	112	Dominant activity	
		˙ Regional growth rate	Regional production.
		˙ Sectors of activities and main sub-sectors	
		˙ Size of firms	
		˙ Change factors (reconversions, extensions, ...)	Importance of projects. Development credits.
	113	Zone for spread of information (local, regional, national, international)	
	114	Geographical locations of urban centres (concentrated or scattered)	No. of activity centres. Distances between activity centres and the regional capital.
12		The organisation of information dissemination:	
	121	Existing means of conveying information	No. of people responsible for information. No. of subscribers and users of information sources. No. of publications and displays.
	122	Adaptation of communication means to the needs of people in the sector	No. of requests from people in the sector. Frequency of use of information services. Observations of information practices of people in the sector. Studies and surveys on information needs.
	12	Capacity for innovation and support provided by the organisation: The management team of the organisation	Examples of projects and achievements with innovative quality. Studies carried out concerning information. Cross-checking with members of other organisations.
		The user's milieu Capacity for accepting innovation	Reactions to CCI initiative. Examples of local initiative.
	14	Ease of experimentation: Technical Economic Institutional.	

Detailed analysis of existing information provision and of the information-communication system within the organisation focused on:

- Information services provided by the organisation

- Information services needed by the users (institutions and individuals).

The second part of the study assessed the information services which regional organisations could supply and the economics of data communication systems. It included:

- Assessment of supply and demand for information:
    - intra-regional information (services and/or users),
    - inter- and extra-regional information (input/output).

- Assessment of useful means of communication and economic analysis of information activities, especially cost/benefit estimates.

- Conditions for trials of data communication services and barriers to introduction of the services in the organisation and in the region.

It went on to consider the implications for regional development. We concluded that data communication services can aid communication access to information as much inside regions as between them. Economic criteria tend to favour uses which cross regional boundaries.

A breakdown of information needs reveals the types and respective size of information services for zones of regional and local common interest, and for groups with specific common interests who are scattered in several regions or over the whole country (transport companies, exporters, teachers, members of the medical professions, and so on).

Attention should be given to the needs of small and medium sized firms, although they are the least aware of the uses for such services.

The main difficulties are connected with the organisation of information services. It seems necessary to bring together a number of organisations and to co-ordinate their contributions. At the regional, as at the national level, moreover, this marketing action challenges the existing habits of the institutions involved. The organisational and management problems of such a service show the difficulties involved in setting it up.

The allocation of an information service between the broadcast medium and the telephone system can be clearly assessed at the regional level, taking into account their relative installation and running costs.

3.3 The Videotransmission System

While the cable television experiments in seven towns authorised by the Government in 1973 have not been carried out very satisfactorily, the 'videotransmission' trials that have been carried out are of some interest. They involved the use of large-screen colour TV projection in public places to display transmissions of interest to special groups.

The experiment was conducted from 12 April to 5 May 1977 in five cities in the Massif Central, the second priority area for regional development strategy. It was undertaken in collaboration between the PTT (Direction Generale des Telecommunications), the distribution carrier for public broadcasting (TDF), and the public programming corporation (SFP), with support from DATAR. Various forms of audience feedback were tried, including audio return channels and, in some cases, video facilities provided from mobile studios (videobuses). Terminals for the system were provided in such places as cinemas, town halls and hospitals.

Of the 46 programmes transmitted, 39 were 'commercial' programmes, a category interpreted widely to include cultural and social programming - such as special programmes for immigrants, the elderly and the handicapped - as well as entertainment. There were also 7 'institutional' programmes for medical/paramedical staff, artisans, farmers and legal professionals.

The organisations running the trial conducted a survey in which questionnaires from 2,000 of the total of 44,000 participants were analysed. 61% of the respondents were under 25 years of age. It was notable that most respondents (57%) saw the medium as mainly a leisure facility; only 27% saw a role for it in professional training.

The potential contribution of such a 'videotransmission' scheme to regional development remains uncertain.

The costs of the system are such that the national distribution of information is perhaps a more likely early role for it than are intra-regional and inter-regional dialogue and exchange of information. The need to make software profitable will cause a trend towards instantaneous and simultaneous broadcasting in as many places as possible, thus reducing the possibility of two-way communication. The possible impact on regional development could be limited to that of a simple broadcasting infrastructure if steps are not taken to encourage the use of cabled premises for two-way communication.

3.4 Overview

A brief analysis of these three new services, at their different
stages of experimentation, shows several key points:

· The coming together of the different public common carriers
  to optimise the use of the 'technological infrastructure'
  is promising, but it will clearly not be easy to create
  the 'institutional infrastructure' needed for the new
  services.

· The principle of public monopoly of transmission is being
  maintained, with the private sector providing content.

· The new services have not been set up as a result of a
  regional initiative, but have been instigated by national
  public bodies: common carriers with the support of the
  regional planning agency. Good means of assessing the
  regional priorities for new telecommunication services
  are still lacking.

· The development of the new services are still severely
  constrained by lack of capital.

· The teleconferencing service employs a national switched
  network which will permit public and private access and
  compatibility with foreign systems if the service is
  extended abroad.

· A general users' data communication service of the
  'Teletext/Viewdata' type could encourage services and
  enterprises of regional interest.

· The videotransmission service could develop gradually
  into a two-way means of communication for intra- and
  inter-regional use, but it may have to compete with
  Pay-TV.

Further work is needed to assess the difficulties retarding
the adoption of these new services by potential users. This involves the systems themselves, the characteristics of the service,
the possible areas of application and experimental conditions.[8]
A very rough comparison shows that data communication services tend
to have more difficulties to resolve than the teleconferencing system, while videotransmission runs up against fewer problems as a
'pay-TV' service than as a more ambitious community communication
service aimed at special applications, often involving two-way
operation.

From the viewpoint of regional development, the analysis of each service and comparisons between them (see Table 2) allow consideration of the effects to be expected from their introduction and of the respective amount of funds to be allocated to them. These criteria should obviously be expressed as quantified indices. They would permit a preliminary assessment and a post-implementation evaluation.

4. TOWARDS A CO-ORDINATED STRATEGY FOR INTRODUCING NEW COMMUNICATION SERVICES INTO THE REGIONS

In this section, we consider how to reduce the gap between the conceptual framework for regional planning, and the field test which evaluates the conditions for introduction of a specific service. A variety of approaches have been used in countries with an advanced telecommunications infrastructure.

4.1 Telecommunications for Regional Development

The introduction of telecommunications systems as a means to encourage regional development has started from several different approaches.

In Canada the need for planning local and regional development was expressed early, with an emphasis on appraisal of capital requirements paying due attention to financial risks. Long term planning studies have distinguished different strategies for urban communication systems and rural communication systems.[5]

In Sweden an important analysis of telecommunications and regional development has led to proposals for "the two-way communication systems on narrowband network where computer aids are of special interest".[6]

Local authorities were invited to submit their views on regional priorities. Then a northern region of primary importance in spatial development strategy was selected and a local seminar for priority discussions was organised. The latter made proposals for projects to demonstrate the application of new telecommunication services. Pilot projects in two areas were to tackle key problems – notably health and education – associated with remoteness and severe weather conditions.

In the U.S.A. a very large amount of work has been carried out in a socially-oriented application of two-way cable television. Development prospects for two-way broadband telecommunications networks in urban areas appear promising, and advanced rural telecommunications projects are also being pursued.

Table 2: The Contribution of New Services to Some
Criteria for Regional Development

Criteria	Teleconfer-encing by audiographic service	Data communication service		Video transmission service
		By telephone	By television	
1) Contribution to over-all regional planning:				
- Experimentation in region with develop-ment priority	+	+	+	+
- Extensions:				
international	+			
national	+	(+)	+	+
inter-regional		(+)		+
intra-regional			+	
- Improvement of inter-regional relations	++	+	+	+
- Decentralisation	++	o	o	o
- Location of offices	++			
- Make small and medium-sized towns more attractive	o	o	o	
- Improve access to serv-ices in rural areas	(+)	+	+	+
- Urban-rural linkage	(+)	o	o	+
Intra-regional:				
- Improve quality of life	o			
- Regionalised informa-tion on collective services and employment	o	++	+	o
- Enhance community serv-ices and neighborhoods				+
2) Contribution to spe-cific regional needs:				
- Priority applications in the region	o	o	o	o
Commercial services:				
Business	++	o	o	o
Public and social services:				
- Inter-city informa-tion				
Health	+	o	o	+
Education	+	o	o	o
Employment	+	o	o	o

+ positive effect
o non-determined

As the past shortage of the means for telecommunication in
France is reduced, the modernisation of the networks with electronic
switching and a digital network will give priority to the less developed regions.

The geographical, sociological and institutional features particular to the country mean that the priorities for application of
new services in France will not be the same as those in other countries. Socio-psychological barriers to the adoption of new services
will be noticeable in certain activities. Group working practices,
management relations, and unfamiliarity with procedures may all
involve problems. Potential users' and decision makers' lack of
information and awareness of telecommunication services and systems
will cause problems in the introduction of these services.

An information and demonstration programme concerning telecommunication services would be helpful in these respects, as the
Swedish initiative indicates.

4.2 Needs for Research in Regional Planning

The development of new services raises several basic questions:

• Which will be considered the basic services, available in
  every region? Which will be considered on-demand services
  (optional) with a variety of installation possibilities
  varying from one area to another?

• How will these systems be extended, and what will be the
  level of organisation - local, regional, or national -
  appropriate to each service?

• What will be the optimal use of existing and projected
  networks bearing in mind the institutional problems involved?

Different alternatives exist for making these services available:

• Offer the same development conditions to all services; or
  give preference to those which tie in with regional priorities. Certain services are supplied automatically when a
  consumer good is bought (a CB Radio for instance) and tend
  to spread quickly. Others imply a more complex organisation
  of operations, as in the case of access to data communication
  by telephone or by television.

• Encourage domestic access through individuals' terminals (cable
  TV); or give preference to the provision of communal access
  points (videotransmission, telecentres, ...), which would
  have a different type of social and economic effect.

· Base the tariff policy on equal treatment, regardless of
  distance, for certain types of service; or have a more closely
  cost-related tariff.

It is important to distinguish between the case where a new
service serves a few local areas intensively, and the case where
potential users are scattered over the whole country in groups with
a common interest, professional or private - as in the case of computer communication for banking. The evolution of public administration, and urban and rural living conditions have changed traditional
living spaces, neighborhood feelings and the sense of belonging, as
well as economic structures, so that 'communities of interest' for
communication have changed. A study has been commissioned in France
to analyse 'communities of interest' for telecommunications in four
typical urban and rural areas chosen from different regions, and to
work with the inhabitants to define the service priorities.

Key areas for future research are:

· Analysis of the obstacles that institutional inertia poses
  for the development of new services.

· Analysis of alternative organisational structures for the
  new services at local, regional, national and international
  levels suited to emerging technologies such as satellites
  and digital networks.

The experience reviewed in this paper suggests that the further
improvement of planning for telecommunications in a regional context
requires an in-depth study of one priority development region, focusing on:

· organisational alternatives
· optimum design of the introduction of new services from an
  experimental and promotional point of view.

The work would aim to provide scenarios for regional development; to promote co-operation and exchange of information between
those who direct regional planning, and involved organisations and
citizen groups; and to guide the selection and implementation of
operational pilot projects.

Three parts of the necessary future work programme are especially
important:

(1) Analysis of the development characteristics of the region,
and of the particular zone being studied in depth. Research
to identify the most important activity sectors.

(2) In collaboration with regional authorities and organisations:
* identification of the main types of services according to an initial 'hypothesis of priorities'
* evaluation of available resources
* assessment of information needs
* choice of services to be implemented
* assessment of likely impacts of the services to be implemented
* assessment of obstacles to implementation,

(3) Identification of priority actions:
* key equipment and infrastructure requirements – networks, terminals, data-bases, etc.
* implications for the planning of the main public network
* actions needed to overcome obstacles (technical, economic, institutional)
* organisational measures required (e.g., by regional economic agency).

CONCLUSION

The development of new services depends on the one hand on the results of the confrontation between multinational corporations and the administrations of the national government, and on the other on the behaviour of the potential user.

It is important to analyse this interaction at the regional level. This allows applications to be defined which will be wide-ranging enough to be economically viable services, and which will be sufficiently small-scale to respond to the needs of local users.

We have assessed the difficulties involved in doing systematic planning of this kind. We have observed the limitations of field tests being carried out at the moment which, though promising, have been too hastily organised to provide much useful information as to their regional application.

This paper is intended to provide some basic information drawn from recent experiments in the field of new services, and to ask several essential questions.

We must now find practical answers and methodologies for the establishment of regional schemes for telecommunication services.

A thorough study of a region designated as a priority area under the 7th Plan, possibly linked to international comparisons, would yield much valuable information for these purposes.

REFERENCES

1) Preparation du VIIème Plan, Commissariat Général du Plan –
   Rapport des Commissions: Amenagement du Territoire et du
   Cadre de Vie; Transports et Télécommunication.

   Rapport d'activité "Amenagement du Territoire" (DATAR, 1976).

2) Services nouveaux de télécommunication. H. Nora et V. Chaumont
   (DATAR November 1973).

3) Décentralization des centres informatiques (DATAR SEGIC, 1973).

   Telecommunications for a decentralized France in the Year 2000
   (MITRE, March 1974).

4) Approche prospective des systèmes de communication (CAF-DATAR,
   1975).

5) Communications et developpement régional. Etude 2d Télécommission
   Ministère des Communications CANADA (1970).

   Telecommunications Experiments in Urban and Regional Planning.
   John DAKIN, May 1974.

6) Telecommunications and Regional Development in Sweden: ERU –
   export board for regional development; STU – national Swedish
   board for technical development. Thomas Ohlin and Bertil
   Thorngren (Stockholm, 1976 and 1977).

PLANNING EXPLORATORY TRIALS OF NEW

INTERPERSONAL TELECOMMUNICATIONS

C. D. Stockbridge

Bell Telephone Laboratories

Holmdel, New Jersey 07733

This paper describes a practical approach to the problems of deciding what locations to include in a field trial of a new interpersonal telecommunications service and of selecting the appropriate sequence in which to install them. This involves balancing the costs of hypothetical networks against their utility for the purposes of the trial.

The formulation requires the different types of application which may occur in the trial to be scored for their relative importance to the user organization. It also requires corresponding estimates of communications traffic levels currently applying for conventional means, and estimates of the probabilities that the new service will in fact be used for each application.

Solutions are obtained by means of a simple (heuristic) procedure.

The paper also shows the curve relating utility to cost which was obtained when the method was applied to the trial for which it had been developed. This contained two clear breakpoints, which were used to divide the trial into three phases.

Some statistics on the usage levels experienced during the trial are provided. These show use rising to a peak, then falling to a plateau.

1. INTRODUCTION

In exploring new applications of interpersonal telecommunications systems there are important places for laboratory studies and a variety of techniques for making market projections. The exploratory field trial is, however, essential to provide convincing answers to the question, "Will the service be used enough in nontrivial ways to justify its cost?"

More generally exploratory field trials are experiments designed to answer questions like:

1) Will the concept work as conceived? Are worthwhile improvements to the concept possible?

2) Can necessary equipment be constructed, installed, and maintained as planned?

3) Are the design features appropriate?

4) Can people in relevant organizations make effective use of the service?

5) Will they actually use it?

In the interpersonal telecommunications field trial the service in question must often be introduced at more than two points, so that a transmission network is required. Two questions then arise: "Which of a set of possible locations should be included within the network and which excluded?" "In what sequence should these points be added to the network?" Answering these questions involves striking a balance between costs and the utility of the network for the purposes of the trial.

The utility of the network is an elusive concept. The assumption made here is that, when the network will be used for a number of purposes (applications), utility depends in a multiplicative way on the relative importance of each application times the frequency with which it is used for that purpose. It is further assumed that each individual application utility can be summed over all applications and locations to give the utility of the network.

One becomes involved in trials in situations where the levels of uncertainty about use are high; nevertheless decisions must be made about the trial network. There is a need for systematic procedures which allow one to make good use of the inevitably subjective information one does have.

This paper describes such a procedure. It also provides some empirical results from its application in the design of a recently terminated, and successful, field trial. (One of these results, the rate at which marginal utility falls off as the network develops towards completion, would be an interesting subject for investigation across a range of field trials.)

It should be stated at the outset that it would be possible to develop mathematically more rigorous formulations of this problem and to use solution techniques which were better in the sense of optimization theory. It is felt however, that it is preferable to describe an approach which has actually been used. It was straightforward to apply and did not require the use of outside experts; hence it had the added advantage of encouraging recalculation as the trial developed and subjective probabilities changed. As all good models do, it provided those implementing the trial with insight into their problem and confidence in the methodology used. Such advantages can be important when weighed against the value of increased sophistication of technique.

A rigorous formulation of the problem has since been developed by others[1], and solutions were obtained with the same data as reported below, by means of procedures guaranteed to be optimal. Comparison showed that the differences were minimal. In a model which must inevitably use imprecise data the dangers involved in the more approximate approach appear to be insignificant.

2. EXPLORATORY TRIAL PLANNING

2.1 Background

For the past several years the Bell System has been exploring the utility and demand for switched visual communications in the United States. The exploration consists of laboratory studies, market studies, and field trials of prototype equipment. This paper is based on the thinking which went into planning one of these field trials prior to rather than after installation of equipment. In large measure actual implementation of the trial followed the plan as modified by ongoing experience.

2.2 The Trial Plan

In 1974 the American Telephone & Telegraph Co (AT&T), a large communications supplier in USA, teamed[2] up with the Law Enforcement Assistance Administration (LEAA), a federal agency, to study the need and use of visual communications in the

Criminal Justice System (CJS). A thorough examination of
potential trial sites started early in 1974, ending in selection
of Maricopa County, Phoenix, Arizona. This selection was based
on the size of the area (about 1 million people), an incidence of
major crimes in excess of 80 per day, a relevant and accessible
local criminal justice database, a high level of cooperation
among the area's criminal justice agencies, a legal environment
receptive to the innovative program, reasonable geographic
dispersal of criminal justice activities and reasonable
implementation cost.

Preliminary work by LEAA and AT&T in 20 candidate cities
had shown that there are 30 frequently occuring tasks in CJS
which are potential applications of visual communications
service (VCS). It was decided to use switched VCS transmission,
as distinct from preplanned dedicated point-to-point transmission,
to interconnect the trial service locations in Phoenix; this
allows unplanned usage to emerge spontaneously during the trial
and thereby gives users an added sense of participation. It was
also decided to narrow the candidate task (befefit) applications
down to the seven "core applications" shown in column one,
Table 1. These and other applications are discussed in
relation to VCS in References 3 and 4.

TABLE 1

CJS CORE APPLICATIONS AND THEIR WEIGHTS

TASK OR BENEFIT APPLICATION (a)	ASSIGNED WEIGHT (w)
1. Remote Testimony	10
2. Bail Bond Interview	8
3. Sentencing Conference	4
4. Booking Review	7
5. Pretrial Conference	8
6. Inmate/Attorney Conference	8
7. Family/Inmate Visit	7

2.3 Location Scoring

The core applications were labeled a_j (j=1 to 7) and each
core application was assigned a weighting score w_j ($1 \leq w_j \leq 10$) of its
perceived relative value to the user; these scores were agreed to
by LEAA and AT&T as representing the best available estimate

of the relative importance of each application to CJS.* They were, of course, to be reassigned any time that substantially better confidence in new weighting factors became available.

Each link between locations in a network over which an application can take place, involves at least two VCS terminals with two sets of participants. Thus, Inmate/Attorney conferences involve inmates and attorneys usually located in a Jail and Court House, respectively. Given a link between the locations and an organization which assures access to the terminals, a four level estimate was made of the relative volumes of traffic (v) for each application which was taking place between each pair of locations by conventional means, i.e., before the VCS trial. The levels chosen were: zero, low, medium, and high which were coded 0, 1, 2, and 3. The values 0, 1, 2, 3 are of course arbitrary. In other applications or when more precise information is available, they could be chosen differently. What is important is that they be consistent with the description "low," "medium," and "high."

To take account of the reality that applications which can take place will not necessarily occur or even be attempted over VCS, an additional factor (e), the subjective probability of use, was introduced. This factor ($0 \leq e \leq 1$) takes into account all the reasons (political, behavioural, perceptual) which may cause a user not to use the system for a particular application. A multiplicative scoring function was then used to assign a value of the application to the trial. For example, when application (j) with weight w_j = 8 took place over link (k) originating at location (ℓ) with relatively high volume $v_{jk,\ell}$ = 3, and it was expected that it would be used 1/2 the number of times it could be used, then $e_{jk,\ell}$ = 0.5, and the score ($s_{jk,\ell}$) which was assigned to the orginating location involved, was $s_{jk,\ell} = w_j v_{jk,\ell} c_{jk,\ell}$ or $s_{jk,\ell}$ = 8×3×0.5 = 12 scoring units.

---

* The combined decision maker would be indifferent between aiming for application i and j, if $w_i p_i = w_j p_j$, where p_i p_j were their estimates of the probabilities of achieving success with the two applications.

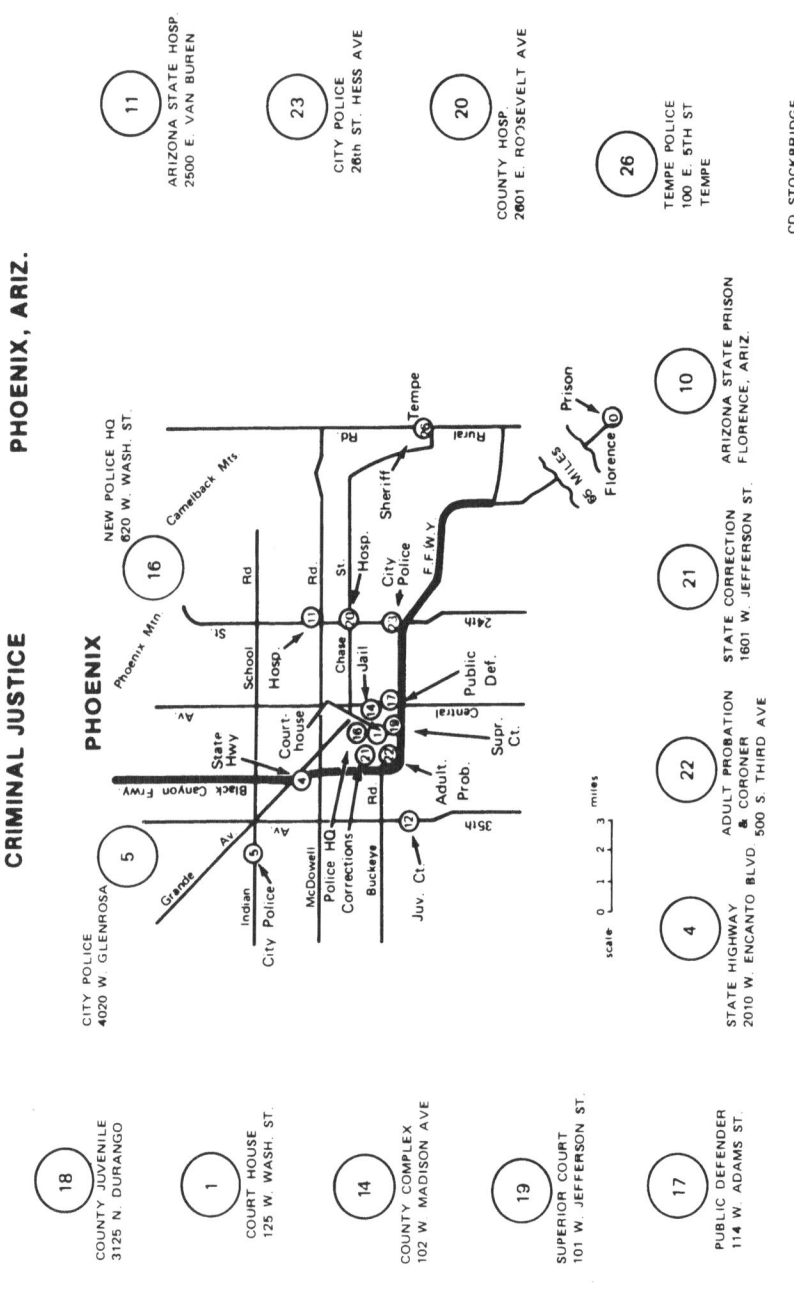

Figure 1. Geographical Locations of 2-Way Visual Communications Terminals

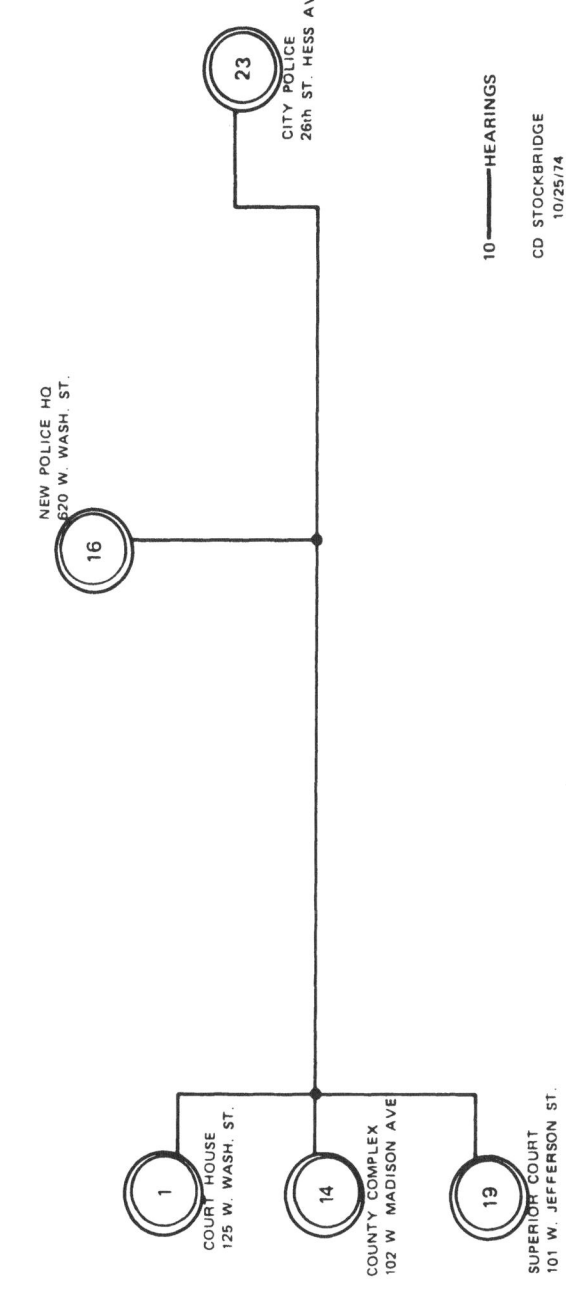

Figure 2:   Applications Network Showing Task Links Between Locations Involved in 'Hearing' of Remote Testimony

TABLE 2

LOCATIONS, CODE NUMBERS OF LOCATIONS
IN FIGURES 3A TO 6B, LOCATION SCORES,
COSTS IN RELATIVE DOLLARS, AND EFFICIENCY MEASURES

LOCATION	CODE # IN FIGURES	LOCATION SCORE	LOCATION COST[†]	EFFICIENCY MEASURE
Superior Court	19	219	3.8	58
Court House	1	194	1.1	176
County Complex	14	164	1.9	86
Public Defender	17	136	1.1	124
Police Station	23	126	2.1	60
Tempe Police	26	92	4.1	22
State Prison*	10	71	10.5	7
City Service Center	5	66	2.5	26
Police HQ	16	56	3.2	17
State H'way Office	4	44	1.1	40
State Hospital	11	16	1.0	16
Juvenile Center	18	16	3.7	4
Coroner	22	12	1.7	7
County Hospital	20	3	0.9	3
Correction Office	21	3	0.8	4
TOTALS		609	39.5	

Notes:

1) [#]These code numbers were assigned sequentially as the locations were considered; the numbers missing from the sequence shown are those of locations which were dropped from the trial plan.

2) [†]An unspecified scaling factor is used in these location costs.

3) *Located in Florence, Arizona, 79 miles by radio link from Phoenix.

The cumulative use score (s) for all the application (a) associated with each location (ℓ) can then be calculated using their weights (w), relative volumes (v) and subjective probabilities (e) by means of the formula:

$$s_\ell = \sum_j \sum_k w_j v_{jk,\ell} e_{jk,\ell} \cdots \begin{array}{l} 1 \leq \ell \leq \text{number of locations} \\ 1 \leq j \leq 7 \\ 1 \leq k \leq \text{number of links} \end{array} \qquad (1)$$

where the summation is over all applications and all links in the network (note that v_j and e_j vary from link to link). Typical cumulative location scores are listed in Table 2, column 3, for the 15 location CJS trial in Phoenix, Arizona.

Each location score approximates the value of the location to the trial as a vehicle for testing the seven tasks (benefit applications). The geographical locations are shown on a map of Phoenix in Figure 1. For convenience we will use the location code numbers shown in the circles and on the street plan, in the following discussion.

The applications which can take place between two or more locations can be visualized graphically as a task link or applications network. Figure 2 shows a simple set of task links for the Hearing or Remote Testimony application listed in Table 1.

When all task links are drawn an applications network can be formed. An example of such a network labeled Phase 1, is shown in Figure 3A; we shall explain the term Phase 1 later, in Section 3.

Figure 3A was first drawn for the CJS trial before any of the corresponding wired network, see Figure 3B, was installed.

Figure 3A is simpler in concept than it at first appears. Each application is indicated as a circle around the location where it occurs, the smaller circles indicate the more valuable applications, while lines between the circles indicate the wired links which will be needed to perform that task by VCS during the trial. Note that the more circles there are around a location the busier and more valuable it is expected to be in the trial.

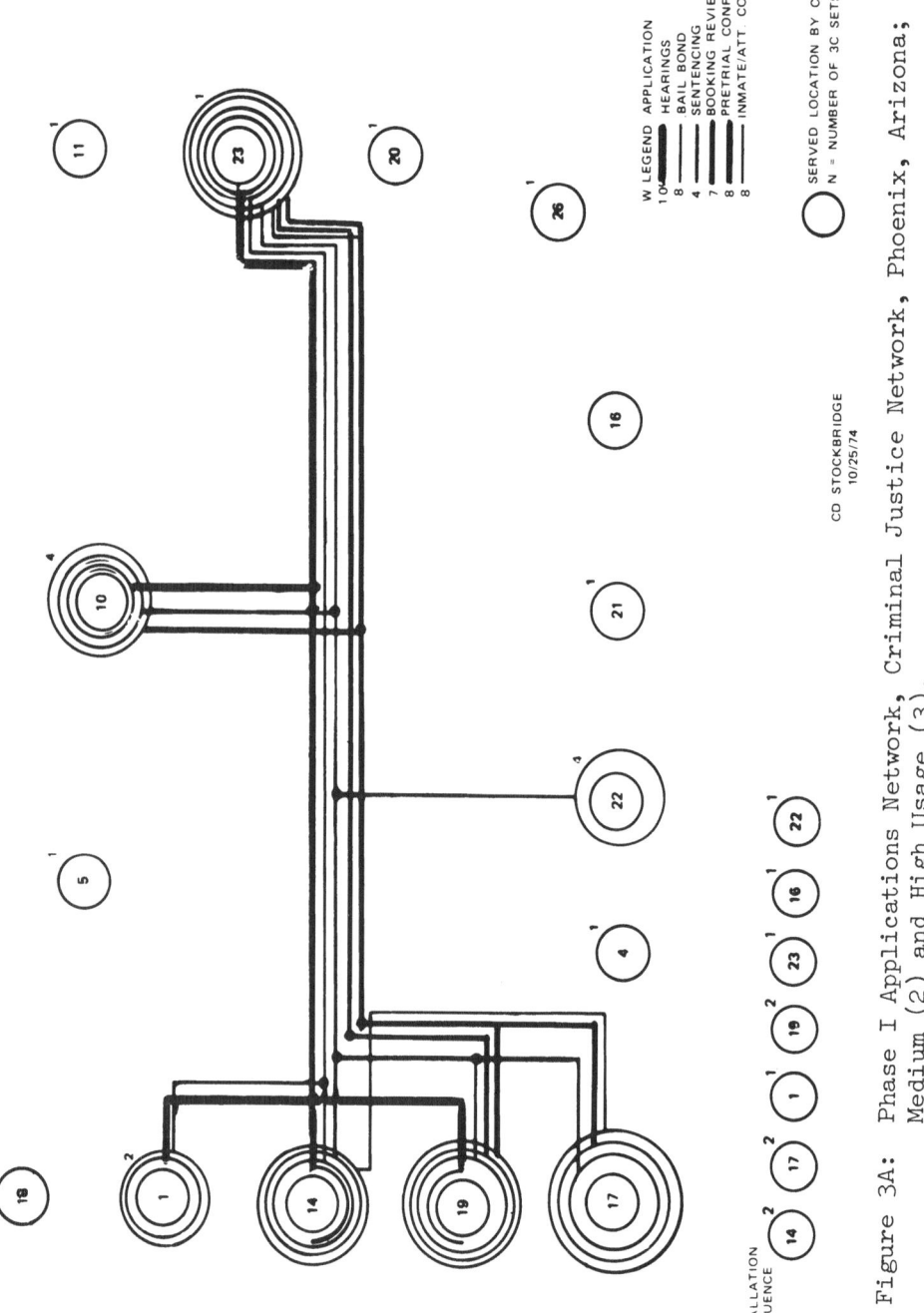

Figure 3A: Phase I Applications Network, Criminal Justice Network, Phoenix, Arizona; Medium (2) and High Usage (3).

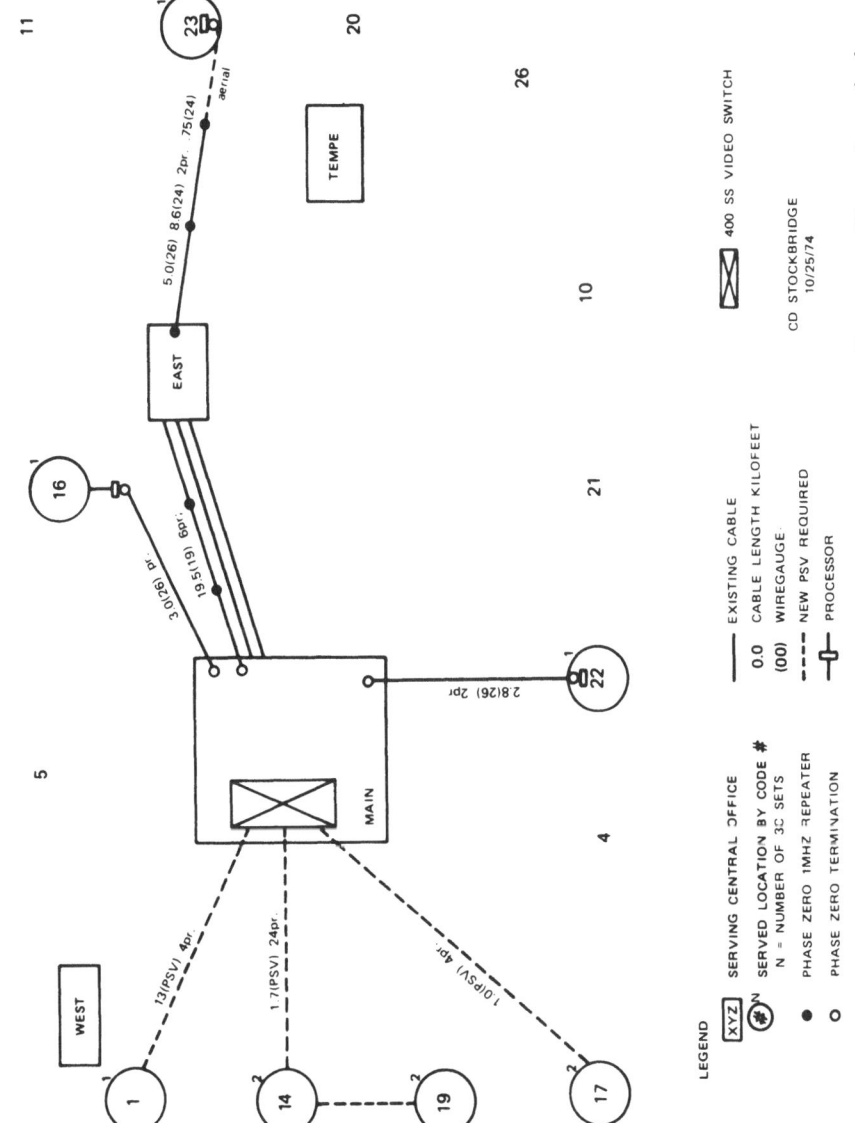

Figure 3B: Phase I VCS Network, Criminal Justice System, Phoenix, Arizona.

2.4 Location Costs

For each candidate location in the proposed trial network there will be a cost for installing the visual communications service. Installations near the center of the city will be relatively inexpensive while remote installations will cost a great deal more. The minimum 'network,' which consists of two basic terminals connected back-to-back, costs approximately $2N$; we note in passing that N for an audio/video terminal is about ten times that for an audio only facility. Some tasks will require special equipment such as graphics transmitters and/or hard copy output devices; these increase the particular location cost substantially.

As soon as a third location is added to the elemental two-location 'network' some form of addressing switch is required so that each location may call either of the others. The switches that were available to the CJS trial could handle 35 locations and were costly.* Since it would be unreasonable to assign the cost of the switch to the third location alone, the following procedure for calculating location costs was adopted:-

1. With knowledge of existing facilities, lay out a complete hypothetical network consisting of terminals, special devices, transmission and switching facilities and estimate the total network cost.

2. When a location is not installed, money will not be spent; equate that number of dollars not spent to the cost of installation at the location.

3. Share the cost of switches and trunks, between those locations directly connected to the switches.

A partial transmission network for the CJS trial in Phoenix is shown in Figure 3B. Terminals are indicated by circles around location code numbers, transmission facilities are shown as lines, and switches are shown as rectangular boxes.

Each location cost (c_ℓ) is composed of a specific location cost (terminal(s), transmission facilities, and hard copy devices, etc.) and shared costs (engineering effort, switches, trunks, traffic measuring equipment, etc.). Thus the location cost (c_ℓ) is given by

---

* Type 400 SS dial addressable telephone switch with video slave adjunct; line selecting key switches could be used in a small trial.

$$c_\ell = \text{specific cost at location}(\ell) + \text{shared costs} \qquad (2)$$

The location costs for the Phoenix CJS network are shown in relative numbers in Table 2, column 4.

When the score contributed by each location is divided by the cost of providing service, an efficiency measure of that location in testing applications, in score units per dollar, can be derived and used to compare the relative value of the locations to the trial. Thus we have

$$m_\ell = s_\ell/c_\ell. \qquad (3)$$

This efficiency measure which is largest for the most worthwhile location and smallest for the least worthwhile is shown in Table 2, column 5.

2.5 Network Scores and Costs

The total network applications score (S_n) and the total network cost (C_n) for the complete network can then be calculated:

$$S_n = \sum_{\ell=1}^{15} s_\ell = 609; \qquad C_n = \sum_{\ell=1}^{15} c_\ell = 39.5 . \qquad (4)$$

Note that the measure $M_n = S_n/C_n$ might sensibly be used for comparison between trials if similar weighting factors, volumes, and subjective probabilities were employed.

3. IMPLEMENTATION

3.1 Implementation in Several Phases

Implementing the above complete network at one time is both a formidable job and a dubious strategy with untested equipment in a new application environment. Personnel may be hostile to the equipment, the equipment may fail and/or require modification; the subjective probability of use may have been grossly misjudged.

A preferred strategy is to implement the network in phases, such as an initial exploratory phase, a utility augmenting phase, and if warranted, an expansive phase. The purpose of the initial exploratory phase is to implement just enough of the network to determine whether an augmented network is likely to be of increased utility.

A heuristic algorithm was used to select a sequence for installation at successive locations. It is believed that the algorithm provides solutions which are either optimal or very close to being so.*

The heuristic algorithm was applied as follows:

1) Given the complete 15 location network,

2) Consider each location in turn as a candidate to be dropped from the network on the basis of its application score and cost,

3) Calculate the network applications score reduction and the dollars saved by removing each location in turn (not succession) from the complete network. Then,

4) Treat the applications score reduction as the value of the location to the network for testing CJS applications.

5) Treat the money saved as the cost of serving that location.

6) Discard the location whose ratio of incremental score added to incremental cost removed, is lowest. This is the location which will be implemented last.

7) Given the resulting 14 location network, repeat steps 2) to 6) above, and discard the location which will be implemented next to last.

8) Repeat the above discarding process iteratively until only two locations remain. These two locations form the bare minimum 'network' which will be implemented first.

The above process yields reduced network scores and reduced network costs. The process is nonlinear since some applications will not be possible when a critical location is discarded. Figure 4 shows scores as a function of costs for the Phoenix CJS network as it was reduced. Locations which lower the network application score relatively little are in general dropped first (i.e., installed last). All factors considered, this heuristic algorithm yields an installation sequence which is near optimal; given the information available, it indicates how to proceed so as to obtain as many score units as possible per dollar spent.

---

* The efficiency of this algorithm has been tested by a computer simulation against a more complex algorithm which guaranteed optimality, see Reference 1.

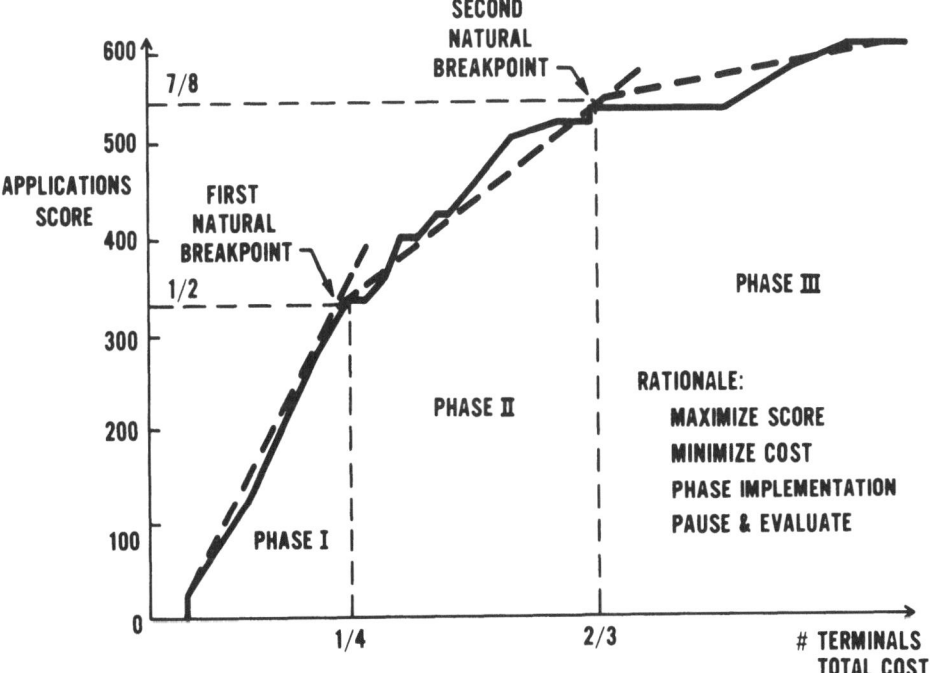

Figure 4:   Plot of Applications Score Versus the Number of
            Terminals or Network Cost Showing Phase Breakpoints.

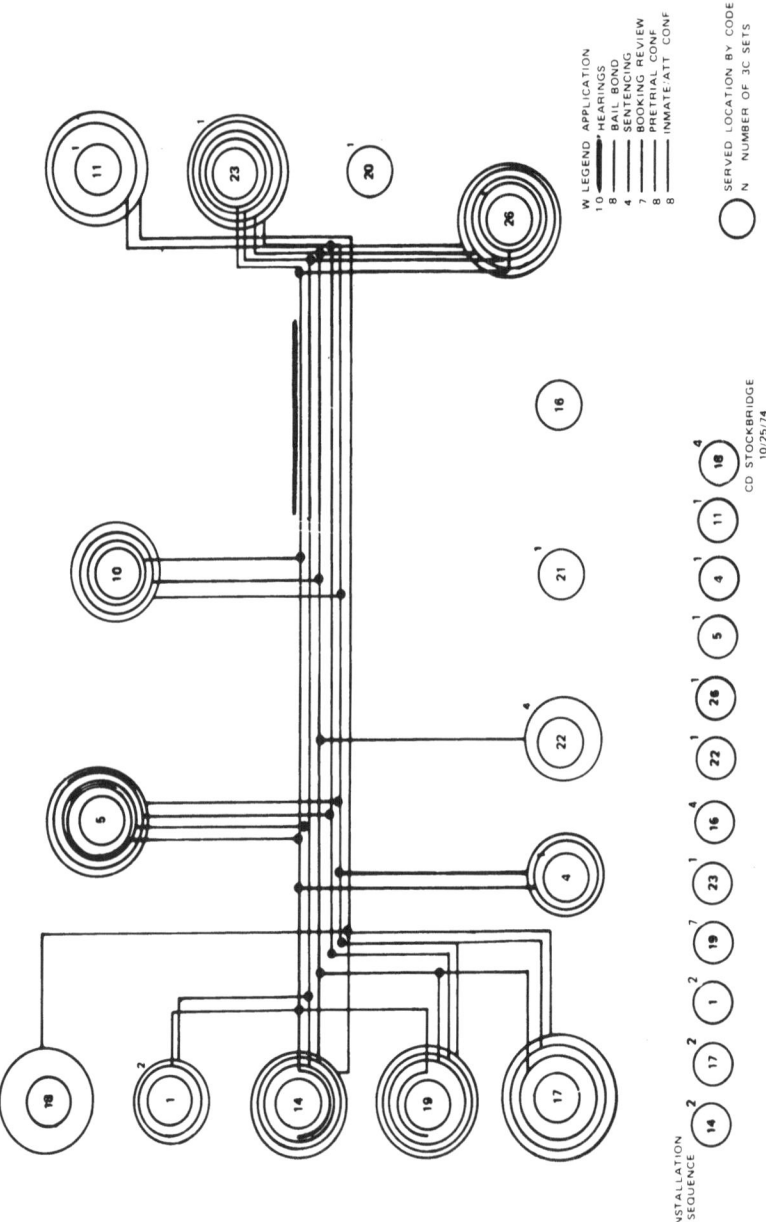

Figure 5A: Phase II Applications Network, CJS Network, Phoenix, Arizona.

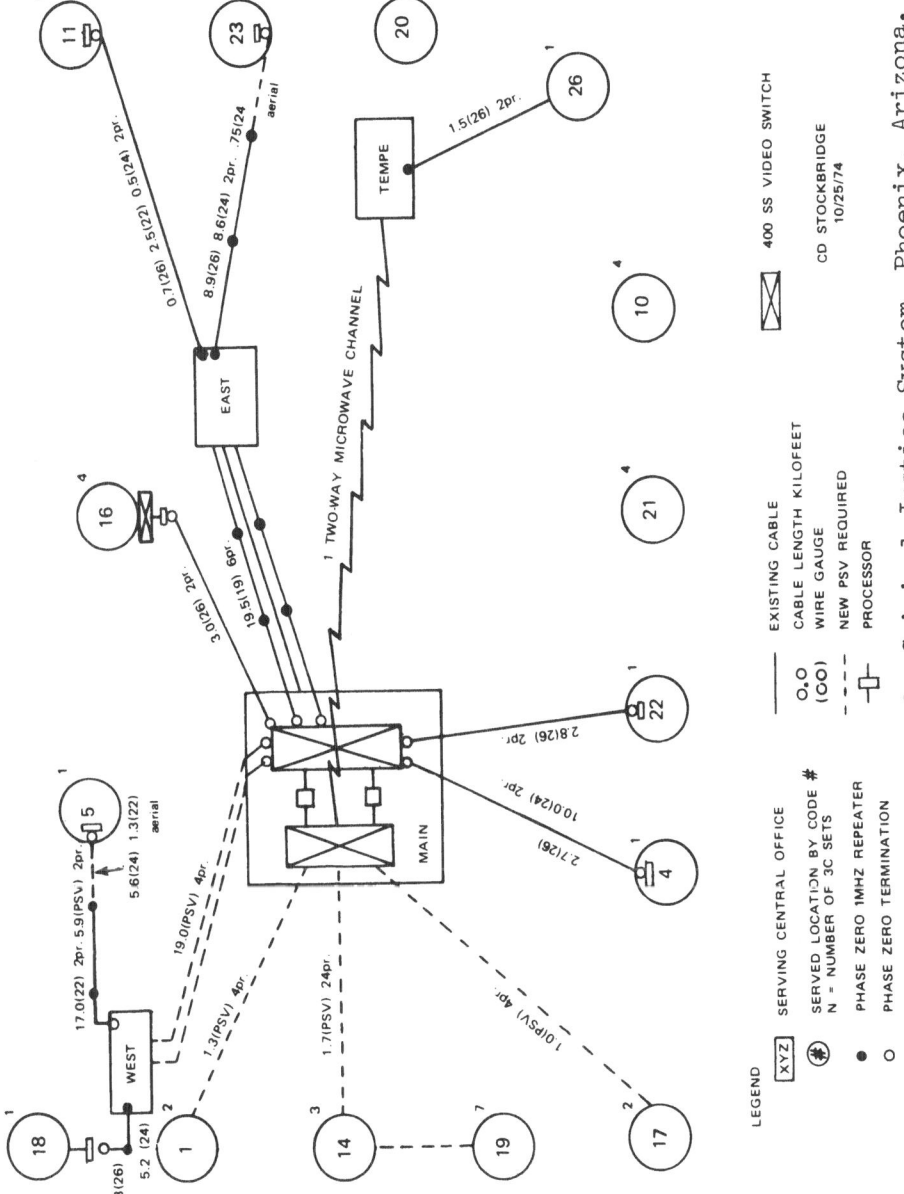

Figure 5B: Phase II VCS Network, Criminal Justice System, Phoenix, Arizona.

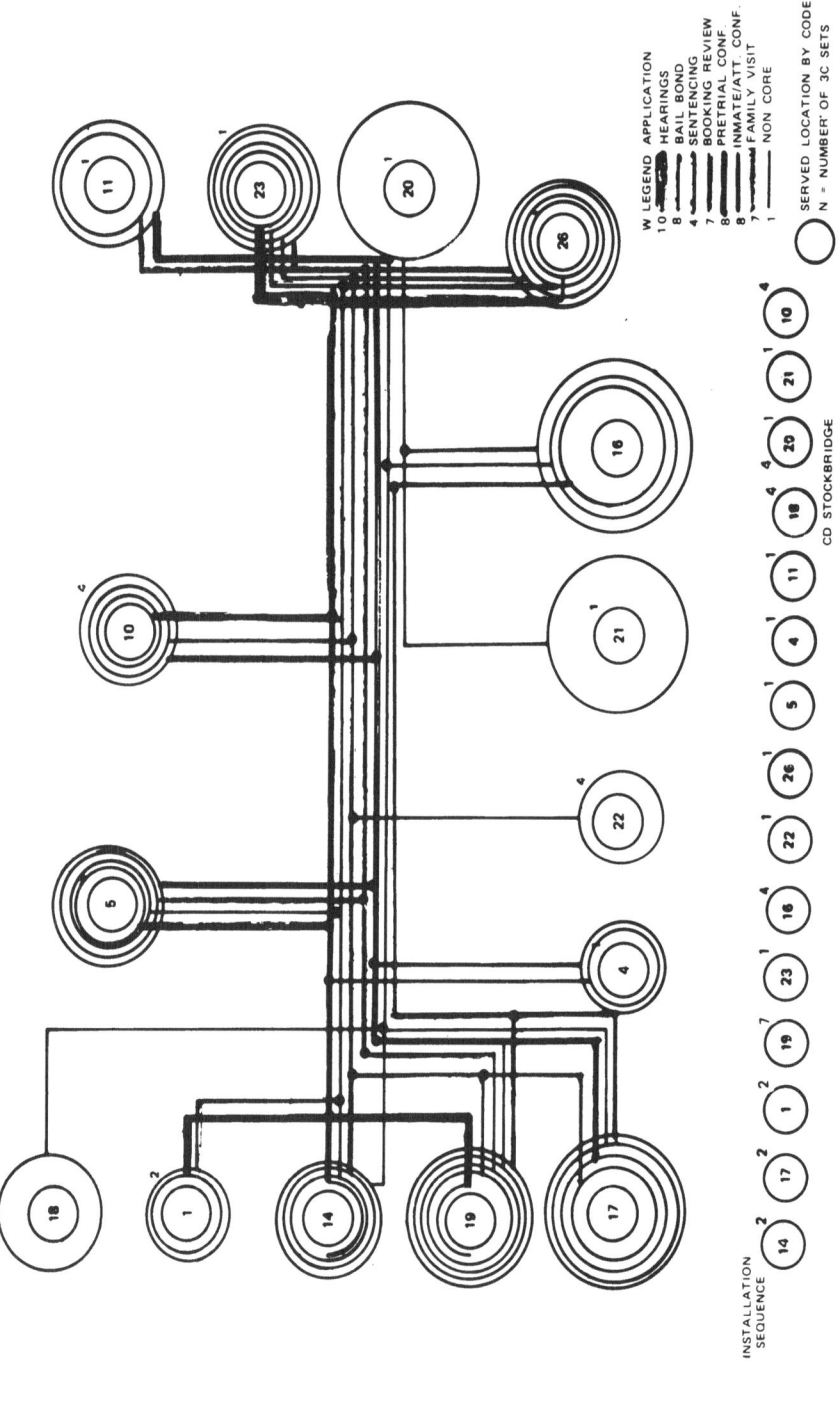

Figure 6A: Phase III Applications Network, CJS Network, Phoenix, Arizona.

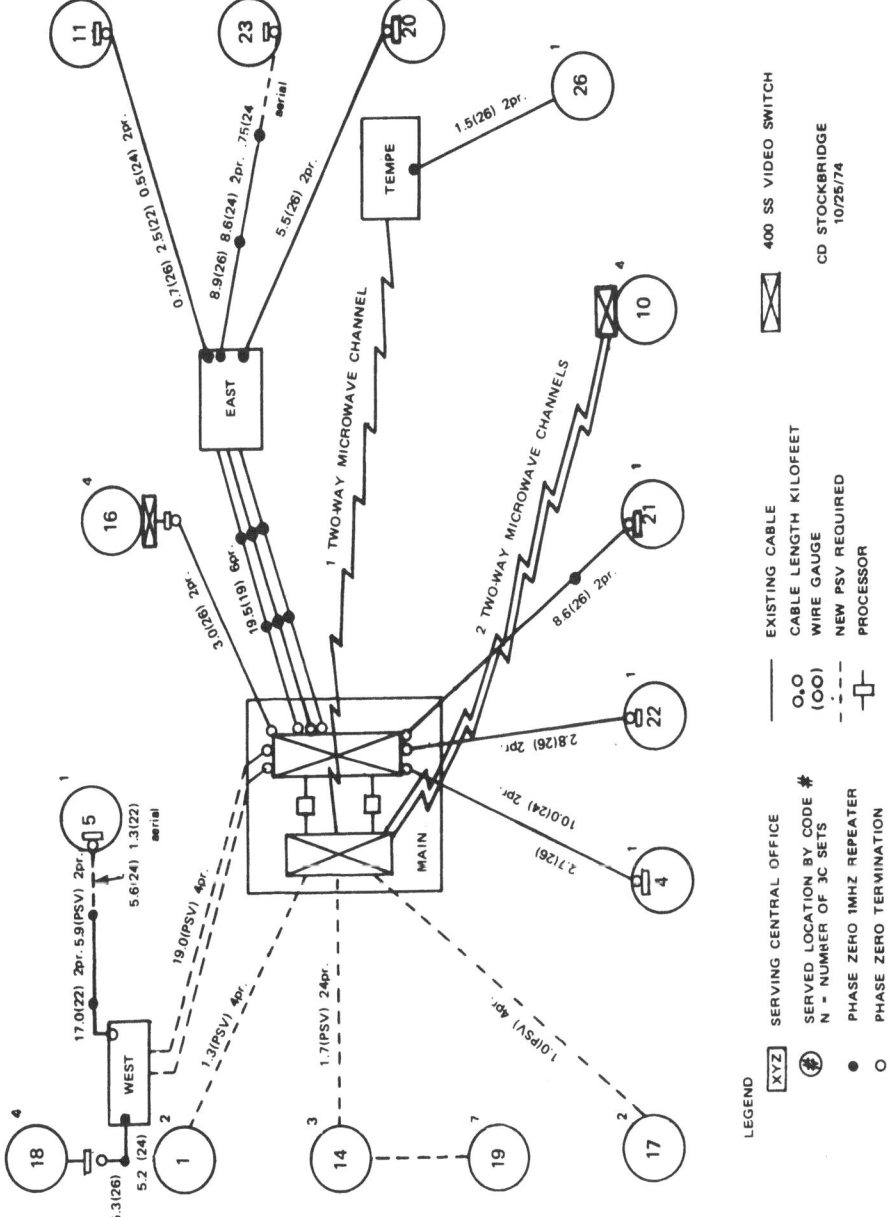

Figure 6B: Phase III VCS Network, Criminal Justice System, Phoenix, Arizona.

Figure 4 shows two clear breakpoints which may be used to define three implementation phases. Phase 1 achieves one-half the total score at one-fourth the total cost; adding locations in Phase 2 increases the score to seven-eighths for two-thirds of the total cost, while adding the last locations in Phase 3 increases the cost greatly while adding little to the expected value of the trial.

If one were to add further locations to the Phase 3 network there would be increasing duplication of the applications tested. Beyond a certain point diminishing returns would set in; this could be taken into account by reducing the applications scores for these later, and harder to serve, links.

At this point the trial would become a service; little additional technical or behavioural knowledge would be gained by increasing the size and cost of the network; however one may at this point begin to obtain statistical information useful for market prediction.

The application and the transmission networks for Phase I, Phase II, and Phase III of the CJS trial in Phoenix are given in Figures 3A and 3B, 5A and 5B, 6A and 6B, respectively.

Pauses for evaluation were planned between the phases to evaluate the merit of implementing the next phase, as well as to refine the factors w, v, and e.

3.2 Network Efficiency

Given a chosen ultimate network which will be realized in time, there exists an optimal sequence of connecting up locations which maximizes the core applications tested for the dollars spent up to that point. This sequence is given in the rows of Table 3 labeled Optimal, and in Figures 3A, 5A and 6A.

The sequence chosen for actual installation in Phoenix, differed slightly from this optimal sequence, since the latter did not take into account equipment availability and political factors which, for example, dictated that location 22 be installed before location 11.

We can define the notion of network installation efficiency,

$$N_E = \frac{(score/cost)\ actual}{(score/cost)\ optimal} \quad \text{where } 0 \le N_E \le 1 \tag{5}$$

It can be seen from Table 3, that N_E = 0.9, for Phase I,
and N_E = 1.0 for Phases II and III. Note that less is spent in
Phase II with the planned installation sequence and that the
money spent per score point increases by 50 percent from Phase I
to Phase III, see column labeled "measure."

TABLE 3

EFFICIENCY OF INSTALLATION

Installation	Locations	Score	Cost[†]	Measure	Efficiency
Phase I Actual	14,17,19,23 1,16,22	304	10.	3.29	0.90
Optimal	14,17,19,23 1,16,19,4	336	10.	2.98	1.0
Phase II, planned	5,26,4,18,11	538	25.6	4.76	1.0
Optimal	5,26,11,18, 22	538	27.2	5.06	1.0
Phase III, planned	20,21,10	609	30.8	5.06	1.0
Optimal	20,10,21	609	30.8	5.06	1.0

[†] See Note 2, Table 2.

3.3 Rationale

The approach described in Section 3.1 provides an
intuitively appealing means of making good use of informed
estimates in order to decide the sequence in which to add nodes
to the network. The rationale is that, in the absence of more
"objective" information, one should make the best use of what one
has, however "subjective." Of course the latter is likely to
change during the trial: the probabilities (e), and possibly the
weights (w) and volumes (v), would need to be updated prior to
recalculating the installation sequence for Phases II and III.

4. RESULTS

4.1 Network Traffic

Figure 7 shows the network call volume which materialized
with the installation of Phase I completed in May, 1975, and
the installation of some Phase II locations modified by Phase I
evaluation, which began in November, 1975 and continued into

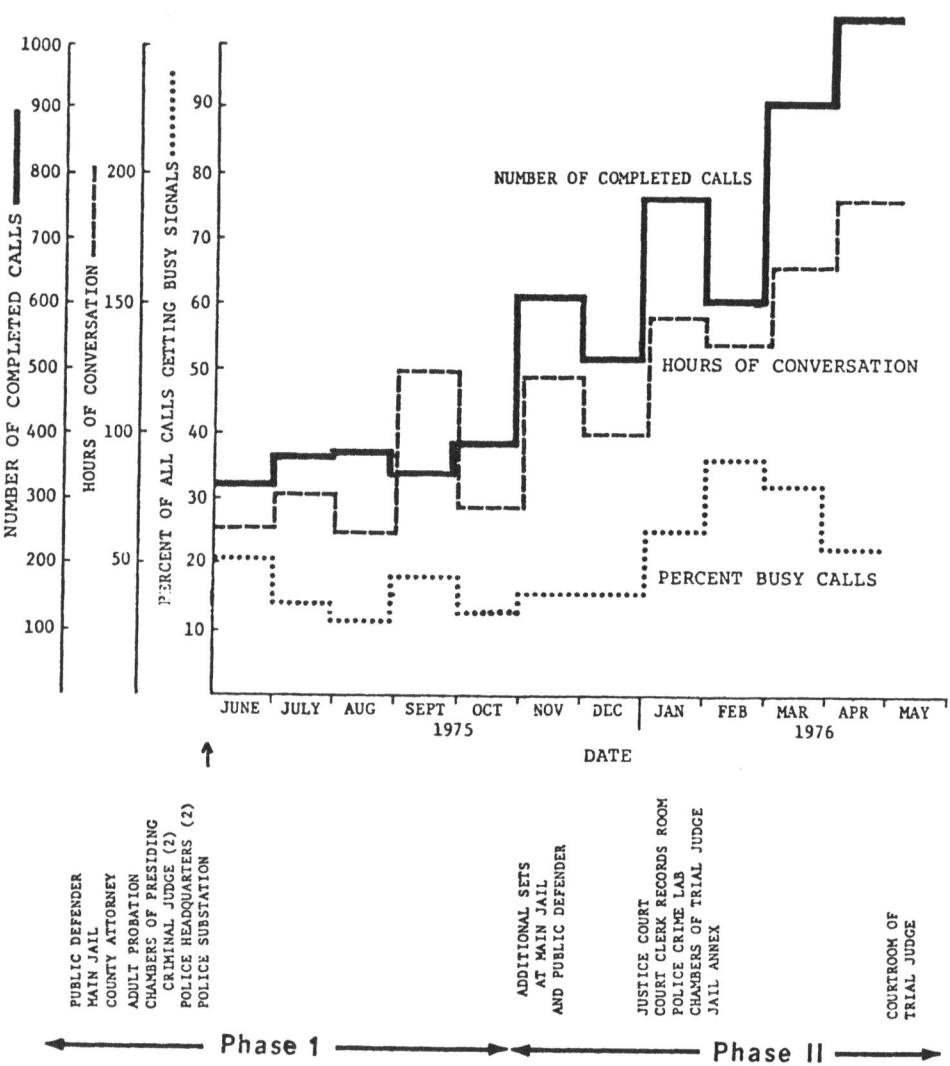

Figure 7:   VCS Network Monthly Call Statistics, As Measured On
             The Phoenix Criminal Justice System in 1975 and 1976.

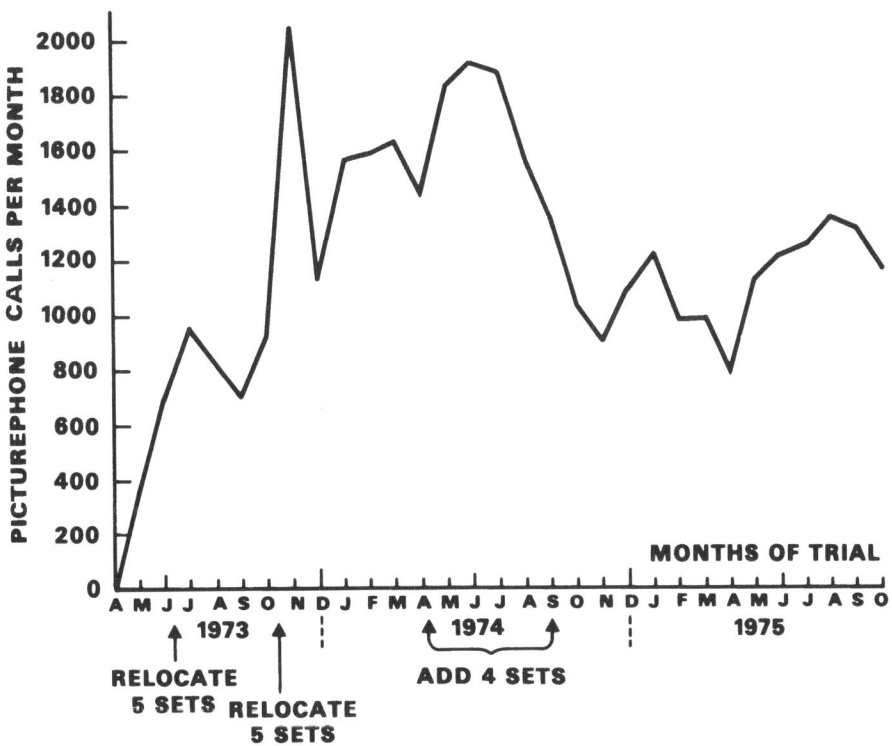

Figure 8:   PICTUREPHONE Network Call Volumes Per Month From
            1973 to 1975 Showing the Initial Rise to a Peak
            of 2050 Calls/Month and Fall to a Plateau of
            Approximately 1100 Calls/Month After 2 Years of Use.

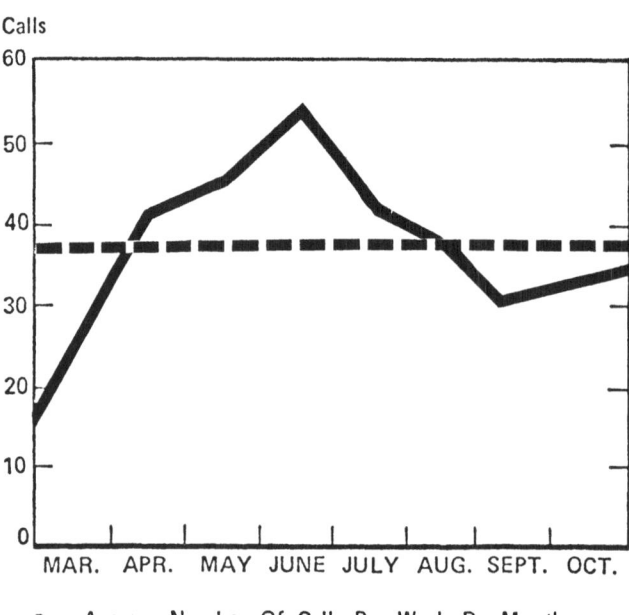

Figure 9:   Public Defender to Jail Inmate VCS Conference Call
            Volume From March to October, 1975; Criminal Justice
            System, Phoenix, Arizona.

TABLE 4

COUNTERPARTS IN ORGANIZATIONAL SYSTEMS

HEALTH CARE	CRIMINAL JUSTICE	BUSINESS
Surgeon	Judge	Vice Pres.
Physicians	Attorneys	Professionals
Patients	Inmates	Products
Operating Theatre	Court Room	Board Room
Emergency Room	Police Station	Receiving
Admissions	Police Desk	Reception
Pathology Lab	Foresic Lab.	Central Files
Exec. Director	Asst. Chief of Police	V. P. Admin.
Isolation Ward	Jail	Warehouse
Family Visitors	Family Vistors	Customers
Clinics	Remote Police Stations	Branch Offices
Admin. Office	Admin. Office	Admin. Office
Nurses	Guards	Service Personnel

1976.[3] These call volumes show the anticipated rise in the presence of advocacy and attention by the trial organizers.

Based on experience with PICTUREPHONE® system trials in health care delivery systems[5,6] whose participants have striking counterparts in criminal justice delivery system, see Table 4, the author expects that the network call volumes will peak and then decline to a more or less steady-state plateau after about 2 years, as is shown in Figure 8 for a health care trial which was similar in many respects to the CJS trial.

It is suggested that when this steady state value has been identified that the most meaningful cost benefit evaluation can be made. There is an indication that the peak and plateau were reached in the Public Defender/Inmate conference applications, see Figure 9. In a health care trial using a network of 23 PICTUREPHONE station sets[5], individual locations plateaued after about 6 months, while the network plateau was only clearly identifiable after 2 years.

On July 1, 1976 approximately one-half of the CJS network was removed (Superior Court Room, Judges' secretary, Superior Court clerk, Crime Lab Criminal Investigation, Information Bureau Justice Court, Police Station (airport) and the Jail Annex); these stations had insufficient use to warrant continuation, their removal marked the transition from a trial to an operational network. The network traffic volume between the remaining stations has since been averaging 500 calls/month for the subsequent 9 months.[7]

5.   ACKNOWLEDGEMENTS

I thank my colleagues in the Bell System, particularly T. M. Potrykus with whom I worked on the CJS project in Phoenix.

6.   REFERENCES

1. Unpublished work by J. A. Hooke and P. S. Unger, January 7, 1975.

2. L. C. Troup, Justice by Phone, Telephony, November 8, 1976.

3. L. L. Stine and L. G. Siegel, The Video Telephone in Criminal Justice, The Phoenix Project, Volume II, Analysis of Applications, MITRE Corporation MTR-7328, August, 1976.

4. G. R. Blakey, Application of the Video Telephone to the Administration of Criminal Justice: A Preliminary Assessment, Journal of Police Science and Administration, Vol. 3, No. 1, March, 1975.

5. C. D. Stockbridge, Experience of PICTUREPHONE System Use in Hospitals, Soc. Photo-Instrumentation Engineers, Application of Optical Instrumentation in Medicine II, Vol. 43, November 29, 1973.

6. C. D. Stockbridge, PICTUREPHONE System Use at Bethany Garfield Hospital, Progress and Prospects in Health Care Distribution Systems, Second Miami International Conference, November 24, 1975. [Traffic on this network of 23 terminals peaked at 2050 calls/month, then declined over 20 months to a steady plateau of 1100 calls/ month until the network was removed in 1976.]

7. H. M. Kroll, private communication, June 1977.

DISCUSSION OF PAPERS BY CHAPUIS, DU ROURE, MANDELBAUM, SHINN AND
BURNS

Discussants and Moderators: Charles Stabell and David Conrath

Stabell began by saying that the papers were provocative in
their attempt to understand the forces that shape the environment
within which we work. Instead of attempting to be comprehensive, he
chose to focus on two themes apparent in the papers by Chapuis,
du Roure, and Mandelbaum.

Chapuis and du Roure addressed the concept of integration. In
Stabell's view their use of the term is unclear, although they view
integration as both in progress and desirable. For Stabell,
however, the concept suggests several different meanings: (1)
integration, together with differentiation, as a means by which
an organization (system) might obtain the necessary variety to be
able to adapt to a complex environment (the law of requisite
variety); (2) integration through the use of standard interfaces
(as in the IBM 360 Series) in order that system components may be
changed without having to change the whole system; (3) integration
as a resurrection of the now defunct Total System concept.

The last point reminded Stabell of the New York blackout - the
city had a totally integrated electric power grid system. A limit
to integration comes with the recognition that control of information and communication is power.

Mandelbaum's paper does not identify the actors in his self-
designing community. Recognizing who they are might provide an
understanding of why the community cannot solve some of the
problems discussed. Stabell pointed to the engineering and natural
science backgrounds of the participants in the self-designing
community. They lack the staying power to overcome their disappointment that behavioral science theories cannot provide the type
of information that engineers are accustomed to obtaining: theory
does not indicate which medium of communication should be used in
a specific situation. They tend to focus on technology without
recognizing that communication and computing devices are not well-
defined theoretical constructs. Finally, engineers approach the
political and scientific components of a problematic as separate
analytical elements - which they clearly are not. In particular,
big research requires big funds which, in turn, are subject to big
political pressures. One way to reduce their interdependence is
obviously to opt for smaller research efforts.

747

Conrath stressed that the issues of policy and methodology are not separate. The du Roure and Chapuis papers ask for a research scenario but ignore the problems of their own research community. The structure of the research community affects research.

The central theme of Shinn's paper was a plea for staging rigorous experimental research. Burns' paper related much more to research experience than to hard numbers. In Conrath's view there is frequent neglect of an ethical question that arises when technology is introduced in field experiments. Since service often cannot be provided beyond the experiments, what should be done about the increased expectations of the participants in them?

For Conrath the points raised by Mandelbaum on robust design did not necessarily indicate flaws. However, the conflict between broad perspective and narrow specialization is a problem. It can be overcome because it is possible to combine depth and integrative studies. But, to date, we have not done an adequate job of handling research to bridge the conflict Mandelbaum points up. Another participant questioned the idea of carrying results from the laboratory environment to the real world. The laboratory, he said, is not representative of the real world. The real question is whether the empirical researcher extracts anything from the laboratory? Shinn distinguished between the purposes of field and laboratory experiments. For hypothesis testing, the laboratory experiment is better. In his view the Reading project served political goals excellently, but scientific goals less well, in large part because these were more difficult to define. He suggested that there is tension between science and politics and one experiment may not be able to resolve questions for both. On the other hand, the Reading project could be considered to have provided an opportunity for much good research which was not itself entirely experimental in nature.

Regarding the Reading experiment, Elton suggested that it does allow us to reject the null hypothesis: the experimental system did work. There was an important hypothesis generated too: spontaneous two-way interaction can be a powerful alternative to traditional TV production. Although the research is still in progress, it is clear that there will be valuable output of a theoretical nature derived from behavioural observations.

Moss said that the burden of proof is greater for the Reading cable project specifically because it has survived beyond the experimental period, unlike many projects which are evaluated on the basis of narrower criteria. The design of the Reading project allows it to be evaluated in terms of the hypothesized effects as well as its broader social and political impact on the community.

There is always a trade-off between the rigidity of an experimental design and its capacity to capture and reflect the full range of effects generated by any treatment. The choice of research strategy should really be a function of the type of questions being asked and the nature of the evidence necessary to answer those questions.

Hiltz stated that a field experiment is doomed to failure if we think we can predict all outcomes and develop good measures ahead of time. In her opinion, researchers must mix "soft" and "hard" methods in analyzing a project. The unanticipated outcomes may be the most important ones; and these are best detected by such ethnographic methods as field observation and unstructured interviews. Brownstein pointed out that the goal for the projects, of which Reading was one, was to be able to say whether two-way cable can provide socially useful services. The answers are unambiguously yes. With regard to Shinn's concern about politics, Brownstein said that major political problems did not materialize for the NSF-sponsored projects; he did not accept major social change as a function of the projects. Goldstein agreed that the laboratory is not the real world, but added that neither is the field experiment the real world. In the real world someone puts real money on the line to introduce a new service. Brownstein agreed that field experiments are not the real world because of the experimenters' stake in the projects. This stake intrudes into the research. The question is, can one generalize on what has been learned?

Shulman said that no one approach to research will provide all the answers. Obviously, each one has its own shortcomings. It is only in their combined use that we obtain a good understanding of a phenomenon we are interested in.

Stabell noted that we don't find theories, we create them. Much of the empirical work that has been done could have been explored theoretically by sitting down and thinking. Referring to Stabell's points on integration, Thorngren said that one may decompose the design system into a number of subsystems. The Reading project might have been done with another type of technology. Stabell felt that research should focus attention on interfaces and should not be overwhelmed by the seeming dissimilarity of systems.

Wish said less support was now available for basic research on interpersonal communication than in the past. Perhaps people are expecting too much from research. Who will support basic research in the future at places beside Bell Labs? How do you choose among types of basic research?

Regarding questions of social science methodology Tyler said that we are sometimes intimidated by physics. The result is that some researchers seem to believe that scientific activities only occur in laboratories or controlled experiments. We need to pay serious attention to other kinds of data. For example, we do not know what the benefits of services are worth to people who have not experienced them, if the services must be offered on such a scale as to make field experiments infeasible. Reading addresses the ethical question of continuity, raised by Conrath, because the community now supports the service. In response to a question whether behavioral laboratory research is on the way out, Tyler suggested that while this is not entirely true, support for such work has waned because much of the action in the real world is not in areas which can be addressed by laboratory research. Lucas added that science should be viewed in terms of uncertainty reduction. All NSF's projects are either rejecting hypotheses or tentatively accepting them. In field experiments we try to reduce uncertainty about the markets for services and other questions that cannot be answered in the laboratory.

Chauche said that the real world is concerned with technological innovation, while field experiments are concerned with social innovation. What are the links between the two? The field test is a link which connects the laboratory to the real world.

Stabell felt that the link between research and practice is method. An important function of the research community is to develop methods and establish their validity and applicability.

Goldstein argued against the view that things happen in a logical progression from the laboratory through field experimentation to the real world. Entrepreneurial activity intervenes: somebody decides to take a risk by committing a large amount of money (compared to the cost of research). There is a need to understand people's willingness to pay for services, not just their willingness to use them. He warned against trying to do everything through research, adding that researchers are needed to clarify concepts and add to understanding.

Jull said that Canadian experiences have identified several factors which he believed are fairly obvious: for example, constraints on travel funds have an impact on teleconferencing. But while we know there is some impact we do not know how much. In his opinion, the usefulness of interpersonal telecommunications cannot be broadly generalized: their value depends on the particular environments of application. Nevertheless, it would be useful for researchers to agree amongst themselves on research procedures to identify constraints and driving forces which influence the use or nonuse of new interpersonal telecommunications services in particular environments. From this it may be possible to provide useful guidance to telecommunications planners.

Goldstein stated that it is the service provider's responsibility (and problem) to develop products which are attractive enough that people will buy them.

Ohlman suggested that useful statements could perhaps be made about impacts if someone were to pull together all the research findings. It might be worthwhile to concentrate upon this. In reply Conrath pointed out that this is difficult when the bases of analysis are so different.

Hiltz did not agree with the position that the "Laboratory to Field" model is always the best for developing research methodology and testing hypotheses in the telecommunications field. In many cases, she said, researchers can use observational data from field trials to suggest what are the most important casual relationships and what appropriate indicators or measures of these variables may be; they may then return to the laboratory to test these hypotheses.

Mandelbaum suggested that it is very hard for large numbers of people to develop careers in the design of telecommunications applications. Do we just "use" people like Red Burns or can they be created? More such people are needed, for we must certainly engage in hard and sustained systems design.

A.S.

M.E.

DISCUSSION OF THEMES RELATING TO POLICY AND METHODOLOGY

POLICY: LAWRENCE H. DAY

This was not designed primarily as a policy conference and there was not any significant focus on particular policy issues. These issues did, however, emerge in the discussions.

I have used a "hit system" in preparing this summary: each time a point arose in the discussions and started to repeat itself, if I thought it was a question of policy (in the broadest sense of the word), I made a note of it. These "hits" are the basis of the following remarks.

The first and most discussed issue (in fleeting shots) can be labeled the productivity (cost-benefit) type of queston: "Why do people use these systems we try to produce, test and sell?" In the discussions of STI, the CTS Satellite, the Reading Cable TV Experiment, Eurosat, EFTS, Telemedicine and so on there was a continuing question of whether any cost-benefit trade-offs were achieved, although we knew we were serving some sort of useful purpose. I think that this highlights a policy problem in that with many of the systems that we put together, we think or hope we are serving useful purposes, but when we start trying to caculate the cost-benefit trade-offs, it gets very tricky. We have to play some very interesting games with the calculations and nobody believes the results when we are through calculating them.

I noticed that this point continually came up and then disappeared again. It is one real problem we have in planning for new telecommunications services: in many cases we do not know if they meet stated or unstated cost-benefit trade-offs. There was one exception to the rule, the usual exception, and that was for military systems. Craig Fields said that cost is no object when the cost of a mistake is so high. That was in marked contrast with what everybody else said.

A related issue was the question, "Whom are we designing these things for? Who are the users?" We seem to be confused about who the real users are. Are they ourselves as academics or as scientists? Are they managers or military officals? The fanatics or "real people?" In the discussion of EFTS it appeared that there were some real people who seemed to know what they wanted to do. In many cases, however, we do not seem to have made the step from the trials, the in-group playing around with new technology, to those who will be the ultimate buyers.

One of my favorite interests arose a few times: the fact that users are exploring as well. They are defining services around the capabilities that are being provided. The subject of user-driven applications brought out what I noted as the rising illegal use of technology: the fact that delivery of certain STI services, computer mail and computer conferencing appears to be - or recently to have been - illegal in various European countries. The things that are really turning users on appear to be against the rules.

I used three different labels for another question that we covered. Each of them has probably been the subject of a previous NATO conference. I refer to the concept of innovation, sometimes called technology transfer - a buzz word a few years ago - and sometimes labeled decision-making. We were often reminded about the "real world" and this brought out questions of how decisions get made, who makes them, who puts the money up to make things happen on an experimental basis and on a real world basis. There is also an issue relating to certain social services where there will not be a sufficient demand if individual users have to put up their money. Who aggregates the demand for social services, in telemedicine, in communities like Reading?

A number of other issues surfaced. Policy can inhibit innovation as Roxanne Hiltz and others mentioned. Jim Cowey brought up the concept of the emotional context which I believe to be important. Inertia was a term used several times: the idea that things do not happen as fast as we would like.

It may be an unfashionable word to use but this business goes through fads; it goes through cycles. We did not talk about some of the neat things we used to talk about in meetings like this. Satellites were hardly mentioned. They used to be very important in the sixties. We mentioned computer communications a bit. That used to be important too. Interactive cable TV is really a "blast from the past." It used to very important though you do not hear much about it now. I think you will start to hear more about it again soon as the experiments start to be reported upon. Well, the old fads did not emerge too much here.

There was a discussion of policy intersections. This was
basically the North Americans versus the Europeans. There seemed
to be a European view which can be expressed as, "You silly idiots
over there are going to cause a lot of problems for us in Europe."
That is true. Many institutions are not really designed so that
they can handle these types of pressures. You cannot <u>contain</u> as
nicely and easily in the European environment as you used to be
able to. That is causing considerable discomfort. I know that if
I worked in Europe I would feel very discomforted too.

There seems to be a "Let's handle things more conservatively"
view versus a "The world is falling apart" approach. (I am sure
that some of our European colleagues were not too thrilled with the
fact that the basic telecommunications policy chaos that exists
particularly in the United States, and to a lesser extent in Canada
is now starting to move eastward across the Atlantic. Frankly I
would not blame you if you are a little upset about it. If you can
slow it down, I think it would be a wise thing to do. That was an
Editorial Remark, not direct reporting of the conference results!)

We discussed the politics or policies of integration, whatever
that term means - too many of these words we never define. Sometimes we talked about the integration of technology, which is one
important concept. Techologies result in services. In telecommunications we basically sell <u>service</u>; we do not sell hardware, though
we sometimes think we do. But the services are starting to become
integrated. The users are starting to integrate things at their
end. They are starting to expand their areas of control. And this
is a new phenomenon.

There was confusion in much of the discussion of integration
because sometimes we were concerned with technological integration
and sometimes with integration of services.

Finally, we did have a wild discussion on the sociology of
doing policy research, which I personally labeled as an "Airlie
Conference East Discussion." (The Airlie Conference is the annual
telecommunications policy conference in the U.S.A.) The discussion
yesterday was very reminiscent of some of those discussions. But
one of the good things about this meeting is that we did not
degenerate, as so often happen there, into a gripe session with a
bunch of people complaining that "nobody loves policy researchers."

POLICY: CHARLES N. BROWNSTEIN

I have tried to cover issues not discussed by Larry Day,
and because we attended different discussion sessions I may have
succeeded. My strategy was to try to report on the themes that
people who were presenting points of view from the audience drove
home most strongly. I noted about a dozen points, and I offer
a few of my own.

An early theme, one that we came back to in considering
telecommunications policy as opposed to research policy, was
the critical importance of industry economics in the development
of services. Along those lines it is important to realize that,
even when one considers industries internally and tries to look
at what impact methods of control have on them, external variables
such as acceptable systems of accounting and tax policy may
have as much to do with industry structure in the end. Or, in
getting from here to there (if you know where you want to go),
they may have at least as much effect as regulatory mechanisms and
other variables of policy research which are rather more
normative. These processes are at least as critical as regulatory
processes in systems development, in the shape of the system
when it is developed, and in the way it can be exploited as it
is being developed. That was a very good point made very clearly.

Another interesting point, relating rather less directly
to telecommunications policy, was that research, in addition to
answering specific questions, whether of a basic nature or, for
example, concerning marketing, can force policy attention in
given directions. In many ways that may be one of the strongest
uses of research; rather than push policy to specific conclusions,
it may just focus attention. In that way, I think, the fads (see
Dr. Day's review) may be a symptom of something else working in
the system, something to do with pushing people's interests
around.

This, of course, is an assumption that must be tested, and
in order to test it one must enter areas of policy which do not involve telecommunications directly. Medical policy, service delivery

policies, development policies, and transportation policies may
push telecommunications policy around as much as vice versa. And
this puts telecommunications researchers interested in these broad
issues squarely into a very messy area of science policy interaction,
which is itself a small industry, at least in the United States.

There are other indirect things that certainly effect what
we do and affect public policy. Satellite research was a good example. Locating socially benefical applications, while of great
importance in itself, may also be considered to some extent an
instrument of foreign policy, in the sense of supplying foreign
aid or reserving slots in the variable parking lot up there. These
functions are often not obvious; at least, some people seem not to
take them into account. Using telecommunications to transfer information was imputed to imply formalizing communication and information transfer, which in turn would force legal questions of ownership, of copyright, of standards of privacy, of access. That is an
old list in telecommunications policy research which was emphasized
several times.

There was for many people here a science policy question
whether one should forecast and act so as to create change, or
whether one is there to really understand it. That depends upon
one's view of oneself as a researcher, along a continuum ranging
from scholar to marketing manager to policy maker. This is less
of a problem in some countries than others, depending, I think, on
who appears to have the most vested interests, the deepest embedded
investments in different services and different technologies. It
was mentioned, for example, that in the UK VIEWDATA is seen as
potientially beneficial to TV manufactures, as a new marketing
opportunity. In the US I would expect VIEWDATA to be accused of
reducing the market shares of commercial TV outlets. So you may
have the same technology forcing different policy issues in different countries depending on industry structures.

There are other interesting differences. For example, in the
satellite analysis area, for some reason the US and Canadian focus
was on public services, while the European focus, at least in the
paper presented here, was on business applications. I had to wonder,
since both of these are for really untested demand, just what was
the policy justification of doing the satellite research.

Another broad policy theme was the use of telecommunications
for non-telecommunications purposes, for grander things: regional
development and creating social change. However, I think there is
some confusion as to whether telecommunications was a symptom of
regional development - something that arises because of need for
communications as regions develop - or something that creates
regional development or perhaps these things work together. In

either case people discuss as a policy issue the problem of overcoming start-up problems, saying, "Well we have no plot of which way
things will go, but we know there is a problem of getting telecommunications in place." The start-up issue seems to be a major
one.

One of the two views on that was that you install a system
and make it available and people will find a way to use it - a kind
of technology push. And the other one was that you await market
pressure, wait for a demand before you go ahead with installations.
Those different points of views are a matter of experience and faith
in what telecommunications can provide. They may also be a matter
of how deep your pockets are in terms of what facilities can be
provided. I think a good instance in the discussion was electronic
funds transfer, in which interest had progressed from technology
through marketing through economics and then to policy. The issue
of what impact electronic funds transfer would have on the telecommunications network, probably the most serious telecommunications
policy issue, came rather late and is still developing.

There was some talk, in papers and in discussion, of national
goals on which policy choices might be made. There was some recognition that individual countries have their own problems and their own
goals. I had a difficult time pinning down what those goals were.
I heard very little discussion of just what a national policy goal
in telecommunications is in any one of the countries, or what it would
be like should such a thing be created. As a result, I think, the
business of economic dominance emerged in the discussions - this
business of people finding uses and these uses pushing the policy
around, at least being stronger than the policy in terms of facilitating development of the different kinds of systems. It is interesting here that we did not discuss telecommunications policy as a
reaction, a reaction to perceived imbalances, problems etc. I think
most of us all the way through the discussion, talked about policy
as some sort of instrumental activity to create something we want.
Yet there is that other area that just was not very much discussed
here, as it would be, for example, in a conference on regulation.

Interests in economic dominance brought out the question of
who loses as being a major policy lynch pin or lever.

Another theme concerned the effects of the integration of
telecommunications systems. This may come via standardization and
may or may not be desirable. It may be undesirable from an economic
point of view, while for technology suppliers it may be desirable in
terms of putting services together so that they can be provided better.
It may have social consequences in changing communication between
regions, possibly between nations. Of course, as was mentioned, it
also has the possibility of creating a rather fragile social network,

not just a fragile telecommunication system; it could become somewhat unglued, if there were problems with the telecommunications
system, if someone pulled the plug. This would be much the same as
the way in which very advanced water delivery systems and sewage
systems make a city much more fragile and open to disaster than
old style ones where there is a great deal of segmentation, even
though the new ones offer certain benefits and efficiencies.

There is a newly identified theme (to which I would not yet
ascribe fad status), the question of integration.

In many ways integration is the critical element for thinking
about the future needs, services, industry structures, etc. Its
ramification driven home today was that the different points of
view of different sectors should be taken more explicitly into
account in designing systems and investigating the way they develop.
The issue of integration has many dimensions: economic, service,
social and technical. It may even be a good organizing concept for
dealing with broad policy problems: how are various demands aggregated? how integrated can systems be? what is the method for
integrating resource allocation for the use of telecommunication
systems? It is probably in·this last area - although it hasn't
been mentioned very much here - that some new serious attention
and concern is being expressed in any sort of applications research
in the United States. The best lesson that has been learned is
that integrating resources, systems design and demand, are the
three critical problems if one is going to do instrumental research
or even if one wants to find out if things work very well.

METHODOLOGY: BERTIL THORNGREN

I would like to start with Mr. Goldstein's remark that somebody else is trying to do something in the system. In my view that
is rather important. This "somebody else" may differ from country
to country; it may also change over time. This makes it difficult
to envisage sweeping generalizations about methods to be applied,
because they may have to depend on these regional and timing differences.

There are clear risks if we do not take into account more of
these "somebody else" types of effect. We may risk turning nails
into screws, to try to make full use of the screwdriver.

In this area, when so often concentrating on new developments,
there is also the danger of forgetting that if something you are
studying goes up, something else may go down. This gets back to
the "birth-death" model I referred to earlier in the symposium.
The phenomenon is much more general when considering movement
through a life-cycle: when something is increasing, somewhere else
in the system something may be decreasing. The "somewhere else"

may be quite nearby, as when a straightforward process of substitution is at work, or it may be quite far away. Wherever it is,
those, "somebodies" may wish to, and even be able to, affect outcomes.

I would, therefore, urge the value of broadening frameworks
to take account of these dynamics. There may be forces acting from
elsewhere, counteracting future movement, in what one is studying.
I would agree with Michael Tyler that, very often in the social
sciences, we borrow methods from the natural sciences a little too
readily, thus not taking account of these types of problem.

Another general observation to be made is that many of the
scientific methods we use are intended to reduce uncertainty one
way or another. One type of uncertainty relates to which, of a
given set of alternatives, may be the better or the best. In
constructing the set there is a danger of cutting away a great deal
of variety. Especially if one is forced to adopt a rather short-
run perspective, a whole set of alternatives may be completely lost
to view. There is a need for methods to counteract this danger, to
increase the variety before one goes into assessment.

I have already mentioned the need to recognize the potentially
active parts of the system. There is also the need to explore
systems in terms of some more general socio-economic framework.
The difference between these two needs was apparent in the studies
presented here.

I would place a very high priority now on trying to achieve the
broader framework which is needed for validation of such studies.

Should this be done specifically within the area of knowledge
explored here? Or should it be a much more general kind of undertaking? I would urge the latter, because many of the most important things which will happen in our area will be initiated outside
its conventional boundaries. For example, we have just heard that
users themselves are taking initiatives, as in some of the computer
applications. Very often in history new developments have entered
from outside particular fields.

Even if we come from very different sectors and have difficulty
in putting the pieces together, it is important to recognize the
need for some integration of our activites. However, we must also
keep in touch with our respective specialities in order to cover
the broader area; they are valuable assets.

What I have experienced in this meeting has certainly been
promising in that we have not been exposed to a very high variety
of methods drawn from different sources. The sort of variety is
not something to be expanded. I think it was Martin Elton's view

from the outset that some communication between us should continue.
While we have had valuable glimpses of these different views, we do
need to continue so as to go deeper into some and recombine them
more than has been possible here.

I would be very glad to hear suggestions on ways and means
of achieving ongoing communication between those of us here,
which would draw on our connections with the rest of the world.
There are many different possibilities; computer conferencing is
one of them. Some kind of very loose, informal system might be
a very worthwhile way to proceed.

METHODOLOGY: DAVID W. CONRATH

An appropriate starting point is a question that was passed
to me five minutes ago, one very relevant to this meeting: given
the constraints on the speaker, the probabilistic nature of the
audience reaction and so on, what has information and communication
theory and analysis taught us about the preparation of a ten
minute talk on the subject? On pondering the question I realized
the answer is, damn little.

I don't think this means that everything we have been doing
over the past two years has had no impact. Though, when one reflects on the comments that were made by Ed Goldstein with respect
to market needs and the problem of decisions which can't wait six
years for a well defined study, one does wonder. Still, I do
think that the basis for inputs into such decisions can be
established. But this first requires a look at the question that
Bertil Thorngren posed and which I, too, was thinking about: to
what extent is an information exchange really taking place between
communication researchers?

Is this meeting going to be a one-shot conference as so many
are? And just because a meeting is held every year, or every
other year, or upon demand, it does not mean that it is other than
one-shot. The Institute of Management Sciences holds three or
four conferences a year, and I fail to see any continuity between
them. There has to be some form of coordinated exchange. One of
the problems which I have seen in conferences like this is that
communication is so frequently one-way. All of us want to say our
piece and be heard. The unfortunate thing is that everybody
sitting in the audience is figuring out, "What is it that I can
say?" rather than listening to what is being said. There is no
interaction; there is no possibility for integration.

Let me reflect on the comments made by Seymour Mandelbaum with respect to the community and on the fact that many of you are living in a seminary, a place which one associates with a monastery. When I visited last night I realized that you had created a community there, even though the concept of a monastery is of individual selves. The concept of community is anything but individual selves. A concept of exchange has, I think, to be incorporated in it. Communication and cooperation have to exist on several levels and it is very difficult to have it working on all levels simultaneously.

Putting this into the context of a community of communication researchers, first and forewmost we need to cooperate among ourselves. This, by the way, may be the most difficult task to achieve. As I mentioned, it is so easy for us to speak and so difficult for us to listen; it is so easy for us to write and, believe it or not, so difficult for us to read. A researcher known to a number of you here faced a question asked by a lady at a cocktail party: "Have you read such and such?" His response was: "In full honesty, no. I haven't had time to read it. I've been busy writing." It was another symptom of the problem of a lot of output with little input.

Another point which arose when I was thinking of research and the problems of its communication is that it's so easy to write and so difficult to think. If I take a look at the literature in general, I wish there was a different reward mechanism, one which more clearly rewarded thought rather than output per se. Certainly our problems of reading would be greatly reduced if much of the material in print were removed, if we had some way of filtering out what was written with little thought. Another aspect of this problem is the great new computer system that is going to develop protocols to separate information for the U.S. Department of Defense. It would be really beautiful if one could somehow distinguish those pieces of input which are germane to a decision from those which are not. Maybe it can succeed; maybe defense or crisis situations are so well defined - though I doubt it - that one can predetermine what one wants to receive.

Not only must we communicate with ourselves, but it has become very obvious that one of the failures of research on the use of communications technology is that we have lost sight of the people in the market who have to make everyday decisions. I wish we had more representatives of them here. We are miserable in our communications with the users of our research. We may ponder, but we don't communicate. We seem to love esoteric discussions among ourselves, and we get uptight with the people who are impatient for answers. Of course the problems are not all one-way. However, those in the market listen a couple of times and they hear what the consider to be pure jiberish. After two or three times feedback mechanisms

cause them to say, "My God, I'm not going to listen to him again.
He has no idea what my problems are." They are right because we
continue to speak, not to listen. As a consequence we don't communicate. Without that communication, quite frankly, what we accomplish is of little value because it won't be used. What will
happen, as Larry Day pointed out, is that there will be a push
from the consumer with the market responding, a process which
completely bypasses what we have to offer.

Here we are, in our monastery, making our community in a
monastery, and having parties late at night, but nevertheless
having nothing to do with what is going on in what Ed Goldstein
refers to as the "real world."

We must also communicate with the technologists, and we
appear to be doing better here. But it seems that all our research,
and this includes all that I've heard up to today, is on existing
technology. The challenge was thrown to us three days ago, "How
can we do research on things as yet undeveloped?" I heard no response and this upset me, because the issues and the questions into
which we can make useful input must deal with tomorrow's technology.
The uses of today's technology are so far down the road that
nothing we can do from a researcher's standpoint is going to make
any real impact.

Well what does this imply? One thing it suggests is that we
need a basis for exchange. To go back to the first point - communication among ourselves - we need some bases of standardization, of
common data. I reflect back to what is required in engineering.
Without standardization, technological integration is not feasible.
So technologists set up committees that do achieve standards both
within a nation and internationally. But behavioral scientists
comment that the engineer's problem is easy. They know when a
standard is good and when it isn't. Nonsense! What they know is
that they have something which will work. Later on they may very
well find that there are other things which work better.

Engineers are generally pragmatists. Their first criterion
is: "Will it work and can we get agreement? Then at least we can
get something accomplished if we use this standard." The questions
asked so frequently among behavioral scientists are: "Isn't something being left out?" "Can't I find something better?" We are
still searching for the Holy Grail. Probably we will be searching
for it for another 500 years unless we start using other criteria,
for example, "Will it work?"

The largest problem of all is our own egos. "I've got my
little cell and I like it. I can grow in it. I can develop and
I can get promoted in my research organization or increase my
publications. Don't mess with it; I've got a nice thing going."

I think that we can do something to overcome this, and I'd like to make a specific suggestion for some ongoing cooperation as an example (however, I really hate to throw out suggestions now with the fear that the very act of making the suggestion will lead to it negation). There is a research program being developed in Sweden to examine a Broadband Video Network. There is research being planned in England on Viewdata, a different use of video technology but nevertheless one which allows some interaction. There is research going on in the States on interactive cable television, yet another use of video technology with different characteristics. It provides for the kind of feedback that Viewdata will have in the United Kingdom. It has broadband video characteristics similar to those the trial system will have in Sweden. No research methodology is universally applicable to all of these, but there are some similarities that can link the experience so that what is learned in one community and one country may have something to say to another community in another country. Without some common measures things will be virtually no different than they are at present. At least a small probability of making some useful comparisons is better than zero.

Back to the fear that a methodology may be non-comprehensive or less than ideal in some other way. There is nothing to prevent us from setting up standards that allow one to probe a few things in depth, and several in breadth. It is not that everyone should measure everything, or avoid using other measures. Some measures should be collected over all three studies so that at least there is a basis for exchange. What I mean by that is standardization by cooperation, by common data. It is not a universal answer; there are none. It permits innovation in the collection of data. It permits all sorts of opportunities for new ways of thinking about the problem. But it also provides some common bases, a common language for discussing common problems.

What are the relevant dimensions? We certainly need technological dimensions. In communications, if we cannot analyze our data in terms of implications for technology, we are wasting our time (unless our purpose is strictly human relations). We certainly need behavioral dimensions. We hear the term "needs research". Unless we are going to live in a future world of robots, needs are evidenced by human behavior and we have to tackle these dimensions directly. Thirdly, we also have to have - and this seems so frequently stated - dimensions of value, or performance. When we identify needs, we have to be able to determine whether in fact these needs are being met. Anecdotes are not sufficient.

Can we cooperate and establish a means for communicating among ourselves? Can it be done? Yes. There is one success story which at its beginning I never thought would succeed. This has taken place in an organization called, interesting enough, the International Communications Association. Some of you are familiar with it. It started out as a speech and rhetoric group. There is a group within it which is interested in Organizational Communication. They set up what they referred to as a committee for organizational communication audits. The purpose was to establish a nationwide data base, with common questionnaires, interview techniques, et cetera, so that there would be common data for work done throughout North America on the problem of organizational communication. The orientation was purely behavioral. There are no data there that can really be converted into technology, which was a disappointment to me. But the lesson to be learned is that cooperation did take place.

What has happened is that researchers take some of the dimensions, some of the questionnaires, some of the methods out of this common data base. They use these, elaborating them and adding to them as they see fit in the light of their own particular research interests. I think we can do this too.

This is a platform for the first stage. Perhaps what is needed now is a small group of researchers who have a vested interest in communications among themselves to commence such an undertaking.

G.Z.

D.K.

PARTICIPANTS IN THE SYMPOSIUM*

CANADA

Red Burns
Alternate Media Center
School of the Arts
New York University
144 Bleecker Street
New York, N.Y. 10012
U.S.A.

Anna Casey-Stahmer
Department of Communications
300 Slater Street,
Ottawa, Ontario, Canada

David Conrath
CEROG
Institut d'Administration des
  Enterprises
13617 Aix-en-Provence
France

John Daniel
Tele-universite
Universite du Quebec
Ste-Foy
Quebec, GIV 2M3 Canada

Larry Day
Bell Canada Business Planning
620 Belmont, Room 1105
Montreal 101, P.Q., Canada

Ric Irving
Dept. of Management Science
University of Waterloo
Waterloo, Ontario, Canada

George Jull
Communications Research Center
P. O. Box #11490, Station H,
Ottawa Ontario K2H 8S2, Canada

Ron McClean
CEROG
Institut d'Administration des
  Enterprises
13617 Aix-en-Provence
France

Nichole Mendenhall
Futures-Studies Division,
Public Service Commission
1725 Woodward Drive,
Ottawa, Ontario, Canada

Percy Tannenbaum
Graduate School of Public Policy
University of California
Berkeley, California 94720
U.S.A.

FRANCE

Jean-Guy De Chalvron
Ecole Nationale Superieure des
  Telecomminications
46 Rue Barrault
75634 Paris Cedex 13
France

Daniel Chauche
18 rue de Villiers
92300 Levallois
France

Robert Chapuis
CCITT
2 Rue Varembe
1211 Geneve
Switzerland

---

*Names are listed alphabetically by country of citizenship.

Nicholas Curien
Ecole Nationale Superieure des
    Telecomminications
46 Rue Barrault
75634 Paris Cedex 13
France

Michel Dormois
Secretariat d'Etat aux Postes
    et Telecommunications
Tour Maine Montparnasse
33 Avenue du Maine
75755 Paris Cedex 15
France

Fabrice Fioux
Secretariat d'Etat aux Postes
    et Telecommunications
Tour Maine Montparnasse
33 Avenue du Maine
75755 Paris Cedex 15
France

Gabriel du Roure
Institut d'Administration des
    Enterprises
29 Avenue Robert-Shuman
13617 Aix-en-Provence
France

GERMANY

H. Gerke
Siemens AG
8000 Munchen 70
Hofmannstr.
- GE -
West Germany

Dierke-Peter Hansen
Erprobungsstelle 71 Der
    Bundeswehr
D 2330 Eckernfoerde
West Germany

Dieter Kimbel
Direction de la Science de la
    Technologie et de L'Industrie
OECD
2 Rue Andre-Pascal
75775   Paris   Cedex 16
France

Gerhard Rahmstorf
6101 Braunshardt
Rheinstrasse 14
West Germany

Jurgen Rottgardt
SEL AG, Hellmuth-Hirth-Str. 42
7000 Stuttgart 40
West Germany

Karl-Heinz Steinhardt
AEG-Telefunken
Fachbereich Weitverkerhr und
    Kabeltechnik
7150 Backnang, Gerberstrasse 33
West Germany

Carl O. Vernimb
Commission of the European
    Communities
Jean Monnet Building B4/022
Kirchberg
Luxembourg

ITALY

Claudio Baudazzi
Direttore Provinciale delle
    Poste & Tellecomunicazioni
Pavia
Piazza della Posta, 1,
Italy

Bruno Drioli
Telespazio
00198 Roma
Corso d'Italia 43
Italy

Giovanni Zambruno
University of Bergamo
Via Salvecchio
24100 Bergamo
Italy

NORWAY

Charles Stabell
Institute of Industrial
   Economics
P. O. Box 3437
5001 Bergen
Norway

TURKEY

Unver Cinar
Command Control and Systems
   Division
SHAPE Technical Centre
P. O. Box 174
The Hague
Netherlands

UNITED KINGDOM

Ronald T. Clark
Inter-Bank Research Organization
Moor House, London Wall
London EC2Y 5ET
England

J. B. Cowie
British Post Office Telecommun-
   ications
Long Range Studies
88 Hills Road
Cambridge, England

Peter Davis
Wharton Applied Research Center
Vance Hall
University of Pennsylvania
Philadelphia, Pennsylvania
U.S.A.   19174

Martin C. J. Elton
Alternate Media Center
School of the Arts
New York University
144 Bleecker Street
New York, N.Y. 10012
U.S.A.

David Gabbitas
Telecommunications Design Ltd.
308a London Road
Stockport SK74RF
England

Hilary Thomas
Communications Studies &
   Planning Ltd.
56-60 Hallam Street
London, WIN 5LH
England

J. J. Thomas
London School of Economics
Houghton Street
London W.C.2
England

Michael Tyler
Communications Studies &
   Planning Ltd.
50-60 Hallam Street
London, WIN 5LH,
England

Cordon Wells
Programmes Analysis Unit
Chilton, Didcot, Oxfordshire
England OX11 ORF

Mick Williamson
Inter-Bank Research Organization
Moor House, London Wall
London EC2Y 5ET, England

J. Stuart Yerrell
Department of the Environment
2 Marsham Street
London SW1, England

Peter I. Zorkoczy
Faculty of Technology
Open University
Milton Keynes, MK7 6AA,
England

UNITED STATES OF AMERICA

Charles Brownstein
National Science Foundation
1800 G Street, N.W.
Washington, D.C. 20550
U.S.A.

Craig Fields
Defense Advanced Research
   Projects Agency
1400 Wilson Boulevard
Arlington, Virginia 22209
U.S.A.

Carole Ganz-Brown
National Science Foundation
1800 G Street, N.W.
Washington, D.C. 20550
U.S.A.

Edward Goldstein
AT&T
295 North Maple Avenue
Basking Ridge, New Jersey
U.S.A.   07920

Roxanne Hiltz
1531 Golf Street
Princeton, New Jersey
U.S.A.   07976

Robert Johansen
Institute for the Future
2740 Sand Hill Road
Menlo Park, California   20550
U.S.A.

Barbara Lucas
National Science Foundation
1800 G Street, N.W.
Washington, D.C. 20550
U.S.A.

William Lucas
Rand Corporation
2100 M Street N.W.
Washington, D.C. 20037
U.S.A.

Seymour Mandelbaum
Graduate School of Fine Arts
University of Pennsylvania
Philadelphia, Pennsylvania
U.S.A.

Robert Mason
Metrics, Inc.
290 Interstate North, Suite 116
Atlanta, Georgia 30339
U.S.A.

Mitchell Moss
N.Y.U. Reading Consortium
87 West 3rd Street
New York, N.Y. 10012
U.S.A.

H. Ohlman
World Health Organization
1211 Geneva 27
Switzerland

Maxine Rockoff
National Center for Health
   Services Research
Hyattsville, Maryland 20782
U.S.A.

Allen M. Shinn
National Science Foundation
1800 G Street, N.W.
Washington, D.C. 20550
U.S.A.

Arthur Shulman
Department of Psychology
Washington University
St. Louis, Missouri 63130
U.S.A.

Myron Wish
Bell Laboratories
Murray Hill, New Jersey 07974
U.S.A.

SWEDEN

R. Bernemyr
Televerkets centralforvaltning
  Ul
S-123 86 Farsta
Sweden

Norman Gleiss
Televerkets centralforvaltning
  Ul
S-123 86 Farsta
Sweden

P. G. Holmov
Stockholm School of Economics
P. O. Box 6501
S-113 83 Stockholm
Sweden

Klaes Jonnson
Central Administration of
  Swedish Telecommunications
Marketing Department,
  Telephones
S-123 86 Farsta
Sweden

Bertil Thorngren
Televerkets Centralforvaltning
Staben for framtidsplanering
Stockholm
12386 Farsta
Sweden